SELECTED TOPICS IN APPLICATIONS OF QUANTUM MECHANICS

Edited by **Mohammad Reza Pahlavani**

Selected Topics in Applications of Quantum Mechanics
http://dx.doi.org/10.5772/58514
Edited by Mohammad Reza Pahlavani

Contributors

Francisco Bulnes, Nelson Flores-Gallegos, Calin Gh Buzea, Daniel Timofte, Lucian Eva, Decebal Vasincu, Maricel Agop, Radu Florin Popa, Aghaddin Mamedov, Jonathan Bentwich, Mahmoud Merad, Boudjedaa Tahar, Benzair Hadjira, Valeriy Sbitnev, Cynthia Whitney, Mohammad Reza Pahlavani, Vasyl Kovalchuk, Jan Jerzy Slawianowski, Sergio Curilef, Maria Burgos, Karlheinz Schwarz, Alina Gavrilut, Lacramioara Ochiuz, Cristina Popa, Gabriel Crumpei, Dan Tesloianu

CBS, Edition **2017**

Published by InTech

Janeza Trdine 9, 51000 Rijeka, Croatia

Publishing Process Manager Iva Lipovic

Technical Editor InTech DTP Team

Cover Designer InTech Design Team

Additional hard copies can be obtained from orders@intechopen.com

Selected Topics in Applications of Quantum Mechanics, Edited by Mohammad Reza Pahlavani
p. cm.
ISBN 978-953-51-2126-8

INTECH
open science | open minds

Contents

Preface

Without a doubt, modern science is not meaningful without quantum mechanics. When combined with mathematical physics, the basic laws of microscopic physics produce quantum mechanics. Despite it being about one century after the discovery of quantum mechanics, its history and foundations are of great interest to researchers, scientists and students of different branches of science and technology. The philosophy of quantum mechanics, which grows on the foundations, history and basic laws of quantum mechanics, such as uncertainty and the correspondence principal of quantum mechanics, is a great branch of philosophy. It forms the foundations of quantum mechanics as an open topic for researchers who are studying the history and philosophy of science. Therefore, most authors of quantum mechanics and its applications introduce it via a historical survey of its early success.

The book consists of two sections. The first section is "Selected Topics in Foundations of Quantum Mechanics", which consists of seven chapters. These chapters provide a clear insight into the foundations of quantum mechanics, through the basic laws.

The title of the first chapter is "Classical or Quantum, What is Reality?" Therefore, as it is clear from the title, the author of this chapter compares classical physics with quantum mechanics. In chapter two, as a basic object of quantum mechanics, photon is compared with the role that is played by signalling information transfer. In chapter three, generalized uncertainty is discussed by combining quantum mechanical uncertainty and the path integral method of Feynman. The subject of chapter four is the unification of quantum mechanics with relativistic theory. In chapter 5 and 6, measurement and its validation in experimental computation are studied as a major problem in quantum mechanics. In the last chapter, the lie algebra of QED and its fermionic- Fock space are presented.

The second section, "Selected Topics in Applications of Quantum Mechanics", consists of seven chapters. These are published with the cooperation of international community authors from different research institutes and universities. The first chapter is dedicated to the application of quantum mechanics, through the mean field method. Thermodynamic entropy and its role in atoms, molecules and matter are presented in chapter two. Chapter three deals with the computation of material properties via atomic structures using quantum mechanics. Quantum information is studied in chapter four. In chapter five, physical vacuum, as a special super fluid, is presented. Husimi distribution and the fisher information are studied in chapter six. The last chapter is dedicated to the application of quantum mechanics in medical cardiography.

The selected topics of the foundations of quantum mechanics are published with the cooperation of international community authors from different research institutes and universities.

I would like to thank all of them for preparing the chapters in time and the InTech publishing group, as well as MSc Iva Lipović, for their valuable efforts in the publication of this book.

We hope that this book becomes a rich reference for researchers, as well as students.

M. R. Pahlavani
Nuclear Physics Department
University of Mazandaran
Babolsar, Iran

Section 1

Selected Topics in Foundations of Quantum Mechanics

Chapter 1

Classical or Quantum? What is Reality?

J. J. Sławianowski and V. Kovalchuk

Additional information is available at the end of the chapter

http://dx.doi.org/10.5772/59115

1. Introduction

Discussion of those questions and the struggle with the well-known paradoxes of quantum mechanics like decoherence and measurement problems is as old as the quantum mechanics itself. This was just the reason for the search of formulations of any of both theories in terms of concepts characteristic for the other one. It would be injustice to condemn or disqualify those attempts. They were fruitful, enabled one to solve many concrete problems and did shed some light onto the mutual relationship of both theories. Nevertheless all mysterious features of quantum theory remained mysterious as they were. There is a whole spectrum of views; let us quote the dominant ones in a very simplified form:

1. In principle everything is "quantum". The "classical" is an approximation of large quantum numbers in large, macroscopic subsystems of the Universe. In a sense it is a kind of "illusion" well working in a restricted range of physical phenomena.
2. Finally everything should be "classical" and the "quantum" model is phenomenological and temporary, in any case incomplete. Very often one formulates the conjecture that it is the linearity of quantum theory that is guilty. In any case: what other physical factor might be responsible for the discrepancy between unitary evolution of the unobserved microsystem and the measurement reduction phenomena?
3. Physical reality is dualistically built of two incompatible elements: quantum and classical. They are joined into a single whole via statistical interpretation of the wave function. This unification is rather mysterious from the purely classical point of view, nevertheless rigorously described mathematically by the standard probabilistic interpretation. This view is relatively popular, although it is not free of the solliptic or even divine ideas. Let us mention that both those ideas are physically justified.

There are many papers concerning the first and third possibility. But the second item, i.e., nonlinearity, is not very popular, although it seems to be a good candidate for explaining quantum paradoxes and exorcising the solliptic or any dualistic ideas. Incidentally, from the point of view of the development of physics and other natural sciences this reluctance

of physicists to nonlinearity in quantum mechanics is rather strange. It contradicts the whole history of physics. Usually one begins, when it is only possible, from linear models, but later on one introduces nonlinear terms to equations. And now we witness the unbelievable development of nonlinear methods, also essentially nonlinear ones, when there is no well-defined linear background term perturbed by a nonlinear correction. It is simply fashionable to formulate and discuss models based on the non-perturbative nonlinearity of geometric origin. Why really to stick to linearity in so fundamental science as quantum physics? Let us also mention that there are objects in the nano-scale like graphenes, fullerenes, and very large molecules the behaviour of which is placed somewhere in the convolution region of the quantum and classical theory. Let us also remind that there are nonlinear computational methods like the Fermi-Dirac procedure in the usual quantum mechanics. It is not excluded that some more fundamental nonlinearity may be formulated and efficiently used.

One of reasons of the reluctance of physicists to nonlinearity in quanta is some non-physical arbitrariness of the nonlinear dynamical models. And really, there were various models introduced "by hand". But recently some essentially nonlinear, non-perturbative models based on deep geometric ideas were suggested. For example, we mean here papers by Doebner and Goldin with co-workers [4, 5], works by Svetlichny [17], and some of our publications [13].

Below we consider two methods of quasi-classical analysis: one based on the limit transition with the Planck constant tending to zero and one based on the analysis of quickly varying wave functions with many modes. They are both connected with the mechanical-optical analogy and both are useful. Namely, the theoretical approach to physics is one based on differential equations, usually partial ones in appropriate variables. But they also contain plenty of so-called physical constants usually experimentally fixed. Let us mention, e.g., $\hbar$, c, G, e, k, etc. (respectively, the Planck constant, velocity of light in vacuum, gravitational constant, elementary electric charge, Boltzmann constant, etc.). The system of natural units like $\hbar = 1$, $c = 1$ is not appropriate. First of all because it is incomparable with the limit transition to zero or infinity. The only solution of the problem is to introduce $\hbar$, c, etc. as "controlling" parameters and to follow quietly (as possible) the consequence of the mentioned manipulations. Let us mention that there are two aspects of the controlling status of those constants. First, one can think about the anthropic principle and perhaps the divine origin of them. The second aspect may be more material. One can, e.g., think on them as fields and even to formulate the corresponding field or motion equations. Let us mention, incidentally, that putting $\hbar = 0$ we get rid of all paradoxes of quantum mechanics, but obtain the non-realistic world with infinities in electromagnetic radiation theory.

In a sense it might seem strange that quantum mechanics was not formulated or at least suggested some hundred years earlier. Though both geometric and physical optics seemed to be known in the deep of XIX century. And one was also aware that the geometric optics is a short-wave asymptotics of the wave theory. And the analogy between mechanics and geometric optics was also known. Mathematically it was based on the similarity of the Hamilton-Jacobi and eikonal equations. But one interpreted it as a formal similarity between description of single particle trajectories and the geometric theory of optical waves moving along the eikonal rays. The mechanical "waves" were not interpreted as the dynamics of a real wave process even in the sense of the Schrödinger wave picture (as a matter of fact also incorrect, as it turned out later on). It seems that the deciding circumstance here was the absence of the Planck constant, the physical quantity of the dimension of

http://dx.doi.org/10.5772/59115

action. It is necessary to replace the Hamilton-Jacobi action $S\left(t,q^i\right)$ by the wave containing factor $\exp\left(i/\hbar\right)S(q)$ with the dimensionless argument. It is strange that the $\hbar$-divisor was discovered by Planck in phenomena of electromagnetic radiation, where it is deeply hidden from the direct observation. And only when it was discovered by Planck, the further procedure was open by Planck himself, Bohr, Sommerfeld, de Broglie, Heisenberg, Schrödinger, and many others.

2. Weyl-Wigner-Moyal-Ville description

It is neither very easy nor automatic to perform the limit transition $\hbar \rightarrow 0$ in quantum-mechanical equations. In any case passing to zero with $\hbar$ in quantum-mechanical wave functions leads to meaningless results. Rather, one should separately use the explicit $\hbar$-dependence of the modulus and phase of the wave function

$$\Psi = \sqrt{D}\exp\left(\frac{i}{\hbar}S\right) = f\exp\left(\frac{i}{\hbar}S\right) \tag{1}$$

and substitute this to the Schrödinger equation (or any other wave equation). And then one should write the system of equations resulting from the comparison of coefficients at the same (but theoretically all possible) powers of $\hbar$. It is important to remember that D, S are power series of $\hbar$, but the divisor $\hbar$ under the exp sign is universally present in equations. The simplest, heuristic way is to use the Weyl-Wigner-Moyal-Ville (WWMV) star product of operators when the phase space is $\mathbb{R}^{2n}$ (or any $2n$-dimensional linear space).

Let $\left(q^1,\ldots,q^n;p_1,\ldots,p_n\right)$ be affine phase space coordinates, $\left(q^1,\ldots,q^n\right)$ — the underlying configuration space variables, and $\left(p_1,\ldots,p_n\right)$ — the induced momentum variables. The operators acting (in principle) in $L^2\left(q^1,\ldots,q^n\right)$ are represented by their kernels-functions or rather distributions when we do not insist on remaining within $L^2(Q)$:

$$\left(\mathbf{A}\Psi\right)\left(\overline{q}\right) = \int A\left[\overline{q},\overline{q}'\right]\Psi\left(\overline{q}'\right)d_n\overline{q}'. \tag{2}$$

What concerns the adjective "distribution-like" let us stress that such important operators as identity, position, and linear momentum are just represented by distributions:

$$\mathbf{1}\left[\overline{q},\overline{q}'\right] = \delta\left(\overline{q}-\overline{q}'\right),\ \mathbf{Q}^a\left[\overline{q},\overline{q}'\right] = q^a\delta\left(\overline{q}-\overline{q}'\right),\ \mathbf{P}_a\left[\overline{q},\overline{q}'\right] = \frac{\hbar}{i}\frac{\partial}{\partial q^a}\delta\left(\overline{q}-\overline{q}'\right). \tag{3}$$

According to the Weyl-Wigner-Moyal-Ville prescription one can represent any function $A\left(q^1,\ldots,q^n;p_1,\ldots,p_n\right)$ by the kernel

$$A\left[\overline{q},\overline{q}'\right] = \int A\left(\frac{1}{2}\left(\overline{q}+\overline{q}'\right),\overline{p}\right)\exp\left(\frac{i}{\hbar}\,\overline{p}\cdot\left(\overline{q}-\overline{q}'\right)\right)\frac{d_n\overline{p}}{\left(2\pi\hbar\right)^n} \tag{4}$$

and conversely, inverting the Fourier transform:

$$A\left(\overline{q},\overline{p}\right)=\int\exp\left(-\frac{i}{\hbar}\,\overline{p}\cdot\overline{\alpha}\right)A\left[\overline{q}+\frac{\overline{\alpha}}{2},\overline{q}-\frac{\overline{\alpha}}{2}\right]d_{n}\overline{\alpha}. \tag{5}$$

Let us remind that this is a consequence of the Weyl-Wigner-Moyal-Ville star product of the pair of phase-space functions on the affine phase space:

$$\left(A*B\right)(z)=2^{2n}\int\exp\left(\frac{2i}{\hbar}\Gamma\left(z-z_{1},z-z_{2}\right)\right)A(z_{1})B(z_{2})d\mu(z_{1})d\mu(z_{2}), \tag{6}$$

$$d\mu(z)=d\mu\left(\overline{q},\overline{p}\right)=\frac{1}{(2\pi\hbar)^{n}}dq^{1}\ldots dq^{n}dp_{1}\ldots dp_{n}, \tag{7}$$

where in the $(\overline{q},\overline{p})$-basis $\Gamma=\begin{bmatrix}O & -I\\ I & O\end{bmatrix}$, I is the $n\times n$ identity matrix, and O is the $n\times n$ matrix composed of zeros. Obviously, the product $A*B$ is isomorphic to the product of operators represented by the "matrix" rule:

$$[AB]\left(\overline{q},\overline{q}'\right)=\int A\left[\overline{q},\overline{q}''\right]B\left[\overline{q}'',\overline{q}'\right]d_{n}q''. \tag{8}$$

And Hermitian conjugate of operators is represented by the complex conjugate of phase-space functions. The above composition of phase-space functions is non-local and in general the positively definite operators (like, e.g., density operators $\boldsymbol{\rho}$) are not represented by non-negative phase-space functions.

It is clear from the above formulae that the action of the operator $\mathbf{A}$ on the configuration space function Ψ is given by

$$\left(\mathbf{A}\Psi\right)\left(\overline{q}\right)=\frac{1}{(2\pi\hbar)^{n}}\int\exp\left(\frac{i}{\hbar}\,\overline{p}\cdot\left(\overline{q}-\overline{q}'\right)\right)A\left(\frac{1}{2}\left(\overline{q}+\overline{q}'\right),\overline{p}\right)\Psi\left(\overline{q}'\right)d_{n}\overline{q}'d_{n}\overline{p}. \tag{9}$$

Let us remind a few properties of the Weyl-Wigner-Moyal-Ville product of the phase-space functions. So, it is bilinear and associative and preserves the complex conjugation:

$$(\lambda A+\mu B)*C=\lambda A*C+\mu B*C,\quad (A*B)*C=A*(B*C), \tag{10}$$

$$C*(\lambda A+\mu B)=\lambda C*A+\mu C*B,\quad \overline{A*B}=\overline{B}*\overline{A}, \tag{11}$$

where $\mathbf{A}^{+}$ is represented by $\overline{A}$. Besides $1*A=A*1$, $\overline{A}*A\neq 0$, if $A\neq 0$ a.e., $\int A*Bd\mu=\int ABd\mu$, but in general $\int A*B*Cd\mu\neq\int ABCd\mu$. Let us notice that

$$\mathrm{Tr}\mathbf{A}=\int A(\overline{z})d\mu(\overline{z}),\quad \langle\mathbf{A},\mathbf{B}\rangle=\mathrm{Tr}\left(\mathbf{A}^{+}\mathbf{B}\right)=\int\overline{A(\overline{z})}B(\overline{z})d\mu(\overline{z}), \tag{12}$$

and $\langle C * A, B\rangle = \langle A, \overline{C} * B\rangle = \langle C, B * \overline{A}\rangle$. Obviously, the star-product of phase-space functions is non-commutative, just as the operator product. Let us quote some formulae concerning non-commutativity and commutativity:

$$q^a * p_b = q^a p_b + \frac{i\hbar}{2}\delta^a{}_b, \quad p_b * q^a = p_b q^a - \frac{i\hbar}{2}\delta^a{}_b, \tag{13}$$

$$q^a * A(q,p) = q^a A(q,p) + \frac{i\hbar}{2}\frac{\partial A}{\partial p_a}, \quad A(q,p) * q^a = A(q,p)q^a - \frac{i\hbar}{2}\frac{\partial A}{\partial p_a}, \tag{14}$$

$$p_a * A(q,p) = p_a A(q,p) - \frac{i\hbar}{2}\frac{\partial A}{\partial q^a}, \quad A(q,p) * p_a = A(q,p)p_a + \frac{i\hbar}{2}\frac{\partial A}{\partial q^a}, \tag{15}$$

and obviously for functions depending only on one kind of variables

$$(A * B)(q) = A(q) * B(q) = A(q)B(q) = (AB)(q), \tag{16}$$

$$(A * B)(p) = A(p) * B(p) = A(p)B(p) = (AB)(p). \tag{17}$$

The star-product is evidently invariant under the action of affine symplectic group, $(U(A,t)f) * (U(A,t)g) = U(A,t)(f * g)$, where t is a translation vector in $\mathbb{R}^{2n}$ and A is a linear symplectic transformation, $\Gamma_{kl}A^k{}_a A^l{}_b = \Gamma_{ab}$, $(U(A,t)f)(z) = f(Az+t)$.

Quantum states are described by density operators ρ which are Hermitian, normalized to unity and positive: $\langle \rho | \mathbf{A}^+\mathbf{A}\rangle = \mathrm{Tr}\left(\boldsymbol{\rho}\mathbf{A}^+\mathbf{A}\right) \geq 0$, $\mathrm{Tr}\boldsymbol{\rho} = 1$. To be honest, one can also live without the last normalization condition. When the condition is satisfied, we have $\mathrm{Tr}\ \boldsymbol{\rho}^2 \leq \mathrm{Tr}\ \boldsymbol{\rho} = 1$. Particularly important are pure states described by projectors, $\widehat{\boldsymbol{\rho}}^2 = \widehat{\boldsymbol{\rho}}$, $\widehat{\boldsymbol{\rho}} = |\Psi\rangle\langle\Psi|$, $\rho\left[\overline{q}, \overline{q}'\right] = \Psi\left(\overline{q}\right)\overline{\Psi}\left(\overline{q}'\right)$. They are related to the corresponding wave function Ψ as follows:

$$\rho\left(\overline{q}, \overline{q}'\right) = \frac{1}{(2\pi)^n}\int \overline{\Psi}\left(\overline{q} - \frac{\hbar}{2}\overline{\tau}\right)\exp\left(-i\overline{\tau}\cdot\overline{p}\right)\Psi\left(q + \frac{\hbar}{2}\overline{\tau}\right)d_n\overline{\tau}. \tag{18}$$

In general it takes on negative values, nevertheless it is positive in the quantum-mechanical sense:

$$\langle \rho | \overline{B} * B\rangle = \int \rho(z)\left(\overline{B} * B\right)(z)d\mu(z) > 0 \tag{19}$$

for all functions B. The exceptional Wigner functions are positive in the literal sense and are exponential:

$$E_{(\overline{\xi},\overline{\pi})}\left(\overline{q}, \overline{p}\right) = \frac{1}{(\pi\hbar)^n}\exp\left(-\frac{1}{\hbar}\left(\left(\overline{q} - \overline{\xi}\right)^2 + \left(\overline{p} - \overline{\pi}\right)^2\right)\right). \tag{20}$$

It is clear that they represent pure states,

$$E_{(\overline{\xi},\overline{\pi})} * E_{(\overline{\xi},\overline{\pi})} = E_{(\overline{\xi},\overline{\pi})}, \qquad \int E_{(\overline{\xi},\overline{\pi})}\left(\overline{q}, \overline{p}\right)d_n\overline{q}\frac{d_n\overline{p}}{(2\pi\hbar)^n} = 1. \tag{21}$$

They are coherent states strongly concentrated in the phase space about the point $(\overline{\xi}, \overline{\pi})$. Therefore, the coarse-grained quantity, so-called Husimi distribution,

$$\widetilde{\rho}(\overline{q}, \overline{p}) = \int E_{(\overline{q}, \overline{p})}(\overline{\xi}, \overline{\pi}) \rho(\overline{\xi}, \overline{\pi}) d_n \overline{\xi} \frac{d_n \overline{\pi}}{(2\pi\hbar)^n} \tag{22}$$

admits an approximate interpretation of the literally positively-definite probability distribution obtained from the Wigner function ρ. Indeed, $E_{(\overline{q}, \overline{p})}$ is in a sense a pure state Wigner function concentrated at $(\overline{q}, \overline{p})$ and therefore the above integral is a probability density for the system to be found in the phase-space cell at $(\overline{q}, \overline{p})$ when it is known to be in a Wigner state ρ. This interpretation is not bad and certainly $\widetilde{\rho}$ is something that in a sense gives an account of the probability distribution to be found in an $\hbar^n$-volume cell about every $(\overline{q}, \overline{p})$. There is only one drawback of this interpretation. Namely, unlike the true Weyl-Wigner-Moyal-Ville distributions, literally non-positive, the Husimi distributions (22) have non-satisfactory, bad marginal properties, because

$$\int \widetilde{\rho}(\overline{q}, \overline{p}) d_n \overline{q} \neq \overline{\widehat{\Psi}}(\overline{p}) \widehat{\Psi}(\overline{p}), \qquad \int \widetilde{\rho}(\overline{q}, \overline{p}) \frac{d_n \overline{p}}{(2\pi\hbar)^n} \neq \overline{\Psi}(\overline{q}) \Psi(\overline{q}). \tag{23}$$

Here, obviously, Ψ, $\widehat{\Psi}$ are the wave functions underlying ρ, respectively in the coordinate and momentum representations. For the ρ itself the above inequalities become exact equalities.

Let us now discuss the problem of the WKB approximation from the point of view of the above remarks. It is clear that from the point of view of the above statements, in the lowest-order approximation of D, S in $\hbar$, we have the following $\hbar$-independent interpretation of the D, S-functions in terms of the expectation values:

$$\langle\Psi|\mathbf{Q}^i|\Psi\rangle = \int D\left(q^1, \ldots, q^n\right) q^i dq^1 \ldots dq^n, \tag{24}$$

$$\langle\Psi|\mathbf{P}_i|\Psi\rangle = \int D\left(q^1, \ldots, q^n\right) \partial_i S\left(q^1, \ldots, q^n\right) dq^1 \ldots dq^n. \tag{25}$$

It is clear that the Planck constant $\hbar$ is absent in those expressions, so really the functions D, S are $\hbar$-independent up to higher orders. In any case it is so at places distant from the turning points. Let us consider the n-dimensional submanifold given by equations $p_i = \partial S / \partial q^i$, $i = 1, \ldots, n$, i.e., $F_i = p_i - \left(\partial S / \partial q^i\right) = 0$. This submanifold, $\mathfrak{m}_S$, is a special case of what is called Lagrangian manifold, because the Poisson brackets of the left-hand sides of its equations vanish; moreover, they vanish after the restriction to $\mathfrak{m}_S$,

$$\left\{F_i, F_j\right\} = \frac{\partial F_i}{\partial q^a} \frac{\partial F_j}{\partial p_a} - \frac{\partial F_i}{\partial p_a} \frac{\partial F_j}{\partial q^a} = \left(\partial^2_{ij} - \partial^2_{ji}\right) S = 0. \tag{26}$$

Let us take the singular probability distribution concentrated on $\mathfrak{m}_S$:

$$\rho_{\rm cl}[D,S] = \lim_{\hbar\to 0}\rho[D,S] = D\left(q^1,\ldots,q^n\right)\delta\left(p_1-\frac{\partial S}{\partial q^1}\right)\ldots\delta\left(p_n-\frac{\partial S}{\partial q^n}\right)$$
$$= \left|\Psi\left(q^1,\ldots,q^n\right)\right|^2\delta\left(p_1-\frac{\partial S}{\partial q^1}\right)\ldots\delta\left(p_n-\frac{\partial S}{\partial q^n}\right). \tag{27}$$

Obviously, it is different from $\rho(q,p)$, nevertheless the expectation values of q^i, p_j and their linear combinations on $\rho_{\rm cl}$ are just the same as those on ρ,

$$\langle\Psi|\alpha_i\mathbf{q}^i+\beta^i\mathbf{p}_i|\Psi\rangle = \int\left(\alpha_i q^i+\beta^i p_i\right)\rho[D,S]d_n\overline{q}\frac{d_n\overline{p}}{(2\pi\hbar)^n}$$
$$= \int\left(\alpha_i q^i+\beta^i p_i\right)\rho_{\rm cl}[D,S]d_n\overline{q}\frac{d_n\overline{p}}{(2\pi\hbar)^n}. \tag{28}$$

Let us mention that all limit transitions here, in particular the one between $\rho[D,S]$ and $\rho_{\rm cl}[D,S]$ are meant in the distribution theory sense.

It is important that the Weyl-Wigner-Moyal-Ville product may be expanded as a power series in $\hbar$ and that the functional coefficients are interpretable in terms of the symplectic geometry of the classical phase space. The first two terms of the expansion are given by

$$A * B \simeq AB + \frac{i\hbar}{2}\{A,B\} + \ldots, \tag{29}$$

the next terms are given by the multiple Poisson brackets. In any case, the Weyl-Wigner-Moyal-Ville product and the corresponding quantum Poisson bracket are given in the limit $\hbar\to 0$ by the following $\hbar$-independent expressions:

$$\lim_{\hbar\to 0} A * B = AB, \tag{30}$$

$$\lim_{\hbar\to 0}\{A,B\}_{\rm QPB} = \lim_{\hbar\to 0}\frac{1}{i\hbar}\left(A*B-B*A\right) = \{A,B\}. \tag{31}$$

Let us mention that these formulae have interesting features and interpretation. Namely, the eigenequation for the wave function implies the following eigenequation for the corresponding density operator: $\mathbf{A}\boldsymbol{\rho} = a\boldsymbol{\rho}$, i.e., in terms of the Weyl-Wigner-Moyal-Ville approach $A*\rho = a\rho$. But this implies $\rho * A = a\rho$ if A is real, i.e., $\mathbf{A}$ is hermitian, and therefore

$$[\mathbf{A},\boldsymbol{\rho}]_{\rm QPB} = \frac{1}{i\hbar}[\mathbf{A},\boldsymbol{\rho}] = 0, \qquad \text{thus,} \qquad \frac{1}{i\hbar}\left(A*\rho-\rho*A\right) = 0. \tag{32}$$

On the quantum level this equation is a direct consequence of the eigenequation $A * \rho = a\rho$. But these equations have quite different qualitative interpretation in physical terms. Namely, $\mathbf{A}\boldsymbol{\rho} = a\boldsymbol{\rho}$ has a purely informational content. It tells us that on the state $\boldsymbol{\rho}$, or ρ in the Weyl-Wigner-Moyal-Ville language, the physical quantity $\mathbf{A}$ (A in the Weyl-Wigner-Moyal-Ville terms) takes spread-freely the value a. This is the purely informational property. But the Poisson bracket property, mathematically following from it, has a qualitatively different interpretation, namely, such a ρ is invariant under the one-parameter group of unitary transformations generated by $\mathbf{A}$ (A),

$$\exp\left(\frac{i}{\hbar}\mathbf{A}\tau\right)\boldsymbol{\rho}\exp\left(-\frac{i}{\hbar}\mathbf{A}\tau\right) = \boldsymbol{\rho}. \tag{33}$$

This is a symmetry property. Therefore, on the quantum level information implies symmetry. But in classical physics Poisson bracket and the pointwise product of functions are algebraically independent. Therefore, information and symmetry of statistical states become logically independent. This implies that in the classical limit Schrödinger equation or the corresponding eigenequation for the density operator must be in the lowest order of approximation replaced by the pair of equations for the phase and modulus of the wave function. Therefore, substituting (1) to (9) and taking the limit $\hbar \to 0$ we obtain:

$$(\mathbf{A}\Psi)(q) \approx A\left(q^i, \frac{\partial S}{\partial q^i}\right)\Psi(q) + \frac{\hbar}{i}\left(\pounds_v f\right)\exp\left(\frac{i}{\hbar}S(q)\right); \tag{34}$$

higher order terms in $\hbar$ are omitted. The symbol $\pounds_v$ denotes the Lie derivative of f with respect to the vector field $v[A, S]$ which equals

$$v^i = \frac{\partial A}{\partial p_i}\left(q^j, \frac{\partial S}{\partial q^j}\right). \tag{35}$$

In spite of the use of analytical symbols, v^i is a well-defined vector field tangent to the manifold $\mathfrak{m}_S$ given by equations $p_j = \partial S/\partial q^j$, $j = 1, \ldots, n$. It is obtained from the Hamiltonian vector field generated by the function A,

$$\mathrm{X}[A] = \frac{\partial A}{\partial p_i}\frac{\partial}{\partial q^i} - \frac{\partial A}{\partial q^i}\frac{\partial}{\partial p_i}. \tag{36}$$

This vector field is tangent to $\mathfrak{m}_S$, so we restrict it to some vector field on this manifold and project it to the configuration space Q, i.e., to the manifold of q^a-variables. It is clear that f geometrically is not a scalar field, but the scalar W-density of weight $1/2$. Therefore,

$$\pounds_v f = v^a \frac{\partial f}{\partial q^a} + \frac{1}{2}\frac{\partial v^a}{\partial q^a} f. \tag{37}$$

D is a scalar density of weight one, thus,

$$\pounds_v D = v^a \frac{\partial D}{\partial q^a} + \frac{\partial v^a}{\partial q^a} D = \frac{\partial}{\partial q^a} \left(D v^a \right). \tag{38}$$

If we consider the Schrödinger equation

$$i\hbar \frac{\partial \Psi}{\partial t} = \mathbf{H}\Psi, \tag{39}$$

then in the quasiclassical limit we obtain the following system of equations:

$$\frac{\partial S}{\partial t} + H\left(q, \frac{\partial S}{\partial q}, t\right) = 0, \qquad \frac{\partial D}{\partial t} + \frac{\partial}{\partial q^a}\left(D \frac{\partial H}{\partial p_a}\left(q, \frac{\partial S}{\partial q}\right)\right) = 0. \tag{40}$$

This is the system composed of the Hamilton-Jacobi equation for S and the continuity equation for D. The second equation is dependent on the solution of the first one. Geometrically it may be written in the form

$$\frac{\partial D}{\partial t} + \pounds_{v[H,S]} D = 0. \tag{41}$$

Let us take a system of n functions A_i on the phase space with pairwise vanishing Weyl-Wigner-Moyal-Ville commutators,

$$A_i * A_j - A_j * A_i = 0. \tag{42}$$

Consider the family of eigenequations for the Weyl-Wigner-Moyal-Ville density function ρ:

$$A_i * \rho = a_i \rho. \tag{43}$$

Obviously, they imply that $(1/i\hbar)\left(A_i * \rho - \rho * A_i\right) = 0$. In the classical limit this system becomes

$$A_i \rho = a_i \rho, \qquad \{A_i, \rho\} = 0. \tag{44}$$

The quantum compatibility condition (42) for (43) implies that in the classical limit the corresponding condition for (44), i.e., $\{A_i, A_j\} = 0$, also holds. The corresponding solution for (44) may be given as:

$$\rho(q,p) = \delta\left(A_1(q,p) - a_1\right) \ldots \delta\left(A_n(q,p) - a_n\right). \tag{45}$$

To be more precise, this holds when $A_1, \ldots, A_n$ is a system of functionally independent analytic functions. This distribution is concentrated on the Lagrangian manifold $\mathfrak{m}_{(A,a)}$ given by equations: $A_i(q,p) = a_i$, $i = 1,\ldots,n$. Solving them with respect to p_i we obtain the transformed equations in the potential form: $p_j = \partial S(q,a)/\partial q^j$. Short calculation shows that ρ may be written as follows:

$$\rho(q,p) = \left|\det\left[\frac{\partial^2 S}{\partial q^i \partial a^j}\right]\right| \delta\left(p_1 - \frac{\partial S(q,a)}{\partial q^1}\right)\ldots\delta\left(p_n - \frac{\partial S(q,a)}{\partial q^n}\right). \tag{46}$$

The quantity $\det\left[\partial^2 S/\partial q^i \partial a^j\right]$ is known as the Van Vleck determinant [20]. The corresponding quasiclassical wave function is given by:

$$\Psi(q,a) = \sqrt{\det\left[\frac{\partial^2 S}{\partial q^i \partial a^j}\right]} \exp\left(\frac{i}{\hbar}S(q,a)\right). \tag{47}$$

This expression is convenient when one of the functions $A_1,\ldots,A_n$ is physically interpretable as a Hamiltonian H. Or when Hamiltonian is a simple function of other quantities A_i — constants of motion, $H = E\left(A_1(q,p),\ldots,A_n(q,p)\right)$. Then the function

$$\Psi(q,t;a) = \sqrt{\det\left[\frac{\partial^2 S}{\partial q^i \partial a^j}\right]} \exp\left(\frac{i}{\hbar}S(q,a) - E\left(a_1,\ldots,a_n,t\right)\right). \tag{48}$$

is an approximate quasiclassical solution of the Schrödinger equation (39) with the continuous spectrum of $A_1,\ldots,A_n$. And here some additional remarks are necessary. The first one is that (47), (48) are valid only far from the turning points. So, they are valid only in the non-compact spaces $\mathbb{R}^n$, $\mathbb{R}^{2n}$, when there is no quantization of A_j at all, or one must modify them so as to admit compact configuration spaces. But then the above version of the Weyl-Wigner-Moyal-Ville formalism does not work and must be replaced by something else. Some way to remain within the framework is to unify the solutions (48) with the quantization of $A_1,\ldots,A_n$ by the Bohr-Sommerfeld quantum conditions. Roughly speaking, the idea is then that only such submanifolds $\mathfrak{m}_{(a^1,\ldots,a^n)}$ are admitted that the periods of $\omega = p_i dq^i$ on $\mathfrak{m}_{(a^1,\ldots,a^n)}$ are integer multiples of the Planck constant. This condition gives rise to the "quantization" of $A_1,\ldots,A_n$. And this is what one really does in the Old Quantum Theory. But in general some difficulties appear on the level of wave functions (47), (48), namely one has to use some Maslov modifications and use the Airy special functions.

Nevertheless, the very heart of idea survives: quasiclassical pure quantum states are represented by probability distributions concentrated on n-dimensional submanifolds of the phase space; let us repeat that n is the number of degrees of freedom. At least locally the expressions (47), (48) are qualitatively correct. This is very interesting from the geometrical point of view. Namely, in spite of using analytic expressions, the Van Vleck determinant is a well-defined, coordinate-independent scalar density of weight two both in the configuration space Q (q^a-variables) and in the $\mathbb{R}^n$-space of the values $a^1,\ldots,a^n$ of constants of motion $A_1,\ldots,A_n$. And its square root is a well-defined scalar W-density of weight one. By its very geometric interpretation, this quantity is a priori the best candidate for the quasiclassical probability distribution of the wave functions (47), (48). Obviously, the care must be taken concerning the mentioned problems, in particular the behaviour at the classical turning points.

In any case, expressions (47), (48) are almost true (in the quasiclassical sense), when the variables q^i are taken modulo 2π, i.e., when the configuration space is topologically a torus, and when there are no turning points at all. Then the Bohr-Sommerfeld quantum conditions work literally (in approximation) and there is no need to introduce the Airy functions. And Van Vleck determinant is a good approximation to the quantum density function.

Summary of Section 2: The main message following from the above study is that the classical limit transition, when correctly carried out, indicates that it is not points of the classical phase space, but rather n-parameter Lagrangian submanifolds in the phase space that is to correspond to the quantum pure states (n is the number of degrees of freedom). Or more precisely, it is probability distributions on those manifolds that are to describe the pure states. When the time variable is taken into account, then it turns out that the pure states evolutions are what J. L. Synge used to call the coherent n-parameter families of classical trajectories. It is in a sense a surprising result that both on the level of wave functions phases and on the level of probability distributions, the corresponding quantities may be a priori guessed on the basis of the classical Hamilton-Jacobi theory and the Van Vleck determinant following from it.

3. Symplectic and contact interpretation

Our arguments above were based on the assumed affine geometry of the phase space. However, it is clear that this fact is not very important. It did not influence our views. Affine geometry and the Weyl-Wigner-Moyal-Ville procedure were merely the auxiliary tools of our analysis. Nevertheless, it is convenient to comment our results in general symplectic terms.

Let (P,γ) be a symplectic manifold, i.e., a differential manifold P endowed with the differential two-form γ satisfying the following conditions: it is closed and non-degenerate. In coordinates ξ^a this means that

$$\gamma = \frac{1}{2}\gamma_{ab} d\xi^a \wedge d\xi^b, \tag{49}$$

where $\gamma_{ab,c} + \gamma_{bc,a} + \gamma_{ca,b} = 0$, $\det\left[\gamma_{ab}\right] \neq 0$ all over P. The comma symbol denotes the partial derivative. Therefore, $\dim P = 2n$, n being natural. As $d\gamma = 0$, then locally $\gamma = d\omega$. Not always, but in majority of applications P is a cotangent bundle over some n-dimensional configuration space,

$$P = T^*Q = \bigcup_{q\in Q} T^*_q Q, \tag{50}$$

where $T_q Q$, $T^*_q Q$ denote as usual the tangent space at $q \in Q$ and its dual — the cotangent space. If q^i, $i = 1,\ldots,n$, are coordinates in an open domain of Q, then the induced coordinates in T^*Q are denoted by $\left(q^i, p_i\right)$, where p_i are components of the canonical momentum attached of $q \in Q$. This structure gives rise to the Cartan one-form ω given locally by $\omega = p_i dq^i$; the coordinate-free definition is easily possible but we do not quote it here. In any case the symplectic form in T^*Q is given by $\gamma = d\omega = dp_i \wedge dq^i$. Being non-degenerate, γ does possess the inverse form $\widetilde{\gamma}$ with coordinates γ^{ab} such that $\gamma^{ac}\gamma_{cb} = \delta^a{}_b$. This gives rise to the Poisson bracket construction

$$\{F,G\} = \gamma^{ab}\frac{\partial F}{\partial \xi^a}\frac{\partial G}{\partial \xi^b}, \tag{51}$$

in the induced coordinates $\left(q^i, p_i\right)$:

$$\{F,G\} = \frac{\partial F}{\partial q^a}\frac{\partial G}{\partial p_a} - \frac{\partial F}{\partial p_a}\frac{\partial G}{\partial q^a}. \tag{52}$$

Canonical transformations preserve the two-form γ, $\varphi_*\gamma = \gamma$, and infinitesimal ones, i.e., canonical vector fields X satisfy $\pounds_X\gamma = 0$. Of course, the identity $\pounds_X\gamma = (X\rfloor d\gamma) + d(X\rfloor\gamma)$ implies that because of $d\gamma = 0$,

$$(d\,(X\rfloor\gamma))_{ab} = (X^c\gamma_{ca})_{,b} - (X^c\gamma_{cb})_{,a} = 0, \tag{53}$$

therefore, at least locally the vector field X is Hamiltonian $(X\rfloor\gamma)_a = X^c\gamma_{ca} = -\partial F/\partial\xi^a$. It is denoted by $X_F = -\widetilde{dF}$ and called the Hamiltonian vector field generated by the local Hamiltonian F. If F is globally one-valued, we say that X_F is a Hamiltonian field generated by F. Therefore, unlike the symmetry group of the symmetric metric tensor on a manifold M, which is a finite-dimensional Lie group of dimension at most $n(n+1)/2$, the group of symplectomorphisms, i.e., one of canonical transformations is always infinite-dimensional, labelled by arbitrary sufficiently smooth functions on P.

An important problem is a classification of submanifolds in a symplectic manifold. This is completely new in comparison to submanifolds in positively definite Riemann spaces. So, let $M \subset P$ be a $(2n-m)$-dimensional submanifold ("constraints") in a symplectic manifold (P,γ), e.g., given by equations

$$F_a(\xi) = F_a(q,p) = 0, \qquad a = 1,\ldots,m. \tag{54}$$

The system of those functions is functionally independent, at least in some neighbourhood of M. Sometimes it is also convenient to take the foliation by submanifolds M_a

$$F_a(q,p) = c_a, \tag{55}$$

where c_a are constants. At every $p \in M$ there is a tangent space T_pM and its symplectic orthogonal (dual) space $T_pM^{\perp}$ which consists of vectors γ_p-"orthogonal" to T_pM:

$$T_pM^{\perp} = \left\{ v \in T_pP : \gamma(p)_{ab} v^b X^a \quad \text{if} \quad X \in T_pM \right\}, \tag{56}$$

or in more sophisticated terms: $\langle v \rfloor \gamma_p, \cdot \rangle | T_pM = 0$. It is a peculiarity of symplectic geometry that $T_pM^{\perp}$ need not be complementary to T_pM. The following class index was introduced to describe this.

If $\dim K_p(M) = \dim \left(T_pM \cap T_pM^{\perp} \right) = k$, then we put $\mathrm{Cl}_pM = (k, m-k)$. We are interested mainly in situation when this does not depend on p, that is, incidentally, a typical situation. Then we write simply $\mathrm{Cl}\ M = (k, m-k)$. If $k = m$, then we write simply $\mathrm{Cl}\ M = \mathrm{I}$ and say that M is co-isotropic. This means that the subspaces γ-orthogonal to M are tangent to M. If $k = 0$, then $\mathrm{Cl}\ M = \mathrm{II}$, and the subspace γ-orthogonal to M are at the same time transversal (complementary) to M. $\mathrm{Cl}\ M = \mathrm{I}$ implies that the functions F_a in (54) or (55) satisfy respectively $\{F_a, F_b\} | M = 0$ or $\{F_a, F_b\} = 0$. Similarly, $\mathrm{Cl}\ M = \mathrm{II}$ implies that $\det[\{F_a, F_b\}] \neq 0$, at least in a neighbourhood of M. If $T_pM \subset T_pM^{\perp}$, then we say that M is isotropic. Then for any pair of tangent vectors at any $p \in M$ we have: $\gamma(p)_{ab} u^a v^b = 0$, when $u, v \in T_pM$. If M is isotropic, then $\dim M \leq n$. If $\dim M = n$, i.e., if $T_pM = T_pM^{\perp}$, we say that M is Lagrangian. It is described by the system of equations $F_a = 0$, $a = 1, \ldots, n$, $\{F_a, F_b\} | M = 0$, or, if we deal with a foliation by Lagrangian manifolds, i.e., with a polarization, then $F_a = c_a$, $\{F_a, F_b\} = 0$. Equations for the Lagrangian submanifold may be solved in the following way with respect to canonical momenta:

$$p_i = \frac{\partial S}{\partial q^i}, \qquad i = 1, \ldots, n, \tag{57}$$

when it is transversal to the fibres of constant q^a, $a = 1, \ldots, n$. Let us denote the corresponding manifold by $\mathfrak{m}_S$. The Hamilton-Jacobi equation

$$\Omega \left(\ldots, q^{\mu}, \ldots; \ldots, \frac{\partial S}{\partial q^{\mu}}, \ldots \right) = 0 \tag{58}$$

means that $\mathfrak{m}_S$ belongs to the zero-valued surface of Ω. We have used here the Greek symbols μ to indicate that the time variable may be included into coordinates. For example, in non-relativistic mechanics:

$$\frac{\partial S}{\partial t} + H \left(t, \ldots, q^i, \ldots; \ldots, \frac{\partial S}{\partial q^i}, \ldots \right) = 0. \tag{59}$$

The integrability condition for the system of Hamilton-Jacobi equations with functions Ω_{ν}, $\nu = 0, 1, \ldots, n$, is given by the equation $\{\Omega_{\mu}, \Omega_{\nu}\} = 0$, i.e., the manifold $\Omega_{\mu} = 0$, $\mu = 0, 1, \ldots, n$, has the class I, i.e., is co-isotropic.

One can show that on every regular submanifold M the assignment $M \ni p \mapsto K_p(M) = T_pM^{\perp} \cap T_pM$ is an integrable distribution, therefore, the quotient manifold $P'(M) = M/K(M)$ carries the canonical symplectic structure γ' such that $\gamma \| M = \pi^* \gamma'$, where $\pi : M \to P'(M)$ is the natural projection. Obviously, $K(M)$ denotes the system of leaves of the distribution. It is clear that $\dim P'(M) = 2\,(n - (m+k)/2)$.

Lagrange manifolds, i.e., isotropic ones of dimension n, are placed, as seen from the formula, only on co-isotropic, i.e., first class submanifolds. And if they are transversal to the configuration Q-fibres (X-fibres), then using the formula (57) we obtain the Hamilton-Jacobi equations (58), (59) or simply

$$A\left(\ldots, q^i, \ldots; \ldots, \frac{\partial S}{\partial q^i}, \ldots\right) = a \tag{60}$$

for the potential S. Every $\mathfrak{m}_S \subset M$ is composed of the foliation of singular fibres $K(M)$. Singular fibres, first of all one-dimensional ones, i.e., integral curves of the Hamiltonian vector fields X_F, X_Ω are classical trajectories. Those of which $\mathfrak{m}_S$ are composed were called by Synge coherent families with the potential S [18, 19]. As seen, they correspond to quasiclassical wave functions. And classically, being dependent on n parameters, they correspond to the complete integrals of Hamilton-Jacobi equation (59). Let us summarize our symplectic interpretation of them.

In quantum mechanics the eigenstates of the physical quantity represented by the Hermitian operator $\mathbf{A}$ are given by density operators $\widehat{\rho}$ satisfying the operator eigenequation $\boldsymbol{\rho}$. Let us stress that in general this is the equation both on $\boldsymbol{\rho}$ and a. Taking its Hermitian conjugate we obtain $\boldsymbol{\rho}\mathbf{A} = a\boldsymbol{\rho}$. One can write these equations as $(\mathbf{A} - a\mathrm{Id})\,\boldsymbol{\rho} = 0$, $\boldsymbol{\rho}\,(\mathbf{A} - a\mathrm{Id}) = 0$. This is the afore-mentioned informative aspect of the eigenequation. But just as it was within the Weyl-Wigner-Moyal-Ville framework, this information context implies the formal consequence, but qualitatively a completely different symmetry property, namely the invariance of ρ under the unitary group generated by $\mathbf{A}$: $(1/i\hbar)[\mathbf{A}, \boldsymbol{\rho}] = (1/i\hbar)\,(\mathbf{A}\boldsymbol{\rho} - \boldsymbol{\rho}\mathbf{A}) = 0$. Therefore, in the classical limit one must assume that the quasiclassical ρ satisfies a pair of mathematically independent, but physically interpretable just as above, conditions: $A\rho = a\rho$ — information, $\{A, \rho\} = 0$ — symmetry.

Let us introduce the set of operators: $E_\rho := \{\mathbf{F} \in B(H) : \mathbf{F}\boldsymbol{\rho} = 0\}$. In principle $B(H)$ denotes the set of bounded operators acting in the Hilbert space H. Although, to be honest, one can weaken this assumption. It is also clear that similarly as in classical statistics, the following holds in quanta:

$$-S(\rho_1) = \mathrm{Tr}\,(\rho_1 \ln \rho_1) \leq \mathrm{Tr}\,(\rho_2 \ln \rho_2) = -S(\rho_2)\,, \tag{61}$$

when $E_{\rho_1} \subset E_{\rho_2}$. In other words, the larger E_ρ, the greater informational content of $\boldsymbol{\rho}$. Of course, we mean here the quantum concept of the Shannon entropy and the mathematical sense of $\mathrm{Tr}\,(\rho \ln \rho)$. Quantum pure states are defined in such a way that E_ρ is a maximal ideal. It answers uniquely the maximal number of experimental questions. There exists then the one-dimensional linear subspace $V \subset H$ such that E_ρ consists of operators which vanish on V,

$$E_{\rho} = \{\mathbf{F} \in B(H) : \mathbf{F}|V = 0\}. \tag{62}$$

This means that the subspace $V \subset H$ given by $V = \bigcap_{\mathbf{F}\in E_{\rho}} \mathrm{Ker}\mathbf{F}$ satisfies conditions: $\boldsymbol{\rho}(H) = V$, $\boldsymbol{\rho}|V = \mathrm{Id}_V$, $\boldsymbol{\rho}\boldsymbol{\rho} = \boldsymbol{\rho}$, therefore, ρ is a projector of H onto V. The entropy (information) takes on ρ the minimal (maximal) value, $\mathrm{Tr}\,(\boldsymbol{\rho} \ln \boldsymbol{\rho}) = 0$. When $P = T^*Q$, then the formulae (47), (48) may be literally applied together with the Bohr-Sommerfeld quantum rules: $\oint \omega = \oint p_i dq^i = nh$ on any closed curve on $\mathfrak{m}_S$. This defines the quantized values of a^i in terms of integers and Planck constant.

Expression for the Van Vleck determinant is correct independently on the additional phase space structures like the affine one. Just because of the structure of this determinant. This is seen from the density formula:

$$\mathcal{V} = \det\left[\frac{\partial S}{\partial q^i \partial a^j}\right] dq^1 \wedge \ldots \wedge dq^n \otimes da^1 \wedge \ldots \wedge da^n. \tag{63}$$

Moreover, it turns out that this determinant is much more general and even the cotangent bundle structure is not necessary for it. Namely, let us assume a pair of polarizations, i.e., a pair of complementary foliations of a general phase space (P, γ) by Lagrangian manifolds. Let us observe that P need not be identical with T^*Q and everything we assume is just a pair of foliations. Lagrangian submanifolds of any foliation have a local affine structure. Introducing coordinates q^i, a^i, we can formally describe them in terms of equations: $p_i = \partial S(q,a)/\partial q^i$, but S is non-unique up to the gauging: $S \mapsto S + \varphi \circ \mathrm{pr}_1 + \Psi \circ \mathrm{pr}_2$, where pr_1, pr_2 are projections from P to the $Q, \mathbb{R}^n$-manifolds. But this gauging does not influence the value of the Van Vleck determinant.

It is interesting to see what follows when we consider Hamiltonian and quantum dynamics in a homogeneous formulation of Hamiltonian/quantum dynamics. So, let us consider the motion of a particle in an $(n+1)$-dimensional space-time manifold X and take a complete integral $S\left(x^{\mu}, a^i\right)$ depending on n arbitrary constants. Then instead of the above quantities we obtain the following vector-density object:

$$\mathcal{V} = D^{\mu} dx^0 \wedge dx^1 \wedge \ldots \wedge \mu \wedge \ldots \wedge dx^n \otimes da^1 \wedge \ldots \wedge da^n, \tag{64}$$

where D^{μ} is a minor of the matrix $\left[\partial^2 S / \left(\partial x^{\mu} \partial a^i\right)\right]$ obtained by removing the μ-th column. The symbol μ in the exterior product means that dx^{μ} is omitted. It is clear that the above expressions imply that

$$\frac{\partial j^{\mu}}{\partial x^{\mu}} = 0, \tag{65}$$

where $j^{\mu} - (-1)^{\mu} D^{\mu}$. This formula is geometrically correct, because j^{μ} is a contravariant vector density of weight one. Therefore, the left-hand side of (65) is well defined in spite of using the usual partial differentiation. One can easily check that it follows from (58). In particular, if Ω in (58) equals the non-relativistic $\Omega = p_0 + H\left(x^0, x^i; p_i\right) = -E + H\left(t, q^i, p_i\right)$ (E denotes the energy variable), then j^μ in (64), (65) equals the formerly written non-relativistic four-current

$$(j^\mu) = \left(\det\left[\frac{\partial^2 S}{\partial q^i \partial a^j}\right], \det\left[\frac{\partial^2 S}{\partial q^i \partial a^j}\right]\frac{\partial H}{\partial p_i}\left(q, \frac{\partial S}{\partial q}\right)\right). \tag{66}$$

This j^μ satisfies the continuity equation (65) in virtue of (40). For the general relativistically written $\Omega\left(x^\mu, p_\mu\right)$, one obtains the four-current density, e.g., for the quasiclassical Klein-Gordon equation. The current (66) corresponds to some choice of the complete integral of Hamilton-Jacobi equations.

Let us mention that for the system of Hamilton-Jacobi equations

$$\Omega_\wedge\left(x^\mu, \frac{\partial S}{\partial x^\mu}\right) = 0 \tag{67}$$

we obtain generalized continuity equations. However, there is no place to stop at this topic here. In any case j^μ is also built of the complete integral of (67).

It is difficult not to be astonished by the fact that the above structures were not discovered some hundred years earlier. They are based on the purely classical and deeply geometric concepts. As mentioned, this may be explained only by the fact that the Planck constant was not known then. To be more precise, it was hidden deeply in the thickest of radiation theory and its thermodynamics.

Let us mention some additional facts. We said that the Van Vleck symbol may be assigned to any complementary pair of polarizations $Q \times \mathbb{R}^n \ni (q, a) \to V(q, a)$. It may be interpreted in a statistical way due to its structure of the double scalar density. Indeed, the quantity $\mathcal{V}(q, a) = \det\left[\partial^2 S / \partial q^i \partial a^j\right]$ may be dualistically interpreted as the density of probability both in Q and in $\mathbb{R}^n$. If $A \subset Q$, $B \subset \mathbb{R}^n$, then

$$P(A, B) = \int_{A \times B} \mathcal{V}(q, a) dq^1 \ldots dq^n da^1 \ldots da^n \tag{68}$$

may be interpreted as the quasiclassical probability that the system with values of integrals of motion in $B \subset \mathbb{R}^n$ will be found in the region $A \subset Q$ of the configuration space. And conversely, it is equal to the probability that the system placed in $A \subset Q$ will show the values of integration constants in $B \subset \mathbb{R}^n$. To be honest, in general they are non-normalized to unity relative probabilities.

When performing pull-backs of probability densities on Q to $\mathfrak{m}_S$, we obtain some probability distributions on the Lagrangian manifold. Therefore, quasiclassical pure quantum states may be interpreted as probability distributions concentrated on submanifolds $\mathfrak{m}_S$. So, their supports are n-dimensional and distinguished by the fact that $\gamma\|\mathfrak{m}_S = 0$. Quasiclassical

http://dx.doi.org/10.5772/59115

mixed states are usually smeared out as $2n$-dimensional probability distributions on $P = T^*Q$.

Let us quote yet some another quasiclassical structures. To do this we begin with the linear symplectic spaces. Let $D(P)$ denote the set of all linear Lagrangian subspaces of P. Let $M \subset P$ be some co-isotropic linear subspace of P and $D(M) \subset D(P)$ denote the set Lagrangian subspaces contained in M. One can show that any $\mathfrak{m} \subset D(P)$ intersects M along some at least $(n-m)$-dimensional isotropic subspace. But the singular fibre of M, i.e., $M^\perp \subset M$ is m-dimensional. Therefore, the subspace

$$E_M(\mathfrak{m}) := \mathfrak{m} \cap M + M^\perp \tag{69}$$

is also Lagrangian and contained in M. Therefore, without any additional structure M gives rise to the mapping $E_M : D(P) \to D(M)$ with the following properties:

1. E_M is a retraction onto the subset $D(M)$, moreover, it is a projection:

$$E_M | D(M) = \mathrm{id}_{D(M)}, \qquad E_M \circ E_M = E_M. \tag{70}$$

2. M, N are co-isotropic and compatible, i.e., $M \cap N$ is also co-isotropic, then E_M, E_N commute and

$$E_M \circ E_N = E_N \circ E_M = E_{M \cap N}. \tag{71}$$

3. If $E_M \circ E_N = E_N \circ E_M$, then M, N are compatible and (71) holds.
4. If $f : P \to P$ is a symplectic (γ-preserving) mapping, then

$$E_{f(M)} = F \circ E_M \circ F^{-1}, \tag{72}$$

where $F : D(P) \to D(P)$ is induced by f.

This is interesting and easily interpretable in terms of quasiclassical wave functions. The relationship with the corresponding quantum relations is also readable. Let us illustrate this with the following simple example.

We consider an affine (or linear) phase space with affine coordinates $\left(q^i, p_i\right)$. Then we have $\omega = p_i dq^i$, $\gamma = d\omega = dp_i \wedge dq^i$. Let us take as M the manifold of states on which the momentum variable p_1 takes on a fixed values b,

$$M = \{p \in P : p_1(p) = b\}, \qquad D(P) \ni \mathfrak{m} = \left\{p \in P : q^i(p) = a^i\right\}, \tag{73}$$

where $i = 1, \ldots, n$, therefore, M is a manifold with fixed values of p_1 equal to b, and $\mathfrak{m}$ is the Lagrangian space (the carrier of a quasiclassical state) with fixed positions a^i. One can easily show that

$$E_M(\mathfrak{m}) = \left\{p \in P : p_1(p) = b, q^2(p) = a^2, \ldots, q^n(p) = a^n\right\}. \tag{74}$$

Therefore, if we fix the value of p_1 with the help of E_M, we result in a complete indeterminacy of q^1. But this is just the classical uncertainty principle. Simply on the classical, or rather semi-classical level, it is not a point in the phase space, but rather Legendre submanifold, or to be more precise, a probability distribution on it, that is a counterpart of the quantum wave state.

P was assumed here to be a linear space endowed with a symplectic structure. But it turns out that the above prescription may be globalized to the general symplectic manifold. Roughly speaking, this follows from its "flatness" which makes it similar to a linear symplectic space in finite domains due to the existence of Darboux coordinates which enable one to write $\gamma = dp_i \wedge dq^i$. This holds in every symplectic manifold, not necessarily cotangent bundle, locally, but in finite domains. Indeed, let $M \subset P$ be a co-isotropic submanifold, $K(M)$ — its singular foliation, and $\mathfrak{m} \subset P$ — Lagrangian submanifold. The manifolds M and $\mathfrak{m}$ need not to intersect; in such situation we say that $E_M(\mathfrak{m}) = \varnothing$. In this way the empty set $\varnothing$ is joined to $D(P)$. It corresponds to the vanishing wave function. We put also $E_M(\varnothing) = \varnothing$. Similarly we do when $\mathfrak{m}$, M intersect in a non-clean way, i.e., otherwise than linear subspaces. Let us assume the generic case, when $\mathfrak{m}$, M intersect in a regular way, i.e., when $T_p\mathfrak{m} \cap T_pM = T_p(\mathfrak{m} \cap M)$ for any $p \in \mathfrak{m} \cap M$. To be more precise, we assume that the subset of points p at which this is satisfied is a Lagrangian submanifold. Obviously, $\mathfrak{m} \cap M$ is an isotropic submanifold. If the intersection $\mathfrak{m} \cap M$ is regular at its every point p, then $E_M(\mathfrak{m})$ is defined as the maximal extension of $\mathfrak{m} \cap M$ by the singular foliation $K(M)$, $E_M(\mathfrak{m}) = \pi^{-1}(\mathfrak{m} \cap M)$. It is evidently Lagrangian and $E_M : D(P) \to D(M)$ satisfies the above properties (70)–(72). One should mention only that (72) is satisfied by every canonical mapping, i.e., every diffeomorphism preserving the two-form γ. The finite-dimensional symplectic group is replaced by the infinite-dimensional group "parameterized" by arbitrary functions.

It is also interesting to know that there are classical counterparts of superpositions and scalar products. We have seen that pure states are represented in a sense by probability distributions concentrated on Lagrange manifolds $\mathfrak{m}_S$. But the function S itself, i.e., the phase of wave functions, is not contained in the corresponding analogy. One should adjoint an additional dimension, action, and consider locally the manifolds $Q \times \mathbb{R}$, $P \times \mathbb{R}$, or rather $Q \times \mathrm{SU}(1)$, $P \times \mathrm{SU}(1)$. To be mathematically more honest, one should use the principal fibre bundles with the bases Q, P and the structure group $\mathbb{R}_{\text{additive}}$ or $\mathrm{SU}(1)_{\text{multiplicative}}$. The corresponding geometry of the contact fibre bundle C over P is locally given by

$$\Omega = p_i dq^i - dz. \tag{75}$$

Take the set of Legendre submanifolds corresponding to the complete integral $\{S_a : a \in A\}$ of the Hamilton-Jacobi equation (40), (58), (59). These solutions may be represented by their diagrams (locally) in $Q \times \mathbb{R}$ or $Q \times SU(1)$:

$$\text{Graph } S_a = \{(q, S_a(q)) : q \in Q\}. \tag{76}$$

The independence of the Hamilton-Jacobi equation on the algebraic presence of the variable S implies that for any value of a the function $S_a + t(a)$ with $t(a) \in \mathbb{R}$ is a solution too. The Hamilton-Jacobi equation imposes only conditions on the tangent elements of functions, therefore, the envelope of diagrams $\{(q, S_a + t(a)) : q \in Q\}$ denoted by

$$\text{Env}_{a \in A} \{(q, S_a(q) + t(a)) : q \in Q\} \tag{77}$$

also represents some solution of (40), (58), (59). The arbitrariness of these solutions corresponds exactly to the arbitrariness of functions $t : A \to \mathbb{R}$. Let us repeat that (77) is a diagram of the set of values $\{(q, S(q)) : q \in Q\}$, where the function S is obtained from the family of S_a-s and t in the following way:

i We start from equations:

$$\frac{\partial}{\partial a^i} (S_a(q) + t(a)) = 0 \tag{78}$$

and solve them, at least in principle, with respect to a. One obtains (also in principle) some q-dependent solution, $a(q)$.

ii This solution is substituted to $S_a(q) + t(a)$ and one obtains the expression denoted by the Stat-symbol,

$$S(q) = S_{a(q)}(q) + t(a(q)) = \text{Stat}_{a \in A} (S_a(q) + t(a)). \tag{79}$$

This follows from the theory of the Hamilton-Jacobi equation. But the same may be shown from "deriving" the continuous superpositions of wave functions satisfying the Schrödinger equation, by performing the WKB-limit transition $\hbar \to 0$ in the following expression:

$$\int \sqrt{w(a)} \exp\left(\frac{i}{\hbar} t(a)\right) \sqrt{D(a)} \exp\left(\frac{i}{\hbar} S(q, a)\right) d_n a. \tag{80}$$

In the WKB-limit $\hbar \to 0$ one obtains just (79) as the limit condition. Let us mention that for any function S on a differentiable manifold A the symbol Stat$f$ denotes the value of S at the stationary point $a \in A$, where $dS_a = 0$. If there are many stationary points, then S in (79) is multivalued. This means that the quasiclassical superposition consists of several waves with various values of phases. We omit the relatively complicated quasiclassical behaviour of moduli, just for simplicity. Similarly, for the pair of wave functions $\Psi_1 = \sqrt{D_1} \exp(i/\hbar) S_1$, $\Psi_2 = \sqrt{D_2} \exp(i/\hbar) S_2$ we can investigate the quasiclassical behaviour of the scalar product:

$$\langle \Psi_1 | \Psi_2 \rangle = \int \overline{\Psi}_1(q) \Psi_2(q) d_n q - \int \sqrt{D_1 D_2} \exp\left(\frac{i}{\hbar} (S_1 - S_2)\right) d_n q. \tag{81}$$

Denoting $\langle \Psi_1 | \Psi_2 \rangle = \sqrt{D} \exp\left(i/\hbar\right) S$ and applying the method of stationary phase we again obtain $S = \text{Stat}\left(S_2 - S_1\right) = \text{Stat}_{q \in Q}\left(S_2(q) - S_1(q)\right)$, where just as previously, $\text{Stat}_{q \in Q}\, \varphi(q)$ denotes the value of φ at the point $q \in Q$, where the differential of φ vanishes:

$$d\varphi_q = 0. \tag{82}$$

The situation is simple when (82) has exactly one solution. If there are a few $q_1, \ldots, q_k$, then the scalar product is a superposition of ones with the corresponding phases

$$\lambda_1 \exp\left(\frac{i}{\hbar}\varphi_1\right) + \ldots + \lambda_k \exp\left(\frac{i}{\hbar}\varphi_k\right), \tag{83}$$

where λ-s are obtained from the quasiclassical limits of D. In any case there is a multivalued phase $\varphi_1, \ldots, \varphi_k$. If (82) has no solutions, then we say that Ψ_1, Ψ_2 are quasiclassically orthogonal.

It would be nice to express those concepts in terms of the contact geometry (75). Let us remind that the Lagrange submanifolds $\mathfrak{m}_S \subset P$ represent rather the density operators of quasiclassical pure states than their wave functions. The latter ones are represented by horizontal lifts of Lagrange submanifolds, i.e., by the maximal, thus n-dimensional horizontal submanifolds in C, i.e., such ones $\mathfrak{M}$ that $\Omega \| \mathfrak{M} = 0$. In particular, they are given as $\mathfrak{M}_S = \text{hor}\, \mathfrak{m}_S$, where

$$\mathfrak{M}_S := \left\{ \left(dS_q, S(q)\right) : q \in Q \right\}. \tag{84}$$

But they need not be so; another extreme example is the horizontal lift of $T_q^* Q$,

$$\mathfrak{M}_q := \left(T_q^* Q, 0\right) = \left\{ (p, 0) : p \in T_q^* Q \right\}. \tag{85}$$

Intermediate examples between (84), (85) are labelled by pairs (M, S) where $M \subset Q$ is a submanifold of Q and $S : M \to \mathbb{R}$ is a real-valued function on M. The corresponding Legendre submanifold in $T^* Q$ is given by

$$\mathfrak{M}_{(M,S)} := \left\{ (p, S(\pi(p))) : \pi(p) \in M,\ p | T_{\pi(p)} M = dS_{\pi(p)} \right\}, \tag{86}$$

$\pi : T^* Q \to Q$ is the natural projection of the co-tangent bundle onto its base.

Let us now translate the above formulae into the language of contact geometry. Some similarities to the rigorous quantum expressions will be obvious. The vertical fibre bundle action of the structural group $\mathbb{R}$ or $\text{SU}(1)$ when operating on the elements $\mathfrak{M}$ of the set of Legendre manifolds $\mathcal{H}(C)$ will be denoted as follows: $[t]\mathfrak{M} := \{g_t(z) : z \in \mathfrak{M}\}$, where g_t is the action of the group element in C. If T is a countable subset of the group elements, then we put:

$$T\mathfrak{M} := \bigcup_{t\in G}[t]\mathfrak{M} = \bigcup_{t\in G}\{g_t(z) : z \in \mathfrak{M}\}. \tag{87}$$

In the case of empty sets (which correspond formally to zero), we have $\varnothing\mathfrak{M} = \varnothing$, $T\varnothing = \varnothing$.

Now let us express the quasiclassical "superposition" and phases of the "scalar products" in terms of the contact geometry. Let $M \subset C$ be a submanifold. Its characteristic subset $\Sigma(M)$ is defined as the set of all points $z \in M$ at which $\Omega_z|T_zM = 0$. In practical applications we often deal with the situation that $\Sigma(M)$, which is always horizontal, is at the same time n-dimensional, therefore, it is a Legendre submanifold, i.e., an element of $\mathcal{H}(C)$. Let us assume that $\{\mathfrak{M}_a : a \in A\}$ is a family of elements of $\mathcal{H}(C)$, i.e., a family of Legendre submanifolds. We say that its superposition is the maximal element of $\mathcal{H}(C)$ which is contained in the determinant set of $\bigcup_{a\in A}\mathfrak{M}_a$. We denote it by $\mathfrak{M} = E_{a\in A}\mathfrak{M}_a$. Without going into details we show with the help of examples below that this superposition is in fact, from the point of view of $Q \times \mathbb{R}$, the envelope or "generalized envelope" of the family of surfaces:

i Let us again consider the contact manifold $C = T^*Q \times \mathbb{R}$ with the natural contact form $p_i dq^i - dz$. We take a manifold A parameterizing functions $S_a(q) = S(q,a)$ and the coefficients function $f : A \to \mathbb{R}$. S gives rise to the following family of Legendre manifolds:

$$\mathfrak{M}_a := \mathfrak{M}_{S(\cdot,a)} = \left\{\left(dS(\cdot,a)_q, S(q,a)\right) : q \in Q\right\}. \tag{88}$$

If it happens (it need not be so) that

$$E_{a\in A}[f(a)]\mathfrak{M}_a = \mathfrak{M}_S = \left\{\left(dS_q, S(q)\right) : q \in Q\right\}, \tag{89}$$

then we obtain $S(q) = \mathrm{Stat}_{a\in A}\left(S(q,a) + f(a)\right)$. And this means that the manifold $\xi_S := \{(q, S(q)) : q \in Q\} \subset Q \times \mathbb{R}$ is really the literal envelope of the family of submanifolds $\xi_a := \xi_{S(\cdot,a)} = \{(q, S(q,a)) : q \in Q\} \subset Q \times \mathbb{R}$.

ii Again we consider $C = T^*Q \times \mathbb{R}$ with the following natural contact form: $\Omega = p_i dq^i - dz$. And we take again the q-localized Legendre manifold $\mathfrak{M}_q = \left(T_q^*Q, 0\right)$, and $\mathfrak{M}_S = \left\{\left(dS_q, S(q)\right) : q \in Q\right\}$. One can show that $\mathfrak{M}_S = E_{q\in Q}[S(q)]\mathfrak{M}_q$; this is a kind of irregular envelope from the point of view of the geometry of Q.

iii We take a linear space V as a manifold Q. Then we have $T^*Q \simeq V \times V^*$ and $C \simeq V \times V^* \times \mathbb{R}$. And then we take as Legendre submanifolds the following ones with well-defined positions, $\mathfrak{M}[x] = \{(x, p, 0) : p \in V^*\}$, and with the fixed canonical momenta, $\mathfrak{M}[p] = \{(x, p, \langle p, x\rangle) : x \in V\}$. One can easily check the next rules of the quasi-classical Fourier analysis: $\mathfrak{M}[p] = E_{x\in V}\left[\langle p, x\rangle\right]\mathfrak{M}[x]$, $\mathfrak{M}[x] = E_{p\in V^*}\left[-\langle p, x\rangle\right]\mathfrak{M}[p]$. And then for any "phase" function $S : V \to \mathbb{R}$ we have $\mathfrak{M}_S = E_{x\in V}\left[S(x)\right]\mathfrak{M}[x] = E_{p\in V^*}\left[\widehat{S}(p)\right]\mathfrak{M}[p]$ with the following translation rules between S and $\widehat{S}$:

$$\widehat{S}(p) = \mathrm{Stat}_{x\in V}\left(S(x) - \langle p, x\rangle\right), \; S(x) = \mathrm{Stat}_{p\in V^*}\left(\widehat{S}(p) + \langle p, x\rangle\right). \tag{90}$$

When we take into account that the analogy between phases of $\Psi(x)$ and $\widehat{\Psi}(p)$ is seriously accepted, we see immediately the obvious quasiclassical relationship between $\Psi(x)$ and $\widehat{\Psi}(p)$. It is based on the concept of generalized envelope.

Let us also notice that all those concepts are invariant with respect to the special contact transformations in C. First of all, let us remind that $u : C \to C$ is a special contact transformation when it preserves Ω, $u^*\Omega = \Omega$. If u is such and $U : \mathcal{H}(C) \to \mathcal{H}(C)$ is the corresponding transformation of $\mathcal{H}(C)$, then $UE_{a\in A}[t_a]\mathfrak{M}_a = E_{a\in A}[t_a]U\mathfrak{M}_a$.

Now let us begin with the concept of the vertical distance, i.e., scalar product of Legendre submanifolds. Let us take a pair of Legendre manifolds $\mathfrak{M}_1, \mathfrak{M}_2 \in \mathcal{H}(C)$ with the property that their Lagrange projections $\mathfrak{m}_1$, $\mathfrak{m}_2$ and also $\mathfrak{m}_1 \cap \mathfrak{m}_2$ are connected and simply-connected. Then there exists exactly one number $t \in \mathbb{R}$ (or $\exp(it) \in \mathrm{SU}(1)$) of the property that $(g_t\mathfrak{M}_1) \cap \mathfrak{M}_2 \neq \varnothing$. This number t or better its unitary exponent $\exp(it)$ is the scalar product of $\mathfrak{M}_1$ and $\mathfrak{M}_2$. To be more precise, the classical scalar product is $\exp(it)$, where t is its phase. More generally, we say that the vertical distance, or the Huygens scalar product $[\mathfrak{M}_1|\mathfrak{M}_2]$ of the pair of Legendre submanifolds $\mathfrak{M}_1$, $\mathfrak{M}_2$ is a subset of $\mathbb{R}$ (or exponentially of $\mathrm{SU}(1)$) such that if $t \in [\mathfrak{M}_1|\mathfrak{M}_2]$, then $\mathfrak{M}_2 \cap g_t(\mathfrak{M}_1) \neq \varnothing$. If $[\mathfrak{M}_1|\mathfrak{M}_2]$ is empty, then we say that $\mathfrak{M}_1$, $\mathfrak{M}_2$ are orthogonal. Their Lagrange projections $\mathfrak{m}_1$, $\mathfrak{m}_2$ onto $P = T^*Q$ are then disjoint. It is clear that any mapping $U : \mathcal{H}(C) \to \mathcal{H}(C)$ generated by a special contact transformation $u : C \to C$ is then "unitary" in the sense of the scalar product $[\cdot,\cdot]$, i.e., $[U\mathfrak{M}_1|U\mathfrak{M}_2] = [\mathfrak{M}_1|\mathfrak{M}_2]$ for any pair of Legendre submanifolds $\mathfrak{M}_1$, $\mathfrak{M}_2$.

Let $M \subset P = T^*Q$ be any co-isotropic submanifold and $\mathcal{H}_M(C) \subset \mathcal{H}(C)$ denote the set of Legendre submanifolds with Lagrange projections to $P = T^*Q$ placed within M. Then the operations E_M on Lagrangian submanifolds introduced above may be canonically lifted to the operations Π_M acting on the horizontal lifts of $\mathfrak{m} \subset D(P) = D(T^*Q)$. Namely, for any co-isotropic M there exists the canonical mapping $\Pi_M : \mathcal{H}(C) \to \mathcal{H}_M(C)$ with the property: $\Pi \circ \Pi_M = \Lambda_M \circ \Pi$, $(\Pi_M\mathfrak{M}) \cap \mathfrak{M} = \left(\pi^{-1}(M)\right) \cap \mathfrak{M}$, where $\Pi : \mathcal{H}(C) \to D(P) = D(T^*Q)$ is the natural projection induced by the fibre bundle projection $\pi : C \to P$. In fact, $\Pi_M\mathfrak{M}$ is the horizontal lift of $\Lambda_M\mathfrak{m}$ which contains $\mathfrak{M} \cap \left(\pi^{-1}(M)\right)$. If $\mathfrak{m}$ intersects M in a regular way, then $\Pi_M\mathfrak{M}$ is a maximal extension of the intersection $\mathfrak{M} \cap \left(\pi^{-1}(M)\right)$ along the fibres of the Ω-horizontal lift $K^{\Omega}(M) = \text{lift } K(M)$. If $\mathfrak{m} \cap M = \varnothing$ or if it is not regular, then it is assumed that $\Pi_M\mathfrak{M} = \varnothing$.

Let us repeat again that Π_M satisfy the properties (69)–(72) modified by the admitted empty-set values: $\Pi_M|\mathcal{H}_M(C) = \mathrm{Id}_{\mathcal{H}_M}(C)$, $\Pi_M \circ \Pi_M = \Pi_M$, $\Pi_M \circ \Pi_N = \Pi_N \circ \Pi_M = \Pi_{M\cap N}$ if Cl $M \cap N$ = I. If $\Pi_M \circ \Pi_N = \Pi_N \circ \Pi_M$, then $M \cap N$-compatible and the both sides equal $\Pi_{M\cap N}$. For any special contact transformation $\Pi_{fM} = F \circ \Pi_M \circ F^{-1}$, where F is a transformation of $\mathcal{H}(C)$ induced by f.

It is clear that every special contact transformation u, i.e., diffeomorphisms of C preserving Ω, projects to P onto canonical transformation $\underline{u}$ preserving γ, $\pi \circ u = \underline{u} \circ \pi$. Let $\{\mathfrak{M}_q : q \in Q\}$ be a system of Legendre submanifolds of C such that $\bigcup_{q\in Q}\mathfrak{M}_q$ is an image of a cross-section of C over P, such that the projections to P, $\mathfrak{m}_q$ form a polarization, i.e., a family of mutually disjoint Lagrange submanifolds of P. Then u acts on $\mathfrak{M}_q$ in such a way that

$$U\mathfrak{M}_q = E_{q'\in Q} U\left(q',q\right)\mathfrak{M}_{q'}, \qquad U\left(q',q\right) = \left[\mathfrak{M}_{q'}|U\mathfrak{M}_q\right]. \tag{91}$$

Then for any superposition-envelope $\mathfrak{M} = E_{q\in Q}\left[S(q)\right]\mathfrak{M}_q$ the following holds:

$$U\mathfrak{M} = E_{q\in Q}\left[S'(q)\right]\mathfrak{M}_q = E_{q\in Q}\left[S(q)\right]U\mathfrak{M}_q, \tag{92}$$

where $S'(q) = \mathrm{Stat}_{q'\in Q}\left(U\left(q,q'\right) + S\left(q'\right)\right)$. This is an obvious analogue and the phase classical limit of the linear rule for superposition of wave functions with the phase factors $\exp\left(i/\hbar\right)S\left(q'\right)$. And $U\left(q,q'\right)$ is just the generating function of the type $W(q,Q) = U\left(q,q'\right)$. And a similar construction may be built for other types of generating functions.

Let us also mention that a similar "quasilinear" representation may be achieved for other operations on Legendre submanifolds, not necessarily ones induced by diffeomorphisms acting in C. For example, let us consider Π_M, i.e., $\mathfrak{M} = E_{q\in Q}\left[S(q)\right]\mathfrak{M}_q$, $\Pi_M\mathfrak{M} = E_{q\in Q}\left[S'(q)\right]\mathfrak{M}_q$. Then we have $S'(q) = \mathrm{Stat}_{q'\in Q}\left(S\left(q'\right) + \Pi_M\left(q',q\right)\right)$, where $\Pi_M : Q\times Q \to \mathbb{R}$ is the Legendre propagator $\Pi_M\left(q',q\right) = \left[\mathfrak{M}_{q'}|\Pi_M\mathfrak{M}_q\right]$. This is again the envelope-like Huygens-quasilinear rule. Using the Stat-symbol one can also write a nice-looking analogue of the Feynman-Stückelberg "sum over paths" rule.

Incidentally, let us remind that by the W-type generating function $W\left(q,q'\right)$ we mean such one that the corresponding canonical transformation $(q,p) \mapsto \left(q',p'\right)$ is given by $p_i = \partial W\left(q,q'\right)/\partial q^i$, $p'_i = -\partial W\left(q,q'\right)/\partial q'^i$. Not every canonical transformation does possess such a function in the usual sense, but it may be replaced by a more general generating function. There is no place here to get deeper into details, cf. e.g. [11].

Let us also stress a few other facts connected with the notion of (generalized) envelope. Consider the compatible system of Hamilton-Jacobi equations: $F_a\left(\ldots,x^\mu,\ldots;\ldots,\partial S/\partial x^\mu,\ldots\right) = 0$. Any fibre of the cotangent bundle may be Λ_M-projected onto $D(M)$ — the set of Lagrange submanifolds of M, $\mathfrak{m}_x := \Lambda_M\left(T^*_x X\right)$. Then, every $(n+1)$-dimensional $\pi^{-1}(\mathfrak{m}_X)$ (where $n = \dim X$) is foliated by the family of Legendre lifts of $\mathfrak{m}_X$. When $C = T^*X\times\mathbb{R}$ or $C = T^*X\times \mathrm{U}(1)$ those lifts are $\mathfrak{M}_{(x,c)} := \Pi_M\left(T^*_x X, c\right)$. Then locally

$$\mathfrak{m}_x \cap T^*_y X = \left\{d\sigma(x,\cdot)_y\right\}, \quad \mathfrak{M}_x \cap \left(T^*_y X\times\mathbb{R}\right) = \left\{d\sigma(x,\cdot)_y, \sigma(x,y)\right\}. \tag{93}$$

For every pair of points $x,y\in X$ we define the quantity $\sigma_M(x,y)$, namely

$$\sigma_M(x,y) = \int_{l(x,y)} \omega = \int_{\ell(x,y)} p_\mu dx^\mu, \tag{94}$$

where $\ell(x,y)$ is any curve placed on a singular fibre containing $x,y\in X$. This σ_M is a fundamental solution, $\Pi_M\mathfrak{M}_S - \mathfrak{M}_{S'}$, $\Lambda_M\mathfrak{m}_S = \mathfrak{m}_{S'}$, where $S'(x) = \mathrm{Stat}_y\left(S(y) + \sigma_M(y,x)\right)$. When $\Sigma \subset X$ is a Cauchy surface for our Hamilton-Jacobi system, then $S(x) =$

$\mathrm{Stat}_{q\in\Sigma}\left(f(q)+\sigma_M(q,x)\right)$, where $f:\Sigma\rightarrow\mathbb{R}$ are initial data. Therefore, the two-point characteristic function is a Hamilton-Jacobi propagator. The idempotence property of Λ_M, Π_M implies that $\sigma_M(x,y)=\mathrm{Stat}_z\left(\sigma_M(x,z)+\sigma_M(z,y)\right)$. Let us quote an interesting example of the free material point in Galilean space-time. Then

$$\frac{1}{\hbar}\sigma_M(x,y)=\frac{1}{\hbar}S(a,z;q,t)=\frac{m}{2\hbar(t-z)}g_{ij}\left(q^i-a^i\right)\left(q^j-a^j\right). \tag{95}$$

When the Van Vleck determinant $\det\left[\partial^2 S/\partial q^i\partial a^j\right]$ is multiplied by some normalization constant, then the Van Vleck solution

$$\sqrt{\det\left[\frac{\partial^2 S}{\partial q^i\partial a^j}\right]}\exp\left(\frac{im}{2\hbar(t-z)}g_{kl}\left(q^k-a^k\right)\left(q^l-a^l\right)\right) \tag{96}$$

becomes

$$\mathcal{K}\left(\overline{\xi},\tau\right)=\left(\frac{m}{2\pi i\hbar\tau}\right)^{n/2}\exp\left(\frac{im}{2\hbar\tau}\overline{\xi}^2\right), \tag{97}$$

where $\tau=t-z$, $\xi^k=q^k-a^k$, and $\overline{\xi}^2=g_{kl}\xi^k\xi^l$. The normalization we have accepted is given by $\lim_{\tau\rightarrow 0}\mathcal{K}\left(\overline{\xi},\tau\right)=\delta\left(\overline{\xi}\right)$, where $\mathcal{K}$ is the usual rigorous quantum propagator for the Schrödinger equations:

$$i\hbar\frac{\partial\Psi}{\partial t}=-\frac{\hbar^2}{2m}\Delta\Psi=-\frac{\hbar^2}{2m}g^{ij}\partial_i\partial_j\Psi, \tag{98}$$

in spite of the fact that it was obtained in a purely classical way.

Summary of Section 3: In this section we have reminded some classical problems concerning classification of submanifolds in the classical phase space. Their classical interpretation in terms of symplectic and contact structures was discussed. Again it turns out that the classical limits are Huygens constructions based on the envelope concepts. The quantum and classical relationships between information and symmetry were discussed. This analysis shows again, without any use of the Weyl-Wigner-Moyal-Ville product that it is probability distributions concentrated on n-dimensional Lagrange manifolds that corresponds to the quantum pure states on the classical level. The homogeneous Van Vleck objects corresponding to the evolution problems were discussed. In particular, this may be used to the analysis of the Klein-Gordon equation. Discussed is the WKB-limit of certain quantum expressions like superpositions of wave functions and their scalar products. Again one obtains expressions based on the envelope concepts and the Huygens-like operations on the set of Lagrangian manifolds. Classical counterparts of the projection operators were found. The envelopes of diagrams of phases of wave functions are geometrically interpreted in terms of contact geometry. This is a geometric picture valid for all types of the eikonal equations. It enables one to interpret also the purely classical concepts like generating functions of canonical transformations in quantum-like form based on the envelopes of diagrams of phases. Roughly speaking, the envelope represents the phase of the classical superposition. It is shown that the quantum propagation for the free evolution Schrödinger equation may be smoothly guessed on the purely classical level, in terms of the Van Vleck determinant.

http://dx.doi.org/10.5772/59115

4. Nonlinearity program in quantum mechanics

Let us consider a finite-level quantum mechanical system. We try to interpret the Schrödinger equation as a usual self-adjoint equation of mathematical physics, just as if it was to be a classical one. If both the first- and second-order time derivatives of the state vector Ψ are to be admitted, the Lagrange function may be postulated as

$$L(1,2) = i\alpha\Gamma_{\bar{a}b}\left(\overline{\Psi}^{\bar{a}}\dot{\Psi}^{b} - \dot{\overline{\Psi}}^{\bar{a}}\Psi^{b}\right) + \beta\Gamma_{\bar{a}b}\dot{\overline{\Psi}}^{\bar{a}}\dot{\Psi}^{b} - \gamma_{\Gamma}H_{\bar{a}b}\overline{\Psi}^{\bar{a}}\Psi^{b}, \tag{99}$$

where α, β, γ are constants and $\Gamma_{\bar{a}b}$ are components of the scalar product. $_{\Gamma}H_{\bar{a}b}$ are components of the Hamiltonian matrix in the covariant form,

$$_{\Gamma}H_{\bar{a}b} = \Gamma_{\bar{a}c}H^{c}{}_{b}, \tag{100}$$

whereas $H^{c}{}_{b}$ are usual mixed tensor components. To be honest, in physics it is this mixed Hamilton operator that is considered as a primary quantity. From the Lagrangian point of view it is a twice covariant form that is primary. Because of this we decided to assume $H^{c}{}_{b}$ as a primary quantity, but the Hermitian matrix $H_{\bar{a}b}$ is assumed as the constitutive element in (99). When we are going to introduce a direct nonlinearity to the treatment, we introduce in addition a real-valued potential $V\left(\Psi,\overline{\Psi}\right)$, e.g.,

$$V\left(\Psi,\overline{\Psi}\right) = f\left(\Gamma_{\bar{a}b}\overline{\Psi}^{\bar{a}}\Psi^{b}\right), \tag{101}$$

where f is a real-valued function of the one real variable.

For the Lagrangian $L = L(1,2) - V$ we obtain the following "Schrödinger", or rather "Schrödinger-Klein-Gordon", equation:

$$2i\alpha\frac{d\Psi^{a}}{dt} - \beta\frac{d^{2}\Psi^{a}}{dt^{2}} = \gamma H^{a}{}_{b}\Psi^{b} + f'\Psi^{a}. \tag{102}$$

The comparison with the usual Schrödinger equation tells us that $\alpha = \hbar/2, \gamma = 1$. The energy function for $L = L(1,2) - V$ is given by $\mathcal{E} = \beta\Gamma_{\bar{a}b}\dot{\overline{\Psi}}^{\bar{a}}\dot{\Psi}^{b} + \gamma_{\Gamma}H_{\bar{a}b}\overline{\Psi}^{\bar{a}}\Psi^{b} + V\left(\Psi,\overline{\Psi}\right)$. Legendre transformation tells us that the corresponding Hamiltonian in the sense of analytical mechanics is:

$$\mathcal{H} = \frac{1}{\beta}\left[\Gamma^{a\bar{b}}\pi_{a}\overline{\pi}_{\bar{b}} + i\alpha\left(\pi_{a}\Psi^{a} - \overline{\pi}_{\bar{a}}\overline{\Psi}^{\bar{a}}\right)\right] + \left[\frac{\alpha^{2}}{\beta}\Gamma_{\bar{a}b} + \gamma_{\Gamma}H_{\bar{a}b}\right]\overline{\Psi}^{\bar{a}}\Psi^{b} + V\left(\Psi,\overline{\Psi}\right), \tag{103}$$

where π_{a}, $\overline{\pi}_{\bar{a}}$ are canonical momenta conjugate to Ψ^{a}, $\overline{\Psi}^{\bar{a}}$. It is clear that $\mathcal{E}$ is always defined all over the state space. Unlike this, $\mathcal{H}$ is defined all over the phase space only when $\beta \neq 0$. If $\beta = 0$, it is defined only on the constraints submanifold.

The possible nonlinearity of quantum mechanics is due to the term $V\left(\Psi,\overline{\Psi}\right)$. As mentioned, this is a rather artificial, perturbative nonlinearity. It is definitely better to use the essential nonlinearity of the geometric, group-theoretic origin. To achieve this, one should follow the idea of transition from special to general relativity. The simplest way is to "de-absolutize" the scalar product. Namely, instead of being fixed once for all, the scalar product will be reinterpreted as a dynamical variable. It is to be self-interacting and free of any fixed absolute background. Therefore, in the finite-level theory, its dynamics will be $\mathrm{GL}(n,\mathbb{C})$-invariant. The only natural Lagrangian will follow the structure of affinely-invariant kinetic energy of affinely-rigid body. So, for the metric $\Gamma_{\overline{a}b}$ we postulate the following Lagrangian:

$$T = L[\Gamma] = \frac{A}{2}\Gamma^{b\overline{c}}\Gamma^{d\overline{a}}\dot{\Gamma}_{\overline{a}b}\dot{\Gamma}_{\overline{c}d} + \frac{B}{2}\Gamma^{b\overline{a}}\Gamma^{d\overline{c}}\dot{\Gamma}_{\overline{a}b}\dot{\Gamma}_{\overline{c}d}. \tag{104}$$

This is the only possibility which is not based on anything absolutely fixed. Obviously, the main term is the first one, controlled by A. The B-term is a correction, not very essential, but acceptable from the point of view of the assumed $\mathrm{GL}(n,\mathbb{C})$-symmetry.

In the Lagrangian (100), (101) for the wave function the scalar product is also replaced by this new, dynamical version. Due to the resulting very essential nonlinearity following from (104) the quantum-classical gap in a sense becomes diffused. One can hope that in the resulting theory the decoherence phenomena may be explained. For example, if for simplicity we fix Ψ^a as constant (non-excited), then we can show that differential equations for Γ obtained from (104) have the following solutions: $\Gamma_{\overline{r}s} = G_{\overline{r}s}\exp\left(Et\right)^{z}{}_{s} = \exp\left(\overline{F}t\right)_{\overline{r}}{}^{\overline{z}}G_{\overline{z}s}$, where $G = \Gamma(0)$ is Hermitian and the forms ${}_{G}E_{\overline{r}s} = G_{\overline{r}z}E^{z}{}_{s}$, $(F_{G})_{\overline{r}s} = \overline{F}_{\overline{r}}{}^{\overline{z}}G_{\overline{z}s}$ are also Hermitian. Let us observe that depending on the initial data G, E, F the scenarios of the evolution of $t \mapsto \Gamma(t)$ may be quite different: oscillatory, exponentially increasing, exponentially decaying. This may suggest that in the rigorous total solutions for $t \mapsto (\Psi(t), \Gamma(t))$ also various phenomena may be predicted, e.g., oscillations, but perhaps also decoherence.

The maximal class of $\mathrm{GL}(n,\mathbb{C})$-invariant Lagrangians $L[\Psi,\Gamma]$ is relatively wide. It seems however that the simplest and at the same time most realistic subclass is given by the following expression:

$$\begin{aligned} L = {} & i\alpha_1\Gamma_{\overline{a}b}\left(\overline{\Psi}^{\overline{a}}\dot{\Psi}^{b} - \dot{\overline{\Psi}}^{\overline{a}}\Psi^{b}\right) + \alpha_2\Gamma_{\overline{a}b}\dot{\overline{\Psi}}^{\overline{a}}\dot{\Psi}^{b} + \left(\alpha_3\Gamma_{\overline{a}b} + \alpha_4 H_{\overline{a}b}\right)\overline{\Psi}^{\overline{a}}\Psi^{b} \\ & + \alpha_5\Gamma^{d\overline{a}}\Gamma^{b\overline{c}}\dot{\Gamma}_{\overline{a}b}\dot{\Gamma}_{\overline{c}d} + \alpha_6\Gamma^{b\overline{a}}\Gamma^{d\overline{c}}\dot{\Gamma}_{\overline{a}b}\dot{\Gamma}_{\overline{c}d} - \mathcal{V}\left(\Psi,\overline{\Psi};\Gamma\right). \end{aligned} \tag{105}$$

The quantities $\alpha_1,\ldots,\alpha_6$ are real constants. They control all the effects mentioned above. The separation of the α_3- and α_4-terms may look artificial. In fact, their superposition is as a matter of fact one term. Nevertheless, it seems reasonable to separate the true Hamiltonian effect from that following from the identity operator type. As mentioned formerly, to obtain the correct Schrödinger behaviour in the Ψ-sector we must put $\alpha_1 = \hbar/2$, $\alpha_4 = -1$. But of course if $\alpha_2 \neq 0$, then in addition to the Schrödinger behaviour we have also as usual in analytical mechanics, the acceleration term in Ψ. The scheme is a bit obscured because of our dealing with a finite-level system. The extension to the usual quantum mechanics, say in $\mathbb{R}^3$, is possible. Besides, let us remind our papers devoted to the study of the $\mathrm{SU}(2,2)$-gauge gravitation theory [12, 13]. There the problem of nonlinearity and the interplay between first- and second-order differential equations for the matter fields appear in a much more evident way.

We do not quote "Schrödinger equation" for the pairs (Ψ, Γ) ruled by Lagrangians (105). Their structure is very readable, nevertheless, their strong nonlinearity prevented us from finding their convincing full solutions, when the mutual interaction between Ψ- and Γ-degrees of freedom is taken into account.

Summary of Section 4: This section was one of the main parts of our study. We are aware that in spite of all similarities and analogies there is still some really quantum kernel of the theory which seems to be incompatible with any attempts of formulating the peaceful coexistence of the unitary "between measurement" and the "reduction-like" phenomena in quantum physics. As usual, the idea of nonlinearity in quantum physics turns out to be attractive. It seems to be the only way to coordinate the "between measurements" unitary evolution and the measurement reduction process. There were various more or less happy ways to introduce nonlinearity; some of them were rather artificial. Our idea resembles the transition from the special to general relativity. Namely, we give up the concept of scalar product fixed once for all and instead consider the scheme in which the wave function and scalar product are both dynamical objects in the mutual interaction. Lagrangian for the scalar product is geometric, invariant under the full linear group and so is the total Lagrangian for the system: wave function and the scalar product. The resulting scheme is nonlinear in an essential, non-perturbative way. There are some indications that the resulting nonlinear system of equations may describe both the "between measurements" evolution and the reduction of state process.

5. Modifications of the WWMV approach

We have seen that the mentioned approach was a very fruitful tool for studying the quasi-classical problems and the relationship between information and symmetry. Unfortunately, its literal version applies rigorously only to systems with affine geometry of the classical phase space. There are various ways to generalize those methods, usually based on group theory and deformation techniques. Some of those methods are applicable also to discrete structures.

Let G be a locally compact topological group. Its Haar measure element will be denoted by dg. To be honest, to avoid problems with the convergence of integrals, we may assume G to be compact. Let us introduce the following non-local product of functions on G:

$$(A \perp B)(g) = \int \mathcal{K}(g; g_1, g_2)\, A(g_1) B(g_2) dg_1 dg_2. \tag{106}$$

To be honest, we think about the multiplication rule for functions not necessarily on G itself, but rather on its affine space, i.e., on the set on which G acts with trivial isotropy groups. Therefore, we assume the translational invariance, so that $\mathcal{K}(g; g_1, g_2) \equiv \mathcal{K}\left(g_1 g^{-1}, g_2 g^{-1}\right)$. The simultaneous assumption of associativity, $A \perp (B \perp C) = (A \perp B) \perp C$ implies that $\mathcal{K}$ must satisfy the following functional equation:

$$\int \mathcal{K}(g_1,g)\mathcal{K}\left(g_2g^{-1},g_3g^{-1}\right)dg = \int \mathcal{K}\left(g_1g^{-1},g_2g^{-1}\right)\mathcal{K}(g,g_3)dg. \tag{107}$$

When G is a locally compact Abelian group, one can try to translate (107) into the language of Fourier transforms. This is suggested by the convolution-like structure of this condition. Let us remind that the dual group $\widehat{G}$ is the multiplicative group of all continuous homomorphisms of G into $\mathbb{T} = SU(1)$ — the group of complex numbers of modulus 1. The Fourier transform of a complex function Ψ on G is the function $\widehat{\Psi}$ on $\widehat{G}$ given by:

$$\widehat{\Psi}(\chi) = \int \overline{\langle\chi|g\rangle}\Psi(g)dg, \tag{108}$$

where $\langle\chi|g\rangle$ denotes the value of $\chi \in \widehat{G}$ at $g \in G$. And conversely,

$$\Psi(g) = \int \langle\chi|g\rangle\widehat{\Psi}(\chi)d\chi. \tag{109}$$

Performing the two-argument Fourier transformation on the equation (107) we obtain the following condition:

$$\widehat{\mathcal{K}}\left(\chi_1,\chi_2\chi_3\right)\widehat{\mathcal{K}}\left(\chi_2,\chi_3\right) = \widehat{\mathcal{K}}\left(\chi_1,\chi_2\right)\widehat{\mathcal{K}}\left(\chi_1\chi_2,\chi_3\right). \tag{110}$$

This is an equation for the factor of ray representations. It is clear that the Fourier representation of (106) in the Abelian case is given by:

$$\left(\widehat{A}\top\widehat{B}\right)(\chi) = \int \widehat{\mathcal{K}}\left(\chi_1,\chi_1^{-1}\chi\right)\widehat{A}\left(\chi_1\right)\widehat{B}\left(\chi_1^{-1}\chi\right)d\chi_1. \tag{111}$$

This is the $\widehat{K}$-twisted convolution of functions. It becomes the usual convolution when $\widehat{K} \equiv 1$. In analogy to (111) one defines the twisted convolution of measures.

A similar operation, i.e., twisted convolution of functions, or more generally, one of measures, may be defined in any locally compact topological group,

$$\left(A\top B\right)(g) = \int \omega\left(h,h^{-1}g\right)A(h)B\left(h^{-1}g\right)dh, \tag{112}$$

again in the sense of Haar measure dh. This product is associative for any group G, not necessarily the Abelian one, if and only if the mentioned functional equation (110) holds, i.e., if ω behaves like the factor of the ray representation, $\omega\left(g_1,g_2\right)\omega\left(g_1g_2,g_3\right) = \omega\left(g_1,g_2g_3\right)\omega\left(g_2,g_3\right)$. One can show that there is a relationship between the twisted convolutions of functions (or measures) over G and the usual ones in some G-extension of the circle group $\mathbb{T} = SU(1)$. The choice of the phase-space group G depends on the particular model.

Let us go back to the situation when G is the group which models the configuration space, not the phase space. And we assume G to be Abelian. The phase space $\mathcal{G}$ will be given by $\mathcal{G} = G \times \widehat{G}$, where the dual group $\widehat{G}$ is to model the "space of momenta". In analogy to the natural symplectic two-form on the linear space $V \times V^*$ we introduce the following two-character on $\mathcal{G}$, $\zeta : \mathcal{G} \times \mathcal{G} \to \mathbb{C}$ (the two-character, because $\zeta(\xi, \cdot)$, $\zeta(\cdot, \xi)$ are characters on $\mathcal{G}$ for any $\xi \in \mathcal{G}$):

$$\zeta\left((x_1, \pi_1), (x_2, \pi_2)\right) = \langle \pi_1 | x_2 \rangle \overline{\langle \pi_2 | x_1 \rangle} = \frac{\langle \pi_1 | x_2 \rangle}{\langle \pi_2 | x_1 \rangle}. \tag{113}$$

It is non-singular in the sense that the mappings $\xi \mapsto \zeta(\xi, \cdot)$, $\xi \mapsto \zeta(\cdot, \xi)$ are isomorphisms of $\mathcal{G}$ onto $\mathcal{G}$. Wave functions in the position and momentum representations are defined as amplitudes on G and $\widehat{G}$ respectively. The group actions of G, $\widehat{G}$ on wave functions are given by the following unitary representations:

$$\left(U(x)\Psi\right)(y) = \Psi\left(x^{-1}y\right), \qquad \left(V(\pi)\Psi\right)(y) = \langle \pi | y \rangle \Psi(y). \tag{114}$$

The second operator is obviously equal to the argument translation when the momentum representation is used: $\left(V(\pi)\Psi\right)^{\wedge}(\lambda) = \widehat{\Psi}\left(\pi^{-1}\lambda\right)$. One can check easily that the following fundamental commutation relation is satisfied:

$$U(x)V(\pi)U(x)^{-1}V(\pi)^{-1} = \overline{\langle \pi | x \rangle} = \langle \pi | x \rangle^{-1}. \tag{115}$$

Following the ideas of the Weyl prescription we define the following unitary operators: $W_p(x, \pi) = \langle \pi | x \rangle^p U(x) V(\pi) = \langle \pi | x \rangle^{p-1} V(\pi) U(x)$. If in G or $\widehat{G}$ there exists a unique square root like in $\mathbb{R}^n$, then we put $p = 1/2$ and then $W\left(x^{-1}, \pi^{-1}\right) = W(x, \pi)^{-1}$. But in general it does not exist and we retain p as a non-defined label. We take the linear closure:

$$\mathbf{A} = \int \widehat{A}(x, \pi) \mathbf{W}_p(x, \pi) dx d\pi, \tag{116}$$

$\widehat{A}$ denoting the Fourier transform of A. The corresponding "multiplication" rule for the functions A, B is based on the kernel:

$$K_p\left((x_1, \pi_1), (x_2, \pi_2)\right) = \int \langle \pi_1 | \xi \rangle \langle \eta | x_1 \rangle \langle \pi_2 | \zeta \rangle \langle \theta | x_2 \rangle \langle \eta | \zeta \rangle^{1-p} \langle \theta | \xi \rangle^{-p} d\xi d\eta d\zeta d\theta. \tag{117}$$

One can ask about the analogue of the "continuous canonical basis" of the usual H^+-algebra over $\mathbb{R}^{2n}$:

$$\rho_{\overline{q}_1, \overline{q}_2}\left(\overline{q}, \overline{p}\right) = \delta\left(\overline{q} - \frac{1}{2}\left(\overline{q}_1 + \overline{q}_2\right)\right) \exp\left(\frac{i}{\hbar}\overline{p} \cdot \left(\overline{q}_2 - \overline{q}_1\right)\right), \tag{118}$$

$$\rho_{\overline{p}_1, \overline{p}_2}\left(\overline{q}, \overline{p}\right) = \delta\left(\overline{p} - \frac{1}{2}\left(\overline{p}_1 + \overline{p}_2\right)\right) \exp\left(\frac{i}{\hbar}\left(\overline{p}_1 - \overline{p}_2\right) \cdot \overline{q}\right). \tag{119}$$

Those bases satisfied:

$$q^i * \rho_{\overline{q}_1,\overline{q}_2} = q_1{}^i \rho_{\overline{q}_1,\overline{q}_2}, \quad \rho_{\overline{q}_1,\overline{q}_2} * q^i = q_2{}^i \rho_{\overline{q}_1,\overline{q}_2}, \tag{120}$$

$$p_i * \rho_{\overline{p}_1,\overline{p}_2} = p_{1i} \rho_{\overline{p}_1,\overline{p}_2}, \quad \rho_{\overline{p}_1,\overline{p}_2} * p_i = p_{2i} \rho_{\overline{p}_1,\overline{p}_2}. \tag{121}$$

It turns out, however, that when there is no square-rooting in $G, \widehat{G}$, there are some problems. Namely, in $\mathbb{R}^{2n}$ we could use both ζ and ζ^2 as kernels. But in a general Abelian group it is essential that we use ζ, not ζ^2 as a kernel. The analogue of (120), (121) reads: $A * \rho_{x,y} = A(x)\rho_{x,y}, \rho_{x,y} * A = \rho_{x,y}A(y)$. We obtain $\rho_{x_1,x_2}(x,\pi) = \delta\left(x_1 x_2 x^{-1}\right)\langle \pi | x_1 x^{-1} \rangle$. If $x_1 x_2$ fails to be a square, then $g_{x_1 x_2} = 0$. It is not yet clear for us if our procedure was improper or if we deal with the real superselection rule.

Let us observe that the group commutator does not feel the choice of p:

$$W_p(x_1,\pi_1)W_p(x_2,\pi_2)W_p(x_1,\pi_1)^{-1}W_p(x_2,\pi_2)^{-1} = \zeta\left((x_1,\pi_1),(x_2,\pi_2)\right)\mathrm{Id}. \tag{122}$$

If $G = \mathbb{Z}^n$ or $\mathbb{T}^n = (\mathrm{SU}(1))^n$, then the mentioned problem with ζ may be connected with what in solid state physics is known as so-called Umklapp-Prozessen.

Let us observe also that in a sense one can use the following kernel of the non-local product of functions over the discrete group $\mathbb{Z}^{2n}$: $K(\overline{n},\overline{m}) = \exp\left(iB_{ab}n^a m^b\right)$, where $[B_{ab}]$ is the real skew-symmetric matrix. Nevertheless, the resulting product will have then some strange features.

Let us finish with some remarks concerning the asymptotics of "large quantum numbers" in the quasi-classical limit transition. For simplicity we consider only the wave functions of the planar rotators, $\Psi_n(\varphi) = \exp(in\varphi)$, $n \in \mathbb{Z}$, where φ is the angular variable. Ψ_n is proportional to the eigenfunction of the angular momentum with the eigenvalue $n\hbar$,

$$\frac{\hbar}{i}\frac{\partial}{\partial\varphi}\Psi_n = \hbar n \Psi_n. \tag{123}$$

Clearly

$$\Psi_n \simeq \exp\left(\frac{i}{\hbar}(n\hbar)\varphi\right); \tag{124}$$

it is just $l_n = n\hbar$ that is interpreted as the physical value of the angular momentum. But the analysis $\hbar \to 0$ does not work directly. We must take superpositions of quickly-oscillating eigenequations,

$$\Psi(\varphi) = \sum_n c_n \exp(in\varphi), \tag{125}$$

where the sequence $\mathbb{Z} \ni n \mapsto c_n \in \mathbb{C}$ is concentrated in a range

$$n_0 - \Delta n \ll n \ll n_0 + \Delta n. \tag{126}$$

It is assumed here that $n_0 \gg \Delta n \gg 1$ and the sequence is assumed to be slowly-varying in the range (126), so that

$$\frac{|c_{n+1} - c_n|}{|c_n|} \ll 1. \tag{127}$$

It follows from the Fourier theory that approximately (125) may be replaced by:

$$\Psi(\varphi) = \int c(k) \exp(ik\varphi) dk, \tag{128}$$

where at the discrete values of $k = n$ $c(k)$ equals c_n and changes slowly, e.g., linearly between them. Then $\Psi(\varphi)$ is well concentrated and one can consider it as a quickly-vanishing at infinity function on $\mathbb{R}$. And then one substitutes $k = p/\hbar$ and further on the previous asymptotics $\hbar \to 0$ may be used. The same when there are more degrees of freedom. Let us remind that it was just this limit transition we have used in the theory of angular momentum [14–16]. By the way, the conditions (126), (127) enabled one to remove artificial picks of the basic wave functions at $k = 2\pi$ in $\mathrm{SU}(2)$. Namely, the subsequent picks have opposite signs and mutually cancel when (126), (127) are satisfied.

Summary of Section 5: We have mentioned here about some generalizations of the Weyl-Wigner-Moyal-Ville procedure. They are based on some group-theoretic models and may be perhaps helpful in the formally "classical", although in fact quantum, approach to dynamics.

6. Conclusions

We have discussed certain problems concerning the relationship between classical and quantum theories. Analyzed are both differences and formal similarities between them. What concerns similarities, we show that in contrast to some popular views, it is not points in the classical phase space but rather n-dimensional Lagrangian submanifolds in the phase space that corresponds to the quantum pure states. More precisely, the classical "pure state" is a probability distribution on the Lagrange manifold, or rather on its horizontal Legendre lift to the contact space. Here n is the number of degrees of freedom and the contact space is, roughly speaking, the Cartesian product of the phase space by $\mathbb{R}$ or $\mathrm{U}(1)$ with geometry given by $p_i dq^i - dz$. It was shown that superpositions, scalar products, etc. are defined in the set of Legendre manifolds and have some formal properties of the corresponding quantum concepts. They are based on the Huygens notion of envelope of the wave fronts. This was shown both directly on the basis of limit transition in the Weyl-Wigner-Moyal-Ville formalism and on the basis of general symplectic language. Nevertheless, it is clear that quantum mechanics with its reduction and decoherence problems is something completely different than the classical theory. We try to show that unlike this view, there is a nonlinear modification of quantum theory which perhaps would be free of the mentioned paradoxes. It is based on the classical language of variational principles and on the concept of dynamical scalar product. The system consisting of wave function and scalar product satisfies an essentially nonlinear, non-perturbative dynamical equation. Its characteristic nonlinearity seems to be able to describe analytically the decoherence process. Finally, we review some generalization of the Weyl-Wigner-Moyal-Ville formalism and discuss the quasi-classical limit in terms of "large quantum numbers".

The general conclusion/hypothesis is that perhaps there is no such a gap between classics and quanta as one commonly believes.

Acknowledgements

This paper partially contains results obtained within the framework of the research project N N501 049 540 financed from the Scientific Research Support Fund in the years 2011-2014. The authors are greatly indebted to the Polish Ministry of Science and Higher Education for this financial support.

Author details

J. J. Sławianowski* and V. Kovalchuk*

Institute of Fundamental Technological Research, Polish Academy of Sciences, Warsaw, Poland

*Address all correspondence to: jslawian@ippt.pan.pl, vkoval@ippt.pan.pl

References

[1] Abraham R., Marsden JE. Foundations of Mechanics (second edition). London-Amsterdam-Don Mills-Ontario-Sydney-Tokyo: The Benjamin-Cummings Publishing Company, Inc.; 1978.

[2] Arnold VI. Mathematical Methods of Classical Mechanics (second edition). Graduate Texts in Mathematics 60, New York-Berlin-Heidelberg-London-Paris-Tokyo-Hong Kong-Barcelona-Budapest: Springer-Verlag; 1989.

[3] Caratheodory C. Variationsrechnung und Partielle Differential-Gleidungen Erster Ordnung. Leipzig: B.G. Teubner; 1956.

[4] Doebner HD., Goldin GA. Introducing Nonlinear Gauge Transformations in a Family of Nonlinear Schrödinger Equations. Phys. Rew. A 1996;54 3764–3771.

[5] Doebner HD., Goldin GA., Nattermann P. Gauge Transformations in Quantum Mechanics and the Unification of Nonlinear Schrödinger Equations. J. Math. Phys. 1999;40 49–63; quant-ph:9709036.

[6] Landau LD., Lifshitz EM. Quantum Mechanics. London: Pergamon Press; 1958.

[7] Mackey GW. The Mathematical Foundations of Quantum Mechanics. New York: The Benjamin-Cummings Publishing Company, Inc.; 1963.

[8] Moyal JE. Quantum Mechanics as a Statistical Theory. Proc. Cambridge Philosophical Society 1949;45 99–124.

[9] Moyal JE. Stochastic Processes and Statistical Physics. Journal of the Royal Society B 1949;11 150–210.

[10] Sławianowski JJ. Uncertainty, Correspondence and Quasiclassical Compatibility. In: Uncertainty Principle and Foundations of Quantum Mechanics. London-New York-Sydney-Toronto: John Wiley & Sons; 1977.

[11] Sławianowski JJ. Geometry of Phase Spaces. Warsaw/Chichester-New York-Brisbane-Toronto-Singapore: PWN — Polish Scientific Publishers/John Wiley & Sons; 1991.

[12] Sławianowski JJ. Order of Time Derivatives in Quantum-Mechanical Equations. In: Pahlavani MP. (ed.). Measurements in Quantum Mechanics. Rijeka: InTech; 2012. p57–74.

[13] Sławianowski JJ., Kovalchuk V. Schrödinger and Related Equations as Hamiltonian Systems, Manifolds of Second-Order Tensors and New Ideas of Nonlinearity in Quantum Mechanics. Rep. Math. Phys. 2010;65(1) 29–76.

[14] Sławianowski JJ., Kovalchuk V., Martens A., Gołubowska B., Rożko EE. Quasiclassical and quantum systems of angular momentum. Part I. Group algebras as a framework for quantum-mechanical models with symmetries. Journal of Geometry and Symmetry in Physics 2011;21 61–94.

[15] Sławianowski JJ., Kovalchuk V., Martens A., Gołubowska B., Rożko EE. Quasiclassical and quantum systems of angular momentum. Part II. Quantum mechanics on Lie groups and methods of group algebras. Journal of Geometry and Symmetry in Physics 2011;22 67–94.

[16] Sławianowski JJ., Kovalchuk V., Martens A., Gołubowska B., Rożko EE. Quasiclassical and quantum systems of angular momentum. Part III. Group algebra su(2), quantum angular momentum and quasiclassical asymptotics. Journal of Geometry and Symmetry in Physics 2011;23 59–95.

[17] Svetlichny G. Nonlinear Quantum Mechanics on the Planck Scale. Int. J. Theor. Phys. 2005;44(11) 2051–2058.

[18] Synge JL. Geometrical Mechanics and de Broglie Waves. Cambridge: Cambridge University Press; 1954.

[19] Synge JL. Classical Dynamics. Berlin: Springer; 1960.

[20] Van Vleck JH. The Correspondence Principle in the Statistical Interpretation of Quantum Mechanics. Proc. Nat. Acad. Sci. USA 1928;14(2) 178–188.

Chapter 2

Photons and Signals in the Age of Information

Cynthia Kolb Whitney

Additional information is available at the end of the chapter

http://dx.doi.org/10.5772/59067

1. Introduction

The history of Physics contains some inexplicable mysteries, and one of them is this: back in the early 20th century, Einstein was working on the idea of the 'Photon' and on the idea of the 'Signal' at essentially the same time [1,2], but he did not relate them to each other. They seem to have arisen totally separate in his mind, and they led to totally separate subsequent developments. The Photon played a central role in the development of Quantum Mechanics (QM), and the Signal played the central role in the development of Special Relativity Theory (SRT).

QM and SRT are now the two great pillars of early 20th century Physics, but they seem to be in conflict over the issue of communication. In QM, Schrödinger's Equation is basically the Fourier transform (just a restatement in terms of different variables) of a statement from Classical Mechanics: Total energy = kinetic energy + potential energy. This statement has no signal propagation speed involved in it. So QM appears to allow instantaneous communication over arbitrary distances. But in SRT, Einstein's Second Postulate limits *all* communication to light speed, c.

Since QM and SRT conflict so dramatically on the issue of communication, at least one of them must, in some sense, be wrong. So we are left unsure about what to believe, or take as a foundation for future development.

Can we find out anything decisive from experiments or observations?

The foundation for QM is usually called the Quantum Hypothesis, and not the Quantum Postulate, because the granularity aspect of QM is experimentally testable in various ways. So far, the granularity aspect of QM never seems to fail. This fact stands in favor of QM. But many people still do not believe in the instantaneous communication aspect of QM that is manifest in Schrödinger's Equation.

The foundation for SRT can be called the Light-Speed Postulate, but not the Light-Speed Hypothesis, because light speed really is *not* experimentally testable, since a test would involve at least two different spacetime points, and the correlation of data from two different spacetime points would involve the Light-Speed Postulate itself. Despite numerous claims to the contrary, SRT has *not* been tested in a way that actually could have falsified it, and only very indirect testing appears even feasible for SRT.

Einstein's General Relativity Theory (GRT) flows from SRT, and GRT is testable, at least observationally, although not experimentally. But the new hypotheses that it offers for observational test are few in number, and great in technical difficulty. So even indirect testing of SRT through GRT does not look very promising.

An altogether different approach therefore seems needed: instead of demanding experiments or observations, we should be reviewing the founding ideas themselves, in light of new insights gathered in the intervening century. Evidently, at least one of the founding ideas, and possibly both of them, need to be updated, or else retired and replaced. This paper aims to identify possible update(s)/replacement(s) that may help.

Some of the needed insights come from engineering practice, rather than from theoretical physics. In the mid 20th century there was a flowering of Information Theory (IT), first in connection with wartime code breaking and code making, and then in connection with the post-war communication industry. All of that development led in turn to our modern computation industry, and our current 'Age of Information'.

IT uses the concept of Entropy, taken from classical Thermodynamics, and with the application of a minus sign, provides a quantitative mathematical measure for Information. This measure can be used in support of all sorts of engineering concept analyses and design decisions, *etc.* A convenient reference about the IT concepts and their general applications is Leon Brillouin's wonderful little book *Science and Information Theory* [3]. Flores Gallegos [4] discusses some particular applications in QM.

Viewed from our vantage point here in the early 21st century, IT actually provides a clear disqualifier for Einstein's Second Postulate. The problem is this: the Second Postulate is based on the behavior of a classical infinite plane wave, and an infinite plane wave cannot convey any information whatsoever!

The reason for this perhaps startling assertion is that an infinite plane wave is to electromagnetic communication what a steady hum is to auditory communication: background at best. There is no music in the monotonous hum, and there is no message in the infinite plane wave.

Information requires structure: amplitude modulation, or frequency modulation, or on-off switching. An infinite plane wave does not have any such structure. Because of this deficit, we certainly need a new Signal model as the foundation for an updated SRT.

Possibly we also need a better Photon model at the foundation of QM. We do not actually have a detailed and universally accepted model for photons in QM. Attention has focused more on material systems, which are said to change state, and in so doing, emit or absorb photons that carry packets of energy and angular momentum. As for electric and magnetic fields, what we

http://dx.doi.org/10.5772/59067

have is the notion from Quantum Electrodynamics of a 'virtual photon'. This terminology reveals the desirability of having some unified, photon-like, approach for both Coulomb-Ampère and radiation fields, but it does not actually provide details.

Without detail to argue against, we are apparently free to develop a Photon model *de novo*, in a way that serves not only as the Photon model for QM, but also provides the more realistic Signal model needed for an updated SRT.

The new Photon/Signal model need *not* involve yet another new Postulate. Remember what Euclid taught the world through his Geometry: use no more Postulates than absolutely necessary. The reason is that unnecessary Postulates can conflict with other Postulates already in place, and so lead to Paradoxes.

In SRT, we do indeed have many Paradoxes, involving rods, clocks, trains, lightening strikes, snakes, barns, twins, and so on, and on. A prime suspect for their root cause is the unnecessary Second Postulate. In QM, Schrödinger's Equation seemingly came full-formed from heaven, and to that extent was also a Postulate, and indeed one that conflicted with Einstein's Second Postulate. The mysterious feel of quantum duality, for example, may suggest a possible QM Paradox yet to be fully articulated. If so, the new Photon/Signal model can offer a candidate approach to solve the problem.

The Photon/Signal model is just very familiar, old-fashioned mathematics: **1)** partial differential equations (Maxwell's first order coupled field equations), **2)** their family of solutions (starting with Gaussian pulses, and generating by differentiations the higher-and-higher order Hermite polynomials multiplying the Gaussian pulses), and **3)** boundary conditions (no backflow of energy behind the source, no overflow of energy beyond the receiver). This formulation is enough to determine the particular solution that fits any particular problem.

One general rule in Science is this: when a venerable and well-tested mathematical approach exists, try it **first**, before abandoning it in favor of a new approach. The irony is that Einstein chose not to apply usual the mathematical approach, and instead to introduce his additional Postulate. But he did it so long ago that, despite the long list of Paradoxes generated by SRT, his Second Postulate has itself become 'venerable', and therefore nearly impossible to unseat.

But with SRT updated with the new Photon/Signal model, based entirely on the old-fashioned mathematical approach, QM no longer needs to conflict with SRT. Atoms can have solution states in which energy loss by radiation is countered by an energy gain mechanism newly identified with the updated SRT. It is no longer necessary to postulate the Schrödinger equation just to prevent atomic death by loss of orbit energy to far-field radiation.

With n new starting point for QM, there can be a new development for QM. Some type of Quantum Gravity has long been sought, but with GRT being founded in SRT, with *c*-speed-only communication, and with QM being founded in Classical Physics, with instantaneous communication, that goal has been hard to reach. But the combination of an updated SRT and traditional Statistical Mechanics (SM) offers a way forward.

The new protocol for treating gravity is this: **1)** First notice that gravitational attraction formally resembles a statistical residue from magnetic interactions between elements of charge-neutral matter that are carrying electrical currents, 'current elements' for short. Current elements were well described by Ampère before Maxwell ever came along. **2)** Then apply SM to pairs of current elements.

QM also connects to modern technology problems. Chemistry is wonderfully rich application area for QM. But current-day Quantum Chemistry (QC) is not very easy to use, mainly because of heavy computation loads associated with numerical integrations. The new photon/signal model leads to a different approach for QC, one that is algebraic, rather than integral, in character. This paper includes some recent results from the application of Algebraic Chemistry (AC) to research on the chemistry of water; namely, the form of water known as 'EZ water', because it excludes positive ions.

Finally, QM may connect to Elementary Particle Physics in a manner yet to be fully developed. Just as the myriad compounds in Chemistry arises from not-very-many chemical elements, some significant part of the myriad of currently understood 'elementary' particles may arise from just two of them: the electron and the positron.

The suggestion for this idea lies in Chemistry's Periodic Table (PT). The way that the electron spin states fill up with increasing nuclear charge suggests the existence of not only electron pairs that have opposing spins, but also electron rings that have aligned spins. Electron rings involve up to three, five, even seven, electrons. Basically, electrons within atoms form, not only opposite-spin couples, but also same-spin teams. Since electron rings apparently do occur in atoms, they may well also occur separated from atomic nuclei, and therefore looking like exotic elementary particles. And the same may be said of positrons. Thus we have a rich array of possibilities yet to explore.

2. The photon / signal model

The first part of the Photon/Signal model consists of the governing partial differential equations. These are Maxwell's four first-order coupled field equations. Jackson [5] gives Maxwell's equations in modern notation and Gaussian units as:

$$\nabla \bullet \mathbf{B} = 0, \nabla \bullet \mathbf{D} = 4\pi\rho, \nabla \times \mathbf{E} + \frac{1}{c}\partial \mathbf{B} / \partial t = 0, \nabla \times \mathbf{H} - \frac{1}{c}\partial \mathbf{D} / \partial t = \frac{4\pi}{c}\mathbf{J}. \tag{1}$$

Here **B** is magnetic field and **E** is electric field. The constant $1/c=\sqrt{\varepsilon_0\mu_0}$, where ε_0 is electric permittivity and μ_0 is magnetic permeability. In free space, $\mathbf{D}=\varepsilon_0\mathbf{E}$, $\mathbf{H}=\mathbf{B}/\mu_0$, and charge density ρ and current density $\mathbf{J}$ are zero. Free space will be the case of interest henceforth in this paper.

The second part of the Photon/Signal model is the family of suitable finite-energy solutions. This is developed as follows. Let the word 'pulse' be the short description for a field profile that is rounded on top and sloping down on the sides, and fading gradually to zero; for example, a Gaussian function.

The mechanism for the waveform development is that the spatial derivatives applied in Maxwell's equations change the original Gaussian pulse into successively longer wavelets consisting of successively higher order Hermite polynomials multiplying the original Gaussian. As a result, the energy in the wavelet gets more and more spread out along the propagation path.

Observe that waveform development is inexorable, just like ever-growing entropy in Thermodynamics. This is, I believe, where Entropy really enters into Physics. That is, Maxwell's first order coupled field equations are what give to Physics its obvious Arrow of Time. In citing electromagnetism as the cause for irreversibility, this idea follows Bentwich [6]. (It does *not*, however, give any hope for *reversing* anything.)

Let the spatial variable argument for the pulse be x. Let the direction of the initial pulse be y. Figure 1 illustrates this Gaussian pulse, along with a snapshot showing how it evolves over one complete cycle through Maxwell's first-order coupled field equations. Series 1 is the input pulse, and Series 2 is the waveform developed from it. A more complicated Figure was given in Whitney [7]. This simplified Figure 1 focuses on just the one issue: waveform development. The wavelet is shown bold because it has not been given sufficient attention before.

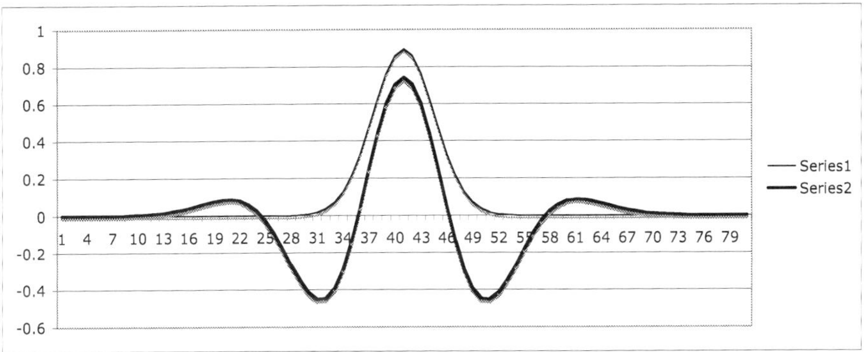

Figure 1. Illustration of waveform development.

Now, to solve the stated propagation problem, one can pose a primary pair of pulses in **E** and **B** to guarantee travel, and, for more realism, add a second such pair, offset a quarter cycle in time and perpendicular in space, to model circular polarization, like a real physical photon exhibits.

The third part of the Photon/Signal model is the pair of propagation boundary conditions: no backflow of energy behind the source, and no overflow of energy beyond the receiver. To guarantee these boundary conditions, one can demand **E**=0 at these boundaries.

One can fulfill the required zero **E** fields by matching the signal leaving the source toward the receiver with **1)** another fictitious signal going from the source in the opposite direction, so as to make the **E** field zero at the source, and **2)** another fictitious signal approaching the receiver from the opposite direction, so as to make the **E** field zero at the receiver. One can even continue this boundary-fixing process, to clean up each tiny new departure from zero **E** at a boundary that each additional fictitious signal creates at the end of the path opposite to the end that it is meant to correct.

Carried to an infinite sum of corrections upon corrections, this is certainly a very complicated picture about electromagnetic fields. How then can we extract from it some simple statement about propagation speed? We have all been trained to just say c. "But relative to what?" we might ask. If we knew the source and receiver were stationary relative to each other, we could be sure that $c = c_{\text{relative to receiver}}$, consistent with Einstein's Second Postulate. Otherwise, we would have to say more.

If we would focus attention to moments when the bulk of the energy is very near the receiver, we could be fairly confident that $c = c_{\text{relative to receiver}}$, again consistent with Einstein's Second Postulate. But if we would focus attention to moments when the bulk of the energy is still very near the source, moments when the receiver is nothing more than a distant phantom, we would be hard pressed to argue against the proposition that $c = c_{\text{relative to source}}$. (This would be consistent with the 1908 Ritz Proposal [8], which was much investigated in the early to mid 20th century as a candidate alternative to Einstein's Second Postulate, but was ultimately rejected.)

If we would need to characterize an entire propagation scenario, we would have to go even further, and consider *all* moments along the way. We would have neither the Einstein Second Postulate, nor the Ritz Proposal, but rather something else more complicated. It certainly must conform to Einstein's Second Postulate late in the scenario, when the bulk of the energy is near the receiver. But early in the scenario, when the bulk of the energy is still near the source, it must conform to Ritz's Proposal. And in between, it must represent in some mathematically appropriate way an idea not previously considered: a transition from one reference for c to the other.

Here is one way to formulate the problem. Let variable x represent distance along the propagation path, and let variable t represent time into the propagation process. At any point x, t there are fields $\mathbf{E}(x, t)$ and $\mathbf{B}(x, t)$ with magnitudes $E(x, t)$ and $B(x, t)$, and from them a local energy density:

$$S(x,t) = \left[E^2(x,t) + B^2(x,t)\right]/2. \tag{2}$$

One way to characterize the propagation is with path integrals. [9-13] Consider the following ratio of two path integrals:

$$r(t) = \int_{\text{path}} xS(x,t)dx \Big/ \int_{\text{path}} S(x,t)dx. \tag{3}$$

This ratio provides a function that begins at the source, and ends at the receiver, and at the temporal midpoint of the scenario, gives equal weight to both the source and the receiver. This behavior captures the proposed transition of the reference to which the light speed c is assumed relative.

3. Using the changing reference for c changes the results

The concept of the changing reference for c to be relative to appears relevant for an important scenario that was considered even before Einstein arrived on the scene. The scenario involves the potentials and fields created by rapidly moving sources, and it was addressed starting in the late nineteenth century and very early twentieth centuries. Researchers then made the same Assumption that Einstein later made his Second Postulate in founding SRT, but they were not attentive enough to see that there indeed *was* an Assumption, and to call it out as his Second Postulate. So Einstein is to be commended for calling attention to this Assumption.

The sources generally cited for this early (1898 to 1901) problem are A. Liénard [14] and E. Wiechert [15]. Although they worked at about the same time, they worked separately. They got the same results, as did all contemporary and subsequent investigators, because all persons working from then up until now have used the same input Assumption; namely, that the speed of light is always c with respect to the receiver of the light.

The Liénard- Wiechert results are given in [5], and in every other standard EM book. They are displayed in [7], and that short passage is quoted again here, for review and subsequent further discussion:

"The standard scalar and vector potentials are:

$$\Phi(\mathbf{r},t) = e\left[1/\kappa R\right]_{\text{retarded}} \text{ and } \mathbf{A}(\mathbf{r},t) = e\left[\boldsymbol{\beta}/\kappa R\right]_{\text{retarded}}, \tag{4}$$

"where $\kappa = 1 - \mathbf{n}\cdot\boldsymbol{\beta}$, $\boldsymbol{\beta}$ is source velocity normalized by light speed c, and $\mathbf{n}=\mathbf{R}/R$ (a unit vector), and $\mathbf{R}=\mathbf{r}_{\text{source}}(t-R/c)-\mathbf{r}_{\text{receiver}}(t)$ (an implicit definition for the terminology 'retarded').

"The LW fields obtained from those potentials are then:

$$\begin{aligned}\mathbf{E}(\mathbf{r},t) &= e\left\{\frac{1}{\kappa^3R^2}(\mathbf{n}-\boldsymbol{\beta})(1-\beta^2) + \frac{1}{c\kappa^3R}\mathbf{n}\times\left[(\mathbf{n}-\boldsymbol{\beta})\times(d\boldsymbol{\beta}/dt)\right]\right\}_{\text{retarded}}, \\ \mathbf{B}(\mathbf{r},t) &= \mathbf{n}_{\text{retarded}}\times\mathbf{E}(\mathbf{r},t).\end{aligned} \tag{5}$$

"The $1/R$ fields are radiation fields, and they make a Poynting vector (energy flow per unit area per unit time) that lies along $\mathbf{n}_{\text{retarded}}$:

$$\begin{aligned}\mathbf{P} &= \frac{c}{4\pi}\mathbf{E}_{\text{radiative}}\times\mathbf{B}_{\text{radiative}} \\ &= \frac{c}{4\pi}\mathbf{E}_{\text{radiative}}\times\left(\mathbf{n}_{\text{retarded}}\times\mathbf{E}_{\text{radiative}}\right) = \frac{c}{4\pi}\left(E_{\text{radiative}}\right)^2\mathbf{n}_{\text{retarded}}\end{aligned} \tag{6}$$

"The $1/R^2$ fields are Coulomb-Ampère fields, and the Coulomb field

$$\mathbf{E}(\mathbf{r},t) = e\left\{(\mathbf{n}-\boldsymbol{\beta})(1-\beta^2)/\kappa^3R^2\right\}_{\text{retarded}}. \tag{7}$$

"does *not* lie along $\mathbf{n}_{\text{retarded}}$ as one might initially expect; instead, it lies along $(\mathbf{n}\text{-}\boldsymbol{\beta})_{\text{retarded}}$. Assume that $\boldsymbol{\beta}$ does not change much over the total field propagation time, in which case $(\mathbf{n}\text{-}\boldsymbol{\beta})_{\text{retarded}}$ is virtually indistinguishable from $\mathbf{n}_{\text{present}}$."

Thus the Coulomb attraction/repulsion and the radiation Poynting vector have distinctly different directions. This result does *not* look right physically. It looks as though advance information is being provided on one, but not the other, of two information channels.

Now consider the same problem using the new and more nuanced definition for the light speed reference. Observe that a line that connects the source position at the temporal midpoint of the scenario to the receiver position at the temporal midpoint of the scenario defines both the distance and the direction that the energy must travel in order to achieve the proposed transmission of the signal from the source to the receiver. This specification implies that we need potentials and fields to be, not *retarded*, but **half**-*retarded*. Now the potentials become:

$$\Phi(\mathbf{r},t) = e\left[1/\kappa R\right]_{\textbf{half}-\text{retarded}} \text{ and } \mathbf{A}(\mathbf{r},t) = e\left[\boldsymbol{\beta}/\kappa R\right]_{\textbf{half}-\text{retarded}}. \tag{8}$$

The fields become:

$$\mathbf{E}(\mathbf{r},t) = e\left\{\frac{(\mathbf{n}-\boldsymbol{\beta})(1-\beta^2)}{\kappa^3R^2} + \frac{\mathbf{n}}{c\kappa^3R}\times\left[(\mathbf{n}-\boldsymbol{\beta})\times d\boldsymbol{\beta}/dt\right]\right\}_{\textbf{half}-\text{retarded}} \text{ and } \mathbf{B}(\mathbf{r},t) = \mathbf{n}_{\textbf{half}-\text{retarded}}\times\mathbf{E}(\mathbf{r},t). \tag{9}$$

The Poynting vector becomes:

$$\begin{aligned}\mathbf{P} &= \frac{c}{4\pi}\mathbf{E}_{\text{radiative}}\times\mathbf{B}_{\text{radiative}} \\ &= \frac{c}{4\pi}\mathbf{E}_{\text{radiative}}\times\left(\mathbf{n}_{\textbf{half}-\text{retarded}}\times\mathbf{E}_{\text{radiative}}\right) = \frac{c}{4\pi}(E_{\text{radiative}})^2\,\mathbf{n}_{\textbf{half}-\text{retarded}}.\end{aligned} \tag{10}$$

The direction of the Coulomb field becomes:

$$(\mathbf{n}-\boldsymbol{\beta})_{\textbf{half}-\text{retarded}} \approx (\mathbf{n}_{\text{present}})_{\textbf{half}-\text{retarded}} \triangleq \mathbf{n}_{\textbf{half}-\text{retarded}}. \tag{11}$$

This direction is the same as the direction of the radiation Poynting vector. That is, the Coulomb field and the Poynting vector are now reconciled to the *same* direction, instead of conflicting with each other.

4. The corrected force direction means the hydrogen atom can survive classically

Many authors have expressed the opinion that really explaining the Hydrogen atom requires some presently-unknown short-range repulsive force between the electron and the proton; see, for example, Lokajicek, *et al* [16]. But given the results just presented, no mysterious new repulsive force is needed. With the direction of the Coulomb field being $\mathbf{n}_{\text{half-retarded}}$, there is a tiny tangential component of Coulomb force aligned with the orbit velocity. So there is a torque on the atom, and the torque pumps energy into the atom, and that process can work to balance the energy loss due to radiation.

That is to say: having a more nearly correct model for potentials and fields created by rapidly moving charges makes it possible to *explain* the immortality of the Hydrogen atom without first *postulating* the immortality of the Hydrogen atom; *i.e.*, postulating Schrödinger's equation.

That is to say: we need not postulate Schrödinger's equation; w can instead just carry out the old-fashioned math.

In my 2012 and 2013 Intech papers, I listed just the pertinent results. Here is more detail and derivation:

Let the masses of the electron and the proton be m_e and m_p. Note that $m_e << m_p$, but m_p is *not* infinite.

Let the orbit radii of the electron and the proton be r_e and r_p. Note that $r_p << r_e$, but r_p is *not* zero.

Let the charges on the electron and the proton be $-e$ and $+e$.

The magnitude of the nominally attractive force within the atom is $F = e^2 / (r_e + r_p)^2$.

Let the orbit frequency be Ω. The orbit speed of the electron is $v_e = r_e\Omega$ and that of the proton is $v_p = r_p\Omega$.

The magnitude of the tiny tangential force on the electron is $F_e = F v_p / 2c = F r_p \Omega / 2c$.

The magnitude of the tiny tangential force on the proton is $F_p = F v_e / 2c = F r_e \Omega / 2c$.

The magnitudes of the torques on the electron and on the proton are $T_e = r_e F_e$ and $T_p = r_p F_p$, both equal to $F r_e r_p \Omega / 2c$.

The total torque on the electron-proton system is $T_{total} = T_e + T_p = 2T = F r_e r_p \Omega^2 / c$.

The torque power delivered to the system is $P_{torque} = T\Omega = F r_e r_p \Omega^2 / c$.

The squared orbit frequency is determined from either $F = m_e r_e \Omega^2$ or $F = m_p r_p \Omega^2$.

The more convenient of the two options is $\Omega^2 = F / m_e r_e$. With that expression, the approximation $r_e \approx r_e + r_p$ yields:

$$P_{\text{torque}} = F r_e r_p \Omega^2 / c = F^2 r_p / m_e c = \left(e^4 r_p / m_e c\right) / \left(r_e + r_p\right)^4 \approx \left(e^4 / m_p c\right) / \left(r_e + r_p\right)^3 . \tag{12}$$

This is a reasonably simple expression. But more important than its simplicity is its very *existence*. The existence of any such expression means that there exists an energy *gain* mechanism to balance against the known energy loss mechanism, *i.e.* radiation. This situation provides a chance for balance, allowing the Hydrogen atom to avoid death by energy loss to radiation.

The P_{torque} is actually quite large, and so it changes the whole emphasis of worry concerning the Hydrogen atom. The question becomes, not why does the Hydrogen atom not radiate and collapse to death, but rather why does the Hydrogen atom not torque itself up and expand way beyond its known size?

The fact is: there exists much, much more radiation than was previously worried about, and it is enough to produce the proper balance between radiation and torque.

In [17], I said that the extra radiation arises from finite signal propagation speed, which results in circular motion of the center of mass of the Hydrogen atom, which in turn produces Thomas rotation, and thereby scales up by a factor of 2 the overall rotation rate generating the radiation, which increases the radiation power by a factor of 2^4.

But what should one say about that center-of-mass circular motion? In Newtonian physics, where signal propagation speed is infinite, there is no such thing. In Maxwell physics, the emphasis is on fields, and the responses of individual charges, but not as much on the responses of whole charge systems, such as atoms. So the issue doesn't come up there. In Einstein's relativity physics, the emphasis is often on the observers of events more than on the events themselves. System center-of-mass circulation seems not to come up, although Thomas rotation does.

In [17], I noted that Thomas rotation is generally believed to be a result of the properties of Lorentz transformations, and hence of SRT. That is the belief because one can think of Lorentz transformations, not only in the usual, passive sense, as conversion from an observer in one inertial coordinate frame to another observer in another inertial coordinate frame, but also in the active sense, as the application of a 'boost' in velocity, the result of an acceleration, the result of a physical force. A series of non-co-linear boosts does indeed produce Thomas rotation.

But in [17] I also remarked that Thomas rotation does arise, not just from Lorentz transformations, but also from Galilean Transformations. That fact can be demonstrated in detail as follows:

For simplicity, let all motion be in the x, y plane. The scenario begins at time coordinate ct_0 with one of the particles, say the electron, at rest at spatial coordinates x_0, y_0. Let an attraction from another particle act in the x direction. Let an increment of velocity $\Delta\mathbf{V}_x = \Delta V$ be imposed, and let an increment of time Δt elapse. The coordinates of the electron then become:

$$t_1 = c(t_0 + \Delta t), x_1 = x_0 + \Delta V \Delta t \text{ and } y_1 = y_0. \tag{13}$$

Now let an attraction from another particle act in the y direction. Let an increment of velocity $\Delta \mathbf{V}_y = \Delta V$ be imposed, and let another increment of time Δt elapse. The coordinates of the electron then become:

$$\begin{aligned} &t_2 = c(t_0 + \Delta t + \Delta t), \\ &x_2 = x_0 + \Delta V(\Delta t + \Delta t) = x_0 + 2\Delta V \Delta t, \\ &\text{and } y_2 = y_0 + \Delta V \Delta t. \end{aligned} \tag{14}$$

Observe that, if the Galilean velocity boosts had been applied in the opposite order, then the ending coordinates of the electron would have been:

$$\begin{aligned} &t_2 = c(t_0 + \Delta t + \Delta t), \\ &x_2 = x_0 + \Delta V \Delta t, \\ &and\ y_2 = y_0 + 2\Delta V \Delta t. \end{aligned} \tag{15}$$

Observe that: the squared incremental length changes have the same magnitude either way: $(2\Delta V\Delta t)^2 + (\Delta V\Delta t)^2 \equiv (\Delta V\Delta t)^2 + (2\Delta V\Delta t)^2 = 5(\Delta V\Delta t)^2$. That fact means the two possible sequences of Galilean boost applications differ only by a rotation. That means each one individually contains a rotation equal to half that total angle difference. This is the Thomas rotation.

Let me now go further, and assert that Thomas rotation will arise from *any* kind of velocity transformation – Lorentz, or Galilean, or any other new kind that may not have a name yet. Thomas rotation is a property of actual reality, not of any particular mathematical model for reality.

With the Thomas rotation included, the total radiation from the atomic system is:

$$P_{\text{total radiated}} = \underline{2^4} \frac{2e^2}{3c^3} a_e^2 = \left(\underline{2^5} e^6 / m_e^2\right) / 3c^3 (r_e + r_p)^4. \tag{16}$$

The value of the separation $r_e + r_p$ for which $P_{\text{total radiated}} = P_{\text{torque}}$ is:

$$r_e + r_p = 32 m_p e^2 / 3 m_e^2 c^2 = 5.5 \times 10^{-9}\,\text{cm}. \tag{17}$$

In the traditional approach to QM, $r_e + r_p = h^2 / 4\pi^2 \mu e^2$, where μ is the reduced mass, defined by $\mu^{-1} = m_e^{-1} + m_p^{-1}$, and very nearly equal to m_e, and h is Planck's constant, 6.626176×10^{-34} Joule-sec, a fundamental constant given by Nature.

The present analysis does not require Planck's constant as an input. Instead, it provides an estimate of Planck's constant as an *output*:

$$h = \sqrt{4\pi^2 \mu e^2 (r_e + r_p)} \approx \sqrt{4\pi^2 m_e e^2 32 m_p e^2 / 3m_e^2 c^2}$$
$$\approx \frac{\pi e^2}{c}\sqrt{128 m_p / 3m_e} \approx 6.77\times 10^{-34}\,\text{Joule-sec} \ .; \tag{18}$$

and, for convenience, also an estimate of the often-seen reduced Planck's constant:

$$\hbar = h / 2\pi \approx (e^2 / c)\sqrt{32 m_p / 3m_e} \approx 1.08\times 10^{34}\,\text{Joule-sec.} \tag{19}$$

These estimates can be improved by taking due account of the fact that the sines of small angles are not exactly equal to the angles themselves, and the cosines of small angles are not exactly equal to unity. The sine corrections are third order in angle, while the cosine corrections are second order in angle, which is more significant. Including those corrections reduces the radiation power slightly, and so reduces the solution $r_e + r_p$ slightly, and so reduces the estimates of h and $\hbar$ slightly – a step in the right direction.

Observe that, in making h an output *from*, rather than an input *to*, a theory, the present work follows both Enders [18] and Ralston [19].

5. EM interactions between neutral atoms: A candidate model for gravity?

Our current best understanding of gravity comes from GRT, which is founded in, and developed from, SRT. The parameter c from SRT appears in GRT, and reveals the lineage. Like electromagnetic signals, gravitational signals must have the finite propagation speed c. So might gravitational signals actually be electromagnetic signals? If we choose to update SRT with a more realistic signal model in place of the number c, does that require a similar update for GRT? This Section explores such questions.

First let it be noted that the Einstein gravitational field equations, like Maxwell's coupled field equations, are a description at the microscopic scale, but the observable phenomena exist at an extremely macroscopic scale.

Figure 2 illustrates a fairly typical barred spiral galaxy. What this image seems to suggest is: there are mass concentrations at the two armpits of this galaxy. Maybe they are mega stars, or black holes. Maybe they orbit another mass concentration at the center of the galaxy, or maybe they just orbit each other. In any event, the two mass concentrations together create a rather structured field of gravitational potential, into which millions, or billions, of smaller stars are entrained, or temporarily detained, as they orbit the galaxy.

http://dx.doi.org/10.5772/59067

Figure 3 shows the skeleton of a potential field created by two super-massive bodies orbiting at half the signal propagation speed. The lines mark minima in gravitational potential as a function of angle around the galaxy. Observe that this skeleton approximately matches this galaxy image.

Figure 2. A typical barred spiral galaxy: "A Barred Spiral Galaxy ngc 1300 hubble photo".

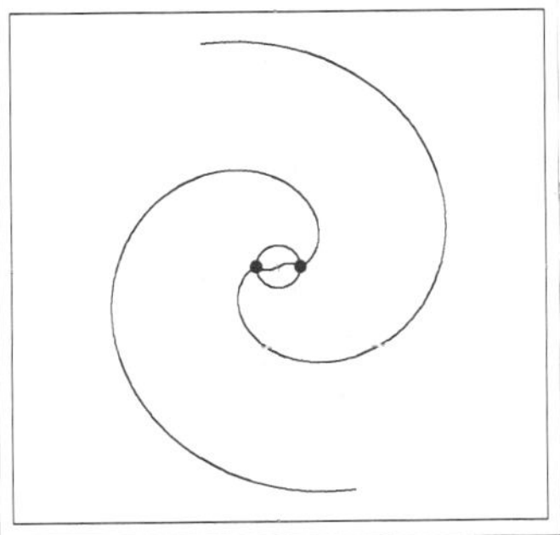

Figure 3. The skeleton for a barred spiral disc galaxy. Originally computed for [20].

Note 1: To persist over time, the orbit speed of the driving two-body system in Figs. 2 and 3 must be such that the rates of energy loss by gravitational radiation and energy gain by torquing balance each other. Like the Hydrogen atom, the galactic-size two-body problem can be solved this way.

Note 2: Since most of the captive stars in Fig 2 orbit at lesser speed than the driving two bodies, and their potential pattern, with its skeleton, Fig. 3, the individual stars in the outer reaches do not keep up with the rotating spiral potential pattern. An individual star sees a recurring 'density wave' of neighbor stars first approaching, then receding. Such density waves have long been known, but not well explained.

Note 3: Fig. 3 is constructed using brute-force calculation of potential for lots and lots of location points, and then using numerical search over angles for local minimum values, and then fitting a function to the minima. Note the slightly sinuous bar between the two driving bodies. Even this detail is suggested in Fig. 2.

Now, in order to model gravity in terms of electromagnetic interactions, we need an expression of electromagnetic interaction that is appropriate for neutral atoms. The best expression appears to be one that was known even *before* Maxwell. André Marie Ampère already had a well-developed theory about forces between what he called 'current elements'. This term referred to charge-neutral material increments in electrical circuits. In modern times, P. Graneau wrote extensively about Ampère's theory and experiments; see for example Graneau [21]

Ampère's theory works perfectly well for ordinary closed circuits, as well as for incomplete broken circuits, such as may exist momentarily in transient situations, like explosive rupture of circuits. Ampère's theory ought not be forgotten solely on the basis that more modern theory also works perfectly well for closed and stable electrical circuits. Indeed, in some technological applications involving transient situations like ruptures, Ampère's theory explains more than the modern theory does.

One reason why Ampère's formulation can sometimes be more powerful than a formulation based on Maxwell's equations is that Ampère's formulation can describe a multi-participant scenario straight away, whereas Maxwell's equations require iteration through successive steps involving both Maxwell's equations, and the Lorentz force law for the force **F** acting on a charge q moving with velocity **v**:

$$\mathbf{F} = q\left[\mathbf{E} + \mathbf{v} \times \mathbf{B}\right]. \tag{20}$$

The iteration goes as follows: first, the input charge positions and motions generate **E** and **B** fields; second, the Lorentz force law tells how each charge q responds to the fields from the other charges; another step through Maxwell's equations tells how all the fields change, and so on.

One particular scenario illustrates the difference between the approaches especially well. Consider a current-carrying wire. It is tedious to use the Maxwell-Lorentz-Maxwell-Lorentz iterative approach to arrive at the understanding that the moving electrons favor the surface of the wire, or even the exterior neighborhood near the wire, leaving the interior of the wire depleted of electrons, and therefore in a state of internal repulsion between the remaining positive nuclei. The Ampère's formulation skips over all this detail, and describes the resulting consequence: the wire experiences internal longitudinal force, and in fact might even rupture. If it does rupture, one can easily tell that the event was *not* due to ordinary resistive heating and melting, since the fragments are found to be neither hot to touch nor melted in appearance.

The Ampère approach looks promising for gravity problems because *any* gravity problem is definitely a multi-participant scenario. And for the same reason, the following analysis also invokes ideas from modern Statistical Mechanics.

http://dx.doi.org/10.5772/59067

Ampère's force formula can be written:

$$\Delta F_{m,n} = +i_m i_n \left[\Delta m \, \Delta n \, / \, (r_{m,n})^2 \right] (3\cos\alpha \cos\beta - 2\cos\gamma). \tag{21}$$

The indices m and n identify two interacting currents. The i_m and i_n are current magnitudes. The Δm and Δn are magnitudes of tiny directed length increments $\Delta\mathbf{m}$ and $\Delta\mathbf{n}$ through which the currents flow. The products of currents and directed length increments, $i_m \, \Delta\mathbf{m}$ and $i_n \, \Delta\mathbf{m}$, are the current elements. The $r_{m,n}$ is the length of the vector separation $\mathbf{r}_{m,n}$ between the current elements. The α, β, and γ are angles with respect to the connecting line between the two current elements, and with respect to each other. Current element $i_m \, \Delta\mathbf{m}$ is at angle α from the connecting line, and current element $i_n \, \Delta\mathbf{n}$ is at angle β from the connecting line. The γ is the angle between the two planes defined by the connecting line and each of the two current elements, as if the distance $r_{m,n}$ did not separate them. The value ranges are all full circle: $0 < \alpha < 2\pi$,

One can get a feel for the general behavior of Ampère's force formula by considering the angle factor $3\cos\alpha\cos\beta - 2\cos\gamma$ for a few special cases:

1. Current elements side-by-side and parallel, as in parallel wires. Both current elements are perpendicular to the connecting line, so α and β are $\pi/2$ and $\cos\alpha$ and $\cos\beta$ are zero. But γ is zero, and $\cos\gamma = 1$, so the angle factor evaluates to -2. The force $\Delta F_{m,n}$ is then negative. The current elements attract each other. If they reside in parallel wires, the wires attract each other. This you know from experience is true. In a plasma, instead of in a solid wire, it is called the 'pinch effect'.

2. Current elements side-by-side, but anti-parallel. This case is just opposite to Case 1 above: now $\gamma = \pi$ and $\cos\gamma = -1$. The current elements repel each other. If they reside in a circuit, that circuit likes to straighten out any kinks and enclose more area. This you may know from experience is true.

3. Current elements end-to-end, as in an electrical circuit. All three angles are zero, all three cosines are unity, and the angle factor evaluates to $+1$, so the force $\Delta F_{m,n}$ is positive. The current elements repel each other.

The $1/(r_{m,n})^2$ aspect of the Ampère force law is just like Newton's law for gravity. Ampère designed his law that way, because, in his time, the greatest prior achievement in Science was Newton's conquest of gravity. Now we wish to return the favor, and exploit the Ampère Force Law to understand something novel about gravity.

What makes the Ampère current element so potentially appropriate for application to gravity? First of all, it is charge-neutral, like the masses in a gravity scenario. Secondly, its electrons are moving, and although its nucleus is moving too, that motion is not anywhere near as fast. So at all times and all places where matter exists, at the microscopic level a net electron current flows.

The concept that current elements generate forces that can attract or repel each other suggest that pairs of current elements – or pairs of atoms – can be regarded as a system that can have positive or negative total energy. The kinetic part of the energy may be disregarded, since the current elements may be essentially static, but the potential part of the energy is worth paying attention to.

The main novel feature that gravity presents is that we usually have, not two current elements, but huge numbers of atoms, and each atom must have some relationship with *all* other atoms. The complexity of the situation naturally conjures up ideas from Statistical Mechanics. Here, ideas from Statistical Mechanics are applied to gravity described in terms of Ampère forces between atoms that are viewed as current elements. Some atom-to-atom relationships are momentarily attractive, and some relationships are momentarily repulsive, and all relationships must vary over time. We can look at the population of atom pairs as a whole, and think of it as a statistical ensemble, in which every condition of attraction/repulsion is represented somewhere.

Every area of physics that has statistical ensembles has Gaussian probability functions. In Classical Thermodynamics, a Gaussian probability density function for a random variable, such as a component of a particle momentum vector, implies maximum entropy, subject to a prescribed value for the standard deviation of that random variable. In Quantum Mechanics, a Gaussian probability density function (the squared wave function amplitude) is associated with minimum uncertainty, meaning minimum product of standard deviations in Fourier conjugate variables, like position and momentum.

Sometimes it is not immediately obvious that a problem involves the equivalent of a Gaussian function, because the Gaussian itself involves a squared variable, such as x^2 or p^2. The squared variable is proportional to some energy E. So one sees probability density functions expressed in the form $\exp(-E/<E>)/<E>$ where $<E>$ is the average value of energy E, usually something like kT, where k is Boltzmann's consistent and T is absolute temperature.

In the case of gravity, that energy E of interest is gravitational potential energy. It can have both positive *and* negative values. The central concept in Statistical Mechanics is that lower-energy states are populated more richly than higher-energy states are. This concept means that any two atoms, viewed as current elements, will be with respect to each other in a state of negative potential energy more often than in a state of positive potential energy. So they will, on average, attract each other more than repel each other. Therefore, $<E>$ is negative. Since temperature cannot be negative, this is something novel.

To deal with negative $<E>$ we really need a probability density function with a Gaussian factor of the form $\exp[-E^2/2<E^2>]$, where $<E^2>$ is another parameter, positive, but also not related to temperature.

A few examples can illustrate how to find the parameters.

Let the two energies be E_{max} and $E_{min}=-|E_{max}|$. With the attractive–force, negative-energy state dominating the scenario, $<E>$ is negative, $<E>=-|<E>|$. The two Boltzmann factors are:

$$\exp(+E_{max} / |< E >|) \text{for the state with negative energy, } E_{min} = -E_{max} \tag{22}$$

and

$$\exp(-E_{max} / |< E >|) \text{for the state with positive energy,} E_{max}. \tag{23}$$

The average energy E_{avg} must satisfy the definition:

$$E_{avg} = \frac{-E_{max} \exp(+E_{max} / |< E >|) + E_{max} \exp(-E_{max} / |< E >|)}{\exp(+E_{max} / |< E >|) + \exp(-E_{max} / |< E >|)}. \tag{24}$$

Simple trial calculations and numerical search of the results works well enough to solve the problem at hand. The solution is approximately:

$$E_{avg} \approx -E_{max} / 1.2 = -0.8333 E_{max} = -\tfrac{5}{6} E_{max}. \tag{25}$$

(There is also, of course, a positive, and presently irrelevant, solution of the same magnitude.)

For this case we have we have line integration in place of point evaluation. E_{avg} becomes:

$$E_{avg} = -\int_{-E_{max}}^{0} E \sinh(E / |< E >|) dE \Big/ \int_{-E_{max}}^{0} \cosh(E / |< E >|) dE. \tag{26}$$

Because the denominator is simpler, begin with that. It is:

$$\begin{aligned} | E_{avg} | \sinh(E / | E_{avg} |) \Big|_{-E_{max}}^{0} &= -\left[| E_{avg} | \sinh(-E_{max} / | E_{avg} |) \right] \\ &= + | E_{avg} | \sinh(E_{max} / | E_{avg} |) \quad . \end{aligned} \tag{27}$$

This is a positive number.

The numerator is more complicated, but it can be evaluated using integration by parts:

$$\int_{-E_{max}}^{0} U dV = UV \Big|_{-E_{max}}^{0} - \int_{-E_{max}}^{0} V dU \quad . \tag{28}$$

where $U = E$ and $dV = \sinh(E / | E_{avg} |) dE$, so that $V = | E_{avg} | \cosh(E / | E_{avg} |)$. The first term in the numerator evaluation is:

$$UV|_{-E_{max}}^{0} = -U(-E_{max})V(-E_{max}) = -[-E_{max} |< E >| \cosh(-E_{max} / |< E >|)] \\ = E_{max} |< E >| \cosh(E_{max} / |< E >|) \ . \tag{29}$$

The second term in the numerator evaluation is:

$$-\int_{-E_{max}}^{0} VdU = -\int_{-E_{max}}^{0} |< E >| \cosh(E / |< E >|) dE = -| E_{avg} |^2 \sinh(E / |< E >|)|_{-E_{max}}^{0} \\ = |< E >|^2 \sinh(-E_{max} / |< E >|) = -|< E >|^2 \sinh(E_{max} / |< E >|) \ . \tag{30}$$

.

The sought numerator divided by denominator for $< E >$ is then:

$$< E > = \frac{E_{max} |< E >| \cosh(E_{max} / |< E >|) - | E_{avg} |^2 \sinh(E_{max} / |< E >|)}{| E_{avg} | \sinh(E_{max} / | E_{avg} |)} \\ = \frac{E_{max}}{\tanh(E_{max} / |< E >|)} + |< E >| \ . \tag{31}$$

Since the sought $< E >$ is negative, equal to $- | < E > |$, we have:

$$2 < E > = -2 |< E >| = \frac{-E_{max}}{\tanh(E_{max} / |< E >|)}. \tag{32}$$

Again, numerical search is a practical approach for finding a solution. We find:

$$< E > \approx -0.485 E_{max} . \tag{33}$$

Observe that, as should be expected, this solution is significantly smaller in magnitude than was the solution with only two energy values, $\pm E_{max}$, to balance between, which came in at $-0.833 E_{max}$.

The continuous Gaussian profile,

$$\exp(-E^2 / 2 < E^2 >) / \sqrt{2\pi < E^2 >}, \tag{34}$$

extends to infinity in both directions. This attribute is inappropriate for the problem at hand, which definitely possesses limits $\pm E_{max}$ beyond which the modeling problem does not extend.

http://dx.doi.org/10.5772/59067

Therefore, let us turn to discrete approximations for a Gaussian. These are based on the binomial expansion for an arbitrary $(a+b)^n$. The binomial coefficients are familiar to many people from Pascal's famous triangle:

$$\begin{array}{c} 1 \\ 1 \quad 1 \\ 1 \quad 2 \quad 1 \\ 1 \quad 3 \quad 3 \quad 1 \\ etc. \end{array} \tag{35}$$

The numbers in Pascal's triangle are constructed with addition of neighboring numbers above. This is easy for small n, but small n means *crude*, and we need *refined*. So we want *large* n. So we need a formula involving multiplication instead of addition. That would be:

$$1,\ n,\ \frac{n(n-1)}{1\times 2},\ \frac{n(n-1)(n-2)}{3!},\ \cdots,\ \frac{(n)!}{(n')!(n-n')!},\ \cdots,\ \frac{n(n-1)(n-2)}{3!},\ \frac{n(n-1)}{1\times 2}, n, 1. \tag{36}$$

Observe that the binomial coefficients are symmetric around the middle of the list, like a Gaussian function is symmetric around zero argument. If n is an even number, the number of binomial coefficients is $2n-1$, odd, and the middle one, the maximum one, is $n!/[(n/2)!]^2$. If n is an odd number, the number of binomial coefficients is $2n$, an even number, and the middle two numbers, the maximum two, are both $n!/\{[(n+1)/2]![(n-1)/2]!\}$.

For our modeling problem, let n be an odd number. Let the binomial coefficients be represented as B_b with $b=-n$ to $b=1$. Let them be associated with equal energy increments $\Delta E = E_{\max}/n$ starting from $-E_{\max}$ and covering the range to zero. Associate the minimum binomial coefficient B_{-n} with the increment starting with $-E_{\max}$, and the maximal binomial coefficient with the increment ending with $E=0$, and associate the other coefficients with the increments between those limits. The problem to solve is:

$$< E > \approx \sum_{b=-n}^{1} B_b E_b \sinh(E_b/<E>) \Big/ \sum_{b=-n}^{1} B_b \cosh(E_b/<E>). \tag{37}$$

Numerical investigations done to date suggest that the solution comes at approximately $<E> = -E_{\max}/2\sqrt{n}$. Here the $\sqrt{n}$ for this discrete model is analogous to the standard deviation σ for the corresponding continuous Gaussian.

The problem of modeling gravity therefore reduces to the problem of determining what value of n should be used. Here is the most pertinent fact: compared to anything electromagnetic, gravity is extremely weak. Consider two Hydrogen atoms at a given separation distance. Let us compare the gravitation force with the maximum Ampère force between them.

The gravitational attraction is proportional to $G(m_p)^2$, where G is the universal gravitation constant, about 6.6×10^{-11} Newton× meter^2 per kilogram^2, and m_p is the mass of the proton, about 1.66×10^{-27}kg, so $(m_p)^2$ is about 2.76×10^{-54} kg^2. Overall,

$$\frac{1}{4\pi\varepsilon_0}e^2(v/c)^2 \approx \frac{2.56\times10^{-38}\times0.45\times10^{-4}}{1.113\times10^{-10}} \approx 1.035\times10^{-32}\text{Newton}\times\text{meter}^2. \quad (38)$$

Clearly, the maximum Ampère force between atoms viewed as current elements is generously larger than the gravitational force between atoms viewed as charge-neutral masses – by about 32 orders of magnitude!

But the typical Ampère force is nowhere near as big as the maximum Ampère force, due to the fact that it depends on three angles, any one of which can spoil it. Occurrence of the maximum Ampère force is very rare indeed. And occurrence of the *minimum* (negative) Ampère force is equally rare. Only the piddling near-zero Ampère forces are common, and even then, each tiny attractive force is mostly cancelled with the tiny repulsive force of equal magnitude but less frequent occurrence. All we have is a tiny residue of attractive force, due to the nature of Boltzmann factors and Statistical Mechanics.

Of course, being tiny does not mean being insignificant. Like the tiny residue that is microwave background radiation, the tiny residue that is gravity is a possible key to understanding something about the Universe in which we live. Here is an example problem: at present, we know the actual particle radius of the electron is something extremely tiny, but we do *not* know what its numerical value is. There exist a number of length-dimensioned quantities associated with the electron, all called 'radius', but distinguished by specific names and numerical values. MacGregor [22] lists seven of them. Most are on the order of 10^{-13} cm, although one is much smaller, and is presently only upper-bounded at $<10^{-16}$ cm.

The radius attributed to the electron can have a role in the gravity problem. The ratio of an atomic radius to the electron radius can imply a candidate level of discretization for the binomial approximation to the Gaussian factor involved in the gravity problem. For an atomic radius, let us consider the first orbit radius of Hydrogen, $r_{H1}=0.529\times10^{-8}$cm. For the electron radius, let us consider two of the possibilities from MacGregor [22].

One of the electron radii is called *classical*. This one captures the Coulomb energy equivalence

$$\frac{1}{4\pi\varepsilon_0}e^2/r_{\text{e classical}} = m_e c^2, \quad (39)$$

which implies

$$r_{\text{e classical}} = \frac{1}{4\pi\varepsilon_0}\left(e^2/m_e c^2\right) = 2.82\times10^{-13}\text{cm}. \quad (40)$$

The ratio $\rho = r_{H1} / r_{e\ classical}$ is then:

$$n = 0.529 \times 10^{-8} / 2.82 \times 10^{-13} = 5.29 \times 10^{-9} / 2.82 \times 10^{-13} \approx 1.86 \times 10^{4}, \tag{41}$$

which implies

$$n = 0.529 \times 10^{-8} / 2.82 \times 10^{-13} = 5.29 \times 10^{-9} / 2.82 \times 10^{-13} \approx 1.86 \times 10^{4}. \tag{42}$$

The square root of this number would then be the dimensionless $\sqrt{n}$ for the discretization:

$$\sqrt{n} = \sqrt{1.86 \times 10^{4}} \approx 1.37 \times 10^{2} = 137. \tag{43}$$

This is a number already famous in Physics, but in a context other than gravity. It is the inverse of the so-called 'fine structure constant' α, defined as:

$$\alpha = 2\pi e^{2} / ch. \tag{44}$$

This number plays a role in spectroscopy, where spectral lines occur in families, closely spaced but clearly distinguishable. There, the explanation comes from QM; clearly, another manifestation of natural discretization.

But if the classical radius of the electron were used in the gravity problem, the ratio of the average Ampère force magnitude to the maximum Ampère force magnitude would be approximately

$$\sqrt{n} / 2n = 1 / 2\sqrt{n} \approx 3.65 \times 10^{-3}. \tag{45}$$

This ratio is *not* appropriately small, so this is *not* the right discretization level for the gravity problem.

Another one of the electron radii given by MacGregor [22] is called *actual*. It characterizes results of scattering experiments, and is the one presently only upper-bounded, at $r_{e\ actual} < 10^{-16}$ cm. No one knows how much smaller it could eventually turn out to be. So how much smaller would it *have* to be, in order to account for the extreme weakness of gravity? Gravity requires $2 / \sqrt{n} \approx 10^{-32}$, or.

$$\sqrt{n} \approx 2 \times 10^{+32}, \text{ or } n \approx 4 \times 10^{64}. \tag{46}$$

That in turn requires:

$$r_{\text{e actual}} = r_{\text{H1}} / \rho = 0.529 \times 10^{-8}\,\text{cm} / 4 \times 10^{64} \approx 10^{-73}\,\text{cm}. \tag{47}$$

At present, such a value for $r_{\text{e actual}}$ certainly looks impossible to test with any kind of measurement. It is smaller than anything we yet know about any elementary particle. But that circumstance may be a good thing, because an extremely small electron makes it easier to understand what data from Chemistry reflects, discussed next. And an extremely small electron, along with a correspondingly small positron, helps explain aspects Elementary Particle Physics, discussed after that.

6. Algebraic chemistry and EZ water

Prof. Gerald Pollack of U. Washington wrote the authoritative book [23] about the physical phenomenon called 'EZ water'. The EZ is short for Exclusion Zone, with the word 'exclusion' referring to a surface phenomenon that expels positive hydronium ions.

Prof. Pollack gave a talk about EZ Water at the 2013 meeting of the Natural Philosophy Alliance at College Park, MD, USA. All the phenomena he described were surprising; some were truly puzzling. EZ water apparently makes extended orderly arrays of hexagonal units. How can *that* behavior comport with our understanding that Nature maximizes entropy? Explanations then available were not at all quantitative. That fact suggested a real need for a more quantitative approach.

I had recently written my book about Algebraic Chemistry (AC). [24] The name reflects the fact that the technique has no integrals or other complicated math operations that would demand capabilities beyond those of a hand calculator. The worst operation is square root. So the AC approach looked promising for quick application to EZ water.

The fundamental idea behind AC is that all atoms share some similarities with Hydrogen atoms: **1)** They have a nucleus that is similar to a proton, but scaled up to nuclear charge Z and nuclear mass M ; **2)** They have a population of electrons that is not entirely unlike a single electron; *i.e.*, an interacting community that is somewhat coherent, and somewhat like one *big* electron orbiting the nucleus; **3)** It is possible for the electron count to be different from the nuclear charge. This last possibility is what characterizes ions, and thereby creates all of Chemistry.

We begin with a clue: Eq. (17) indicates that the radius of the Hydrogen atom scales with the mass of the proton. This fact suggests that the base orbit energy of the Hydrogen atom scales with the inverse of proton mass. It further suggests that for element with nuclear charge Z and mass M, the base orbit energy may scale with Z/M. If so, then when first-order ionization potentials for all elements are scaled by the inverse factor, M/Z, then the scaled first-order ionization potentials (called $IP_{1,Z}$) might fall into some pattern.

We proceed with an observation: A pattern indeed emerges: the rise on *every* period in the Periodic Table is *exactly the same factor*, 7 / 2.

We make a Hypothesis: All $IP_{1,Z}$ contain information valuable for all other elements: *population generic information*. Each $IP_{1,Z}$ contains a universal baseline contribution $IP_{1,1}$ about interaction between the nucleus and the population of electrons as a whole. For all elements beyond Hydrogen, there is also a contribution $\Delta IP_{1,Z}$ about interactions among the electrons.
The $\Delta IP_{1,2}$ can be very significant. For Helium, $\Delta IP_{1,2}$ is *huge*, meaning that two electrons bond together very *strongly*. And for Lithium, $\Delta IP_{1,3}$ is *negative*, meaning that two electrons actively work together to try to exclude a third electron.

Over the periods, there is obvious detail about the electron-electron interactions. Within each period, there are obvious sub-periods keyed to the nominal angular-momentum quantum number that is being filled. Plotted on a log scale, all sub-period rises are straight lines. The slopes all appear to be rational fractions. We can display these rational fractions in a Table, as was done in [24] and [25]:

period	N	l	*fraction*	l	*fraction*	l	*fraction*	l	*fraction*
1	1	0	1						
2	2	0	1/2	1	3/4				
3	2	0	1/2	1	3/4				
4	3	0	1/4	2	5/18	1	2/3		
5	3	0	1/4	2	5/18	1	2/3		
6	4	0	1/4	3	7/48	2	5/16	1	9/16
7	4	0	1/4	3	7/48	2	5/16	1	9/16

A non-traditional parameter N is included in the display because, for $l > 0$, it is possible to write a simple formula for the fraction:

$$\text{fraction} = \left[(2l+1)/N^2\right]\left[(N-l)/l\right]. \tag{48}$$

Also, all periods in the Periodic Table have length $2N^2$.

All this numerical regularity suggests that there really is a reliable pattern here, and we can reasonably seek to exploit it. Here is the first exploitation that suggests itself: Given first-order ionization potentials of many elements, we can estimate the additional energy required to remove a second electron from each, and then a third, and so on. This was first done in [24]. Formulae were given for each *individual* electron removal or addition, and evaluated for a large number of elements.

One point that Ref. [24] emphasized was that the energy to remove a second electron, or a third, and so on, is *not* the same thing as the so- called 'second-order ionization potential', 'third-order ionization potential', and so on. Those energies are very *large*, which implies that those events are very *violent*: ripping two, or three, or more, electrons off an atom *all at once*.

Those energies do exhibit a lot of numerical regularity, but that isn't important for understanding typical lab-bench chemistry, which is all about *gentle* events that occur *one-at-a-time*.

Ref [25] presented the equivalent summed formulae for removal of, or addition of, one, two, and three electrons, removed or added *one-at-a-time*. Basically, use of these formulae save the user some repetitive arithmetic that would be incurred using the formulae from Ref. [24].

Development of More Formulae:

For the present paper, the formulae from [25] are extended from the illustrative cases of removing, or adding, one, two, and three electrons, to the general case of removing, or adding, N electrons.

Ref. [25] used symbols W and H to distinguish between energy increments associated with electron-nucleus interaction, and energy increments associated with electron-electron interactions. That distinction is analogous to the distinction between work and heat in thermodynamics: the work part is something a human can control, and the heat part is something that Nature simply does, regardless of what the human does.

We had:

$$W_{\text{removing e}_1\text{ from the neutral atom}} = IP_{1,1}(Z/M_Z). \tag{49}$$

The ion produced has a little less attraction between the nucleus and the now reduced electron cloud. So removing another electron should take a little less work:

$$W_{\text{removing e}_1\ \&\ \text{e}_2} = IP_{1,1}\left[Z + \sqrt{Z(Z-1)}\right]/M_Z. \tag{50}$$

And then:

$$W_{\text{removing e}_1,\ \text{e}_2,\ \&\ \text{e}_3} = IP_{1,1}\left[Z + \sqrt{Z(Z-1)} + \sqrt{Z(Z-2)}\right]/M_Z. \tag{51}$$

This pattern generalizes to:

$$W_{\text{removing e}_1\text{ through e}_N} = IP_{1,1}\left[\sum_{i=1}^{N}\sqrt{Z(Z+1-i)}\right]/M_Z. \tag{52}$$

We also had:

$$H_{\text{removing e}_1\text{ from the neutral atom}} = (\Delta IP_{1,Z} - \Delta IP_{1,Z-1})(Z/M_Z). \tag{53}$$

We inferred in [25] that:

$$H_{\text{removing e}_1,\ \&\text{e}_2} = \left[\Delta IP_{1,Z}Z - \Delta IP_{1,Z-2}(Z-2)\right]/M_Z. \tag{54}$$

and

$$H_{\text{removing e}_1,\text{e}_2,\ \&\text{e}_3} = \left[\Delta IP_{1,Z}Z - \Delta IP_{1,Z-3}(Z-3)\right]/M_Z. \tag{55}$$

This pattern generalizes to:

$$H_{\text{removing e}_1\text{ through e}_N} = \left[\Delta IP_{1,Z}Z - \Delta IP_{1,Z-N}(Z-N)\right]/M_Z. \tag{56}$$

Finally, in [25] we had:

$$\begin{gathered}(W+H)_{\text{removing e}_1,\ \&\text{e}_2} = \\ IP_{1,1}\left[Z+\sqrt{Z(Z-1)}\right]/M_Z + \left[\Delta IP_{1,Z}Z - \Delta IP_{1,Z-2}(Z-2)\right]/M_Z \quad ,\end{gathered} \tag{57}$$

and

$$\begin{gathered}(W+H)_{\text{removing e}_1,\text{e}_2,\ \&\text{e}_3} = \\ IP_{1,1}\left[Z+\sqrt{Z(Z-1)}+\sqrt{Z(Z-2)}\right]/M_Z + \left[\Delta IP_{1,Z}Z - \Delta IP_{1,Z-3}(Z-3)\right]/M_Z \quad ,\end{gathered} \tag{58}$$

This pattern generalizes to:

$$\begin{gathered}(W+H)_{\text{removing e}_1\text{ through e}_N} = \\ IP_{1,1}\left[\sum_{i=1}^{N}\sqrt{Z(Z+1-i)}\right]/M_Z + \left[\Delta IP_{1,Z}Z - \Delta IP_{1,Z-N}(Z-N)\right]/M_Z \quad ,\end{gathered} \tag{59}$$

Now let us turn to *adding* electrons. First, use the formula for the energy for removing an electron from a neutral atom of element Z to describe instead *removing* an electron from the *singly charged negative ion* of element Z, which has $Z+1$ electrons to start with:

$$(W+H)_{\text{removing e}_1\text{ from negative ion}} = \left[IP_{1,1}\sqrt{Z(Z+1)} + \Delta IP_{1,Z+1}(Z+1) - \Delta IP_{1,Z}Z\right]/M_Z. \tag{60}$$

Reversing the direction of the operation:

$$(W+H)_{\text{adding } e_1 \text{ to neutral atom}} = -\left[IP_{1,1}\sqrt{Z(Z+1)} + \Delta IP_{1,Z+1}(Z+1) - \Delta IP_{1,Z}Z \right] / M_Z . \tag{61}$$

This means the work for adding one electron into the nuclear field is:

$$W_{\text{adding } e_1 \text{ to neutral atom}} = -IP_{1,1}\sqrt{Z(Z+1)} / M_Z . \tag{62}$$

And the heat for re-adjusting the electron population is:

$$H_{\text{adding } e_1 \text{ to neutral atom}} = -\left[\Delta IP_{1,Z+1}(Z+1) + \Delta IP_{1,Z}Z \right] / M_Z . \tag{63}$$

Now let us add a second electron. This will require additional work:

$$W_{\text{adding } e_2 \text{ after } e_1} = -IP_{1,1}\sqrt{Z(Z+2)} / M_Z . \tag{64}$$

And it will cause another heat adjustment:

$$H_{\text{adding } e_2 \text{ after } e_1} = -\Delta IP_{1,Z+2}(Z+2) / M_Z + \Delta IP_{1,Z+1}(Z+1) / M_Z . \tag{65}$$

This means total energy involved in adding two electrons is:

$$(W+H)_{\text{adding } e_1 \text{ \& } e_2} = -IP_{1,1}\left[\sqrt{Z(Z+1)} + \sqrt{Z(Z+2)} \right] / M_Z + \left[\Delta IP_{1,Z}Z - \Delta IP_{1,Z+2}(Z+2) \right] / M_Z \quad . \tag{66}$$

Likewise, the total energy involved in adding three electrons is:

$$(W+H)_{\text{adding } e_1, e_2 \text{ \& } e_3} = -IP_{1,1}\left[\sqrt{Z(Z+1)} + \sqrt{Z(Z+2)} + \sqrt{Z(Z+3)} \right] / M_Z + \left[\Delta IP_{1,Z}Z - \Delta IP_{1,Z+3}(Z+3) \right] / M_Z \quad . \tag{67}$$

This patten generalizes to:

$$(W+H)_{\text{adding } e_1 \text{ through } e_N} = IP_{1,1}\left[\sum_{i=1}^{N} \sqrt{Z(Z+i)} \right] / M_Z + \left[\Delta IP_{1,Z}Z - \Delta IP_{1,Z+N}(Z+N) \right] / M_Z \quad , \tag{68}$$

http://dx.doi.org/10.5772/59067

Numerical Data to Insert in Formulae:

Numerical data for elements up to number 118 are given in [24]. The numerical analysis of EZ water requires at most the data for the first ten elements. Expressed in electron volts, eV, these numerical data are:

Hydrogen:	$Z=1$,	$M=1.008$,	$IP_{1,1}=14.250\text{eV}$,	$\Delta IP_{1,1}=0\text{eV}$.
Helium:	$Z=2$,	$M=4.003$,	$IP_{1,2}=49.875\text{eV}$,	$\Delta IP_{1,2}=35.625\text{eV}$.
Lithium:	$Z=3$,	$M=6.941$,	$IP_{1,3}=12.469\text{eV}$,	$\Delta IP_{1,3}=-1.781\text{eV}$.
Beryllium:	$Z=4$,	$M=9.012$,	$IP_{1,4}=23.327\text{eV}$,	$\Delta IP_{1,4}=9.077\text{eV}$.
Boron:	$Z=5$,	$M=10.811$,	$IP_{1,5}=17.055\text{eV}$,	$\Delta IP_{1,5}=2.805\text{eV}$.
Carbon:	$Z=6$,	$M=12.011$,	$IP_{1,6}=21.570\text{eV}$,	$\Delta IP_{1,6}=7.320\text{eV}$.
Nitrogen:	$Z=7$,	$M=14.007$,	$IP_{1,7}=27.281\text{eV}$,	$\Delta IP_{1,7}=13.031\text{eV}$.
Oxygen:	$Z=8$,	$M=15.999$,	$IP_{1,8}=27.281\text{eV}$,	$\Delta IP_{1,8}=13.031\text{eV}$.
Fluorine:	$Z=9$,	$M=18.998$,	$IP_{1,9}=34.504\text{eV}$,	$\Delta IP_{1,9}=20.254\text{eV}$.
Neon:	$Z=10$,	$M=20.180$,	$IP_{1,10}=43.641\text{eV}$,	$\Delta IP_{1,10}=29.391\text{eV}$.

Ordinary Water:

Here are some example calculations concerning possible ionic configurations of ordinary, normal water.

Most people would guess that water is $2\,H^{+}+O^{2-}$. But let us evaluate that ionic configuration. The transition $H\rightarrow H^{+}$ takes:

$$IP_{1,1}\,/\,M_{1}=14.250\,/\,1.008=14.1369\,\text{eV}. \tag{69}$$

So $2H^{+}$ takes $2\times 14.1369=28.2738$ eV.

The transition $O\rightarrow O^{2-}$ takes:

$$\begin{aligned}
&-IP_{1,1}\left[\sqrt{Z(Z+1)}+\sqrt{Z(Z+2)}\right]\Big/M_{Z}+\left[\Delta IP_{1,Z}Z-\Delta IP_{1,Z+2}(Z+2)\right]\Big/M_{Z}\\
&=-14.250\left[\sqrt{8\times 9}+\sqrt{8\times 10}\right]\Big/15.999+\left[\Delta IP_{1,8}\times 8-\Delta IP_{1,10}\times 10\right]\Big/15.999\\
&=-14.250\left[\sqrt{72}+\sqrt{80}\right]\Big/15.999+\left[13.031\times 8-29.391\times 10\right]\Big/15.999\\
&=-14.250\left[8.4853+8.9443\right]\Big/15.999+\left[104.248-293.910\right]\Big/15.999\\
&=-14.250\left[17.4296\right]\Big/15.999+\left[-189.662\right]\Big/15.999\\
&=-15.5242-11.8546=-27.3788\,\text{eV}\quad .
\end{aligned} \tag{70}$$

So the ionic configuration $2\,H^{+}+O^{2-}$ requires $28.2738-27.3788=0.8950$ eV. This is a positive energy requirement, which implies that some external assistance is needed to create this ionic configuration. So normal water may *not* be $2\,H^{+}+O^{2-}$ after all!

Another possibility is readily at hand though. The ionic configuration for normal water could be $2\,H^{-}+O^{2+}$. The transition $H\rightarrow H^{-}$ takes:

$$\begin{aligned}&-IP_{1,1}\sqrt{Z(Z+1)}/M_Z-\Delta IP_{1,Z+1}(Z+1)/M_Z+\Delta IP_{1,Z}Z/M_Z\\&=-IP_{1,1}\sqrt{1\times 2}/M_1-\Delta IP_{1,2}\times 2/M_1+\Delta IP_{1,1}\times 1/M_1\\&=-14.250\times 1.4142/1.008-35.625\times 2/1.008+0\times 1/1.008\\&=-19.9924-70.6845+0=-90.6769\,\text{eV}\ .\end{aligned} \tag{71}$$

So $2\,H^{-}$ takes $2\times(-90.6769)=-181.3538$ eV. Notice that this is a *huge* negative energy. It reflects the fact that electrons *really* like to make pairs. Indeed, their propensity to do so motivated the invention of the so-called 'spin' quantum number. Without spin, many electrons in atoms would be violating the 'Pauli Exclusion Principle', which says that only *one* electron can be in any particular quantum state. Electron pairs are famous in Condensed Matter Physics too, under the name 'Cooper pairs'.

Proceeding now to the transition $O\rightarrow O^{2+}$, that takes:

$$\begin{aligned}&IP_{1,1}\left[Z+\sqrt{Z(Z-1)}\right]/M_Z+\left[\Delta IP_{1,Z}Z-\Delta IP_{1,Z-2}(Z-2)\right]/M_Z\\&=IP_{1,1}\left[8+\sqrt{8\times 7}\right]/M_8+\left[\Delta IP_{1,8}\times 8-\Delta IP_{1,6}\times 6\right]/M_8\\&=14.250\left[8+\sqrt{56}\right]/15.999+\left[13.031\times 8-7.320\times 6\right]/15.999\\&=13.7907+3.7707=17.5614\,\text{eV}\ .\end{aligned} \tag{72}$$

Thus the creation of the water molecule in the ionic configuration $2\,H^{-}+O^{2+}$ demands altogether $-181.3538+17.5614=-163.7924$ eV. This energy is solidly negative, which means that ordinary water is overwhelmingly in this ionic configuration, $2\,H^{-}+O^{2+}$. This *is* normal water.

However, in situations where more than one version of anything can exist, both generally *do* exist, in proportions determined by their so-called Boltzmann factors, $\exp(-E/kT)$. Here E is energy, k is Boltzmann's constant, and T is absolute temperature. Boltzmann factors are the result of entropy maximization at work. Because of non-zero Boltzmann factors, there will exist a tiny, tiny fraction of the first ionic configuration, $2\ H^{+}+O^{2-}$.

This analysis of normal water shows how quantitative approaches can sometimes unseat long-standing, but never-justified, assumptions in Chemistry.

About EZ Water:

The following transition is generally thought to represent the creation of EZ water:

3 normal water molecules $\rightarrow$ 1 EZ water ion + 1 hydronium ion

That is,3(H_2O)®(H_3O_2)$^-$+(H_3O)$^+$.

Here parentheses are used to avoid implying anything about what charge the individual atoms within any ion or radical may carry. A full numerical analysis should consider all possible, or at least all plausible, ionic configurations of every molecule or radical involved.

One possible ionic configuration for the EZ water ion $(H_3O_2)^-$ is $3H^+ + 2O^{2-}$. The $3H^+$ takes total energy $3 \times IP_{1,1}/M_z = 3\times 14.250/1.008 = 42.4107$ eV. The $2O^{2-}$ takes total energy $2\times(-27.3788) = -54.7576$ eV. So the ionic configuration $3H^+ + 2O^{2-}$ altogether takes $42.4107 - 54.7576 = -12.3469$ eV. This is a negative energy, so this ionic configuration certainly can occur.

But there is also another possibility for the EZ water ion $(H_3O_2)^-$. It could have the ionic configuration $3H^- + 2O^+$. An H^- takes -90.6769 eV, so $3H^-$ takes $3\times(-90.6769) = -272.0307$ eV. An O^+ takes energy:

$$\begin{aligned}&(IP_{1,1}\times 8+\Delta IP_{1,8}\times 8-\Delta IP_{1,7}\times 7)/M_8\\&=(14.250\times 8+13.031\times 8-13.031\times 7)/15.999\\&=(114.000+104.248-91.217)/15.999=7.9399\,\text{eV}\end{aligned} \tag{73}$$

So $2O^+$ takes $2\times 7.9399 = 15.8798$ eV. Then the ionic configuration $3H^- + 2O^+$ takes $-272.0307 + 15.8798 = -256.1509$ eV. This energy is much more negative than that of the first candidate ionic configuration for EZ water, $3H^+ + 2O^{2-}$. This fact means EZ water is nearly always in this second candidate ionic configuration, $3H^- + 2O^+$.

One possible ionic configuration for the hydronium ion $(H_3O)^+$ is $3H^+ + O^{2-}$. This, I believe, is what most people would guess. But from the study of regular water, we know the candidate $3H^+$ would take $3\times 14.250 = 42.4107$ eV, and that the candidate O^{2-} would take -27.3788 eV, so the candidate ionic configuration $3H^+ + O^{2-}$ for hydronium would take $42.4107 - 27.3788 = 15.0319$ eV. This energy is positive, *so this ionic configuration for the hydronium ion is not promising.*

However, as was the case with the EZ water ion, there is another possibility for the hydronium ion. It $(H_3O)^+$ could have the ionic configuration $3H^- + O^{4+}$. The $3H^-$ would take $3\times(-90.6769) = -272.0307$ eV, and the O^{4+} would take:

$$\begin{aligned}&IP_{1,1}\left[Z+\sqrt{Z(Z-1)}+\sqrt{Z(Z-2)}+\sqrt{Z(Z-3)}\right]/M_Z+\left[\Delta IP_{1,Z}Z-\Delta IP_{1,Z-4}(Z-4)\right]/M_Z\\&=14.250\times\left[8+\sqrt{56}+\sqrt{48}+\sqrt{40}\right]/15.999+\left[13.031\times 8-9.077\times 4\right]/15.999\\&=14.250\times 28.7361/15.999+67.940/15.999\\&=25.5947+4.2465=29.8411\,\text{eV}\end{aligned} \tag{74}$$

So for the hydronium ion $(H_3O)^+$, the second candidate ionic configuration $3\,H^- + O^{4+}$ would take $-272.0307 + 29.8411 = -242.1895$ eV. This very negative energy explains why the reaction product that accompanies EZ water *is* a hydronium ion, rather than a naked proton plus a normal water molecule, which would take $14.1369 \ -163.7925 = -149.6556$eV, which is not *as* negative.

The EZ water ion and the hydronium ion together take $-268.6204 - 242.1895 = -510.8099$ eV. Compare this energy to the energy taken by three normal water molecules: $3 \times (-163.7925) = -491.3775$ eV. The EZ water ion with the hydronium ion has lower energy than the three normal water molecules. That means that Nature will take any opportunity to make EZ water ions and hydronium ions.

It appears that what creates the opportunity is a surface, plus a little energy to separate the ions. Any material body provides some gravity to create a surface, and if there is also some small energy source, such as sunlight, to help separate ions, and if there is also some normal water, the situation will automatically create EZ water too. Even an icy comet might be able to create some EZ water.

7. Microphysics

Just as the myriad compounds in Chemistry arise from not-very-many chemical elements, some significant part of the m

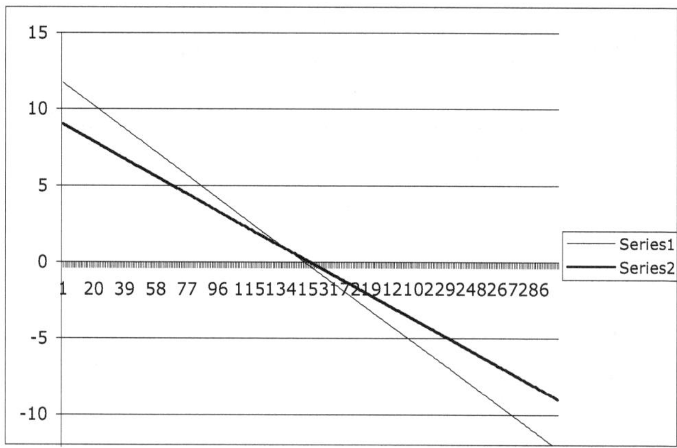

Figure 4. Log energy loss and gain rates *vs.* log system radius.

Observe that radiation dominates for radii below the crossing point, whereas torquing dominates for radii above the crossing point. This means the balance between the two effects is *unstable*: a small excursion from balance in either direction causes more excursion in the same

http://dx.doi.org/10.5772/59067

direction. This is interesting. It means that Hydrogen does not like to exist as an isolated atom. It wants to engage in chemical reactions. In the Universe at large, you will find Hydrogen in H_2 molecules, or other molecules, or you will find Hydrogen plasma, consisting of naked protons and free electrons, but you will *not* find many isolated Hydrogen atoms.

1. We can discover even more about the Hydrogen atom if we include the appropriate angle sine and cosine factors in the calculations. These factors are oscillatory. So negative numbers sometimes occur with the torquing curve, and they cannot be plotted on a logarithmic scale. However, we are mainly interested in the points of balance between torquing and radiation, and the radiation curve is always non-negative, so the torquing curve is non-negative too at the balance points.

Figure 5 shows that we have not just one balance point, but *many* balance points. The balance points occur in close pairs, one stable and one unstable. The two pairs furthest left on the plot do not show as crossings because of the finite resolution of the plot, but they are certainly present.

More solution pairs are to be found, off the Figure to the left. Indeed, the solution pairs continue *indefinitely*, into smaller and smaller system radii. So Hydrogen has an infinite family of 'sub-states'. Mills discusses these in [27].

The smaller and smaller radii of the balance points in Fig. 5 correspond to higher and higher orbit speeds. This idea conflicts with a prohibition imposed by SRT: no physical particle possessing mass is allowed to move at a speed matching or exceeding light speed c. What does this conflict mean? I believe it means the prohibition should be understood more precisely to say: no physical particle possessing mass *can be* **perceived** to move at a speed matching or exceeding light speed c, **if** *we agree to process all received data in accord with Einstein's Second Postulate.* If we do not agree to that, then there is no particle speed limit.

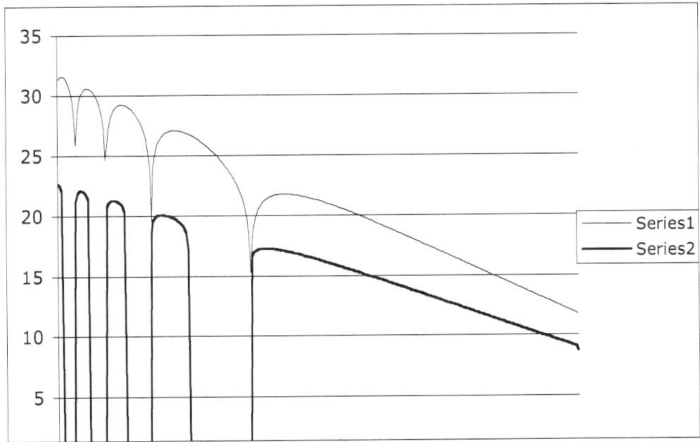

Figure 5. More and more balance points below the Hydrogen ground state.

2. We can also study positronium (a system consisting of one electron and one positron). See Fig. 6. As with Hydrogen, the oscillatory angle factors create a family of solutions for this system too. Half of them are stable, and half are unstable, and uncountably many of them occur at small radii and high speeds well in excess of c. Due to the finite resolution of the plot, only one pair of solutions is clearly visible in Fig. 6. But two more, at smaller radius and higher speed, are also certainly present. We can characterize these, and all high-speed solutions, without even knowing exactly what the radiation curve is like - its exact amplitude, or its r^{-4} dependence. The one low-speed solution is just $v_0 \approx 0.02 \times c$. The many high-speed solutions have to occur in pairs just above and below orbit speeds of the form $v_n = v_0 + n \times 2\pi c$, where n is an arbitrary positive integer.

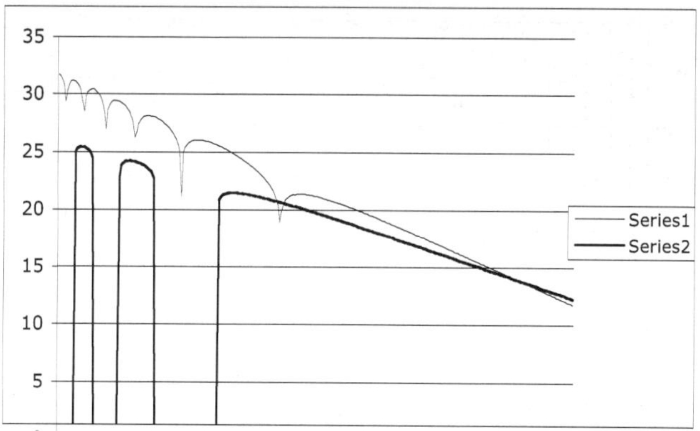

Figure 6. Positronium solutions.

A parenthetical note applies for Fig. 6: for same-mass systems, the amplitude of the radiation curve should be less by a factor of 4 because there is no center-of-mass motion. This sort of numerical detail does not significantly affect where the solutions fall. That is determined almost entirely by the cosine factors that produce the deep dips that intersect the peaks in the curve for rate of energy gain due torquing.

3. The oscillatory nature of the angle factors can turn a situation of seeming repulsion into a situation of actual attraction. This phenomenon of sign reversal due to signal delay is well known to engineers, who often deal with oscillating signals in feedback control systems For the present application, consider two electrons in a circular orbit, and suppose they move at speed πc. One electron launches its signal radially outward. By the time this electron has executed half an orbit, this signal has expanded a distance equal to the orbit diameter. By then, the two electrons have exchanged places. So the expanding signal first contacts the second electron at exactly the signal launch point. Then the two electrons complete their orbit. At the end, the second electron finally understands its signal: it is to move radially. But by now, the two electrons have changed places again, and for the second electron, the direction commanded is *inward*. That situation is equivalent to *attraction*.

Given this mechanism for attraction, we can also study homogeneous systems: two electrons, or two positrons, for example. Again, there exist both stable and unstable solutions, and there are infinitely many of each, corresponding to orbit speeds of the form $v_n = \pi c + n \times 2\pi c$ for arbitrary positive integer n. See Fig. 7.

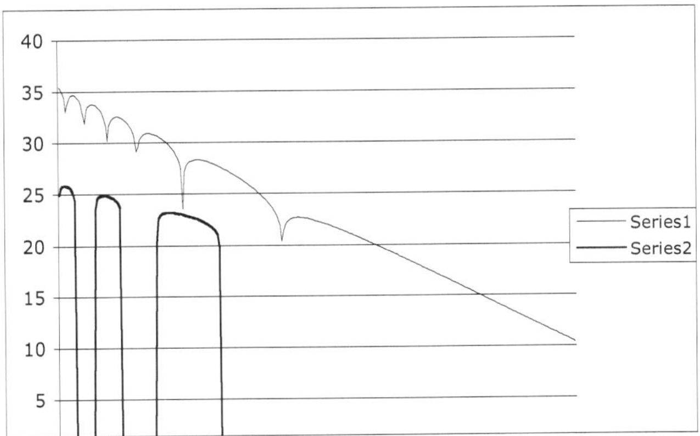

Figure 7. Same-charge solutions.

An additional parenthetical note applies for Fig. 7: for same-charge systems, the angular pattern of the radiation is quadrupole, rather than dipole, so the amplitude of the radiation energy loss curve should decline as r^{-6} rather than r^{-4}. Again, this sort of numerical detail does not significantly affect where the solutions fall, which is determined almost entirely by the cosine factors that produce the deep dips that intersect the peaks in the curve for rate of energy gain due torquing.

The stable solutions for two electrons bring to mind the situation that is so well known in Chemistry: electron pairs. They are everywhere in Chemistry. The most famous case occurs for Helium. Helium is a noble gas, and it reacts with other elements only under extreme duress. Helium has two electrons, and pulling one electron away is very costly: Helium has the highest ionization potential of any element. The message is: two electrons definitely do form a stable subsystem within an atom. The standard QM explanation for this invokes the concept of electron spin, with two possible values, $\pm\hbar/2$, allowing two electrons in the same overall energy state. Electron pairing also occurs famously in Solid State Physics, under the name of Cooper Pairs.

8. Conclusion

In the present Chapter, the new concept applied is the more realistic signal model for use in an improved version of SRT. The realistic signal model is based on Information Theory,

The new concept is implemented with very standard mathematics: differential equations, their family of solutions, and the particular problem boundary conditions. These mathematical ingredients for a proper signal model were all available in 1905, but they were not used in SRT. Why? I believe the fundamental reason is the history: Information Theory was not yet available, so no researcher at that time would have been likely to detect the inadequacy of the infinite plane wave as a signal model.

The present paper has shown that there are rewards for instead using the realistic photon/signal model. They include more insight into Quantum Mechanics, and into Gravity Theory, and potentially into Elementary Particle Physics. These are all subjects to be studied much more fully in the future.

Many textbook treatments of SRT devote a lot of space to Lorentz Transformations (LT's). The present work has not mentioned LT's at all. To this author, LT's just seem to describe the wrongly informed opinions of different observers. So I don't really want to focus on LT's. But I have to mention them, because repairing SRT to take proper account of the concepts of IT casts doubt on Einstein's SRT, and hence on LT's. Therefore, I hereby relegate the unavoidable discussion of coordinate transformations, LT's and others, to the following Appendix.

9. Appendix

The situation in the late nineteenth century included the following fact: Maxwell's first order coupled field equations appeared *not* to be invariant under Galilean transformation of coordinates (GT's). Phipps [28] has written extensively about this apparent conflict between Maxwell's Electromagnetic Theory and Newton's Mechanics. In the early twentieth century, SRT brought in LT's, and the conflict seemed to be resolved: Maxwell's electromagnetic theory was clearly invariant under LT's. This fact was taken as evidence in favor of Einstein's SRT over Newton's Mechanics.

But there is a puzzle left to resolve: Maxwell's first order coupled field equations appear to qualify as *tensor* equations. Mathematicians had developed tensors in the first place to enable the articulation of mathematical statements that would be *coordinate-free*. So tensor equations are *by definition* invariant to *all* invertible coordinate transformations.

So what had happened here? I believe two circumstances had collided to create a very bad situation. One circumstance was that Mathematics had such a long history of developing *eternal* truths: the focus had been on arithmetic, geometry, and trigonometry – all of them eternal in character. Even archeo-astronomy was largely about the eternal *repetition* of events, and *not* about temporal *evolution* of events. Eternal truths really need not have a time dimension. They can, however, have as many spatial dimensions as may be desired, and that became the focus for much of tensor analysis. The other circumstance was that time became a really significant variable with the advent of modern Physics: Kepler, Galileo, Newton, and Maxwell. And time is a really different kind of variable than space is. Maxwell was very well aware of the difference, as he developed his electromagnetic theory in terms of Hamilton's quaternions. The modern equivalent of the quaternion tool is the set of four 2×2 complex Pauli spin matrices:

$$\sigma_{ct} = \begin{bmatrix} 1 & 0 \\ 0 & 1 \end{bmatrix}, \sigma_x = \begin{bmatrix} 0 & 1 \\ 1 & 0 \end{bmatrix}, \sigma_y = \begin{bmatrix} 0 & -i \\ i & 0 \end{bmatrix}, \sigma_z = \begin{bmatrix} 1 & 0 \\ 0 & -1 \end{bmatrix}. \tag{75}$$

The first one, the time-like one, is the identity matrix. The other three, the space-like ones, produce the identity matrix when squared. When two of them are cross-multiplied, they generate a factor of $i=\sqrt{-1}$ times the third one, corresponding to the vector cross product in three-dimensional space.

The collision between circumstances came in the formulation of differential operators. People were familiar with the scalar chain rule,$\boldsymbol{d/dt}=\partial / \partial_t + v\partial / \partial_x$, and did not realize that more information was needed in the context of vector and tensor applications with time as well as space dimensions.

The 2×2 Pauli matrices are easy to appreciate visually, so I will also discuss transformations of coordinates in terms of 2×2 matrices also - but only real ones, not complex ones. Let s stand for any spatial coordinate. A general coordinate transformation involving ct and s has the form:

$$\begin{bmatrix} ct' \\ s' \end{bmatrix} = \frac{1}{\sqrt{1-AB}} \begin{bmatrix} 1 & B \\ A & 1 \end{bmatrix} \begin{bmatrix} ct \\ s \end{bmatrix}. \tag{76}$$

For the familiar Lorentz transformation, $A=B=-v/c$, where v is the speed of the new coordinate frame relative to the old one. The letter v is lower case to remind us that $v<c$; *i.e.* $v/c<1$. We have:

$$\begin{bmatrix} ct' \\ s \end{bmatrix} = \frac{1}{\sqrt{1-v^2/c^2}} \begin{bmatrix} 1 & -v/c \\ -v/c & 1 \end{bmatrix} \begin{bmatrix} ct \\ s \end{bmatrix}. \tag{77}$$

For the long discarded Galilean transformation, $A=-V/c$ and $B=0$. The letter V is upper case to remind us that V is *not* limited, and might exceed c. So we have:

$$\begin{bmatrix} ct' \\ s \end{bmatrix} = \frac{1}{1} \begin{bmatrix} 1 & 0 \\ -V/c & 1 \end{bmatrix} \begin{bmatrix} ct \\ s \end{bmatrix}. \tag{78}$$

For all such general coordinate transformations, there also exists a complement transformation:

$$\begin{bmatrix} ct'' \\ -s'' \end{bmatrix} = \frac{1}{\sqrt{1-AB}} \begin{bmatrix} 1 & -A \\ -B & 1 \end{bmatrix} \begin{bmatrix} ct \\ -s \end{bmatrix}. \tag{79}$$

Its purpose is to preserve inner products; for example:

$$\begin{bmatrix} ct & s \end{bmatrix} \begin{bmatrix} ct \\ -s \end{bmatrix} = (ct)^2 - s^2. \tag{80}$$

Observe that:

$$\begin{aligned} \begin{bmatrix} ct' & s' \end{bmatrix} \begin{bmatrix} ct'' \\ -s'' \end{bmatrix} &= \begin{bmatrix} ct & s \end{bmatrix} \frac{1}{\sqrt{1-AB}} \begin{bmatrix} 1 & A \\ B & 1 \end{bmatrix} \frac{1}{\sqrt{1-AB}} \begin{bmatrix} 1 & -A \\ -B & 1 \end{bmatrix} \begin{bmatrix} ct \\ -s \end{bmatrix} \\ &= \begin{bmatrix} ct & s \end{bmatrix} \frac{1}{1-AB} \begin{bmatrix} 1-AB & 0 \\ 0 & -BA+1 \end{bmatrix} \begin{bmatrix} ct \\ -s \end{bmatrix} \equiv (ct)^2 - s^2 \ . \end{aligned} \tag{81}$$

For Lorentz transformation, the complement transformation is the inverse, or equivalently, the *re*verse transformation:

$$\begin{bmatrix} ct'' \\ -s'' \end{bmatrix} = \frac{1}{\sqrt{1-v^2/c^2}} \begin{bmatrix} 1 & +v/c \\ +v/c & 1 \end{bmatrix} \begin{bmatrix} ct \\ -s \end{bmatrix}. \tag{82}$$

But for Galilean transformation, the complement transformation is:

$$\begin{bmatrix} ct'' \\ -s'' \end{bmatrix} = \frac{1}{1} \begin{bmatrix} 1 & +V/c \\ 0 & 1 \end{bmatrix} \begin{bmatrix} ct \\ -s \end{bmatrix}. \tag{83}$$

This *is* the *in*verse transformation, but *not* the *re*verse transformation. It is rather the *transpose* of the reverse transformation. It looks so very strange because, for more than a century now, *only* Lorentz transformations of velocity have been used in mainstream theoretical Physics, and transposition does not change them.

The vital role for this strange new thing lies with the differential operators. The story is much like it was for the coordinates: there are *two* complementing transformations, and they involve, not only inversion/reversal, but also transposition. Let ∂_{ct} represent differentiation with respect to the time-like coordinate, and ∂_s represent differentiation with respect to the spatial variable. Let us demand invariance of inner products involving differential operators; for example, like:

$$\begin{aligned} &\begin{bmatrix} \partial_{ct} & \partial_s \end{bmatrix} \begin{bmatrix} ct \\ -s \end{bmatrix} = 1 - 1 = 0 \text{ and } \begin{bmatrix} \partial_{ct} & -\partial_s \end{bmatrix} \begin{bmatrix} ct \\ s \end{bmatrix} = 1 - 1 = 0 \\ &\text{or } \begin{bmatrix} \partial_{ct} & \partial_s \end{bmatrix} \begin{bmatrix} \partial_{ct} \\ -\partial_s \end{bmatrix} = \partial_{ct}^2 - \partial_s^2 \text{ and } = \begin{bmatrix} \partial_{ct} & -\partial_s \end{bmatrix} \begin{bmatrix} \partial_{ct} \\ \partial_s \end{bmatrix} = \partial_{ct}^2 - \partial_s^2. \end{aligned} \tag{84}$$

http://dx.doi.org/10.5772/59067

i.e., always *two* statements – not just *one* statement. This level of detail was missing from the *scalar* chain rule, and that omission caused people to believe that Maxwell's equations could not be shown to be invariant under GT. And so they welcomed LT instead. This is not to say we should now revert to using GT again. Indeed, because of the half-retardation issue discussed in Sect. 3, the best transformation to use may involve, not V / c, but rather $V/\underline{2}c$. This question needs detailed future study.

The use of 2×2 matrices can make the detail needed in such future study very clear. However, many mathematicians tend to prefer tensor notation. But current-day tensor notation uses only two index positions, both on the right: down called 'covariant', up, called 'contravariant'. To represent the transformations needed for Physics, it would be helpful, and maybe necessary, to add two more index locations, up and down on the *left*, to acknowledge transposition, and using words like 'trans-covariant' and 'trans-contravariant' to emphasize what putting indices in those positions means.

Acknowledgements

This Chapter is dedicated to the memory of a most courageous researcher in theoretical and applied electrodynamics: Dr. Peter Graneau, 1921-2014. He encouraged me, and many other researchers, to give serious attention to the History behind Physics.

Author details

Cynthia Kolb Whitney*

Address all correspondence to: galilean_electrodynamics@comcast.net

Galilean Electrodynamics, USA

References

[1] Einstein, A., On Thr Electrodynamics of Moving Bodies, Annelen der Physik 17, 891-921 (1905).

[2] Einstein, A., Annelen der Physik On the Theory of Light Production and Light Absorption, Annelen der Physik 20, 199-206 (1906).

[3] Brillouin, L., *Science and Information Theory*, Second Edition, Dover Publications, Inc., Mineola, NY, (2013).

[4] Flores-Gallegos, N., Shannon Informational Entropies and Chemical Reactivity, Chapt. 29, pp. 683-722 in *Advances in Quantum Mechanics,* Intech (2013).

[5] Jackson, J.D., *Classical Electrodynmics,* Second Edition, John Wiley & Sons, New York, NY. (1975).

[6] Bentwich, J., The Theoretical Ramifications of the Computational Unified Field Theory, Chapt. 28, pp. 671-681, *Advances in Quantum Mechanics,* Intech (2013).

[7] Whitney, C.K., Better Unification for Physics in General Through Quantum Mechanics in Particular, Chapter 7, pp. 127-160 in *Theoretical Concepts of Quantum Mechanics,* Ed. M.R. Pahlavani, InTech. (2012).

[8] Ritz, W., Researches critiques sur l'electrodynamique generale, Ann. Chim. et Phys. 13, pp. 145-275, (1908).

[9] Putz, M.V., Path Integrals for Electronic Densities, Reactivity Indices, and Localization Functions in Quantum Systems, *International Journal of Molecular Sciences* 10, pp. 4816-4940 (2009).

[10] Bracken, P., Quantum Mechanics Entropy and a Quantum Version of the H-Theorem, Chapt. 21, pp. 469-488 in *Theoretical Concepts of Quamtum* Mechanics 469-488 (2012).

[11] Bracken, P., The Schwinger Action Principle and its Applications to Quantum Mechanics, Chapt. 8, pp 159-182 in *Advances in Quantum Mechanics,* Intech (2013).

[12] Spitnev, V., Generalized Path Intergral Technique: Nanoparticles Incident on a Slit Grating, Matter Wave Interference, Chapt 9., pp. 183-212 in *Advances in Quantum Mechanics,* Intech (2013).

[13] Bulnes, F., Correction, Restoration and Re-Composition if Quantum Mechanical Fields of Particles by Path Integrals and Their Applications, 489-514 in *Advances in Quantum Mechanics,* Intech (2013).

[14] Liénard, A., Champ Electrique et Magnétique produit par une Charge Electrique Concentrée en un Point et Animée d'un Movement Quelconque, *L'Eclairage Electrique, XVI,* pp. 5-14, 53-59, and 106-112 (1898).

[15] Wiechert, E., Elektrodynamische Elementargesetze, *Archives Néerlandesises des Sciences Exactes et Naturelles, série II, Tome IV,* pp. 549 – 573 (1901).

[16] Lokajicek, M., V. Kundrát, and J. Procházka, Schrödinger Equation and (Future) Quantum Physics, Chapter 6 in Advances in Quantum Mechanics, Ed. Paul Bracken, InTech. (2013).

[17] Whitney, C.K., On the Dual Concepts of 'Quantum State' and 'Quantum Process', Chapter 17 in *Advances in Quantum Mechanics,* Ed. Paul Bracken, InTech. (2013).

[18] Enders, P., Quantization as Selection Rather than Eigenvalue Problem, Chapt 23, pp. 543-564, *Advances in Quantum Mechanics*, Intech (2013).

[19] Ralston, J.P., Emergent un-Qusntum Mechanics, Chapt. 19, pp. 437-475, in *Advances in Quantum Mechanics*, Intech (2013)

[20] Whitney, C.K., Reasonable doubt: Cosmology's gift for physics, *Journal of Computational Methods in Science and Engineering* 2013, *13*, pp. 291-302 (2013).

[21] Graneau, P., *Ampère–Neumann Electrodynamics of Metals*, Hadronic Press, Nonantum, MA, USA, pp. 7 – 22 (1987).

[22] MacGregor, M., *The Enigmatic Electron- A Doorway to Particle Masses*, 2nd ed; El Mac Books, Santa Cruz, CA, USA; pp. 3 –11 (2013).

[23] Pollack, G., *The Fourth Phase of Water, Beyond Solid Liquid Vapor*; Ebner & Sons, Seattle (2013).

[24] Whitney, C.K., *Algebraic Chemistry: Applications and Origins*, Nova Science Publishers (2013).

[25] Whitney, C.K., EZ Water: A New Quantitative Approach Applied – Many Results Obtained, Water 5, January 5, pp. 105-120 (2014).

[26] Whitney, C.K., The Neutron: a Challenge for Post-Maxwell Physics, *Hadronic Journal*, 3, pp. 201 – 250 (2008).

[27] Mills, R.L., The Nature of the Chemical Bond Revisited and an Alternative Maxwellian Approach, *Physics Essays* 17, 3, pp. 342 – 389 (2004).

[28] Phipps, T.E., *Heretical Verities: Mathematical Themes in Physical Description*; classic non-fiction library, Urbana, IL, USA, pp. 102 – 108 (1986).

Chapter 3

Path Integral Methods in Generalized Uncertainty Principle

Hadjira Benzair, Mahmoud Merad and
Taher Boudjedaa

Additional information is available at the end of the chapter

1. Introduction

As we known the modern physics is based on the two fundamental pillars of physics. The first is the general relativity theory, discovered by Albert Einstein, which gave us a detailed explanation of the macro-dimension world; for example, planets, stars, galaxies and clusters of galaxies and even the extra-universe, that explains the force exerted by the gravitational field of a massive object on any body within the vicinity of its surface. It mainly uses the Riemannian geometry as a mathematical formalism. The second perspective is the quantum mechanics, that describes the micro-dimensions, such as; molecules, atoms and even the smallest components of the latter, like the electrons and quarks, which explains the three principal forces in the micro-world, (like the weak force, electromagnetic force and strong force). It uses the operator theory acting on a Hilbert space algebra (von Neumann algebras). After the mid-twentieth century a new theory in physics has been emerged called the Non-commutative (NC) geometry. It came to unify the four fundamental forces, And its roots go back to the inability of classical physics to explain certain macroscopic phenomena. Mathematically described by a Poisson manifold M, and denoted by $F(M)$ algebra (commutative) regular functions on M, called observable. In this case, it is important to quantify these Poisson varieties (quantum mechanics) in order to obtain results more "precise" than classical mechanics. Many studies have focused on the possibility of quantification of such varieties and the idea of using the theory of algebraic deformations, called "deformation of quantization" is due to (Bayen et al, 1978). And as has been creativity in this mathematical aspect, through a group of researchers (for example, (Bordemann et al., (2005); Makhlouf, 2007)). The motivations to the occurrence of this deformation theory are multiple, in a string theory, (see for example (Veneziano, (1986); Amati et al, (1987); Konishi et al, (1990); Kato, (1990) and Guida et al, (1991) also Gross et al, 1988) in a quantum gravity, (Garay, 1995) in a non-commutative geometry, (Capozziello et al, 2000) and in a black hole physics (Scardigli, (1999); Scardigli & Casadio, (2003)).

In the same context, the innovations physicists were prominent by inserting this study algebra on several applications in physics. The first of these applications is the papers of (Kempf et al, 1995), which is based on introducing a parameter of deformation β in the Heisenberg incertitude principle, given by:

$$\Delta x \Delta p \geqslant \frac{\hbar}{2}\left[1+\beta\left(\Delta p\right)^{2}+\beta\left\langle p\right\rangle^{2}\right], \tag{1}$$

Where the commutation between the position and the momentum operators in one dimension, can be written as:

$$\left[\widehat{x},\widehat{p}\right]=i\hbar\left(1+\beta p^{2}\right), \tag{2}$$

This example, we have found several applications in the non-relativistic quantum mechanics; such as, the harmonic oscillator of arbitrary dimensions (cf., e.g., Refs (Chang et al, (2002); Hinrichsen & Kempf, (1996); Kempf, (1994), Kempf et al, (1995) and Kempf, (1997)), the problem of the cosmological constant has been studied (Pet & Polchinski, (1999)), the effect of the minimal length on the 3-D Coulomb potential has also been studied in (Brau, (1999) and Akhoury, (2003)), the one-dimensional box (Nozari & Azizi, (2006)), the study of the dynamics of a non-relativistic particle with mass variable $m(t)$ (cf., e.g., Ref Merad & Falek, (2009)), and also in the relativistic extension of this problem has some limited attempts, among them we mention: the Dirac equation in the presence of a minimum length in (Nouicer, (2006)), where the Dirac oscillator in one dimension has been solved exactly, the generalized Dirac equation was recently studied by Nozari (Nozari & Karami, (2005)), the bosonic oscillator DKP (spin 0 and 1)-dimensional and three-dimensional that were treated respectively in (Falek & Merad, (2009); (2010)).

On the other hand, the path integral is an alternative technical of Heisenberg and Schrödinger methods. This approach is based on the Lagrangian form, which offers an alternative view of quantum mechanics, that has quickly established itself in theoretical physics, with its extension on quantum field theory and gauge theories. The extension of this technique within the framework of deformed algebra was applied to the relativistic and non-relativistic quantum mechanics. For example, the harmonic oscillator in one dimension (Nouicer, 2006) and in D dimensions (Chergui et al, 2010) and the (1+1)-dimensional Klein–Gordon equation with mixed vector-scalar linear potentials (Merad et al, 2010) but recently it is shown that the problem concerning the choice of point discretization in the path integral is not yet resolved and this arbitrariness is fixed by comparing the discrete action in its infinitesimal form with the corresponding wave equation by which judicious choice of the discretization parameter is indicated by the order of operators (cf., e.g., Refs Benzair et al, (2012); (2014)). This resembles the case of curved spaces in which the mid-point (i.e. $\bar{x}=\left(x_{j}+x_{j-1}\right)/2$) was privileged to have correct quantum correction due to the curvature. Similar arguments in the case of space-time transformations (cf., e.g., Ref (Khandekar et al, 1993)) are presented. In (Ref (Kleinert, 1990)), an outcome considers all points of the interval in an equivalent manner but unfortunately with minimal length deformation. The problem is raised and we will say that it is more like that of the quantization with constraints (see, e.g., Ref (Lecheheb et al, 2007)).

In this chapter we propose to construct the path integral formalism in the momentum space representation to adapt this type of deformation, defined in Eqs. (1) and (2). Then, we describe in detail the method of calculating the quantum corrections according to Feynman approach (cf., e.g., Ref (Khandekar et al, 1993)). As it is shown in (Benzair et al, (2012); (2014)), different methods gave different results, where the quantum correction C_T depends on the α-point discretization interval and there are specific options for the choice of the discretization $\alpha-$parameter, which coincides with the equation method, and this leads to the vanishing of the term C_T and this corresponds to $\alpha = 0$ and $\alpha = 1/2$ within the method of (Kleinert, 1990) and to $\alpha = \frac{1}{2}\left(1 \pm 1/\sqrt{2}\right)$ within the standard method of (Khandekar et al, 1993).

2. Brief review of a minimal length relation

As it seems that in the Kempf's work (cf., e.g., Ref (Kempf et al, 1995)), there is a minimal value of $(\Delta x)_{\min}$ different zero which is given by:

$$(\Delta x)_{\min}(\langle p\rangle) = \hbar\sqrt{\beta}\sqrt{1+\beta\langle p\rangle^2}. \tag{3}$$

$$= \hbar\sqrt{\beta} \text{ corresponds to } \langle p\rangle = 0. \tag{4}$$

The operators $(\widehat{x}, \widehat{p})$ that verifies the commutation relation amended (2) may be considered as the functions of q and p operators, satisfying the relationship of canonical commutation: $[\hat{q}, \hat{p}] = i\hbar$, as follow

$$\widehat{x} = i\hbar\left(1+\beta p^2\right)\hat{q},\ \hat{p} = p. \tag{5}$$

In the momentum space representation, we define the expressions of $\hat{x}$ and $\hat{p}$ act on the functions $\Psi(p)$ defined by:

$$\widehat{p}.\Psi(p) = p\Psi(p),\ \ \widehat{x}\Psi(p) = i\hbar\left(1+\beta p^2\right)\tfrac{\partial}{\partial p}\Psi(p). \tag{6}$$

The most important condition to be satisfied by the representation (2), is the preservation of the operators symmetry x and p, where their values are real. Despite the fact that p is not modified, then its symmetry is obvious; it is not the case for the x operator. Indeed, the symmetry condition is written

$$(\langle\Psi|\,\widehat{x})\,|\Phi\rangle = \langle\Psi|\,(\widehat{x}\,|\Phi\rangle). \tag{7}$$

The scalar product should be defined as

$$\langle\Psi \mid \Phi\rangle = \int_{-\infty}^{+\infty}\tfrac{dp}{1+\beta p^2}\Psi^*(p)\,\Phi(p). \tag{8}$$

The modification of this product implies a new closure relation, which is written as

$$\int_{-\infty}^{+\infty} \frac{dp}{1+\beta p^2} \left| p \right\rangle \left\langle p \right| = 1. \tag{9}$$

Inserting the latter relation in the scalar product of two momentum eigenvectors operator, we get:

$$\langle p \mid p' \rangle = \left(1 + \beta p^2\right) \delta \left(p - p'\right), \tag{10}$$

also is given by

$$\langle p | p' \rangle = \delta \left(\frac{1}{\sqrt{\beta}} \arctan \sqrt{\beta} p - \frac{1}{\sqrt{\beta}} \arctan \sqrt{\beta} p' \right). \tag{11}$$

In this case the Schrödinger equation for the particle in the harmonic oscillator of momentum space representation in one dimension, can be written as

$$\hat{H} = \left(\frac{\hat{p}^2}{2m} + \frac{m\omega^2}{2} \hat{x}^2 \right) = \left[\frac{p^2}{2m} + \frac{m\omega^2}{2} \left(i\hbar \left(1 + \beta p^2\right) \frac{\partial}{\partial p} \right)^2 \right]. \tag{12}$$

Exact solutions of spectrum energy and the normalized eigenfunctions of the bound states are defined in (Chang et al, (2002):

$$E_n = \hbar\omega \left[\left(n + \frac{1}{2}\right) \sqrt{1 + \left(\frac{\beta\hbar m\omega}{2}\right)^2} + \left(n^2 + n + 1\right) \left(\frac{\beta\hbar m\omega}{2}\right) \right]. \tag{13}$$

and

$$\Psi_n(p) = \sqrt{\frac{2^{2\lambda-1}(\lambda+n)n!\sqrt{\beta}}{\pi\Gamma(2\lambda+n)[\Gamma(\lambda)]^{-2}}} \left[\frac{1}{\sqrt{1+\beta p^2}} \right]^{\lambda} C_n^{\lambda} \left(\frac{\sqrt{\beta} p}{\sqrt{1+\beta p^2}} \right), \tag{14}$$

with C_n^{λ} are Gegenbauer polynomials.

3. Construction propagators with generalized Heisenberg principle

The purpose of this section is to discuss the propagators and the quantum corrections via the standard Feynman approach, for a non-relativistic quantum mechanics in the context of the deformed and non-deformed space at α-point discretization. Then we will circulate this study on the relativistic problems through the two applications chosen below.

3.1. Ordinary quantum mechanics case

In this subsection, we will illustrate the spatio-temporal technique and the method of calculating the quantum corrections according to the standard Feynman approach in the ordinary quantum mechanics. So, in one dimension, we consider the propagator expression of ordinary non-relativistic quantum mechanics to the discontinuous form path integral

$$K_N = A_N \int \exp\left\{ \frac{i}{\hbar} \sum_{n=1}^{N+1} S_n \right\} \prod_{n=1}^{N} dz_n, \tag{15}$$

with $A_N = \left(\sqrt{\frac{m}{2\pi i\hbar\varepsilon}}\right)^{N+1}$ and S_n is the discrete action into intervals $[n-1, n]$, takes the form follows:

$$S_n = \frac{m}{2\varepsilon}\left(z_n - z_{n-1}\right)^2 - \varepsilon V\left(x_n\right). \tag{16}$$

According to the standard method of Feynman (cf., e.g., Ref (Khandekar et al, 1993)), we will apply the spatio-temporal method of processing $\bar{z}_n^{(\alpha)} = f(\bar{q}_j^{(\alpha)})$ to α-point discretization defined as:

$$\bar{z}_n^{(\alpha)} = \alpha z_n + (1-\alpha) z_{n-1}. \tag{17}$$

Two terms of quantum corrections have appeared (C_{act}, C_{mes}). Let's start by calculating the correction to the C_{act} action. Developing Δz_n to α-point discretization, we have:

$$\Delta z_n = \Delta q_n \bar{f}_n^{(\alpha)\prime}\left(1 + \frac{(1-2\alpha)}{2!}\frac{\bar{f}_n^{(\alpha)\prime\prime}}{\bar{f}_n^{(\alpha)\prime}}\Delta q_n + \frac{(1-\alpha)^3+\alpha^3}{3!}\frac{\bar{f}_n^{(\alpha)\prime\prime\prime}}{\bar{f}_n^{(\alpha)\prime}}\Delta q_n^2\right), \tag{18}$$

where the prime on the function $\bar{f}_j^{(\alpha)}$ indicates the derivative $\bar{f}_j^{(\alpha)}$ over $\bar{q}_j^{(\alpha)}$. So, the kinetic energy term in the action is:

$$\frac{\Delta z_n^2}{4\varepsilon} =$$

$$\frac{\Delta q_n^2}{4\varepsilon}\left(\bar{f}_n^{(\alpha)\prime}\right)^2\left(1 + (1-2\alpha)\frac{\bar{f}_n^{(\alpha)\prime\prime}}{\bar{f}_n^{(\alpha)\prime}}\Delta q_n + \left[\frac{(1-2\alpha)^2}{4}\left(\frac{\bar{f}_n^{(\alpha)\prime\prime}}{\bar{f}_n^{(\alpha)\prime}}\right)^2 + \frac{(1-\alpha^3)+\alpha^3}{3}\frac{\bar{f}_n^{(\alpha)\prime\prime\prime}}{\bar{f}_n^{(\alpha)\prime}}\right]\Delta q_n^2\right). \tag{19}$$

The potential energy takes a simple form:

$$\varepsilon V(z_n) = \varepsilon V(\bar{q}_n^{(\alpha)}) + O(\varepsilon^2) = \varepsilon V(\bar{q}_n^{(\alpha)}). \tag{20}$$

In addition, we note that the transformation $z = f(q)$ made the path integral rather complicated, where the mass parameter is transformed into $m\bar{f}_n^{(\alpha)\prime}$. At this point, we would apply the transformation over time parameter in order to overcome this difficulty:

$$\varepsilon = \sigma_n f'\left(q_n\right) f'\left(q_{n-1}\right), \quad \text{where } \sigma_n = s_n - s_{n-1}. \tag{21}$$

Development $f'(q_n)$ and $f'(q_{n-1})$ at α-point discretization in order two of Δq_n, is written as:

$$\varepsilon=\sigma_n\left(\bar{f}_n^{(\alpha)\prime}\right)^2\left(1+(1-2\alpha)\frac{\bar{f}_n^{(\alpha)\prime\prime}}{\bar{f}_n^{(\alpha)\prime}}\Delta q_n+\left(\frac{(1-\alpha)^2+\alpha^2}{2}\frac{\bar{f}_n^{(\alpha)\prime\prime\prime}}{\bar{f}_n^{(\alpha)\prime}}-\alpha(1-\alpha)\left(\frac{\bar{f}_n^{(\alpha)\prime\prime}}{\bar{f}_n^{(\alpha)\prime}}\right)^2\right)\Delta q_n^2\right). \quad (22)$$

From expressions (19) and (22), we can deduce the quantum correction from the action:

$$\exp\left[\frac{i}{\hbar}\frac{m}{2\varepsilon}(\Delta z_n)^2\right]=\exp\left[\frac{i}{\hbar}\frac{m}{2\sigma_n}(\Delta q_n)^2\right](1+C_{act}), \quad (23)$$

with

$$C_{act}=\frac{i}{\hbar}\frac{m}{8\sigma_n}\Delta q_n^4\left[\left(-16\alpha^2+16\alpha-3\right)\left(\frac{\bar{f}_n^{(\alpha)\prime\prime}}{\bar{f}_n^{(\alpha)\prime}}\right)^2-\frac{2}{3}\frac{\bar{f}_n^{(\alpha)\prime\prime\prime}}{\bar{f}_n^{(\alpha)\prime}}\right]. \quad (24)$$

Turning now to calculate the second correction C_{mes}, we have:

$$A_N\prod_{n=1}^{N}dz_n=\left(\sqrt{\frac{m}{2\pi i\hbar\varepsilon}}\right)^{N+1}\prod_{n=1}^{N}dq_nf'(q_n). \quad (25)$$

This can be achieved by rewriting:

$$A_N\prod_{n=1}^{N}dz_n=\left[f'(q_b)f'(q_a)\right]^{-1/2}\prod_{n=1}^{N+1}\left(\sqrt{\frac{mf'(q_n)f'(q_{n-1})}{2\pi i\hbar\varepsilon}}\right)\prod_{n=1}^{N}dq_n.. \quad (26)$$

Then, we develop $f'(q_n)$ and $f'(q_{n-1})$ to the second order of Δq_n as follows:

$$\left[f'(q_n)f'(q_{n-1})\right]^{1/2}=\bar{f}_n^{(\alpha)\prime}\left(1+\frac{(1-2\alpha)}{2}\frac{\bar{f}_n^{(\alpha)\prime\prime}}{\bar{f}_n^{(\alpha)\prime}}\Delta q_n+\left(\frac{(1-\alpha)^2+\alpha^2}{4}\frac{\bar{f}_n^{(\alpha)\prime\prime\prime}}{\bar{f}_n^{(\alpha)\prime}}-\frac{\alpha(1-\alpha)}{2}\left(\frac{\bar{f}_n^{(\alpha)\prime\prime}}{\bar{f}_n^{(\alpha)\prime}}\right)^2\right)\Delta q_n^2\right) \quad (27)$$

From the expressions (26), (27) and (22) we can deduce C_{mes},

$$A_N\prod_{n=1}^{N}dz_n=\prod_{n=1}^{N+1}\left(\sqrt{\frac{m}{2\pi i\hbar\sigma_n}}\right)(1+C_{mes})\prod_{n=1}^{N}dq_n., \quad (28)$$

to the form

$$C_{mes}=\frac{(1-2\alpha)^2}{4}\left(\frac{\bar{f}_j^{(\alpha)\prime\prime}}{\bar{f}_j^{(\alpha)\prime}}\right)^2\Delta q_n^2. \quad (29)$$

To calculate the total quantum corrections C_T, we use the following expression

$$\left\langle (\Delta q)^{2n} \right\rangle = \left(\tfrac{i\hbar\sigma}{m}\right)^n (2n-1)!!, \tag{30}$$

The result is

$$\left\langle (\Delta q)^{2} \right\rangle = \left(\tfrac{i\hbar\sigma}{m}\right), \left\langle (\Delta q)^{4} \right\rangle = \left(\tfrac{i\hbar\sigma}{m}\right)^2 (3)!! = -3\left(\tfrac{\hbar\sigma}{m}\right)^2. \tag{31}$$

By combining all these results, we arrive at:

$$C_T = V_{eff} = -\sigma_n \tfrac{i\hbar^2}{4m}\left[\left(\tfrac{11}{2} - 28\alpha^2 + 28\alpha\right)\left(\frac{\bar{f}_n^{(\alpha)\prime\prime}}{\bar{f}_n^{(\alpha)\prime}}\right)^2 - \frac{\bar{f}_n^{(\alpha)\prime\prime\prime}}{\bar{f}_n^{(\alpha)\prime}}\right]. \tag{32}$$

When we used the standard formalism of Feynman (cf., e.g., Ref (Khandekar et al, 1993)) to α-point discretization, a single case of $\alpha-$point gave the same result of the method equation, this value is $\alpha = 1/2.$, where the effective potential is

$$V_{eff} = \sigma_n \tfrac{i\hbar^2}{4m}\left[\frac{3}{2}\left(\frac{\bar{f}_n^{(\alpha)\prime\prime}}{\bar{f}_n^{(\alpha)\prime}}\right)^2 - \frac{\bar{f}_n^{(\alpha)\prime\prime\prime}}{\bar{f}_n^{(\alpha)\prime}}\right]. \tag{33}$$

But in the presence β parameter deformation, the value of α-point discretization is different according to what has been explained by the Feynman approach.

3.2. The non-relativistic QM with minimal length case

We illustrate the use of the path integral formalism of the transition amplitude in the momentum space representation for a quantum time-independent quadratic systems with the presence of nonzero minimum position uncertainty. We start with the propagator expressed as:

$$K^{(\beta)}(p_b, t_b, p_a, t_a) = \lim_{N\to\infty}\left\langle p_b \left| \prod_{j=1}^{N} \exp(\tfrac{i\varepsilon}{\hbar}\hat{H}) \right| p_a \right\rangle. \tag{34}$$

Inserting the closure relation for the momentum states (9) between each pair of infinitesimal evolution operators ($U(j, j-1) = \exp(\frac{i\varepsilon}{\hbar}\hat{H})$), we obtain

$$K^{(\beta)}(p_b, t_b, p_a, t_a) = \lim_{N\to\infty}\prod_{j=1}^{N}\int \frac{dp_j}{(1+\beta p_j^2)}\prod_{j=1}^{N+1}\left[1 + \frac{i\varepsilon}{\hbar}\hat{H}\right]\langle p_j \mid p_{j-1}\rangle, \tag{35}$$

where the projection relation ($\langle p_j \mid p_{j-1}\rangle$) is defined in eq.(10). Then we inject the Hamiltonian operator of a particle with nonzero minimum position uncertainty on the projection relation for any systems studied. It is clear that there are only few cases where it is exactly solvable; namely, the case of a linear potential $(V(x) = gx)$ and the case of a harmonic potential $\left(V(x) = \frac{m\omega^2}{2}x^2\right)$. In this chapter, we will study only the form quadratic, for example, the harmonic oscillator potential in one dimension and the spinorial relativistic particle.

The construction of momentum space path integral representation of the transition amplitude for a particle moving in the potential of the harmonic oscillator in one dimension with nonzero minimum position uncertainty. Following the well-known steps to construct a quantum propagator $K^{(\beta)}$, we write:

$$K^{(\beta)}\left(p_b,t_b;p_a,t_a\right) = \lim_{N\to\infty}\prod_{j=1}^{N}\int\frac{dp_j}{\left(1+\beta p_j^2\right)}\prod_{j=1}^{N+1}\int\frac{dq_j}{2\pi\hbar}\left(1+\beta p_j^2\right)\exp\left\{\frac{i}{\hbar}\sum_{j=1}^{N+1}\left[q_j\Delta p_j - \varepsilon\frac{p_j^2}{2m}\right.\right.$$
$$\left.\left.-\varepsilon\frac{m\omega^2}{2}\left(\left(1+\beta p_j^2\right)^2 q_j^2 - 6i\hbar\beta p_j q_j\left(1+\beta p_j^2\right) - 2\hbar^2\beta\left(1+3\beta p_j^2\right)\right)\right]\right\}. \tag{36}$$

The form of expression (36) shows that the path integral over the variables q_j is Gaussian, so the result is simply written as:

$$K^{(\beta)}\left(p_b,t_b,p_a,t_a\right) = \lim_{N\to\infty}\prod_{j=1}^{N}\int\frac{dp_j}{\left(1+\beta p_j^2\right)}\prod_{j=1}^{N+1}\frac{1}{\sqrt{2\pi i\hbar m\omega^2\varepsilon}}$$
$$\exp\left\{\frac{i}{\hbar}\sum_{j=1}^{N+1}\left[\frac{\left(\Delta p_j\right)^2}{2m\omega^2\varepsilon\left(1+\beta p_j^2\right)^2} + \frac{3i\hbar\beta p_j\Delta p_j}{\left(1+\beta p_j^2\right)} - \varepsilon\frac{p_j^2}{2m} - \varepsilon\frac{m\omega^2\hbar^2\beta}{2}\left(-2+3\beta p_j^2\right)\right]\right\}. \tag{37}$$

This latter expression shows that the kinetic term is similar a system of space dependent mass and can be removed by an $\alpha-$point coordinate transformation method (see, Ref (Khandekar et al, 1993)), where the α-point discretization interval defined by

$$\bar{p}_j^{(\alpha)} = \alpha p_j + (1-\alpha)\,p_{j-1}. \tag{38}$$

So, we will introduce the function $f(p)$, where the first derivative of $f(\bar{p}_j^{(\alpha)})$ is equal to $1/(1+\beta\bar{p}_j^{(\alpha)2})$. Thus, there are three quantum corrections obtained in expression (37).

- The first related to the action $C_{act}^{(1)}$
- the second correction related to the measure $C_{mes}^{(1)}$
- and the third correction related to the f-factor C_f^T

Expanding the exponential $\left(\frac{i}{\hbar}\frac{\left(\Delta p_j\right)^2}{2m\omega^2\varepsilon\left(1+\beta p_j^2\right)^2}\right)$ of the $\alpha-$point discretization interval, we find

$$\exp\left[\frac{i}{\hbar}\sum_{j=1}^{N+1}\left(\frac{(\Delta p_j)^2}{2m\omega^2\varepsilon\left(1+\beta p_j^2\right)^2}\right)\right]=\exp\left[\frac{i}{\hbar}\sum_{j=1}^{N+1}\left(\frac{\left(\bar{f}_j^{(\alpha)\prime}\right)^2(\Delta p_j)^2}{2m\omega^2\varepsilon}\right)\right]\left(1+C_{act}^{(1)}\right), \tag{39}$$

where

$$C_{act}^{(1)}=\frac{i}{2\hbar m\omega^2\varepsilon}\left[\frac{2(1-\alpha)\bar{f}_j^{(\alpha)\prime\prime}\left(\bar{f}_j^{(\alpha)\prime}\right)^2}{\bar{f}_j^{(\alpha)\prime}}(\Delta p_j)^3+(1-\alpha)^2\left[\left(\frac{\bar{f}_j^{(\alpha)\prime\prime}}{\bar{f}_j^{(\alpha)\prime}}\right)^2+\frac{\bar{f}_j^{(\alpha)\prime\prime\prime}}{\bar{f}_j^{(\alpha)\prime}}\right](\bar{f}_j^{(\alpha)\prime})^2(\Delta p_j)^4\right]$$
$$-\frac{2(1-\alpha)^2}{(2\hbar m\omega^2\varepsilon)^2}\left(\frac{\bar{f}_j^{(\alpha)\prime\prime}}{\bar{f}_j^{(\alpha)\prime}}\right)^2\left(\bar{f}_j^{(\alpha)\prime}\right)^4(\Delta p_j)^6, \tag{40}$$

and the measure term will be developed as.

$$\prod_{j=1}^{N}\int\frac{dp_j}{1+\beta p_j^2}=\sqrt{\left(1+\beta p_b^2\right)\left(1+\beta p_a^2\right)}\prod_{j=1}^{N}\int dp_n\prod_{j=1}^{N+1}\frac{1}{\sqrt{\left(1+\beta p_j^2\right)\left(1+\beta p_{j-1}^2\right)}}$$
$$=\left[\frac{1}{f_b^{\prime}f_a^{\prime}}\right]\prod_{n=1}^{N}\int dp_j\prod_{n=1}^{N+1}\bar{f}_j^{(\alpha)\prime}\left(1+C_m^{(1)}\right), \tag{41}$$

where

$$C_m^{(1)}=\frac{(1-2\alpha)}{2}\frac{\bar{f}_n^{(\alpha)\prime\prime}}{\bar{f}_n^{(\alpha)\prime}}\Delta p_n+\left(\frac{(1-\alpha)^2+\alpha^2}{4}\frac{\bar{f}_n^{(\alpha)\prime\prime\prime}}{\bar{f}_n^{(\alpha)\prime}}-\frac{\alpha(1-\alpha)}{2}\left(\frac{\bar{f}_n^{(\alpha)\prime\prime}}{\bar{f}_n^{(\alpha)\prime}}\right)^2\right)\Delta p_n^2. \tag{42}$$

and the f-factor term will be developed as

$$\exp\left(-\frac{3\beta p_j\Delta p_j}{1+\beta p_j^2}\right)=1+C_f^T, \tag{43}$$

where

$$C_f^T=\frac{3}{2}\left(\frac{\bar{f}_j^{(\alpha)\prime\prime}}{\bar{f}_j^{(\alpha)\prime}}\right)\Delta p_j+\frac{9}{8}\left(\frac{\bar{f}_j^{(\alpha)\prime\prime}}{\bar{f}_j^{(\alpha)\prime}}\right)^2(\Delta p_j)^2+\frac{3}{2}(1-\alpha)\left(\frac{\bar{f}_j^{(\alpha)\prime\prime\prime}}{\bar{f}_j^{(\alpha)\prime}}-\left(\frac{\bar{f}_j^{(\alpha)\prime\prime}}{\bar{f}_j^{(\alpha)\prime}}\right)^2\right)(\Delta p_j)^2, \tag{44}$$

Now, in order to convert this expression to the usual form of Feynman path integral, let us bring the kinetic term to the conventional one, with a constant mass term by using the following coordinate transformation $p_j = g(k_j)$, this transformation generates two corrections:

- the first related to the action $C_{act}^{(2)}$

- and the other correction related to the measure $C_m^{(2)}$

The α-point expansion of Δp_j reads at each (j)

$$\Delta p_j = g(k_j) - g(k_{j-1}) = \Delta k_j \bar{g}_j^{(\alpha)\prime}\left(1+\frac{(1-2\alpha)}{2!}\frac{\bar{g}_j^{(\alpha)\prime\prime}}{\bar{g}_j^{(\alpha)\prime}}\Delta k_j + \frac{(1-\alpha)^3+\alpha^3}{3!}\frac{\bar{g}_j^{(\alpha)\prime\prime\prime}}{\bar{g}_j^{(\alpha)\prime}}\Delta k_j^2\right). \tag{45}$$

The choice of g is fixed by the following condition $((\partial g/\partial k) = (\partial f/\partial p)^{-1})$, that makes the transformation $g(k) = \frac{\tan\sqrt{\beta}k}{\sqrt{\beta}}$ where the region $p \in [-\infty, +\infty]$ is mapped to $k \in [-\pi/2\sqrt{\beta}, \pi/2\sqrt{\beta}]$. Thereafter, we develop the exponential of the kinetic term as

$$\exp\left[\frac{i}{\hbar}\sum_{j=1}^{N+1}\left(\frac{(\Delta p_j)^2}{2m\omega^2\varepsilon(1+\beta p_j^2)^2}\right)\right] = \exp\left\{\frac{i}{\hbar}\sum_{j=1}^{N+1}\left[\frac{\Delta k_j^2}{2m\omega^2\varepsilon}\right]\right\}\left[1+C_{act}^{(1)}\right]\left[1+C_{act}^{(2)}\right], \tag{46}$$

where $C_{act}^{(2)}$ is given by

$$\begin{aligned} C_{act}^{(2)} = &\left\{\frac{i}{2\hbar m\omega^2\varepsilon}\left[(1-2\alpha)\frac{\bar{g}_j^{(\alpha)\prime\prime}}{\bar{g}_j^{(\alpha)\prime}}\left(\bar{g}_j^{(\alpha)\prime}\right)^2\left(\bar{f}_j^{(\alpha)\prime}\right)^2\Delta k_j^3\right.\right. \\ &+\left[\frac{(1-2\alpha)^2}{4}\left(\frac{\bar{g}_j^{(\alpha)\prime\prime}}{\bar{g}_j^{(\alpha)\prime}}\right)^2+\frac{(1-\alpha)^3+\alpha^3}{3}\frac{\bar{g}_j^{(\alpha)\prime\prime\prime}}{\bar{g}_j^{(\alpha)\prime}}\right]\left(\bar{f}_j^{(\alpha)\prime}\right)^2\left(\bar{g}_j^{(\alpha)\prime}\right)^2\Delta k_j^4\Bigg] \\ &\left.-\frac{(1-2\alpha)^2}{2(2\hbar m\omega^2\varepsilon)^2}\left(\frac{\bar{g}_j^{(\alpha)\prime\prime}}{\bar{g}_j^{(\alpha)\prime}}\right)^2\left(\bar{g}_j^{(\alpha)\prime}\right)^4\left(\bar{f}_j^{(\alpha)\prime}\right)^4\Delta k_j^6+\ldots\right\}. \end{aligned} \tag{47}$$

The measure induce also a correction

$$\begin{aligned} \prod_{j=1}^{N}\int\frac{dp_j}{1+\beta p_j^2} &= \sqrt{\frac{1}{f_b' f_a' g_b' g_a'}}\prod_{n=1}^{N}\int dk_j\prod_{j=1}^{N+1}\bar{f}_j^{(\alpha)\prime}\bar{g}_j^{(\alpha)\prime}\left(1+C_m^{(2)}\right)\left(1+C_m^{(1)}\right) \\ &= \prod_{j=1}^{N}\int dk_j\prod_{j=1}^{N+1}\left(1+C_m^{(1)}\right)\left(1+C_m^{(2)}\right), \end{aligned} \tag{48}$$

where $C_m^{(2)}$ is given by

$$C_m^{(2)} = \frac{(1-2\alpha)}{2}\frac{\bar{g}_j^{(\alpha)\prime\prime}}{\bar{g}_j^{(\alpha)\prime}}\Delta k_j + \left[-\alpha\frac{(1-\alpha)}{2}\left(\frac{\bar{g}_j^{(\alpha)\prime\prime}}{\bar{g}_j^{(\alpha)\prime}}\right)^2+\frac{(1-\alpha)^2+\alpha^2}{4}\frac{\bar{g}_j^{(\alpha)\prime\prime\prime}}{\bar{g}_j^{(\alpha)\prime}}\right]\Delta k_j^2. \tag{49}$$

By combining all these corrections as follows:

$$1+C_T=\left(1+C_{act}^{(1)}\right)\left(1+C_{act}^{(2)}\right)\left(1+C_{m}^{(1)}\right)\left(1+C_{m}^{(2)}\right)\left(1+C_{f}^{T}\right). \tag{50}$$

where the corrections terms are evaluated perturbatively, using the expectation values

$$\left\langle (\Delta k)^{2\ell}\right\rangle=\left(i\hbar m\omega^2\varepsilon\right)^{\ell}(2\ell-1)!!. \tag{51}$$

So through all of these corrections, one can conclude the correction total C_T depending on the α-point discretization can be obtained as

$$C_T=\frac{i}{\hbar}\hbar^2 m\omega^2\varepsilon\beta\left[-1+\left(8\alpha^2-8\alpha+\frac{5}{2}\right)\tan^2\sqrt{\beta}k_j\right] \tag{52}$$

Furthermore, a judicious choice of the discretization parameter is indicated by the order of operators present in the wave equation method. The result coincides with the path integral approach when C_T equals

$$C_T=i\hbar\varepsilon m\omega^2\beta\left[-1+\frac{3}{2}\tan^2\sqrt{\beta}k\right], \tag{53}$$

We note that the different α-point discretization value (i.e. $\alpha=\frac{1}{2}\left(1\pm 1/\sqrt{2}\right)$) are obtained compared with the ordinary quantum mechanics case ($\alpha=1/2$). So the propagator (37) becomes,

$$K^{(\beta)}\left(p_b,t_b;p_a,t_a\right)=\lim_{N\to\infty}\prod_{n=1}^{N}\int dk_n\prod_{n=1}^{N+1}\frac{1}{\sqrt{2\pi i\hbar m\omega^2\varepsilon}}\exp\left\{\frac{i}{\hbar}\sum_{n=1}^{N+1}\left[\frac{(\Delta k_n)^2}{2m\omega^2\varepsilon}-\varepsilon\frac{\tan^2\left(\sqrt{\beta}k_n\right)}{2m\beta}\right]\right\}, \tag{54}$$

This expression is exactly the path integral representation of the transition amplitude of a particle, moving in the symmetric Pöschl-Teller potential (cf., e.g., Ref (Grouche & Steiner, 1998)):

$$K^{(\beta)}=\lim_{N\to\infty}\prod_{n=1}^{N}\int dk_n\prod_{n=1}^{N+1}\frac{1}{\sqrt{2\pi i\hbar m\omega^2\varepsilon}}\exp\left\{\frac{i}{\hbar}\sum_{n=1}^{N+1}\left[\frac{(\Delta k_n)^2}{2m\omega^2\varepsilon}-\varepsilon\frac{\beta\hbar^2 m\omega^2}{2}\lambda(\lambda-1)\tan^2\left(\sqrt{\beta}k_n\right)\right]\right\} \tag{55}$$

with ($\lambda=\left(1+(1+(2/\beta\hbar m\omega)^2)^{1/2}\right)/2$). The solution of this path integral is given by:

$$K^{(\beta)}\left(p_b,t_b;p_a,t_a\right)=\sum_{n=0}^{\infty}\frac{2^{2\lambda-1}(\lambda+n)n!\sqrt{\beta}}{\pi\Gamma(2\lambda+n)[\Gamma(\lambda)]^{-2}}\exp\left[-\frac{i}{\hbar}\frac{\beta\hbar^2m\omega^2(t_b-t_a)}{2}\left(n^2+2\left(n+1\right)\lambda\right)\right]$$
$$\cos^{\lambda}\left(\sqrt{\beta}k_b\right)\cos^{\lambda}\left(\sqrt{\beta}k_a\right)C_n^{\lambda}\left(\sin\left(\sqrt{\beta}k_b\right)\right)C_n^{\lambda}\left(\sin\left(\sqrt{\beta}k_b\right)\right). \tag{56}$$

We finally obtain the spectral decomposition of the transition amplitude for the one-dimensional harmonic oscillator with nonzero minimum position uncertainty

$$K^{(\beta)}\left(p_b,t_b;p_a,t_a\right)=\sum_{n=0}^{\infty}\Psi_n\left(p_b\right)\Psi_n^*\left(p_a\right)e^{-\frac{i}{\hbar}E_n(t_b-t_a)}. \tag{57}$$

The energy spectrum is obtained from the poles of the Green function (57):

$$E_n=\frac{\beta\hbar^2m\omega^2}{2}\left(n^2+2\left(n+1\right)\lambda\right). \tag{58}$$

Using the expression of λ, one finds:

$$E_n=\hbar\omega\left[\left(n+\frac{1}{2}\right)\sqrt{1+\left(\frac{\beta\hbar m\omega}{2}\right)^2}+\left(n^2+n+1\right)\left(\frac{\beta\hbar m\omega}{2}\right)\right]. \tag{59}$$

Also, the normalized eigenfunctions of the bound states can be easily deduced

$$\Psi_n\left(p\right)=\sqrt{\frac{2^{2\lambda-1}(\lambda+n)n!\sqrt{\beta}}{\pi\Gamma(2\lambda+n)[\Gamma(\lambda)]^{-2}}}\left[\frac{1}{\sqrt{1+\beta p^2}}\right]^{\lambda}C_n^{\lambda}\left(\frac{\sqrt{\beta}p}{\sqrt{1+\beta p^2}}\right). \tag{60}$$

We note that equations (59) and (60) coincide exactly with those obtained in (Chang et al, 2002). Also we can verify these results when $\beta\to 0$, which transform to these results:

$$E_n\underset{\beta\to 0}{=}\hbar\omega\left(n+\frac{1}{2}\right),\quad \Psi_n\left(p\right)\underset{\beta\to 0}{=}\left[\frac{1}{2^n n!\sqrt{\pi}}\right]^{1/2}\left(\frac{1}{m\hbar\omega}\right)^{1/4}e^{-\frac{p^2}{2m\hbar\omega}}H_n\left(\sqrt{\frac{1}{m\hbar\omega}}p\right). \tag{61}$$

For the one dimensional harmonic oscillator in the framework non-commutative geometry represented by Eqs. (1) and (2), the quantum corrections from the viewpoint of Feynman (cf., e.g., Ref (Khandekar et al, 1993)) at the α-point discretization interval, we found only two points discretization ($\alpha=\frac{1}{2}\left(1\pm 1/\sqrt{2}\right)$) consistent with differential equation (see, Ref (Chang, et al, 2002)) which gives different value of ordinary quantum mechanics.

So, in the following subsection we aim to expand this type deformation for relativistic systems.

3.3. The relativistic QM with minimal length

Our interest in the following section is to construct the propagator for two applications relativistic quantum mechanics in the presence of a minimal length, the first is (1+1)-dimensional Dirac oscillator, where the momentum component is shifting p by $p - im\omega\gamma_0 x$ (cf., e.g., Ref (Szmytkowski & Gruchowski, 2001), and the second is a spinorial relativistic particle under the action of a Lorentz potential $\left(V(x), \vec{\mathbf{A}} = 0\right)$ plus a scalar potential $S(x)$, described by the (1+1)-dimensional Dirac equation:

$$\left(\gamma^{\mu}\hat{\Pi}_{\mu} - m + \imath\epsilon\right)\hat{S}^{(\beta)} = I, \tag{62}$$

where $\mu = 0, 1$, γ_{μ} are the Dirac matrices in the 2-dimentional Minkowski space. So, via the same procedure in the our previous work (Benzair et al, 2012 and 2014), we can obtain the standard propagator result for both systems, where there are two types of propagation, one with positive energy $(+E_n^{(\beta)})$ propagation and the other with negative energy $(-E_n^{(\beta)})$ propagation

$$S^{(\beta)}\left(p_b, p_a, t_b - t_a\right) = -\sum_{n=0}^{\infty}\left[\begin{array}{c}\Theta\left(t_b - t_a\right)\Psi_n^{(\beta)+}\left(p_b\right)\bar{\Psi}_n^{(\beta)+}\left(p_a\right)e^{-\imath E_n^{(\beta)}\left(t_b - t_a\right)} + \\ \Theta\left(-\left(t_b - t_a\right)\right)\Psi_n^{(\beta)-}\left(p_b\right)\bar{\Psi}_n^{(\beta)-}\left(p_a\right)e^{\imath E_n^{(\beta)}\left(t_b - t_a\right)}\end{array}\right], \tag{63}$$

For one-dimensional Dirac oscillator in the momentum space representation with the presence of minimal length uncertainty, can be expressed the energy spectrum as follows

$$E_{n,\pm}^{(\beta)} = \pm\sqrt{m^2 + \beta\left(m\omega\right)^2 n^2 + 2n\left(m\omega\right)}. \tag{64}$$

and the corresponding wave functions

$$\Psi_n^{(\beta)+}\left(p\right) = \begin{pmatrix} f_n^{(\beta)+}\left(p\right) \\ g_n^{(\beta)+}\left(p\right)\end{pmatrix}, \text{ and } \Psi_n^{(\beta)-}\left(p\right) = \begin{pmatrix} f_n^{(\beta)-}\left(p\right) \\ g_n^{(\beta)-}\left(p\right)\end{pmatrix}, \tag{65}$$

where the components of the wave functions $f_n^{(\beta)\pm}\left(p\right)$ and $g_n^{(\beta)\pm}\left(p\right)$ are respectively

$$f_n^{(\beta)+}\left(p\right) = \sqrt{\Gamma\left(\eta\right)^2 \frac{2^{2\eta-1}(n+1)!(n+\eta)\sqrt{\beta}\left(E_n^{(\beta)} + m\right)}{\pi\Gamma(n+2\eta)2E_n^{(\beta)}}}\left(\frac{1}{1+\beta p^2}\right)^{\eta} C_n^{\eta}\left(\frac{\sqrt{\beta}p}{1+\beta p^2}\right). \tag{66}$$

$$g_n^{(\beta)+}\left(p\right) = \frac{2\imath}{\sqrt{\beta}}\sqrt{\Gamma\left(\eta\right)^2 \frac{2^{2\eta-1}n!(n+\eta)\sqrt{\beta}}{\pi\Gamma(n+2\eta)2E_n^{(\beta)}\left(E_n^{(\beta)} + m\right)}}\left(\frac{1}{1+\beta p^2}\right)^{\eta+1} C_{n-1}^{\eta+1}\left(\frac{\sqrt{\beta}p}{1+\beta p^2}\right). \tag{67}$$

and

$$f_n^{(\beta)-}(p) = \sqrt{\Gamma(\eta)^2 \frac{2^{2\eta-1} n!(n+\eta)\sqrt{\beta}\left(E_n^{(\beta)}-m\right)}{\pi\Gamma(n+2\eta)2E_n^{(\beta)}}} \left(\frac{1}{1+\beta p^2}\right)^{\eta} C_n^{\eta}\left(\frac{\sqrt{\beta}p}{1+\beta p^2}\right). \tag{68}$$

$$g_n^{(\beta)-}(p) = \frac{2\imath}{\sqrt{\beta}}\sqrt{\Gamma(\eta)^2 \frac{2^{2\eta-1} n!(n+\eta)\sqrt{\beta}}{\pi\Gamma(n+2\eta)2E_n^{(\beta)}\left(E_n^{(\beta)}-m\right)}} \left(\frac{1}{1+\beta p^2}\right)^{\eta+1} C_{n-1}^{\eta+1}\left(\frac{\sqrt{\beta}p}{1+\beta p^2}\right). \tag{69}$$

and in the second application we can express these results; the energy spectrum are

$$E_n^{(\beta)\pm} = -\frac{m_0 V_0}{S_0} \pm \omega_n^{(\beta)}; T = t_b - t_a. \tag{70}$$

with

$$\omega_n^{(\beta)} = \frac{\left(S_0^2-V_0^2\right)}{S_0}\sqrt{\beta(n^2 + \frac{2n}{\beta\sqrt{S_0^2-V_0^2}})}. \tag{71}$$

and the wave functions appropriate to the energy spectrum $E_n^{(\beta)\pm}$:

$$\Psi^{(\beta)\pm}(k) = \exp\left(i\frac{\left(E_n^{(\beta)\pm}V_0+m_0S_0\right)}{\left(S_0^2-V_0^2\right)}k\right)$$
$$\times\begin{pmatrix} \sqrt{\frac{N_n\left(S_0^2-V_0^2\right)}{4S_0\left(E_n^{(\beta)\pm}S_0+m_0V_0\right)}}\left[\sqrt{\frac{\left(E_n^{(\beta)\pm}S_0+m_0V_0\right)}{(S_0+V_0)}}v^{\eta}C_n^{\eta}(u) + \frac{2i}{\sqrt{\beta}}\sqrt{\frac{(S_0-V_0)}{\left(E_n^{(\beta)\pm}S_0+m_0V_0\right)}}v^{\eta+1}C_{n-1}^{\eta+1}(u)\right] \\ \sqrt{\frac{N_n\left(S_0^2-V_0^2\right)}{4S_0\left(E_n^{(\beta)\pm}S_0+m_0V_0\right)}}\left[-\sqrt{\frac{\left(E_n^{(\beta)\pm}S_0+m_0V_0\right)}{(S_0-V_0)}}v^{\eta}C_n^{\eta}(u) + \frac{2i}{\sqrt{\beta}}\sqrt{\frac{(S_0+V_0)}{\left(E_n^{(\beta)\pm}S_0+m_0V_0\right)}}v^{\eta+1}C_{n-1}^{\eta+1}(u)\right]\end{pmatrix} \tag{72}$$

To use the old variables, we need the following relations

$$u = \frac{p\sqrt{\beta}}{\sqrt{1+\beta p^2}}, v = \frac{1}{\sqrt{1+\beta p^2}} \text{ and } k = \frac{\arctan}{\sqrt{\beta}}\left(\sqrt{\beta}p\right). \tag{73}$$

In the end, in order to separate the β dependent contribution, let us consider a very small β. The form of (71) can easily expand to first-order in β, be written as

$$\omega_n^{\ll\beta} = \sqrt{2n}\frac{\left(S_0^2-V_0^2\right)^{3/4}}{S_0} + \beta\frac{\left(S_0^2-V_0^2\right)^{5/4}(n^2)}{2S_0\sqrt{2n}} + O\left(\beta^2\right). \tag{74}$$

Then we get

$$E_n^{(\beta)} = -\frac{m_0 V_0}{S_0} \pm \sqrt{2n}\frac{\left(S_0^2-V_0^2\right)^{3/4}}{S_0} \pm \beta\frac{\left(S_0^2-V_0^2\right)^{5/4}(n^2)}{2S_0\sqrt{2n}} + O\left(\beta^2\right), \tag{75}$$

The first term in (75) is the energy spectrum of the ordinary Dirac equation in the presence of electromagnetic field and the second term represents the correction due to the presence of the minimal length.

3.4. Resolution of (1+1)-dimensional Dirac equation in position space representation

In this subsection, we'll examine the same above second system in the position space representation, and using the properties of the Hermite polynomial. We can calculate the corrections in the values of spectrum energy and this will be seen in this regard. This system is described by the (1+1)-dimensional Dirac equation

$$\left\{\sigma_2 \hat{p} + \sigma_3 \left(m_0 + S\left(\hat{x}\right)\right) - \left(i\partial_t - V\left(\hat{x}\right)\right)\right\} \Psi(x,t) = 0, \tag{76}$$

where σ_2 and σ_3 are the standard Pauli matrices

$$\sigma_2 = \begin{pmatrix} 0 & -i \\ i & 0 \end{pmatrix}, \quad \sigma_3 = \begin{pmatrix} 1 & 0 \\ 0 & -1 \end{pmatrix}. \tag{77}$$

We note that in (1+1) dimensions, the Dirac algebra is represented by the Pauli matrices. These reflect the invariant character of the parity of the Dirac equation. In fact in this dimension, there are no spin properties. This looks meaningless. However, in second quantization, we are obliged to use anticommutation relations to take into account the statistics of particles and to have a stable theory. At this level and even in (1+1) dimensions, the spin is an intrinsic characteristic of the particles in connection with the Wigner representation of relativistic particles.

As the potentials are time-independent, we have then to find the stationary states of this equation. Accordingly, let us choose for $\Psi(x,t)$ the form $\exp\left(-iEt\right)\, \Phi(x)$; we then get the following eigenvalue equation

$$\left\{\sigma_2 \hat{p} + \sigma_3 \left(m_0 + S\left(\hat{x}\right)\right) - \left(E - V\left(\hat{x}\right)\right)\right\} \Phi(x) = 0. \tag{78}$$

In the position space acts as

$$\hat{x} = x, \qquad \hat{p} = -i\partial_x \left(1 - \tfrac{\beta}{3}\partial_x^2\right). \tag{79}$$

Then, a modified Dirac equation can be written as,

$$\left[-i\sigma_2 \partial_x \left(1 - \tfrac{\beta}{3}\partial_x^2\right) + \sigma_3 \left(m_0 + S\left(x\right)\right) - \left(E - V\left(x\right)\right)\right] \Phi(x) = 0. \tag{80}$$

By using the following ansatz

$$\Phi = \left\{ -i\sigma_2 \partial_x \left(1 - \tfrac{\beta}{3}\partial_x^2\right) + \sigma_3 \left(m_0 + S\left(x\right)\right) + \left(E - V\left(x\right)\right) \right\} \chi, \tag{81}$$

where $\chi = \binom{\chi_1}{\chi_2}$ is a two-component function spinor, Eq.(80) becomes a differential equation of fourth order whose solution is very complicated in the presence of potentials.

Now, by suggesting that the system is subjected to the action of linear vector plus scalar potentials,

$$V(x) = V_0 x, \quad S(x) = S_0 x, \tag{82}$$

with S_0 and V_0 being arbitrary constants, we find that the Dirac spinor satisfies:

$$\left\{ -\tfrac{2}{3}\beta\partial_x^4 + \partial_x^2 + (E - V_0 x)^2 - (m_0 + S_0 x)^2 - \left(1 - \beta\partial_x^2\right)\left(S_0\sigma_1 + iV_0\sigma_2\right) \right\} \chi = 0, \tag{83}$$

We note that the linear potential, such a uniform external electromagnetic field plays a significant role in various domains of physics. For example, in particle physics, it can be regarded as a model to describe quark confinement (cf., e.g., Ref (Ferreira et al, 1971)). Further, the linear potential well has potential applications in electronics (in semiconductor devices), where the electrons are confined in almost linear quantum wells (cf., e.g., Ref (Singh, 1997)). Quantum Mechanics-Fundamentals and Applications to Technology).

Now in order to decouple the system (83), we introduce the following canonical transformation

$$\chi\left(x\right) = U\xi\left(x\right), \tag{84}$$

where U is given by

$$U = \begin{bmatrix} \frac{(V_0 + S_0)}{\sqrt{(S_0^2 - V_0^2)}} & \frac{(V_0 + S_0)}{-\sqrt{(S_0^2 - V_0^2)}} \end{bmatrix}. \tag{85}$$

Then the function $\xi\left(x\right)$ satisfies the following equation

$$\left[-\tfrac{2\beta}{3\alpha}\partial_x^4 + \tfrac{(1+\varepsilon\beta\alpha)}{\alpha}\partial_x^2 - \alpha\left[x^2 + 2\tfrac{(m_0 S_0 + E V_0)}{\alpha^2}x\right] + \tfrac{(E^2 - m_0^2)}{\alpha} - \varepsilon \right] \xi_\varepsilon\left(x\right) = 0, \tag{86}$$

with $\alpha = \sqrt{(S_0^2 - V_0^2)}$ and $\varepsilon = \pm 1$.

As it has been mentioned previously that the solution is complicated, we try to find via the usual perturbation method of quantum mechanics the first energy correction at order 1 in β and point out how the introduction of the modified Heisenberg algebra affects the physical results. To do this, let us first suppose in this case that $\alpha^2 > 0$ so as to avoid complex eigenvalues and arrange equation (86) as a sum of two terms, one of which being the perturbative term, as follows,

$$\left[H^0\left(z,\partial_z\right)+H^{pert}\left(\partial_z\right)\right]\xi_\varepsilon\left(z\right)=0, \tag{87}$$

by setting

$$z=\left(S_0^2-V_0^2\right)^{1/4}\left(x+\frac{\left(m_0S_0+EV_0\right)}{\left(S_0^2-V_0^2\right)}\right). \tag{88}$$

and

$$\begin{aligned} H^0 &= \partial_z^2 - z^2 + z_1, \\ H^{pert} &= -\frac{2}{3}\beta\alpha\partial_z^4 + \varepsilon\beta\alpha\partial_z^2, \end{aligned} \tag{89}$$

where

$$z_1=\frac{\left(m_0S_0+EV_0\right)^2}{\left(S_0^2-V_0^2\right)^{\frac{3}{2}}}+\frac{\left(E^2-m_0^2\right)}{\sqrt{\left(S_0^2-V_0^2\right)}}-\varepsilon. \tag{90}$$

Now, in case where $H^{pert}\left(\partial_z\right)$ vanishes, (i.e. when $\beta\rightarrow 0$), equation (87) becomes that of the harmonic oscillator whose solution is known,

$$\xi_\varepsilon^{\beta=0}\left(z\right)=C_{n'}\exp\left(-\frac{1}{2}z^2\right)H_{n'}\left(z\right),\;\; n'=n+\frac{1}{2}+\frac{\varepsilon}{2}, \tag{91}$$

with z_1 verifying

$$z_1=2n+1.\qquad n=0,1,2,\dots, \tag{92}$$

where $n'=(n+1,n)$. Hence from (90) and (92), we obtain the following energy levels for our Dirac equation:

$$E_{n,\pm}^{\beta=0}=-\frac{m_0V_0}{S_0}\pm\sqrt{2n}\frac{\left(S_0^2-V_0^2\right)^{3/4}}{S_0}. \tag{93}$$

We note the existence of the two signs in (93) which is a characteristic property of energies in relativistic quantum mechanics. Now, to find the first correction in the energy levels, we take the expectation value of the perturbation operator by using eigenfunctions (91)

$$\Delta z_{n1}=\frac{\left\langle\xi^{\beta=0}(z)\left|H^{pert}\right|\xi^{\beta=0}(z)\right\rangle}{\left\langle\xi^{\beta=0}(z)|\xi^{\beta=0}(z)\right\rangle}. \tag{94}$$

With the help the properties of Hermite polynomial (Gradshteyn & Ryzhik, 2000), we obtain this result:

$$\Delta z_{n1} = \frac{\beta\alpha \int \exp\left(-\frac{1}{2}z^2\right)H_n(z)\left[-\frac{2}{3}\partial_z^4+\varepsilon\partial_z^2\right]\exp\left(-\frac{1}{2}z^2\right)H_n(z)dz}{\int \exp\left(-\frac{1}{2}z^2\right)H_n(z)\exp\left(-\frac{1}{2}z^2\right)H_n(z)dz} = -\frac{\beta\alpha}{2}\left(n^2\right), \tag{95}$$

From the relation (90), we derive the expression of ΔE_{n1} as a function of Δz_{n1}, and we write,

$$\Delta E_n^1 = \frac{\left(S_0^2-V_0^2\right)^{3/2}\Delta z_{n1}}{2S_0\left(E_{n,\pm}^{\beta=0}S_0+m_0V_0\right)}. \tag{96}$$

Then, by substituting (95) and (93) in (96), we find,

$$\Delta E_n^1 = \pm\beta\frac{\left(S_0^2-V_0^2\right)^2\left(n^2\right)}{2S_0\sqrt{(2n)}\left(S_0^2-V_0^2\right)^{3/4}}. \tag{97}$$

The energy spectrum of this study at order 1 in β can be rewritten as

$$E_n(\beta) = E_{n,\pm}^{\beta=0} + \Delta E_n^1 + O\left(\beta^2\right), \tag{98}$$

which is equal to

$$E_n(\beta) = -\frac{m_0V_0}{S_0} \pm \sqrt{(2n)}\frac{\left(S_0^2-V_0^2\right)^{3/4}}{S_0} \pm \beta\frac{\left(S_0^2-V_0^2\right)^{5/4}\left(n^2\right)}{2S_0\sqrt{(2n)}} + O\left(\beta^2\right). \tag{99}$$

We note that the same correction spectrum energy obtained where using the momentum space representation defined in eq. (75).

4. Conclusion

We have discussed in this chapter the path integral formalism in the case of the appearance of the parameter of deformation β in the generalized Heisenberg principle (1), where we calculated the quantum corrections according to the Feynman approach (cf., e.g., Ref (Khandekar et al, 1993)) for Harmonic oscillator particle in one dimension. And we have shown that the different α values obtained for the ordinary quantum mechanics, that make us wonder about these results. In addition, we have generalized the study of relativistic particles which have one half (1/2) spin for example Dirac oscillator and relativistic spinning particle subjected to the action of combined vector and scalar linear potentials, with a deformed commutation relation for the Heisenberg principle. We have obtained the same α values for Harmonic oscillator. This has been explained in our previous works (Benzair et al, 2012; 2014). Using the residue theorem, the energy spectrum and corresponding eigenfunctions expressed in terms of Gegenbauer polynomials are then deduced as a function of the deformation parameter β. It has been noted above the energy spectrum of the relativistic spinorial particle is dependent on term quadratic in n that is similar to the energy levels of a particle confined in a potential well. In addition, we studied in the second relativistic application, the energy spectrum of the Dirac equation for a spin $1/2$ subjected to the action of combined vector and scalar linear potentials, with a deformed commutation relation for the Heisenberg principle, where we have used the old variable p in the position space representation, we have obtained a differential equation of fourth order whose analytic solution is complicated. We have calculated the energy correction in first order for β by using a usual approximation technique of quantum mechanics. We note that the two methods gave the same results for the first energy correction at order 1 in β. In this study, we did not take the case $S_0^2 - V_0^2 < 0$, so as to avoid the complex eigenvalues. But when $S_0^2 - V_0^2 = 0$, the calculation is very simple and we can obtain physical results.

Finally, let us signal that the problems of choosing $\alpha-$point discretization in the case of deformed space are under consideration.

Author details

Hadjira Benzair[1], Mahmoud Merad[2] and Taher Boudjedaa[3]

1 Laboratoire LRPPS, Faculté des Sciences et de la Technologie et des Sciences de la Matière, Université Kasdi Merbah Ouargla, Ouargla, Algeria

2 Laboratoire SDC, Département des sciences de la matière, Université de Oum-El-Bouaghi, Algeria

3 Laboratoire de Physique Théorique, Département de Physique, Université de Jijel, BP 98, Ouled Aissa, Jijel, Algeria

References

[1] Akhoury, R. & Yao, Y. P. (2003). Minimal length uncertainty relation and the hydrogen spectrum, Phys. Lett. Vol B 572: 37-42.

[2] Amati, D., Ciafaloni, M. & Veneziano, G. (1987). Superstring collisions at planckian energiesOriginal Research Article, Phys. Lett. Vol. B 197: 81-88.

[3] Bayen, F., Flato, M., Fronsdal, C., Lichnerowicz, A. & Sternheimer, D. (1978). Defromation Theory and Quantization I: deformations of symplectic structures, Annals Phys.Vol 111: 61-110 (1978) and II: physical applications," Annals of Physics, Vol. 111: 111–151

[4] Benzair, B., Boudjedaa, T. & Merad, M. (2012). Path Integral for Dirac oscillator with generalized uncertainty principle, J. Math. Phys. Vol. 53: 123516.

[5] Benzair, H., Merad, M. & Boudjedaa, T. (2014). Path integral of a relativistic spinning particle in (1+ 1) dimension with vector and scalar linear potentials in the presence of a minimal length, Int. J. Mod. Phys. A 29: 1450037

[6] Bordemann, M.,.Makhlouf, A. & Petit, T. (2005). Déformation par quantification et rigidité des algèbres enveloppantes, Journal of Algebra, Vol 285: 623-648.

[7] Brau, F. (1999) Minimal length uncertainty relation and the hydrogen atom, J. Phys. Vol A 32: 7691-7696

[8] Capozziello, S., Lambiase, G. & Scarpetta, G. (2000). Generalized uncertainty principle from quantum geometry, Int. J. Theor. Phys. Vol. 39: 15-22.

[9] Chang, L. N., Minic, D., Okamura, N. & Takeuchi, T. (2002). Exact solution of the harmonic oscillator in arbitrary dimensions with minimal length uncertainty relations, Phys. Rev. Vol. D 65: 125027-125035.

[10] Chargui, Y., Chetouani, L. & Trabelsi, A. (2010). Path integral approach to the D-dimensional harmonic oscillator with minimal length, Phys. Scr. Vol. 81: 015005.

[11] Falek, M. & Merad, M. (2009). Bosonic oscillator in the presence of minimal lengths, J. Math. Phys. Vol 50: 023508.

[12] Falek, M. & Merad, M. (2010). Generalization of Bosonic oscillator in the presence of minimal lengths, J. Math. Phys. Vol 51: 033516.

[13] Ferreira, P. L., Helayel, J. A. & Zagury, N. (1971). Il Nuovo Cimento Vol: A 2 215.

[14] Garay, L. J. (1995). Quantum gravity and minimum length, Int. J. Mod. Phys. Vol. A 10: 145-165.

[15] Gross, D. J. & Mende, P. F. (1988). String theory beyond the Planck scale, Nucl. Phys. Vol. B 303: 407–454.

[16] Guida, R., Konishi, K. & Provero, P. (1991) On the short distance behavior of string theories,. Mod. Phys. Lett. Vol. A 6: 1487-1504.

[17] Gradshteyn, I. S. & Ryzhik, I. M. (2000). Table of Integrals Series and Products (Academic Press, New York).

[18] Grosche, C. & Steiner, F. (1998). Handbook of Feynman Path Integrals (Springer, Berlin).

[19] Hinrichsen, H. & Kempf, A. (1996). Maximal localization in the presence of minimal uncertainties in positions and in momenta, J. Math. Phys. Vol. 37: 2121-2137.

[20] Kato, M. (1990). Particle theories with minimum observable length and open string theory, Physics Letters B, vol. 245, no. 1, pp. 43–47, 1990.

[21] Kempf, A. (1994). Uncertainty relation in quantum mechanics with quantum group symmetry, Journal of Mathematical Physics, J. Math. Phys. Vol. 35: 4483-4495.

[22] Kempf, A., Mangano, G. & . Mann, R. B. (1995). Hilbert space representation of the minimal length uncertainty relation, Phys. Rev. Vol. D 52: 1108-1118.

[23] Kempf, A. (1997). Non-pointlike particles in harmonic oscillators, J. Phys. Vol. A 30: 2093-2102.

[24] Khandekar, D. C., Lawande, S. V. & Bhagwat, K. V. (1993). Path Integral Methods and their Applications (World Scientific, Singapore).

[25] Kleinert, H. (1990). Path Integral in Quantum Mechanics, Statistics and Polymer Physics (World Scientific, Singapore).

[26] Konishi, K., Pauti, G. & Provero, P. (1990) Minimum physical length and the generalized uncertainty principle in string theory, Physics Letters B, Vol. 234: 276-284

[27] Lecheheb, A., Merad, M. & Boudjedaa, T. (2007). Path integral treatment for a Coulomb system constrained on D-dimensional sphere and hyperboloid, Ann. Phys. Vol. 322: 1233-1246.

[28] Makhlouf,. A (2007). A Comparison of Deformations and Geometric Study of Varieties of Associative Algebras, International Journal of Mathematics and Mathematical Sciences, 1-24

[29] Merad, M. & Falek, M. (2009). The time-dependent linear potential in the presence of a minimal length, Phys. Scr. Vol 79: 015010.

[30] Merad, M., Zeroual, F. & Benzair, H. (2010). Spinless Relativistic Particle in the Presence of A Minimal Length, Electron. J. Theor. Phys. Vol. 7: 41-56.

[31] Nouicer, K. (2006). An exact solution of the one-dimensional Dirac oscillator in the presence of minimal lengths, J. Phys. A: Math. Gen. Vol 39: 5125–5134

[32] Nozari, K. & Karami, M. (2005). Minimal length and the generalized Dirac equation, Mod. Phys. Lett. Vol A 20: 3095-3104.

[33] Nozari, K. & Azizi, T. (2006). Some Aspects of Minimal Length Quantum Mechanics, Gen. Rel. Grav. Vol 38: 735-742

[34] Nouicer, Kh. (2006). Path integral for the harmonic oscillator in one dimension with nonzero minimum position uncertainty, Phys. Lett. Vol. A 354: 399-405

[35] Pet, A. W. Polchinski, J. (1999). UV-IR relations in AdS dynamics, J. Phys. Rev. Vol D 59: 065011.

[36] Scardigli, F. (1999). Generalized uncertainty principle in quantum gravity from micro-black hole gedanken experiment, Phys. Lett. Vol. B 452: 39-44.

[37] Scardigli, F. & Casadio, R. (2003). Generalized uncertainty principle, extra dimensions and holography, Classical and Quantum Gravity, Class. Quantum Grav. Vol. 20: 3915-3926.

[38] Singh, J. (1997). Quantum Mechanics-Fundamentals and Applications to Technology (New York: A Wiley Interscience).

[23] Kempf, A. (1997). Non-pointlike particles in harmonic oscillators, J. Phys. Vol. A 30: 2093-2102.

[24] Khandekar, D. C., Lawande, S. V. & Bhagwat, K. V. (1993). Path Integral Methods and their Applications (World Scientific, Singapore).

[25] Kleinert, H. (1990). Path Integral in Quantum Mechanics, Statistics and Polymer Physics (World Scientific, Singapore).

[26] Konishi, K., Pauti, G. & Provero, P. (1990) Minimum physical length and the generalized uncertainty principle in string theory, Physics Letters B, Vol. 234: 276-284

[27] Lecheheb, A., Merad, M. & Boudjedaa, T. (2007). Path integral treatment for a Coulomb system constrained on D-dimensional sphere and hyperboloid, Ann. Phys. Vol. 322: 1233-1246.

[28] Makhlouf,. A (2007). A Comparison of Deformations and Geometric Study of Varieties of Associative Algebras, International Journal of Mathematics and Mathematical Sciences, 1-24

[29] Merad, M. & Falek, M. (2009). The time-dependent linear potential in the presence of a minimal length, Phys. Scr. Vol 79: 015010.

[30] Merad, M., Zeroual, F. & Benzair, H. (2010). Spinless Relativistic Particle in the Presence of A Minimal Length, Electron. J. Theor. Phys. Vol. 7: 41-56.

[31] Nouicer, K. (2006). An exact solution of the one-dimensional Dirac oscillator in the presence of minimal lengths, J. Phys. A: Math. Gen. Vol 39: 5125–5134

[32] Nozari, K. & Karami, M. (2005). Minimal length and the generalized Dirac equation, Mod. Phys. Lett. Vol A 20: 3095-3104.

[33] Nozari, K. & Azizi, T. (2006). Some Aspects of Minimal Length Quantum Mechanics, Gen. Rel. Grav. Vol 38: 735-742

[34] Nouicer, Kh. (2006). Path integral for the harmonic oscillator in one dimension with nonzero minimum position uncertainty, Phys. Lett. Vol. A 354: 399-405

[35] Pet, A. W. Polchinski, J. (1999). UV-IR relations in AdS dynamics, J. Phys. Rev. Vol D 59: 065011.

[36] Scardigli, F. (1999). Generalized uncertainty principle in quantum gravity from micro-black hole gedanken experiment, Phys. Lett. Vol. B 452: 39-44.

[37] Scardigli, F. & Casadio, R. (2003). Generalized uncertainty principle, extra dimensions and holography, Classical and Quantum Gravity, Class. Quantum Grav. Vol. 20: 3915-3926.

[38] Singh, J. (1997). Quantum Mechanics-Fundamentals and Applications to Technology (New York: A Wiley Interscience).

Chapter 4

Unification of Quantum Mechanics with the Relativity Theory, Based on Discrete Conservations of Energy and Gravity

Aghaddin Mamedov

Additional information is available at the end of the chapter

http://dx.doi.org/10.5772/59169

1. Introduction

The history of physics has two great revolutionary theories, such as relativity and quantum physics. However, the new discoveries of the particle physics do not fit with the principles of both theories. One of the main questions is how to explain the break of symmetry in proton-antiproton collision experiments and formation of more matter particles than that of anti-matter ingredients. The theory, which can explain this phenomenon, should answer to the question how matter in the beginning of universe formed and what is the space-time structure of the universe.

In accordance with the modern physics, the small-scale experiments of particles physics have to be described by the Standard Model of quantum mechanics. However, the phenomenon of matter/antimatter symmetry breaking appearing at subatomic scale requires formulation of the new dynamical laws. Presently it is not clear that the mystery of the small-scale dynamics is due to the incompleteness of the mathematic formulation of dynamics of physical events at small scale or to the change of mathematics. The problem is that our present knowledge on mathematical description of change at small scale of space-time frame and physical theories do not distinguish what special features should have the initial state (position) of universe.

Quantum mechanics suggests that at sub –atomic scale the features of initial sale, the position and velocity of a particle cannot be measured. At this scale, the differential equations of dynamical evolution do not work and the change of the state of a system cannot be found by its velocity and position where takes place appearance of wave-like performance.

The problem is that the concept of "rest energy" or "rest mass" can not be explained on the basis of the our present knowledge of static energy conservation law. The static state of "rest

mass" requires application of uncertain amount of energy to keep a body at certain constant position within infinite time duration.

Quantum mechanics relates the problem of "rest" to the uncertainty principle. By quantum mechanics, a particle cannot be at rest because at the "rest" the position and velocity both have to be certain, which cannot be possible.

In classic formulation of Neother theorem, the concept of energy conservation is uncertain due to the continuous feature of the energy conservation. The symmetry principle, revealed from this theory became also uncertain. Therefore, the scientific basis of symmetry breaking, observed at small scale, cannot be explained because of classic energy conservation principle.

The energy as an identity may be conserved discretely because continuous energy conservation in the form "energy can not be destroyed and created" leads to the arbitrary and infinite feature of the dynamical events: the amount and origin of conserved energy within this formulation is uncertain.

The non-continuous energy conservation in the non-arbitrary frame requires existence of boundary of the energy conservation, which has to be localized within space-time framework. But our present understanding of space-time frame does not describe what the space and time are.

There are many concepts, related to the nature of space and time variables but the true nature of time (space as well) is not known. Particularly it is not clear why space has tree dimensions while time has only one. Due to this reason all the physical theories, describing motion distinguish difference between the past and future. The second law of thermodynamics, correlating irreversible time arrow with the increase of randomness does not explain evolution of biological events in direction of more ordered states.

By relativity, space and time are not absolute variables.Unfortunately, the theory of relativity is the theory of geometric space-time and it does not explain why the space-time unit connects the parameters having different dimensions.

Based on these problems, the aim of our chapter is to give analysis of the quantum mechanics and relativity within the principles of classic energy conservation law and to describe main features of these theories within alternative concept of discrete energy conservation. In the frame of discrete energy conservation the classic symmetry principles gets entirely new feature which occupies the special part of the present work.

In the present chapter, we also discuss one of the mysteries of the physics related to the transformation of the space-time variables. Physics accepts invariance in time, which does not fit with the CP violation. We will discuss also Paolo Scaruffi analysis [1] on the laws of classic physics that "why do objects in accordance with the first law of motion have a preference for travel in a straight line at constant speed is not clear. Where does this property come from? It is not clear also, why a body in motion tends to remain in motion and a body at rest tends to say at rest. These questions are open to interpretation".

In our work, we have suggested a new space-time dynamic boundary mapped discrete energy/ momentum conservation law for unification of quantum mechanics and relativity scales interactions, which is based on a new non-Lagrangian mathematical foundation of non-invariance action principle and commutation of frequency domain differential operators of conjugated canonical space-time variables. We showed that the phenomenon called space is the materialization of background "vacuum" energy in the momentum of dark matter, while time phase destroys everything material, reversing the alignment of dark matter momentum to the initial state of "vacuum" energy. Therefore gravity is not the geometry of space-time, it is the dynamic order to hold discrete energy conservation cycles and its power source is the background coupling of space-time phases.

The mechanism of mass generation has been described in the chapter in detail. It is shown that generation of correlated mass and its conservation takes origin by consumption of energy of discrete space-time background frame, moving in direction of space phase expansion. The "continuous parity" conservation may result only by discrete commutation of **SU (2)** group of interactions with the coupling energy of background virtual space-time ingredients, recycling its discrete uniform energy to the **2:1** confined interactions of nuclear. The uniform exchange of energy, generated by not-yet observed background **[T-B]** bi-meson field through interaction with the **uu**$^{-}$quarks, regulates discrete energy conservation within nuclear and background polarization state.

The new non-Lagrangian time/space dependent function shows that the "resulting weak response force" gets its origin from the action-advanced response non-invariant parity of three-component interactions **(V-A)/A,** providing a new deterministic description of nature and fundamental dynamic symmetry. With the static, single non-correlated existence of the identity and physical parameters, the energy and momentum (mass) are not conserved. The physical reality and its parameters as the resulting (not passive relativistic) quantities, do not exist independently of the advanced response interactions: in formulations of classic and quantum physics, the fundamental Lagrangian action is not conserved.

2. The physics of Einstein's relativity

The question "What is the wrong with the relativity theory?" appears frequently in discussions related to the relativity theory. Of course, the relativity theory is not a complete theory. Particularly, the energy –matter equivalence in $E=mc^2$ formulation is not complete from two reasons: the concept of rest mass and the conditions of energy-mass equivalence of the special relativity is not clear. What mass is equivalent to the energy and what features has the rest mass as an equivalent to the energy? What is the performance of this equivalency at dynamical conditions out of the vacuum? What is the connection of the mass with the dynamics of the space-time frame? There is no physical meaning of the increase of the inertia (mass) with the increase of the velocity even with application of Lorentz transformations.

The special statement of relativity that space and time variables are not absolute variables and they form the non-separable space-time unit is very important statement of this theory. From this statement follows that description of the change by classic differentiation where change of the function (space function, such as length) has a relation with the independent time input leads to the appearance of the uncertainty. Therefore, change of space has to be correlated with the response of the time to the change. Unfortunately, this concept, which follows from the analysis of relativity, was not realized in the mathematic formulation of the special relativity.

General relativity is not a complete theory because it cannot describe space-time structure at small scale. However, general relativity, as the special relativity, also has an important statement that space-time structure is determined by the energy and mass in it, which makes it non-linear. The problem is that general relativity describes this relation by geometry, which cannot show change of the structure of space-time with the change of energy and matter. Due to this incompleteness, the general relativity leads to the singularity at small-scale physics (atomic scale) and at high space scale called black holes.

Another big problem of the relativity is its concept, based on continuous dynamics, which leads to the relation of the motion to the relative reference frames. Elimination of continuity and description of the natural events by discrete dynamics do not need application of "reference frames" for relation of the motion. In discrete dynamics, the motion is related to the action force: in the presence of the action, an event changes its state, but when there is no action force an event returns to its initial state.

Based on the discrete dynamics, the Newton's laws of motion and gravitation could be unified. The motion is the result of the action force, therefore the Newton's first law that' if a body is not affected by any force it will keep its initial uniform motion in straight line" eliminates that the body before "affect of any force" had a motion due to the result of some force.

Einstein showed that space and time are simply different dimensions of the same space-time continuum. All events and quantities decompose into time and space components, which depend on the observers. Einstein connected the curvature of space-time of an event with the energy and momentum of the objects. By Einstein opinion the energy and momentum are the same quantities of space-time which has four dimensions. That is why space-time is the same in relation for all reference frames and change of the event is realized through change of the space and time components of this frame. The relative quantity of energy and momentum depends on the observer.

The problem of this approach is that the dynamical space-time variables were connected within continuum framework, which did not allow distinguish personal properties of time and space identities. Later we will show that the difference in performance of space and time variables do not involve dimension of these parameters but deeply is connected with the asymmetric phases, binding the boundary of the space-time phases. The relative quantity of energy and momentum has to be determined by the asymmetry in the boundary of these variables. The contraction of the length and time dilation cannot be described without specification of the boundary of these variables. Therefore, contraction of space and time delay cannot be without relation of these parameters to their initial boundary. The relative ratio of space-time variables in the dynamic space-time frame (by relativity-decomposition of space-time continuum to its ingredients) may be different for the observers, participating differently in the space-time event and sharing directly or indirect the energy recourses with the event.

http://dx.doi.org/10.5772/59169

In Newtonian physics, time flows at a constant rate for all observers. This statement of Newtonian physics appears from the concept of "independent time". Therefore, in accordance with the Einstein's relativity flow of time has different rate for different observers. Therefore, by relativity time is the personal property of an event: different observers, giving different time flow for the same event, present their "personal time". The problem is that relativity equations do not clarify why different observers could measure the same speed of an event by their own measurements. It is not known also how to unify the measurements of the different observers.

Einstein determines the dynamics of matter by the geometry of space-time and that geometry is determined by the distribution of matter. As in the case of special relativity, in general relativity many questions remain open. Particularly general relativity does not explain the origin of mass and energy, which curves the structure of space-time. The question related to the Newton physics, how the moving body responses to the action, in the relativity theory also remains open. If this energy has an external origin, how this energy is generated and how it interacts with the space-time remains open. But if this energy has an internal origin within space-time frame, the mechanism of the energy generation and the energetic features of space is the subject of interpretation. It is not clear also if the universe is a single system, what energy makes its expansion.

The main question here is the energy content of the space. What is the property of the vacuum, is it the empty unit? How the empty space may have energy? Unfortunately, the basic formulations of special and general relativity do not provide answer to these questions without which the relativity theory itself became the "observer" between Newton's physics and quantum mechanics.

3. Fundamental basis of quantum principles and their problems

Quantum mechanics did not solve problem of classic mechanics by modification of its causal dynamical laws but applied entirely new concept of probability, which was a new philosophy for description of nature. Description of the causal nature by mathematical formulation of classic differentiation had boundary problems, which lead to the uncertainty [2]. Description of the position and velocity by classic approach through application of boundary of variables to the mathematical formulation could solve the problems. Dirac's relativistic quantum mechanics concept is the realization one of these ways but Dirac's concept kept the probability feature of the quantum mechanics.

The fundamental basis of the quantum mechanics was very important due to the application the uncertainty in the description of the "change phenomenon" instead of mathematic formulation of classic physics. Unfortunately, the derivative of the function without involving of the function itself leads to the uncertainty. Therefore, the physical observables, such as momentum and velocity of the classic formulation were replaced by the possible states of operators on a wave function. The starting point of quantum mechanics was that it transformed the physical parameters to the differential operators: the time operator replaced energy while momentum by spatial operator.

In accordance with the quantum mechanics, the physical state of a body cannot be described by classic state and it has to be described by "quantum state", presented by wave function. The change of the wave function describes evolving of a system with time. It is clear that during change of a system with time the evolving wave function cannot be merged with the law of relativity.

It is necessary to note that Schrödinger equation is not free from the continuous energy conservation principle due to the involvement of single time derivative as an energy operator.The another problem of the Schrödinger equation is that it involves Hamiltonian as the linear operator of the wave function. The Schrödinger equation is a single particle's equation that is why Hamiltonian does not describe the interaction of the particle with its surrounding medium. Due to this features this equation has a problem of "locality" of a body, described by the wave function.

Due to the absence of "absolute frame of reference", description of the state of a particle was one of the difficult subjects of the relativistic and non-relativistic concepts. Shrodinger equation contains only first order derivative with respect to the time. Dirac tried to give description, such as a differential equation first order for space and time, which may lead to the simple relativistic relation between energy and momentum.

To eliminate the problems of the Schrödinger equation, Dirac suggested the relativistic quantum mechanical wave function in order to fit the concept of relativity. Dirac equation is the first attempt for generation of theory consistent with the principles of quantum mechanics and relativity theory.

Dirac equation was similar to the Schrödinger equation, but his proposal suggested existence of the anti-matter. The main principle of Dirac equation [3, 4] is that Hamiltonian, correlated with the input wave function, was described through complex value, involving the sum of energy-momentum consistent:

$$\left(\beta mc^2 + \sum_{k=1}^{3} \alpha_k p_k c\right)\psi(\mathbf{x},t) = i\hbar\frac{\partial\psi(\mathbf{x},t)}{\partial t} \tag{1}$$

Dirac assumed that presenting Hamiltonian through energy-momentum sum might describe the atomic spectra and discrete angular momentum of an electron.

It is clear that the form of the wave function and its evolution is determined how the total energy and energy-momentum has been described. Hamiltonian of Schrödinger wave function involves the total energy of a system. Making the Schrödinger equation relativistic, Dirac used some additional transformation expressions and got a equation

$$-\frac{\hbar^2}{2m}\nabla^2\phi = i\hbar\frac{\partial}{\partial t}\phi. \tag{2}$$

http://dx.doi.org/10.5772/59169

Dirac equation in this form still is not a complete dynamical model because it presents the dynamic behavior of a free particle and cannot explain why an electron has a spin angular momentum of half a quantum. The Dirac's relativistic wave equation [3, 4] was the first try to communicate relativity with the quantum mechanics.The left side of the equation (1) describes momentum ingredients:

$$\left(\beta mc^2 + c\left(\alpha_1 p_1 + \alpha_2 p_2 + \alpha_3 p_3\right)\right)\psi\left(x,t\right) = i\hbar\frac{\partial\psi\left(x,t\right)}{\partial t} \tag{3}$$

It is necessary to note that Dirac proposal, suggesting existence of anti-matter (particularly positron) does not explain properly physical state of negative energy, introducing it as a "sea" of negative particles. It could not explain also how matter may degenerate into negative-energy states.

The relativistic feature of Dirac's proposal was explained by his suggestion that relativistic description of a particle should involve multiple wave functions for other potential particles.

But Dirac's equation does not involve first order space-time derivatives which in the form of symmetric ingredients have to describe energy and momentum parts of the space-time frame. Dirac's equation also does not involve any information about initial values of the wave function at every discrete time instants; therefore, it could not lead to the definite solutions for communication of relativity with the quantum mechanics. The initial values of wave function of Dirac equation are freely chosen.

The special feature of Dirac's concept is the "Dirac sea" which is related to the negative energy solutions.The "negative energy" of Dirac's sea is explained with the principle that every quantum state can only be occupied with one electron. This concept accepts that the total energy and total charge are infinite. Therefore, this principle does not explain the source of infinite energy and infinite charge.The Dirac's sea is possible only with the infinitely many particles.

In accordance with the Dirac concept, when one electron is lifted form "Dirac's sea" via a high energy of γ-particle, there forms "a hole in the Dirac sea". By this concept, the "hole" represents the anti-particle to the electron with the positive charge, namely the positron, which has been proven experimentally.

Dirac equation can be obtained from the lagranjian action principle, but lagranjian action cannot give first order derivatives for the event simultaneously in space and time variables.

O.Klein and W.Gordon proposed [5, 6] the concept, which is applied to describe "particle" behavior in the relativistic mode and an equation, which has to simulate the behavior of a spineless free particle which has a spinless wave solutions. The Klein-Gordon equation is a relativistic version of Shrodinger equation, it is the second order in time, and it describes spinless particles in towards and backwards in time.

The Klein–Gordon equation has the following description:

$$\frac{1}{c^2}\frac{\partial^2}{\partial t^2}\psi - \nabla^2\psi + \frac{m^2c^2}{\hbar^2}\psi = 0. \tag{4}$$

The Klein-Gordon equation is the second order for time expression and by separation of positive and negative parts describes time-independent case. In some versions, Klein-Gordon equation was introduced with the second order for time and space but in these expressions have boundary value problems.

Although Klein-Gordon equation claims to be relativistic equation, it does not give complete relativistic picture of a "particle" dynamics simultaneously in space and time coordinates, which in relativistic mode needs to be connected with the space-time frame, required by relativity theory.

4. Discrete conservation of energy as the basis for unification of physics

4.1. Formulation of discrete space-time field theory

The concept, which we are planning to use, has to eliminate the small-scale phenomenon-point particle concept of classic physics and quantum mechanics probability of location of subatomic particles. The new concept is the discrete dynamical structure of space-time phases. The position of an elementary particle, located within space-time phases is not a point; it exists as a time carrying identity within minimum space frame, called elementary space-time manifold commuting space and time phases.

The problem of classic physics is the description of the events by displacement of a point in space (**Δx**) within certain displacement of time (**Δt**), presented in the form of relative simple intervals. Two point particles having the same displacement in space and time could pass the observer's reference position differently being in "phase" or in "opposite phase" due to the having, phase differences starting from different initial conditions.

Feynman [7] showed that elimination of the infinity in dynamical formulations could be done through substitution of interval of function (**Δf**) by the value of the function itself (**f**) but by his opinion in this case the dynamical event will be static.

It is clear that the effect of scale phenomenon to the behavior of dynamical systems can be analyzed by Hamilton canonical coordinate transformations. Unfortunately, a genetic Hamilton's transformation (**g, p, t**) → (**Q, P, T**) has no explicit time dependence and does not preserve this transformation.

We used a new principle of canonical transformation where the transform is the time – frequency representation. The new approach involves commutation of frequency with the time domain through conjugation of the change of function (**Δf**) with the local function (**f_n**) itself. In this case, the reciprocal discrete transform within (Δf) and function (f_n) itself can be generalized to the Abelian group due to the generation of dynamic translation within boundary of canonical variables.

We did not replace the interval of the function (**Δf**) with the function itself (**f_n**), as Feynman suggested, but we conjugated them together to form entirely new mathematical operator, eliminating arbitrary performance of the coupling constant: the evolution of an event will follow to the discrete dynamic transformations of local function **(Δf)/f_1** such as **(f_2/f_1-1)**. This operator has important features because the field theories involve the differences of the parameters, but not input values itself, such as electricity theory involves the difference of voltages but not voltages themselves.

This formulation does not lead to the confusing approach; it is like frequency of repeating of some action in relation to the previous ground instant, giving the product of time phase **Δt/t_1**. This function is the frequency domain product, which is different from the classic physic's frequency.

Now we have to apply this concept for space and time ingredients of the space-time unit. The mathematical unit "time phase" as a frequency domain describes the displacement from a specified reference point at the **t=t_1**. The time and space phases have a commutation through product of certain position in space within some interval of time **(S_1Δt)** and with interval of space at a moment of time (t_1ΔS).

It is easy to show that the time domain in the form of time phase is equivalent and commutative to the space phase of an event and describes the relation of passage of time (**Δt**) to some local instant state or an event local boundary, such as **Δt/t_i** where **t_i** is the local time boundary of an event. Similarly, we can present the space phase with the same way **ΔS/S_i**, describing the change of space in relation to some starting local boundary of the event. On this basis, the event dynamics can be described by the shift in phase within space-time frame giving the event history equivalently in the form of direction of time or displacement of space phase. Therefore, the commuted phases may be equivalent through commutation of variables, mapped within their boundaries.

The new operator in the form (**Δf**)/**f_1** describes change (vibration) of function around its origin with certain rhythm in the form of fluctuation density to repeat its origin. Similarly, the operator ΔS/S_1 describes fluctuation of change of space energy density around space origin while operator Δt/t_1 describes fluctuation time of change around instant origin in the form of frequency.

The initial values of space and time in the form of **S_1** and **t_1** are the local dynamic boundary states: if the event within some region has ended, **t_1** transforms to the end t_2, but if the event is continued, **t_2** became a new initial state. On this basis, the local boundary is the mixture of past and future, which may have uniform and non-uniform states. Therefore, without correlation of initial and end states the action is not conserved. The non-unitary time domain **Δt/t_1** describes discrete time integer numbers of the non-continuum event. In the same way the ΔS/S_1 describes the non-continuum discrete changes of space phase with dynamic (i=1, 2, 3) boundary conditions.

The space phase in the form of integer multiplication **ΔS/S_i** involves the numbers of the "pieces" (grains) located in the space medium, while **Δt/t_i** describes the duration in relation to the instants corresponding to the change of position of these "pieces". It is easy to see that relation of the space and time intervals to the dynamical local state presents the transformation of continuum intervals to the discrete integers of subintervals at points $\mathbf{S}_i$ and $\mathbf{t}_i$ forming definite canonical variables of the differentiation.

This approach leads to the binding of space and time phases to each other. The relation of space and time intervals to each other usually forms the classic velocity vector $\mathbf{\Delta S/\Delta t}$. However, this derivative of uncertain intervals represents the singularity at boundary conditions. The infinity in this case arises due to the absence of the commutation of the change with the local boundary of the **S(t)** function.

Interaction of time $[\mathbf{t}_1, \Delta\mathbf{t}/\mathbf{t}_1, \Delta\mathbf{t}]$ and space phase $[\mathbf{S}_1, \Delta\mathbf{S}/\mathbf{S}_1, \Delta\mathbf{S}]$ matrices leads to the formation of space-time field. Field as the correlation of the end of the phase with the initial boundary eliminates the infinity due to the conjugated displacements in space $\mathbf{t}_1\Delta\mathbf{S}$ and time $\mathbf{S}_1\Delta\mathbf{t}$. On this basis an event can be described through change of the action energy applied to the space-time field instead change of position or coordinates of the system. The $\Delta\mathbf{S}/\mathbf{S}_1$ shows distribution of energy in space: when $\mathbf{S}_1$ is small the space gets the same property within all regions of the field while at high scale, the space field is non-invariant.

By relativity, the space-time frame is curved due to the presence of energy and mass in it. But by our concept space-time field is the resulting inner product of action of the energetic field $[\mathbf{E}_{act}, \Delta E/\mathbf{E}_s, E_s]$, formed from exchange interaction of action and event's energies:

In accordance with the Lagrangian mechanics, "an object subjected to external influence will choose a path which makes the action minimum". By our concept, this phenomenon is due to the discrete exchange of the action with the response force of the system's field, which forms the action – response non-invariance parity of the interaction. The effect of action of the applied force to the space-time field of a body can be formulated in the form of exchange of energetic fields-$(\mathbf{E}_{act} - E_s)/E_s$:

$$\frac{\frac{\Delta S}{S_1}}{\frac{\Delta t}{t_1}} = \left(\frac{E_{act}}{E_s} - \frac{E_s}{E_s}\right) = \frac{(E_{act} - E_s)}{E_s} \tag{5}$$

$$\frac{\Delta S}{\Delta t} = \frac{S_1}{t_1} \cdot \frac{(E_{act} - E_s)}{E_s} \tag{6}$$

where, $\mathbf{S}_1$ and $\mathbf{t}_1$ are the space and time variables corresponding to the dynamic local boundary, $\mathbf{E}_{act}$ and $\mathbf{E}_s$ are the energies of action and under action systems of interaction at conditions corresponding to the local boundaries of $\mathbf{S}_1$ and $\mathbf{t}_1$. The interval of time $\mathbf{\Delta t}$ describes the duration of coupling of the event with the action, while $\mathbf{t}_1$ presents the dynamic time instant or time boundary of correlation with the action. In the same way, the interval of space ΔS describes the expansion, while $\mathbf{S}_1$ presents the local boundary of space of an event at the instant. The quantity of energy, available for change (scattered energy) and the quantity, determining the response (consumed energy-momentum) have different signs (5) therefore, leads to the generation of direction and causality between local past and local future.

The equation (6), which we generated has based on the space-time consistents of energy. Model (6) although is very simple description of event dynamics but it allows to give a new look to the interaction, localized within triangle matrix of space-time-energy boundaries. In accordance with the model (6), the change is the "tree body action" of the space-time-energy operators: space derivative of energy produces momentum, time derivative conservation of identity is the energy and the product of energy derivative in conjugated space-time phases is an event. The product of energy–time multiplication is the identity of an event, localized in the observed space.

As can be seen, the change of a system's state is the result of interaction of two space-time frames, which presents the change of one field in relation to the action "field". Model (6) treats the matter field with space phase and antimatter with time phase. The time phase describes the change of event instants over time duration, but frequency domain shows how much of the time instants lie within each given frequency.

Model (6) may be re-written in the form:

$$\frac{\Delta S}{\Delta t} = \frac{S_1}{t_1} \cdot \left(\frac{E_{act}}{E_s} - 1 \right) \tag{7}$$

$$\lambda = \left(\frac{E_{act}}{E_s} - 1 \right) \tag{8}$$

In the classic formulation the change of the space-time frame **ΔS/Δ**t cannot be the precise Eigen state of the dynamical event but in the model (6) it gives the precise Eigen state of identity due to its conjugation with the Eigenvector of dynamical local position $\mathbf{S_1/t_1}$ and with the exchange energy of the space-time variables (E_{act} / E_{s}-1). The reflection function (**$\mathbf{E_{act}}$ / $\mathbf{E_{s}}$-1)** became the Eigen value of the ΔS/Δt, which covers all the values of correlation of space-time variables. The $\mathbf{E_{act}}$ /$\mathbf{E_s}$ describes the density of energy, distributed in time phase as an action of force, while $\mathbf{E_s}$ has a relation to the density of energy distributed in space phase in the form of matter. Therefore, **ΔS/Δ**t, conjugated with the dynamical local position $\mathbf{S_1}$/ $\mathbf{t_1}$ in the form of speed, became the function of state holding conservation of energy within space-time phases. On this basis, the model (6) became the equation of state. The equation of state (6) in its basic form (7) gives the numbers due to the resulting non-unity of the parameters.

Model (4) shows that particles do not have the fixed position during action and measurement, but they possess the change of space phase by coupling with the action energy. In this case, the measurement of the position is the measurement of the velocity and impossibility of the measurements of these parameters in different order is not the subject of uncertainty in nature.

The minimum portion of energy, which by quantum physics called "quanta", is the elementary space-time "field". The energy in similar way also is "quantized" within space-time field. The total energy is conserved discretely; it comes in discrete amounts, localized within space-time phases. The energy is the inner product of coupling of space and time fields and exists in the form of resulting Eigen value.

Model (6) shows that the space-time phases in the form of two different inner products of action –response parity display the frequency of energy consumption/restoration cycles. When there is no energy ($\mathbf{E_{act}}$=0) the coupling of virtual space and time phases generates energy, then the generated energy produces the non-virtual space-time frame of identity which conserves the produced energy within momentum. On this basis, time is the product and measure of the consumption of energy. The different frequency of energy consumption for different events leads to the separation of the non-continuum periodic events by real numbers forming the "Time phenomenon". Time takes its origin only from discrete energy conservation cycle and due to the relation to the initial state, cannot be described only by intervals. Time is the "proper time" only in connection with the Eigen value (proper value). The notion that" time slows down by distance" is the correlation of variables and this correlation is the result of discrete energy conservation.

The interaction of action and response is affected by the initial action in the form of relations of the action $(\mathbf{E_{act}} - \mathbf{E_s})/ E_{act}$ and response $\mathbf{E_s}/ \mathbf{E_{act}}$ strengths, but the outcome is determined by the advanced response given by the mathematical structure of Eigen value, which regulates energy distribution within space-time frame.

In accordance with the model (6), time appears as the personal product of exchange interaction of action-response parity and is the quantity, which holds discrete energy conservation within this parity. Therefore, relation of time to the external system of reference having no energetic parity with the system does not make sense for description of the event dynamics. On this basis, if an observer does not apply the force for interaction with an event, it became the local "photographic plate" of an event. In accordance with the model (6), the instant of time is the dynamic phenomenon and when energy is applied it produces the response in the form of inertia to regulate the continuity of the event.

In accordance with our concept, the description of a "change" in magnitude and in direction is possible only through correlation of space-time canonical variables regulated by Eigen value $(\mathbf{E_{act}} /\mathbf{E_{s-}1})$.

The classic mathematical tool of differentiation does not describe the boundary of function and the non-boundary concept of time presently is the common acceptable concept of physics. The differential operator in the absence of boundary function leads to the approximation of the action conservation due to the lost of the original function during production of the outcome of the operation.

In accordance with the model (6), when the velocity $\mathbf{\Delta S/\Delta t}$ commutes with the dynamical boundary of space-time it became the non-relativistic operator. The concept of velocity is to be tied to the space-time boundary while description of velocity without boundary leads to the "problem of different observers ". The boundary-mapped space-time eliminates the uncertainty in the continuity and the singularity of boundary conditions. This is the mathematical background of our model based on the new commutation concept.

The conditions $\mathbf{\Delta S} /\mathbf{\Delta t}=\mathbf{S_1}/\mathbf{t_{1,}}\ \lambda_{=}1$ ($\mathbf{E_{act}}=\mathbf{2E_s}$) are similar to the mathematical principle of involution. The inversion of the action by the same coordinate line does not produce the same action. That is why action-response parity should involve one more dynamic intermediate state,

http://dx.doi.org/10.5772/59169

which leads to the three body interactions. The additional intermediate is the "second material observer" of (**S_2**, **$2S_1$**) interaction. On this basis the "color based strong interactions" is due to the "three sigma" color phases (colors), correlation of which in **1:2** coupling creates formation of three body interactions: the energy of two merged colors is balanced with the third color with formation of "color based elementary space-time unit". The local state with the dynamic conjugation of two merged colors with the third color leaves Eigen value the same, leading to the conservation of energy and formation of constant interactions. In accordance with our concept, virtual particle appears when the duration of change is equal to the instant of generation.

By prediction of model (6), the action of a force is to change the state of an event, but response of system (negative sign) is appeared to make the action minimum and maintain the initial state of the event.

The action of the force gradually became minimum that is why the force to be needed to maintain the initial state of action. The inversion of an event in non-virtual space-time frame in spatial and time reversal manner cannot eliminate the asymmetry of the action-effect parity. The difference of action-effect quantities determines positive time direction and magnitude of a motion (momentum).

The action-response parity is the interaction of the two fields, which "shakes" each other with the selection of the direction to move: the Eigen value of the space-time structure determining momentum of the system became the degree of freedom of the resulting Eigenvector field.

By Feynmann analysis, Dirac showed that in quantum mechanics there is important quantity, similar to the differential equation, which carries the wave function from one time **t** to another, such as **t+ε**, **t+2ε**, **t+3ε**. [7]. Feynman developed Dirac analysis and succeeded in representing quantum mechanics directly by the Lagrangian action.

Giving characterization of his research Feynman showed "that the important issue in his development of the space-time views of quantum electrodynamics is that he connected the Lagrangian with the quantum mechanics. But Lagrangian for strong interactions still needs renormalization". The problem of Feynman's Lagrangian is the non-conservation of the action. Our concept of gradually advanced response is opposite to the Feynman action and Weyl concepts where has been used the advanced action wave which violates principles of causality. The advanced response as the resulting quantity is the "hidden correlation" of space-time variables to realize discrete energy conservation.

In accordance with the model (6), space-time without coupling with the discrete energy, conservation law can give only uncertain position. This feature of space-time explains why string theory suggests additional dimensions to describe a position. The extra dimension in reality is the correlation of space-time variables with the coupling energy, which through coupling with the Eigen value rotates the space-time vectors in the form of curled up dimensions. Particle without coupling is not observable and has negative existence.

Model (6) connects position-momentum and time-energy relation and shows that these relations within space-time boundary frame cannot be subject of uncertainty because position as a spatial variable does not have existence, independent of time. Model (6) involves the commutation of first order derivatives of space and time variables from each other.

4.2. The action-effect parity of the "Change" phenomenon

By prediction of the model (6), the action of a force is to change the state of an event, but response of system (negative sign) is appeared to make the action minimum and maintain the initial state of the event. From model (6) also follows that the system, applying the energy (E_{act}) is the origin of the causal effect.

The action of the force gradually became minimum that is why the force to be needed to maintain the initial state of action. The inversion of an event in non-virtual space-time frame in spatial and time reversal manner cannot eliminate the asymmetry of the action-effect parity. The difference of action-effect quantities determines positive time direction and magnitude of a motion (momentum).

The action-response parity is the interaction of the two fields, which "shakes" each other with the selection of the direction to move: the Eigen value of the space-time structure determining momentum of the system became the degree of freedom of the resulting Eigenvector field.

With the discrete energy conservation, the forward action decreases by one while backward response increases by one (7). The advanced response force regulates conservation of energy, mapped within space-time frame. If the response (axial vector) does not change it can not limit the action. This is the non-Lagrangian least action, produced as the resulting exchange quantity.

In accordance with the model (6), the space-time frame without coupling with the discrete energy conservation can give only uncertain position. This feature of space-time explains why string theory suggests additional dimensions to describe position. The extra dimension in reality is the correlation of space-time variables with the coupling energy, which through coupling with the Eigen value rotates the space-time vectors in the form of curled up dimensions. Particle without coupling is not observable and has negative existence (6).

The Heisenberg's uncertainty and relativistic reference frame will have the same nature, if relativity's observer has an energetic interaction with an event. The observers of both theories during the measurement will have variable action (variation of consumed energy), therefore Einstein's and Heisenberg's observer's measurements are not invariant and affected by an "external force of a body under measurement". On this basis, it is obvious that the Heisenberg's and Einstein's observers cannot measure the fixed static position.

The Heisenberg's discovery on non-commutation between velocity and position is due to the problem of classic differentiation, which produces non-conservation of local velocity. Position is the integral product while the velocity is the differential outcome therefore these parameters are not commutative and are origin of the non-invariance of the action-response parity (λ). The action-response parity generates discrete energy distribution, which is only the way to eliminate infinity from space-time frame.

In the case of Heisenberg's thought experiment the measurement is an application of energy $\mathbf{E}_{act}$ to change the position for getting the information. In accordance with the model (6), the action energy portion cannot be less than energy (E_s) of the elementary space-time unit of the light. The space dimension in this case is related to the distance between the wave crests of space and time phases of the light photon. The difference of the phases cannot be smaller than wavelength of the light waves. This is the limit of phase difference of classic space-time frame, which has been called Planck's scale.

On this basis, we can unify Einstein's observers and Heisenberg's measurement. If the observers do not have energetic correlation with the event, they are distant "photographic plates" but if correlate with the event with the same Eigen value, they are invariant observers. Therefore, the Eigen value (the numbers of the scattered energy portions) "is the reference frame which determines the invariance": time is relative to an observer if the observer affects the Eigen value of an event. The Eigen value determines the intrinsic property of a system in the form of identity.

This principle of simultaneity explains Copenhagen interpretations, which states that the outcome of an experiment is only revealed when the quantum system interacts with a macroscopic apparatus of measurement resulting only one outcome.

The measurement changes the energy $\mathbf{E}_{act,}$ applied to the system and correlated with the space-time framework (6). In accordance with the model (6), you cannot measure the parameters at their fixed states, which have no independent existence. The entanglement concept of quantum mechanics may be explained also based on these principles. You cannot measure one parameter determined by space-time, fixing another one because there is no independent existence of the parameters, correlated within space-time boundary. In the absence of measurement, the system has its own deterministic space-time frame, regardless how "quantum physic's and relativity observers will observe an event on their photographic plate".

By quantum mechanics, the action is "quantized", but by our concept, the quantization is the energetic discreteness of the classic action and existence of reality by action-response parity. Conservation laws, mapped within Lagrangian framework of classic mechanics results only approximate conservation due to the non-invariance of the action. The invariance requires constant energy supply, which is not possible in continuous mode therefore the underlying mechanism of reality is the non-linearity of the physical events, realized through non-invariance action-response parity and discrete energy conservation.

5. The Non-Noether's concept of symmetry and energy conservation law

In accordance with the model (6), the classic principle "the total energy of a dynamical system involving kinetic and potential energies is conserved" can be the true concept only if the system's dynamical coupling with the action is conserved. The statement that "the energy is not created and not destroyed, but transforms from one form to another in self-sufficient system" should lead to the non-invariant energy transformation, disappearance of action and momentum. By this principle the action, kinetic and potential energies individually are not conserved and even cannot be conserved.

In our concept, we replaced Hamiltonian static sum of energy by the dynamic scattering energy where total energy, distributed within two systems, presents interaction term. The interaction term of Schrödinger equation is linear while the Eigen value of model (6), correlated with the change of local dynamic space-time boundary, determines the proper states instead of Schrödinger's probability.

Model (6) shows that the problem of Hamiltonian in canonical coordinate's transformation is directly related to the principles of energy-momentum conservation. Hamiltonian operator of total energy involves independence existence of kinetic and potential energies but in discrete energy conservation, concept energy has no independent existence. The expression of resulting energy of the system through simple gradient of kinetic and potential energies may produce only continuous displacement therefore leads to the observed problem of Hamiltonian conservation. Hamiltonian sum of energy describes the state of a system as an independence existence but in accordance with the discrete energy conservation concept, the identity cannot be described by its own existence.

The question is why we need discrete energy conservation law is very important which determines all the features of the new physics. The static continuous energy conservation described by Noether's theorem does not limit the boundary of the conserved quantity, therefore leads to the singularity.

By complementary principle of quantum mechanics, "the classical concepts such as space time location and energy-momentum can not be combined into a single picture. One classical concept excludes the simultaneous application of other classic concept. The uncertainty suggests that this reciprocal limitation is due to the uncertainty and uncontrollable exchange of momentum of a particle with the space-time frame of the object where a particle is located".

In accordance with the discrete energy conservation concept, momentum in the form of two reciprocal parameters appears for generation of the space phase of energy conservation. Conservation of energy in the space phase generates a "mass of space", as transformation of energy to mass, to generate the opposite time phase of energy conservation.

In accordance with the model (6), energy as an identity can be conserved only discretely and discreteness is realized with the alternation of the two opposite appearance –disappearance phases: energy disappears in space phase and appears in time phase. When energy is conserved in space phase, it leads to the appearance of mass. Therefore, the "mass" phenomenon is the property of the energy conservation but not the property of a body "affected by a force".

In accordance with the quantum mechanics, forces are manifestations of exchange of discrete amounts of energy. Without locality in space, exchange of energy quanta is not possible. That is why quantum physics concept that "any field of force manifests itself in the form of discrete particles" may be realized only in the presence of space medium, which display discrete existence.

From model (6), we can get Newton's first principle: Eact=0 describes the body, which is not affected by any force and localized in the inertial frame. This is the initial state where events move with the uniform motion. Therefore, E_{act} is the generator of space-time frame, which appears discretely. The state of appearance of E_{act} is the Planck scale where difference between space and time disappears. Energy is the generator of space-time and energy quanta itself is the space-time cell.

Space-time unit cannot exist alone and has to be interacted with the event where s_1/t_1 is the elementary cell of the space-time frame, $(\mathbf{E_{act} - E_s)/E_s})$ is the event. If there is no action energy Eact=0, everything is going to the initial cell s_1/t_1. In this case, the initial state and change have different sign and initial state became resources of inertia. In the absence of the action energy, the structure of the space-time changes: correlation of space and time variables t_1 $\Delta S=S_1\Delta t$ disappears which lead to the separation of t_1 from Δs. Change of space-time frame is associated with the change of energy-matter structure of an event. This leads to the separation of electron-positron and neutrino-antineutrino pairs and formation of e-/e+and v-/v pairs. The e-/e+and v-/v pairs form the pre-existing form of matter. The matter is composed from a fundamental cell of space-time frame, presented in the form of minimum S_1/t_1 cell.

6. Analysis of quantum mechanics based on discrete conservation of energy

By quantum mechanics, the vacuum energy is to be the virtual particles, which as vacuum fluctuations are created out of the vacuum. This concept does not explain what the nature of lowest vacuum energy is. The important question is why during removal of matter from the vacuum it does not reverses the energy back to the background state.

Quantum mechanics describes energy of empty vacuum in the form of virtual particles giving little push to the start of the universe, which can continue acceleration. The problem is that the calculations of quantum field theories predict 10^{120} times more quantum vacuum energy than that of any possible value.

By quantum mechanics, vacuum cannot have zero energy because of uncertainty principle, which can be violated: the zero energy value is certain. Uncertainty does not describe how the lower energy state may have high zero point energy state.

In accordance with our concept, the zero point energy means that the total energy of space-time is zero (Eact=0) but this energy is accumulated within asymmetric space-time boundaries. The total energy is distributed within different phases, which are in asymmetric state of opposite boundaries.Vacuum and black hole are two asymmetric boundaries of space-time variables. The theories of quantum physics suggest that due to the uncertainty principle, quantum field cannot have zero value, therefore became the origin of vacuum fluctuations. In accordance with our concept the vacuum and "quantum fluctuations" is due to the discrete conservation of energy within asymmetric space-time phases. At small scale of space, the frequency of fluctuations is generated by time phase and is high while at high space scale the giant size of fluctuations is created by the energy consumed in space phase.

In accordance with our model (6), creation of a particle is the result of discrete conservation of energy, which appears within space and time "fields". Conservation of energy is not static, therefore mass also is not static and changes with the expansion of space. This feature of mass is the necessary factor for conservation of energy; with the constant mass, energy cannot be conserved. On this basis, vacuum is the asymmetry of space-time boundaries, therefore light cannot exist in vacuum with the constant static speed. Vacuum appears when the energy of the background state is removed for expansion of space. When energy is completely consumed in space, time returns it back to the initial state by negative gravitational force. That is why we replaced the spatial and time coordinates by space and time phases having performance of the energetic fields.

By quantum field theory, vacuum has properties as a particle and these properties cancel out on average, leaving the vacuum empty. By our concept vacuum is the asymmetric boundary of space-time where space identity is very small which is connected with the high amount of time particles –antiparticles.

The classic physics relates local symmetry to the Lorentz invariant quantity, which is connected with the helicity phenomenon, describing the projection of a spin of a particle in the direction of momentum. In the case of discrete energy concept, local invariant inversions are not allowed.

Discrete conservation of energy in space –time phases cannot be realized without boundary of theses phases which constrains the expansion of space through left handed neutrino (involved in Es) and right handed anti-neutrinos (involved to E_{act}). That is why the right handed neutrino and left handed antineutrino never was observed.

In accordance with the Newton's physics, mass is the inertial rest energy of a particle and the measure of the resistant to the applied force (a=F/m). By special relativity, a massless particle cannot exist at rest it must always move at speed of light. Quantum mechanics suggests that a "massive fermions should have both right and left hand states because field operators that yield a non zero mass for fermions are bilinear products of fields that flip the particles handedness".

In accordance with the discrete energy conservation, the quantum mechanic's mass operator, which annihilates left handed neutrino and creates right handed antineutrino can not be described by the sum of doublet due to the continuous nature of this interaction. Therefore, the mass term that changes a particle into antiparticle, which in quantum mechanics called as Majorana mass term, has to be described by discrete term, such as (E_{act}/E_s-1). The sign in this formula changes (such as quantum mechanic's fermions' number, which changes from n=-1 to n=+1) right-handed particles to the left handed.

The uncertainty of a position and the future motion of a particle can be accepted classically obvious, because position is not a static quantity. It is known that measurement of the particle position involves the scattering of light, which can give only probabilistic exchange of energy and cannot be described by known classic physics. Quantum mechanics explains, "If scattering energy is not uniform the measurement devices has no possibility to measure the position and momentum".

The problem is we cannot measure position with instant of time only and measurement should involve instant of time and its change in the form of time phase $\Delta t/t_1$. In accordance with the model (6), the change of time in relation to the instant may describe the change of a position and velocity. The change of time phase has no independent existence and its changes in correlation with the change of space phase. That is why measurement of the space and time as the independent coordinates leads to the uncertainty. Therefore, the mathematics of the description of the position and velocity should involve them not as an independent coordinates but correlated phases.

Therefore, the statement that classic concept does not fit at the quantum level is not true. If time will involve two parameters as an instant and duration, the ratio of this parameters may describe any level of an event: at small scale the duration is small and instant has high frequency while at high scale instant is small duration of an even is long. At "quantum level", the high frequency instants are not quantum jumps and they are discrete phases, appearing in space phase and disappearing in time phases. The alternation of these phases and exchange of energy between these phases can not be described by probabilistic energy distribution, because discrete exchange of energy is very causal deterministic process.

Heisenberg uncertainty implies that "any two variables that do not commute can not be measured simultaneously". It is possible to give classic explanation to this phenomenon. Two variables that have no independent existence cannot be measured simultaneously. The position and velocity have no independent existence similarly as change of energy is not independent of time. Therefore, uncertainty of quantum mechanics is the non-conservation of energy in the formulations of classic physics. That is why "uncertainty of non-commuted variables" is the uncertainty in conservation of energy in the farme of classic physics.

The question is how to describe displacement of space boundary within certain time interval or displacement of time boundary within certain space boundary. On this basis, the main concept of quantum mechanics is the "quantization" of space-space period.

In accordance of our concept, quantization of space-time frame is the classic phenomenon and it is related to the discrete conservation of energy, mapped within two opposite phases. "Quantization" of space-time frame requires correlation of space-time variables, appearing as a field, realizing the dynamic state of energy conservation. Model (6) shows that connection of space and time together in one unit, is the quantization of space-time, which puts limitation to the boundary of space-time. However, for connection of space and time in space-time unit, free coordinates cannot present these variables.

The non-independent change of space-time variables leads to the displacement of space and time phases by the alternation, which generates a wave function carrying the portions of conserved energy in different phases.

In the case of discrete conservation of energy, the particle-antiparticle annihilation is not a symmetric event and cannot be described by symmetric model. Therefore, the symmetric model leads to the continuous energy conservation which results "Ultraviolent Catastrophic" phenomenon of the background energy. Due to this problem, renormalization is an open question in the quantum field theory where the total energy becomes infinite.

By quantum mechanics, the zero point energy is the "lowest" quantized energy level of a quantum mechanical system. The zero point energy is "the energy which remains when all the energy is removed from a system". The common opinion is that the origin of the zero point energy is the uncertainty principle. By quantum physics a quantum fluctuations is the temporary change for energy in a point in space, arising from Heisenberg's uncertainty principle, "such as temporary change of amount of energy in a point in space without certain time". The uncertainty accepts that conservation of energy can be violated but for small times. In accordance with the energy-time uncertainty, with the decrease of time, violation of energy conservation increases. Therefore, the uncertainty principle allows existence of a virtual particle with the barrowed energy, but the sources of the borrowed energy are not clear.

Quantum mechanics suggests that pair production appears as a pop into existence and then annihilation each other. On this basis there is suggestion that [8] energy in classic physics is conserved properly but in the quantum micro world energy can appear and disappear in a spontaneous fashion. The uncertainty principle implies that particles came into existence for short periods even when there is not enough energy to create them. They borrow energy for a short time and then they return and disappear again. In vacuum, pair of virtual particles is constantly being created and destroyed. The energy of matter is positive which appears spontaneously out of empty space while the energy of anti matter is negative. The matter made of positive energy and matter particles has attracting gravitational energy, which is negative. The total energy is zero. By quantum mechanics, fluctuations are random and have no causal nature.

Quantum theory of the vacuum suggests that the zero state of vacuum energy is negative. Based on expansion of the space it concludes that vacuum ground state has non-zero energy. The vacuum energy, described by quantum field theory without renormalization is mathematically infinite.

It can be thought that discrete energy conservation is obvious concept due to the discrete dynamics of quantum mechanics and discrete energy radiation. But the question is how to describe the discrete energy conservation in mathematical formulation. In accordance with our model (6), without correlation of space-time variables, it is impossible to describe discrete conservation of energy. With the frame of discrete energy conservation, it is impossible to describe spatial or temporal inversion without boundary space-time framework. In accordance with the discrete energy conservation, the space-time dynamics appears as the particle-antiparticle pair, which carries a conserved quantity of energy within these phases.

Quantum mechanic's theories suggest that zero point energy has positive sign for bosons and negative for fermions, which cancel each other for perfect symmetry. The model (6) describes the energy of bosons and fermions with the similar way but shows that they cancel each other at Eact=2Es.

For uniform speed of light, the change of space (ΔS) and time (Δt) have to be correlated, generating invariance of these intervals from the boundary of space-time frame and uniform change of these variables. The energy of light at uniform speed should follow the Noether's

http://dx.doi.org/10.5772/59169

theorem of symmetry. It is known that Maxwell equation is not independent from the property of distance and the speed of light is a function of properties of space. Model (6) shows that property of space (ΔS) and time (Δt) intervals changes with the change of boundary of these variables.

In accordance with the quantum mechanics when matter and antimatter collide, they annihilate in a flash of energy. In accordance with our model, the energy, produced during interaction of asymmetric space-time phases has to be conserved in space phase to form "particle" which disappears in the opposite time phase, forming "virtual particles". The observance of these particles depends from frequency of the discrete energy conservation.

7. The singularity problem of relativity and uncertainty of the quantum measurements

Due to the lost of the initial boundary frame, all space-time models break down at the singularity. To eliminate the singularity, the space-time should have internal constraint holding the dynamics at boundary, because the initial position is not an independent free coordinate. The initial state of minimum space-time identity is the internal constraint of the space-time, which determines how to go from contraction phase to the expansion phase without singularity. General relativity has fused discrete space-time frame to the continuous "geometrical manifold" which with the decrease of space has to move to the infinite space-time curvature. The space-time frame by this concept has no internal constraint therefore has to move to the singularity and the initial state of universe should start from a single point having infinite density. This is the unproved conjecture of general relativity.

At continuous energy conservation, an event with any size has a trend to move to the equilibrium: to remove a system from the equilibrium one additional parameter is needed which can be the application of an external energy. With the continuous energy conservation law a system, consuming any amount of external energy has to move to the equilibrium of "infinite black hole".

The asymmetry of space-time boundaries is the fundamental basis for discrete existence of universe. In some theories, dynamical space-time has been discussed but there was not shown that this dynamics is due to the asymmetry of boundaries. The discrete coupling constant $\mathbf{E}_{act}$ (magnitude is finite) is the boundary constraint of space-time which at $\mathbf{E}_{act}$=0 is different from Einstein's $\mathbf{R}_{ab}$=0 constraint of empty space. The discrete conservation of matter and energy within discrete space-time frame gives a non-geometrical status to the space –time phenomenon.

In accordance with the model (6), inverse transformation appears when all the energy of space-time is consumed by space ("black hole phase") and all the events have the same dynamic characteristics of "free fall "to the state of energy generation".

8. Particle physics based on discrete energy conservation

Presently why the particles acquire the mass is not known. The mass –energy relation which first time has been introduced by Einstein in the form $\mathbf{m=E/c^2}$ does not describe mechanism of mass generation. The problem of this formulation is invariance of mass-energy relation, fixed by speed of light.

As follows from the model (6), conservation of mass cannot be separated from the discrete energy/momentum conservation. Therefore, in discrete conservation of energy there is no invariant rest mass, mapped within discrete space-time frame. The problem of mass appears with the Newton's force. Newtonian law does not explain the origin of force that is why the nature of mass remains unanswered. Newton's invariant mass during change of velocity leads to the approximate energy conservation. If force is needed to change the velocity, then Newtonian inertia appears to describe the magnitude of change in the form of mass. If the action of force, as is shown in (6), is needed to hold the initial state of a body then it is needed to hold the existence of matter in space-time frame in the form of mass. The problem of Newtonian law is that mass, which is invariant during velocity differentiation but affects to the acceleration.

In accordance with the model (6), the entity called mass is the resulting quantity, generated from discrete energy conservation. This feature seems makes it as the "electromagnetic mass". With cyclic discrete energy conservation there is no any preferred reference frame, which could display inertial or relativistic mass. The non-virtual matter particle is created simultaneously with the generation of non-virtual space-time frame: mass is the part of energy to be conserved in space phase: the dense space phase, formed through change of frequency of correlation of space-time variables leads to the generation of different masses responsible for different particles. The density of space phase is determined by the Eigen value (6), therefore mass is not constant, it changes with the change of frequency. Due to this phenomenon electron has no constant mass, it transforms to other families as it moves through space.

Discrete energy conservation mapped within discrete space-time phases has two consecutive steps: a) re-alignment of minimum space phase, "b) accumulation of mass in space phase.

Discrete energy conservation eliminates the infinity of the energy, produced during contraction of space phase to minimum size (6) through re-alignment and coupling with the time phase. The time reversal asymmetry of space-time boundaries arises from discrete energy conservation where forward and backward directions have different initial energetic conditions. Fraction of time is needed to complete a full cycle when the minimum portion of the space phase passes through inversion frame from negative to the positive direction. The time reversal transformation of matter/antimatter particles can be described as follows:

$$\gamma/\gamma = -(e^+/e^- + \nu_e/\nu_e^-) \tag{9}$$

Formation of two pairs of particles from one pair (9) is due to the" three body" identity conservation, described by condition (9).The equation (9) describes total spin symmetry of matter-antimatter interactions. The spin of right-handed photons is equal to the sum of the two half spins of left-handed particles which hold conservation of energy, charge and momentum. The super partner of gamma rays is two pair of (four) fermions (9) which generates re-alignment of the particles in space-time reversal cycles. The ingredients of the equation (9) form the triplet spectrum generating re-alignment of light and dark photons on intermediate $\mathbf{e}^{+}/\mathbf{e}^{-}$frame.

As can be seen, without composite photon it is impossible to describe generation of mass. When neutrinos interact with the charged particles, fermions are generated but when they interact with each other "gamma ray bosons" are formed. The three families of particles participate in generations of right side" pseudo-bosons" and left side composite photons.

The mechanism of equation (9) is in agreement with the results of Stefano Profumo[9], who found that electron, reacting with the surrounding dark matter could fuse into a heavy version of electron and when it returns to the initial state radiates gamma rays.

It is necessary to note that the quantum mechanics concept of transformation between photons and $\mathbf{e}^{+}/\mathbf{e}^{-}$does not preserve energy conservation due to the non-locality of energy transformation. Light quanta do not have independent existence and cannot hold fixed amount of energy without space-time structure.

In accordance with the scheme (9), discrete inversion between light photons and neutrinos (dark photons) is to be realized through electron/positron pair playing a role of "intermediate mixing space-time frame of matter-antimatter pairs" which leads to the generation of charges. On this basis, it is clear why matter-carrying ingredient of Eigen value is negatively charged. In accordance with the scheme (9), merging and co-vibration of light and dark (neutrino) photons in mixed frame produces charges $\mathbf{e}^{-}/\mathbf{e}^{+}$. On this basis, photons and $\mathbf{e}^{-}/\mathbf{e}^{+}$pair show behavior in a similar fashion being origin of each other and light photons have no motion without alignment of electrons with the neutrinos. The inversion model (9) of the matter/antimatter transformation is different from Dirac vacuum, containing only electron/position "see".

The scheme (9) involving inversion operations describe charge and spin conservations. In accordance with the equation (9), electrons are the light photons, localized in the virtual space-time frame that is why all electrons have the same charge. When a particle form non-virtual space-time frame it became a different particle.

In accordance with the condition (9), photon has a virtual leptonic structure and the same time the pair of leptons $\mathbf{e}^{-}/\,\mathbf{e}^{+}+\nu_{e}/\,\nu_{e}^{-}$has a performance of virtual bosons (called Nambu Goldstone bosons). The sum of $\mathbf{e}^{-}/\,\mathbf{e}^{+}+\nu_{e}/\,\nu_{e}^{-}$ (9) in the form of four fermions describes the "fermionic quanta". Due to the discrete energy conservation, the amount of "bosons", distributed in space-time condensate is not unlimited and the condensate should radiate photons, similar to the "black hole" radiation of space energy.

When $\mathbf{E}_{s}$ increases, it absorbs more gamma rays and leads to the formation of "black hole structure" which cannot settle to the stationary state. From model (6) follows that at $\mathbf{E}_{act}=0$ "black hole" has to radiate its energy back to the minimum space state. The radiation of black hole energy happens with transformation of $\mathbf{e^+}/\nu_e$+$\mathbf{e^-}/\nu_e^-$ to the longitudinal wave of neutral ingredients ν_e/ν_e^-+e^+/e^-.

From model (6) and scheme (9) follows, that removal of last portion of matter from vacuum needs more energy than background state. Therefore, the boundary space-time unit participates in the coordinate inversion and grows of particles, which leads to the expansion of space.

In discrete energy conservation concept the initial and end states are vector fields. Absorption of photons by $\mathbf{e^-}/\mathbf{e^+}$+$\nu_e/\nu_e^-$ pairs leads to the transition of "bosonic" structure to the frame of $\mathbf{e^-}/\nu_e^-$+$\mathbf{e^+}/\nu_e$ virtual quarks with transformation of photons to "gluons" existing in the form of meson field (pions). We suggest that the phenomenon called "spontaneous symmetry breaking" is the change of space-time structure with transition of energy within asymmetric space-time boundaries while symmetry is the derivative of energy conservation. The consumed energy is not equivalent to the scattered energy, which leads to the asymmetric space-time boundaries. The space phase (virtual fermions) is less symmetric than time phase, therefore energy conservation has handiness in direction of energy consumption and the consumption of energy is accelerated.

The interference of space-time phases at background state (9) is not linear and coupling of the matter-anti matter particles forms intermediate loop boundary of mixed matter-anti matter frame: electron is produced from neutrino side while positron is generated from photonic side. Spin of the matter-antimatter loop to the minimum size leads to the re-alignment of the phases with generation of $\mathbf{e^-}/\nu_e^-$+$\mathbf{e^+}/\nu_e$ quark/antiguark structure. Due to the different spin for bosons and fermions, the alignment is directed to the matter side with the shift of handiness from the left to the right (consumption of energy in space phase). This is the origin why matter spin has to be twice less than bosons. Therefore, the asymmetry in spin and in the boundary of space-time variables is the origin of the matter of universe. The transformation from spin one to half spin has to lead to the energy loss but this energy is accumulated in the pion and kaon fields.

Leptons are products of virtual space-time while baryons are location of quarks in the non-virtual space-time frame, which makes radiation of matter and energy non-invariant. The quantum theories treat right and left-handed particles symmetrically which leads to the invariance of Lagrangian action. This is the problem all of physical theories. Background state (9) connects all conservation laws except asymmetry of space-time boundaries. The general principle of super symmetry and action invariance can be realized when the change of the energy of bosons, relative to its initial state, is equal to the change of the energy of under action fermions relative to the applied energy. In this case, the laws can remain the same for both systems.

Quantum mechanics considers that empty space is a dynamic medium, which is full of virtual particles where "lowest energy state of the system is not zero and an isolated particles traveling in empty space interacts with the vacuum and produces virtual particles/antiparticles pair, coming from the vacuum itself. The original particle then reappears when the particle and antiparticle meet and anhillate each other". Uncertainty principle suggests that particle/antiparticle pair live on borrowed energy within short time therefore violation of energy conservation takes place only in a short time. The question, which has to be answered here, is the nature and origin of the borrowed energy. The unanswered question of quantum physics is how to describe energy of empty space if contains any kind of energy.

In accordance with the model (6), the space-time can exist only through interaction with other space-time field. The baryonic space-time structure of matter exists in interaction with the non-baryonic matter field forming leptonic matter. Light photons move through sea of e-/v...e+/v waves with constant speed and expand this boundary through generation of new portions of dark energy. When it expands the boundary of space, (e-/v…e+/v) it leads to the formation of red shift.

The existence of neutrinos in three flavors requires the existence of light photons also in three flavors similar to the equation (9):

$$G/G= - \left(\mu^{+} / \mu^{-} + \nu_{\mu} / \nu_{\mu}^{-}\right) \tag{10}$$

$$Z/Z= - \left(\tau^{+} / \tau^{-} + \nu_{\tau} / \nu_{\tau}^{-}\right) \tag{11}$$

The **Y**, **G** and **Z** describe different flavors of photons having different frequency. At this state, the energy is not dispersive (Eigen value is minus one) and the virtual photon particles are indistinguishable.

Based on scheme (9-11) we may extend our analysis on the wave –particle duality of the light. Light is wave, but presently it is not known how it is waving. In accordance with the special relativity light travels through vacuum with constant speed but this concept does not explain what medium has a vacuum, which carries the light with the constant speed. Maxwell equations also do not involve any medium to carry light waves.

Due to the discrete energy conservation, light travels in space-time phases through polarization of wave and particle properties. In accordance with the equations (6) and (7), light does not propagate without momentum which constraints propagation of light wave in space. The space-time frame and light energy radiation are symmetric in discrete mode. In accordance with the scheme (7-9), light has virtual quark's structure and does not travel in empty space.

Connection of discrete energy conservation with the boundary mapped space-time frame explains the mystery of "increase of mass with the velocity": increase of mass with the velocity was suggested by Einstein to hold constant speed of light. Discrete energy conservation and boundary mapped space-time concept describe this problem differently: light is the discrete identity, which produces its derivative space-time frame, and any observable identity is the interaction with this frame. Therefore, light does not exist independent of space-time frame.

In accordance with the discrete energy conservation concept, it is impossible to shift space-time phases of a particle having the same spin unit of its constituents. Spin is the result of energy conservation in different phases, which leads to the generation of direction.

At discrete conservation of energy, the background state of annihilation (9) does not lead to the divergence due to the conservation of finite amount of energy within asymmetric boundary of space and time phases. The alignment of the ingredients of the scheme (9) is the polarization of neutral leptons to fermions. Coupling of γ/γ rays with the **e^-/ e^++ν_e** $^-$/ν pairs leads to the alignment of scalar field to the vector bosons which conjugates left handed neutrino with the left handed electron and right handed antineutrino with the right handed positron with formation of e^-/ν_e^- and e^+/ν quark/anti quark pairs. The alignment of three e^-/ν_e^- and e^+/ν quark/anti quark pairs in 2:1 (λ=1) mode leads to the non-integer electric charge carrying baryon's space-time structure. The energy of gamma rays in the form of "gluons" connects quark/anti quark pairs within meson field of pions. The electric charge keeps photon in the form of gluon, when charge is annihilated gluon also disappears. At decay process (E_{act}=0), the energy of gluons is consumed for re-alignment of e^-/ν_e^-, that is why electron in the resulting process appears with less energy The continuous spectrum of beta decay is due to the absence of correlation which generates left handed helicity of neutrinos. In real world due to the discrete correlation with the applied energy, the anti neutrino has right handed helicity.

Coupling of ingredients of the equation (9) at boundary loop leads to the formation of dense space particle-top quark/anti quark pair which in accordance with the model (6) leads to the energy-matter interaction with generation of virtual ingredients of baryons (**tt-+e^-/ e^++ν_e^-/ν → bb$^-$+uu$_-$**). This is the reason why top quark's mass is very close to the energy scale of alignment (symmetry breaking). Coupling of top quark with up quark takes place at **λ=1**. In this exchange interaction, the top quark is conjugated with its weak isospin partner of bottom quark. The boundary energy of top quark does not limit frequency of decay during its coupling with the up quark. Due to this function, top quark does not form baryon's space-time frame. In accordance with the scheme (9), without boundary space the alignment γ / γ→ **W^-/W^+→ TT^-**is not possible. Generation in the mixing loop (9) gives to the electron mass more than neutrino where the absorbed neutrino became the space locality of the mass-carrying electron **e^-/ν_e^-**.

Therefore, particles, which exchange energy, are bosons while particles exchanging momentum are fermions. With other definition particles carrying energy in time phase are bosons but conserving energy in space phase are fermions. The exchange of energy does not affect the wave function (Eigen state) while exchange of momentum changes the sign of the wave function. The phenomenon, which make light to move and generate discrete radiation is the discrete energy conservation within space and time phases.

9. Generation of quarks — Why the top quark does not form baryon?

The space-time frame (6) at **$E_{act=}$**0 transforms to the virtual space-time frame which by radiation of matter contracts to the minimum space size :

$$\frac{dS}{dt} = \frac{S_1}{t_1(-E_s / E_s)} \tag{12}$$

At (**$E_{act=}$0, λ=-1**) re-alignment of matter ingredients leads to the radiation of mass, consumed in the space giving the same "Femi weak coupling constant G_v "of neutron and muon decay.

The "non-correlated" space matter decay (Kaon and other mesons) generates neutral current similar to the photons. The model (6) in this case explains why correlated electromagnetic interaction of proton with electron is stronger than the reverse weak force of neutron decay to proton and electron. The alignment of the fermionic super partners $\mathbf{e^+/\nu_e{+}e^-/\nu_e^-}$ to the bosonic super partners $\mathbf{\nu_e/\nu_e^-{+}e^+/e^-}$ leads to the formation of neutral current of decay

$$\gamma^*\left(\mathbf{Gluons}\right)+\mathbf{e}^+/\nu_e+\mathbf{e}^-/\nu_e^-=-\ \gamma/\gamma+(\mathbf{e}^+/\mathbf{e}^-)+\nu_e\ /\nu_e^- \tag{13}$$

which correlates the helicity of elementary particles with the negative Eigen value.

In classic way, a matter is the quadratic product of the two fields and coupling of space and time phases gives the potential of the observable frame. This principle is similar to the quantum mechanics concept that the square modulus of wave function is associated with the probability of observing the object.

In quantum field theory, the energy of vacuum is zero, but formation of mass requires the positive sign of energy between the vacuum and the next lowest energy state.

The generation of energy as a quadratic product of coupling of opposite space-time vector fields can be described as follows

$$\left(\frac{dS}{S_1}\right)^2=\left(-\frac{dt}{t_1}\right)^2 \tag{14}$$

The quadratic quantity of space phase describes the energy density in space while the quadratic quantity of time phase describes frequency of the change of this phase. Coupling of opposite charge carrying space-time vector fields is similar to the coupling of matter-antimatter particles (6):

$$(\gamma/\gamma)^2=[-(\mathbf{e}^+/\mathbf{e}^-+\ \nu_e\ /\nu_e^-)]^2\longleftrightarrow(\gamma^*+\mathbf{e}^+/\nu_e\ +\mathbf{e}^-/\nu_e^{-)2}\ll(\mathbf{t}/\mathbf{t}^--\mathbf{bb}^-) \tag{15}$$

Coupling of matter-antimatter particles of the equations (9) is similar to the scheme (13). Here arises very important question why background space-time coupling should leads to the generation of top-bottom quarks. The top-bottom quarks are similar to other quarks but due to the high energy, they decay fast and do not form non-virtual space-time structure of baryons. Therefore, **t /t⁻-bb⁻** interactions may exist within high frequency annihilations generating in time phase **t/t⁻** particles and in space phase **bb⁻** particles with the realization of **t/t⁻-bb⁻** meson structure. But energy-momentum frame in simple background **t/t⁻-bb⁻** transitions is not invariant, therefore coupling of this pair leads to the formation of another pair of quarks, for example **uu⁻-dd⁻** interactions with exchange of vector bosons. The magnitude of the background energy (9) is fixed within boundary of space-time that is why top quark mass does not move to the infinity.

In accordance with the energy conservation (9), **d-u** transformation of quarks cannot be realized by simple emission of the W^- bosons because in this case the energy is not conserved. Therefore, energy and identity conservations in quarks transformations have to be realized through coupling of **t-b** transformations with one side with **u-d,** from other side with **s-c** quarks where the total energy and momentum are discretely conserved:

$$\mathbf{tt^- + uu^- \rightarrow dd^- + bb^-} \tag{16}$$

$$\mathbf{ss^- + bb^- \rightarrow tt^- + cc^-} \tag{17}$$

$$\mathbf{dd^- + cc^- \rightarrow uu^- + ss^-} \tag{18}$$

In accordance with the equation (9), the identity of a particle can be determined with the charged particles but electromagnetism is carried by neutral photons. On this basis the particle performance of baryon in space-time frame, is carried by charged particle that is why the electromagnetic force cannot hold alone the identity conservation of baryons.

During energy transfer (16-18), the bottom quark travels as a part of jet stream but then returns back to generate cycle as "the Feynman loop". In accordance with the conditions (16-18), conservation of causality of action is realized with correlation of end and initial states through [**T-B**] bi-mesons [(**ss⁻-bb⁻**) – (**tt⁻-uu⁻**)] which regulates discrete exchange of energy within nucleons and background polarization state.

Top meson has to be created from anti top and up quark while anti top meson forms from top quark and anti up quark. **B** meson simultaneously is produced from b-anti quark (**b-bar**) and **d**-quark. Its antiparticle **B**-anti meson is formed from **b**-quark and anti **d**-anti quark. These transformations take place at $(\mathbf{E_{act}}\text{-}\mathbf{E_s})/\mathbf{E_s}=1$. The discrete performance of two nuclear frames is the requirement of discrete identity conservation of nuclear within **uud-udd** and **ssc-ccs** nucleons coupled with the **t-b** transformations. The **ssc-ccs** play a role of isotopic nucleon.

The exchange of the two nucleons generates a field in direction $\pi\rightarrow$ Kaon$\rightarrow$ **B** mesons, which decays back at $\mathbf{E_{act}}=0$. This is the inverse spontaneous symmetry breaking. The stability of nuclear against to the repulsion by electromagnetic force is due to the discrete coupling of the recycled energy ($\boldsymbol{\lambda}$**=1**). This is the explanation why nuclear force is short ranged. Depending from Eigen value, meson field can be real or virtual particle. At none zero $\mathbf{E_{act}}$, pion is a particle, but at $\mathbf{E_{act}}=0$ it became field of gamma rays. At this condition, there is no difference between electromagnetic force and gravitation. The heavier the meson the shorter is time allowed for the exchange process therefore only asymmetry of space-time boundaries allows existing of heavy top mesons. The mystery of conventional physics why the pion should have mass is explained with the discrete energy conservation within matter-antimatter particles, located within space-time phases.

The constant coupling with the background energy resources leads to the internal isospin symmetry SU of baryon quarks existing in three flavor of space-time structure $\mathbf{E_{act}=2\ E_s}$. The baryon alone is not stable therefore cannot be the fundamental matter: two nucleons are coupled with the background bi-meson field **[T-B]** to form three flavor of space-time structure $\mathbf{E_{act}=2\ E_s}$. The spin of quarks within baryons is aligned to conserve the condition $\mathbf{E_{act}=2\ E_s}$ through quark/ant quark mesons. Separation of quark-anti quark pair is impossible due to the impossibility of separation of energy, distributed within space and time phases.

It is very interesting that the identical quarks do not obey the normal rules of quantum statistics as spin ½ particles: quarks should be fermions with anti symmetric wave functions but the pattern of observed baryons shows that they have symmetric wave function. In accordance with the quantum principles, quarks would radiate energy of interaction and dissipate their motion. Model (6) explains this phenomenon: the constant motion of quarks in baryons is in invariant coupling with the background state: at constant coupling with the background, state uniform transformation within **p-n** pairs gives to the quarks identity conservation with the symmetric wave function. At these conditions, the quarks have the same symmetric performance with the particles of background photons.

By quantum physics, it is hard to keep electric charge in a small pack because it repels itself. The problem of this concept is that electron /positron pair cannot be created from photons by two body reverse interactions $\mathbf{2Y=e^-/e^+}$due to the violation of energy conservation. The three-body model shows that the mixed "electron /positron bubble" is a composite frame (9) where the energy, momentum and spin are balanced properly. In the mixed frame the three components form the "bubble space-time" with generation of electron with negative sign while anti particle gets positive charge. Photon and electron ingredients of the background state (9) are the dynamic products of the geometric object – space-time vectors where generation of space $\mathbf{S_1}$ transforms antiparticle to particle while generation of time instant t_1 transforms particle to antiparticle. The three body flavor interaction $\mathbf{E_{act}=2\ E_s}$ with the discrete non-Noether's symmetry leads to the generation of ordered structure of baryons.

The momentum of an electron is generated from E_{act}/E_s interactions therefore there is no static mass and photon alone is not the electrodynamics force mediator.

10. Why the weak force is needed?

In accordance with the model (6), the phenomenon called parity does not describe inversion of a position of a static body through spatial coordinate but the parity of two dynamical space-time systems. Violation of parity is the result of non-invariance of the action-response of two systems, which generates direction of time: without local violation of parity, the global symmetry is not conserved and there cannot be energy conservation and direction of time. That is why P violation is associated with the **CP** violation. When the energy of the initial state is completely consumed in space phase ($E_{act}=0$), parity transformation changes the algebraic sign of the space coordinate to return the system back to the energy generation state.

In accordance with the gauge filed theory, the gauge bosons have a parity **P=-1,** but spin ½ particles have **P=+1**. The vector bosons and fermions to acquire masses the gauge invariance Lagrangian and new scalar boson Higgs field is predicted to exist. We suggest that with the non-invariance Lagrangian the intermediate **W** bosons participates in decay process and decay leptonically when λ=-1.

The model (6) connects strong and with weak interactions as the resulting quantity from action –response parity while Yukawa formula [10] of the strong force does not reveals the resulting quantity.Yukawa suggested that nuclear force could not be reduced to the electromagnetic interactions between charged particles. The same conclusion follows from Gell-Mann mechanism of quark interactions locating quarks in baryon octets within certain quantum numbers. The important conclusion of Gell-Mann [11] is that the strong force between quarks will degenerate if the strong force is the only force holding the quarks within nucleons.

In accordance with the model (4), one force cannot conserve the strong nuclear force and this force has to be uniformly coupled with the energy, recycled to the nucleons. In this case, the coupling constant does not diminish with the change of the space within the meson field.

By quantum field theory "when coupling constant is much smaller than one (**g < 1**) then the theory is said is weakly coupled. If the coupling constant is of order of one or larger the theory is said to be strongly coupled". The model (4) meets this statement of quantum field theory. When the Eigen value is equal to one the system is strongly coupled with the action. If $\mathbf{E_{act}}$ is smaller than $\mathbf{E_s}$ the system has a trend to undergo to the β-decay. This condition corresponds to the negative value of Eigen value. When Eigen value is plus one the particle's momentum, spin is aligned, and a particle is right handed. If the Eigen value is minus one the spin and momentum counter aligned, therefore the particle is left handed.

The important question here is why fermions are half integer particles and should obey Pauli principle. In accordance with the model (6), this question is related to the realization of constant action in three jet events and formation of space-time frame: $E_{act}=2E_s$. The three jet events are the result of space-time frame and one of the three jets has unique property under the strong interactions to realize coupling with the constant energy. Therefore, coupling holds the hadronization of quarks and without coupling hadrons' space-time frame transforms to the virtual manifold. That is why strangeness is conserved during creation but is not conserved in decay process. Based on the mathematical structure of the model (6), at constant coupling the quarks can never be liberated from the hadrons.

As follows from the equation (6), four fermion ingredients at $\mathbf{E_{act=}}0$ transforms to the neutral bosonic particle with total spin 2 which in the form of "graviton" leads to the conservation of the background energy.

11. Strong interactions and behind the standard model

Standard Model describes the origin of forces in term of local symmetries. But energetic field has to change discretely which leads to the discrete change of local space-time frame where the energetic radiation field ($\mathbf{E}_{act}$ /$\mathbf{E}_{s\text{-}}1$) is associated with the asymmetric local space-time field. Model (6) shows that every local space-time frame is commuted with the energetic field, which constrains each other from the linear displacement. On this basis, the space-time frame and energy field have no independent existence and their distribution is determined by discrete energy conservation through background state. This eliminates the linear effect of the Newton's force and leads to the rotational dynamical motion (spin) of a particle around its generation frame. Model (6) shows that radiation field ($\mathbf{E}_{act}$ /$\mathbf{E}_{s\text{-}}1$) and particle can be classic commutative variables which by Dirac were accepted as a commutation of quantum variables.

The space-time model based on discrete energy conservation produces symmetric and anti-symmetric dynamical functions describing the direction of decay and the nature of produced particles. In accordance with the model (6), the symmetric and anti-symmetric functions can be transformed to each other depending from energy and identity conservation laws: when the Eigen value is positive, the function leads to the hadronic transformations. When the $\mathbf{E}_{act}$=0, the function (6) being anti symmetric leads to the leptonic decay.

In accordance with the scheme (9), the alignment of the photons with the ½ spin particles ($\mathbf{e}^{+}$/ $\mathbf{e}^{-+}\nu_{e}$ / ν_{e}^{-}) moves the resulting energetic field in direction of space expansion with formation of hadronic ingredients. The resulting hadrons form "jet" along the direction of parent "electron particles".

The action-response parity of the model (6) shows that all the interactions (four forces) can be divided into two groups: two dimensional two body performance (matter –anti matter annihilation) with **SU (2)** symmetry group and three body interactions with commutation of **SU (2)** symmetry with **U(1)** group giving three body **2:1** confined resonance particles. The" two body" performance with **λ=-1** describes the interaction of virtual space-time annihilations (11) where the input function produces the opposite outcome. The total energy, distributed in **2:1** and **1:2** combinations of space-time waves generates 1/3 and 2/3 fractional frequency of waves which is displayed in the form of charges. Due to the operation of space-time waves within discrete symmetry, all electric charges have the symmetry in their unit.

Very important question, which may arise from three family 2:1 resonance performance, is Pauli Exclusion Principle, which states that two fermions such as quarks cannot occupy a quantum state at a given time. Model (6) shows that the existence of the two quarks with the symmetric wave function in classic way requires discrete correlation one of the quarks in proton-neutron baryons with the discrete conservation of the action.

Yukawa showed that [10] for some reason nuclear forces are saturated which lead to the appearance of different concepts for explanation of this phenomenon. Particularly Yukawa's meson theory appeared to explain this problem.

By prediction of the model (6), the symmetry between different spin particles such as bosons and fermions is possible if to accept that the force-carrying particle is not the single boson but is a complex "particle" having the sum of two half-order particles. On this basis the **SU (2)** discrete symmetry of proton-neutron transition (kaon meson field) is coherent with the **SU (2)** symmetry of background transitions. The mechanism of commutation of left handed doublets and right handed singlet $2E_s=E_{act}$ (6) is similar to the **SU (2) xU (1)** symmetry group of fermions. At Eigen value, one the strong interaction of the quarks is confined.

The **2:1** flavor interactions (9) show that the kaon field is necessary to realize **SU (2)** symmetry and parity conservation within proton-neutron discrete transformations to realize discrete parity conservation within **uud-duu** and **ssc-ccs** nucleons.

Here is necessary to explain why three generation of particles, particularly strange **(s)** and charm **(c)** quarks are needed in particle physics[11]. Before we showed that if, correlation of different events with the energy resources has the same coupling they are simultaneous in time having different space phase. The invariance of Eigen value needs simultaneity in different phases. Therefore, identity conservation of the nuclear requires its discrete invariance within different space phases, formed by discrete existence of **uud-duu** and **ssc-ccs** frames.

Recently was observed that quarks treat the right and left hand differently and pick out a direction in space while in empty space they treat the left and right hand without any difference. On this basis, the strong interaction should violate the parity as well. Model (6) explains this phenomenon and shows that strong interactions can hold the parity only through discrete energy conservation while in the absence of constant coupling the strong interaction has to select one-handed direction.

The discrete energy conservation and action-response parity are the necessary laws of nature to give different shapes to the different events: without discrete conservation of energy and non-invariant action all the events and bodies would form non separable mass without any shape and structure.

The action principle (6) shows that the certain frame to be stable its outcome after the change should be the same (Eigen value **λ=1**). Therefore, formation of stable non-virtual matter frame needs application of additional intermediate force. Due to the conservation of energy in discrete mode, the intermediate force is necessary to recycle the energy to the nuclear.

It is known that the square of the parity transformation is the identity conservation. Applying two parity transformations is equivalent to no transformation. Therefore, the condition (6) meets the requirement of identity conservation generating discrete invariance of action-response parity. Model (6) involves non-invariance action instead of Yang-Mills gauge Lagrangian.

Model (6) shows that at constant Eigen value quark cannot radiate energy (**$E_s < E_{act}$**). Decay of proton by radiation of energy can be realized at inversion of the universe back to the initial state (**λ=-1**). In accordance with the model (6) proton decay has to be observed globally when **E_{act}=0.** This is the mathematical proof why quarks can never be elaborated from the hadrons.

By conventional particle theory when matter and antimatter collides, they should destroy each other, leaving behind nothing but only energy. In this concept energy as the resulting quantity is not localized with the lost of matter carrying momentum. That is why matter/antimatter annihilation should produce again other matter/antimatter couples to conserve momentum and space-time frame.

The description of the " change" phenomenon by independent intervals of classic differentiation in the form of "continuous move from one point to another "gives only positive sign (such as Fermi's golden rule) and leads to the lost of discreteness and reversibility of dynamical events due to the elimination of the boundary of the initial function from the commutation. In this case, the identity conservation also is smeared out due to the absence of conjugation of the "energetic jump" with the dynamic boundary of space-time variables.

With the similar way, the second law of thermodynamics smeared out the discrete energy conservation by replacing discrete space-time dynamics by the one parametric thermodynamic arrow.

15. Conclusion

The analysis presented in our chapter shows that the concept of discrete energy/momentum conservation and their commutation within boundary mapped discrete space-time phases allow unification of the forces and interactions within unified classic field theory which completely changes our views on the fundamental interactions and symmetrical laws of nature.

The space and time are the products of discrete energy/momentum conservation and in reverse order, energy/momentum identities of antimatter/matter are the inner products of space-time discrete dynamics [16-18].

Unification of discrete energy conservation and discrete space-time dynamics has the same basis as electromagnetic unification of light as the coupling product of the space-time phases.

The principles that everything is "relative" or "uncertain " are getting different look with the concept that "every space-time identity is the discretely conserved energy/momentum packet within space and time phases" which became a new fundamental concept for description of nature and its physical laws regardless of dimensions and scale.

Model (6) shows that the non-conservation of parity is the result of non-arbitrary process of discrete energy conservation therefore the local CPT non-invariance is to be the fundamental deterministic law of nature. The discrete energy conservation treats its ordered outcomes-space and time through correlation of their asymmetric boundaries which allows energy to perform its discrete conservation within dynamic space-time framework. The non-invariance of action-response parity is more fundamental concept than unification of forces due to the generation of all the forces and events from discrete commutation of this relation.

The reality is not created by observation, as quantum mechanics suggests, but as the resulting quantity, it is created from exchange action-response interactions of space-time frame with the discrete energetic action.

Author details

Aghaddin Mamedov*

Address all correspondence to: amamedov@americas.sabic.com

SABIC Technology Center, Sugar Land, TX, United States

References

[1] P. Scarruffi. The nature of consciousness. 2006

[2] http://wikipedia. org/wiki/calculus, Access date missing (format:Accessed on dd/mm/yyyy)

[3] Dirac, P. A. M. (1958 (reprinted in 2011)). *Principles of Quantum Mechanics* (4th ed.). Clarendon. p. 255

[4] Dirac, P. A. M. (1928). "The Quantum Theory of the Electron". *Proceedings of the Royal Society A: Mathematical, Physical and Engineering Sciences* 117 (778):

[5] Vladimirov, V. S., Mikhailov, V. P., Vasharin A. A., et al., *Collection of Problems on Mathematical Physics Equations* [in Russian], Nauka, Moscow, 1974.

[6] Hazewinkel, M. Klein–Gordon equation", *Encyclopedia of Mathematics*. 2001

[7] P. Feynman. Nobel Lecture. 1965

[8] P. Pantell, H. E. Putoff. Fundamentals of Quantum Elrectrodynamics, New York, 1969

[9] S. Profumo. Lectures for TASI 2012, University of Colorado, June, 2012;ArXiv:1301. 0952v1, 5 Jan 2013

[10] H. Yukawa. Meson theory in its development. Nobel Lecture. 1949

[11] M. Gell Mann. Nobel lecture. Symmetry and currents in particle physics. 1969

[12] T. M. Liss, P. L. Tipton. The Discovery of Top Quark. Scientific American. Sep 1997

[13] L. D. Landay, and E. M. Lifshitz. Quantum Mechanics (Non-relativistic Theory) (Beijing World Publishing Corporation: P290 (1999)2.

[14] M. Koboyashi and T. Maskawa. Nobel lecture. 2005

[15] T. D. Lee. Nobel Lecture. 1957

[16] A. K. Mamedov. European Journal of Scientific Research. v. 36. No. 4. P. 570-584. 2009

[17] A. K. Mamedov. European Journal of Scientific Research. v. 42. No. 3. P. 359-384. 2010

[18] A. K. Mamedov. European Journal of Scientific Research. v. 57. No. 2. P. 314-365. 2011

Chapter 5

The Measurement Problem in Quantum Mechanics Revisited

M. E. Burgos

Additional information is available at the end of the chapter

http://dx.doi.org/10.5772/59209

1. Introduction

In a paper entitled "Against 'measurement'," J. Bell points out [1]: "Here are some words which, however legitimate and necessary in application, have no place in a *formulation* with any pretention to physical precision: *system, apparatus, environment, microscopic, macroscopic, reversible, irreversible, observable, information, measurement*... On this list of bad words... the worst of all is measurement."

To begin with, let us recall that none of the words of the previous list is included in classical theories. Classical Mechanics, Electromagnetism, Statistical Physics and other *classical theories speak about what happens, not about what is observed or measured*; they assume the behavior of apparatuses or measuring devices ruled by the same laws which govern processes where man-made objects are absent; they treat on the same footing microscopic and macroscopic objects; and they offer no room for notions such as environment or information.

The situation is radically different in quantum mechanics. Orthodox Quantum Mechanics, the theory formulated by J. von Neumann in the late 1920, consists of five axioms, and two of them refer to measurements. One of them, which is a generalization of Born's Postulate, refers to the possible results of a measurement, and their corresponding probabilities. The other one, the Projection Postulate, refers to the system's state once the measurement process is completed. In addition to these issues, present since the quantum mechanical formalism was established, in the following years other problems were unveiled and the Projection Postulate became the principal target of criticisms. In particular, it was pointed out that this postulate introduces a subjective element into the theory; it conflicts with the Schrödinger equation; and it implies a kind of action-at-a-distance.

This chapter critically reviews quantum measurement, starting with contributions by E. Schrödinger and M. Born dating from 1926. Schrödinger proposed an electromagnetic interpretation and Born a probabilistic interpretation of the wave function; the latter implies that quantum mechanics has to be considered a probabilistic theory. In 1927 Einstein objected the idea that quantum mechanics is a complete theory of individual processes. Several ways to face the measurement problem are reported and discussed, among them: Dirac's notion of observation; Bohr's point of view; von Neumann's theory of measurement; Margenau's rejection of the Projection Postulate; the Many Worlds Interpretation; and Decoherence. Brief references are made to Schrödinger cat, EPR paradox, Bell's inequalities and quantum teleportation. A comparison between the characteristics of spontaneous processes and those of measurement processes highlights why so many scientists are disappointed with Orthodox Quantum Mechanics formalism, and in particular with its Projection Postulate.

In the last sections of the chapter we deal with the following items: (i) Conservation laws are strictly valid in spontaneous processes and have only a statistical sense in measurement processes; (ii) Ad-hoc use of the Projection Postulate; (iii) Introduction of the essential concepts involved in the Spontaneous Projection Approach; and (iv) Formal treatment of the ideal measurement scheme in the framework of this approach.

2. Born's probabilistic interpretation of the wave function

The mathematical formalism of quantum mechanics was completed in 1926, the theory already exhibiting a spectacular success in accounting for nearly every spectroscopic phenomena. E. Schrödinger largely contributed to the achievement of these goals by showing that his own formalism and Heisenberg's matrix calculus are mathematically equivalent. The only major problem left seemed to be interpreting the function $\psi(x, y, z, t)$ which satisfies a wave equation (at present called the Schrödinger equation). Then, in view that *e.g.* the hydrogen atom emits electromagnetic waves whose frequency and polarization should be related to the initial and final atom states, Schrödinger thought it necessary to ascribe to the function ψ an electromagnetic character. A further elaboration of this idea led him to interpret quantum theory as a simple classical theory of waves. In his view, *physical reality consists of waves and waves only.* [2] This interpretation of quantum theory did not convince many physicists and soon several objections were raised, among them that ψ undergoes a discontinuous change during a process of measurement.

M. Born proposed also in 1926 a *probabilistic interpretation of the wave function* (sometimes called statistical interpretation of the wave function) and, as a consequence, that *quantum mechanics should be considered a probabilistic theory.* Summarizing his interpretation, one could say that $|\psi|^2\, d\tau$ measures the probability of *finding* the particle within the volume $d\tau$, the particle being as a mass point having at each instant both a definite position and a definite momentum. In 1954 Born was awarded the Nobel Price "for his fundamental work in quantum mechanics and especially for his statistical interpretation of the wave function." When explaining why he did not follow Schrödinger's interpretation of the function ψ, he pointed out that "every experiment by Franck [a physicist working close to Born's institute] and his assistants on electron collisions appeared to me as a new proof of the corpuscular nature of the electron." [3]

Born's interpretation was adopted by most leading physicists, included W. Heisenberg. In a letter sent to Born's wife on March 3, 1926, A. Einstein said: "The Heisenberg-Born concepts leave us all breathless, and have made a deep impression on all theoretically oriented people. Instead of dull resignation, there is now a singular tension in us sluggish people." [4] Einstein's conception of quantum mechanics, expressed in this letter, pleased Heisenberg and Born. A few months later, however, Einstein wrote to Born: "Quantum mechanics is certainly imposing. But an inner voice tells me that it is not yet the real thing. The theory says a lot, but does not really bring us any closer to the secret of the 'old one'. I, at any rate, am convinced that He is not playing at dice." [4] Note that *Einstein was not rejecting the probabilistic interpretation of the wave function* ψ; he was expressing dissatisfaction with loss of determinism. In his view, a probabilistic theory cannot be complete since "the real thing" should be described by a deterministic theory. His position became more explicit in the Fifth Solvay Congress (October 24 to 29, 1927).

During this congress, Born and Heisenberg presented a paper on matrix mechanics and the probabilistic interpretation of the wave function ψ. At the end of their lecture they made this provocative statement: "We maintain that *quantum mechanics is a complete theory*; its basic physical and mathematical hypotheses are not further susceptible of modifications." [2, emphases added] During the discussion which followed, H. Lorentz objected the rejection of determinism in atomic physic, as proposed by the majority of speakers. Although admitting that Heisenberg's indeterminacy relations impose a limitation on observation, he objected the notion of probability as an axiom a priori, at the beginning of the interpretation, instead of putting it at the end, as a conclusion of theoretical considerations. Finally, he declared: "*Je pourrais toujours garder ma foi déterministe pour les phénomènes fondamentaux… Est-ce qu'un esprit plus profond ne pourrait pas se rendre compte des mouvements de ces électrons? Ne pourrait-on pas garder le déterminisme en faisant l'objet d'une croyance? Faut-il nécessairement exiger l'indéterminisme en principe*?" [2]

After the intervention of other speakers, A. Einstein intervened to point out that the theory of quanta may be considered from two different viewpoints. To illustrate his assertion he referred to the following experiment; see Figure: "A particle (photon or electron) impinges normally on a diaphragm with slit O so that the ψ-wave associated with the particle is diffracted in O. A scintillation-screen… in the shape of a hemisphere is placed behind O so as to show the arrival of a particle…" [2] Then Einstein asserted:

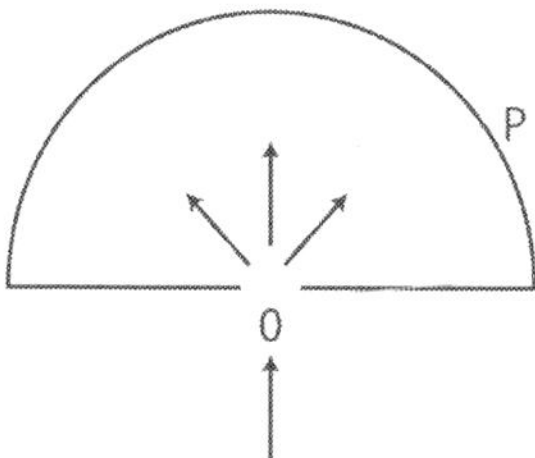

Figure 1. The experiment analyzed by Einstein in the Fifth Solvay Congress

According to viewpoint I, the... waves do not represent one individual particle but rather an ensemble of particles distributed in space. Accordingly, the theory provides information not on an individual process but rather on an ensemble of them. Thus $|\psi(r)|^2$ expresses the probability (probability density) that there exists at r *some* particle of the ensemble [*une certain particule du nuage*].

According to viewpoint II quantum mechanics is considered as a complete theory of individual processes... each particle moving toward the screen in the shape of a hemisphere is described as a wave packet which, after diffraction at O arrives at a certain point P on the screen, and $|\psi(r)|^2$ expresses the probability... that at a given moment one and the same particle shows its presence at r... If $|\psi|^2$ is interpreted according to II, then, as long as no localization has been effected, the particle must be considered as potentially present with almost constant probability over the whole area of the screen; however, as soon as it is localized, a peculiar action-at-a-distance must be assumed to take place which prevents the continuously distributed wave in space from producing an effect at *two* places on the screen.

"It seems to me"-Einstein concluded-"that this difficulty cannot be overcome unless the description of the process in terms of the Schrödinger wave is supplemented by some detailed specification of the localization of the particle during its propagation... If one works only with Schrödinger waves, the interpretation II of $|\psi|^2$, I think, contradicts the postulate of relativity." [2]

Many years after this memorable meeting Born and Einstein continued to discuss about this subject. Their divergence is illustrated in Einstein's letter to Born dated September 7, 1944: "We have become Antipodean in our scientific expectations. You believe in the God who plays dice, and I in complete law and order in a world which objectively exists, and which I, in a wildly speculative way, am trying to capture. I firmly believe, but I hope that someone will discover a more realistic way, or rather a more tangible basis than it has been my lot to find. Even the great initial success of the quantum theory does not make me believe in the fundamental dice-game, although I am well aware that our younger colleagues interpret this as a consequence of senility. No doubt the day will come when we will see whose instinctive attitude was the correct one." [4]

Original Born's probabilistic interpretation of the wave function ψ enjoyed great success in the analysis of atomic scattering, but soon it was evident that it confronted serious difficulties to explain other phenomena. [2] In any case, a generalization of Born's probabilistic interpretation of the wave function ψ was included by J. von Neumann as the third postulate of his quantum mechanics formulation; see next section.

3. The formalism of orthodox quantum mechanics

Von Neumann formulated quantum mechanics as an operator calculus in Hilbert space; the German version of his "Mathematical Foundations of Quantum Mechanics" was published for the first time in 1932. [5] A couple of years earlier P. A. M. Dirac published his celebrated treatise "The Principles of Quantum Mechanics." [6] The essential of the theory was presented there and, even though von Neumann admitted that Dirac's formalism could 'scarcely be surpassed in brevity and elegance,' he criticized it as deficient in mathematical rigor." [2] Many other versions of quantum mechanics followed these pioneer works, most of them motivated by the desire of solving "the measurement problem." But, in general, von Neumann's formulation continued to be preferred to other approaches and, at present, it is frequently the only one taught at the academy. We shall refer to it as Orthodox Quantum Mechanics (OQM).

The primitive (undefined) notions of OQM are *system, physical quantity* and *state*; and the formalism can be summarized in the following way [2]:

a. To every system corresponds a Hilbert space H whose vectors (state vectors, wave functions) completely describe the states of the system.

b. To every physical quantity A corresponds uniquely a self-adjoint operator A acting in H. It has associated the eigenvalue equations

$$\mathbf{A}\left|\mathbf{a}_j^v\right\rangle = \mathbf{a}_j\left|\mathbf{a}_j^v\right\rangle \tag{1}$$

(ν is introduced in order to distinguish between the different eigenvectors that may correspond to one eigenvalue a_j), and the closure relation

$$\sum_{j,v}\left|\mathbf{a}_j^v\right\rangle\left\langle\mathbf{a}_j^v\right| = I \tag{2}$$

is fulfilled (here I is the identity operator). If j or ν is continuous, the respective sum has to be replaced by an integral.

c. For a system in the state $|\Phi\rangle$ the probability that the result of a measurement of A lies between a' and a'' is given by $\|\Psi\|^2$, where $\|\Psi\|$ is the norm of $|\Psi\rangle = (I_{a''} - I_{a'})|\Phi\rangle$ and I_a is the resolution of identity belonging to A.

d. The evolution in time of the state vector $|\Phi\rangle$ is determined by the Schrödinger equation.

e. Projection Postulate: If a measurement of A yields a result between a′ and a″, then the state of the system immediately after the measurement is an eigenfunction of $(I_{a''} - I_{a'})$.

Many prominent authors of quantum mechanics textbooks adopt the primitive notions *system, physical quantity* and *state*, either in an explicit or implicit way. They also take as valid *the first four postulates* of the previous formalism with little or none modification. The exact formulations of these axioms due to some conspicuous authors can be found in: [7-10].

4. The measurement problem and the statistical interpretation of quantum mechanics

The problem pointed out by Einstein at the Fifth Solvay Congress has been considered one of the most serious flaws that quantum mechanics confronts. Some years later, in 1935, he published with B. Podolsky and N. Rosen their celebrated paper "Can Quantum-Mechanical Description of Physical Reality Be Considered Complete?" [11] This article prompted H. Margenau to consider the Projection Postulate as indicative of a defect in the formalism of quantum mechanics and to suggest that it should be abandoned [2]; one of the main reasons to do so being that this postulate contradicts the more fundamental Schrödinger equation of motion. As an example, Margenau considered the measurement of the coordinate of a particle which initially has a definite momentum and argued: as the value of the position (and then the state ψ after the measurement) cannot be predicted, the Hamiltonian of interaction between the particle and the measuring device cannot be a unique operator as usually encountered in the formalism.

In the following we reproduce his argument in case the operator A representing the physical quantity A to be measured has a discrete non-degenerate spectrum, the eigenfunctions of A being $\psi_j (j=1, 2, \cdots)$. Let t be the time the measurement process starts and $t+\Delta t$ the time such a process is over. We shall call H_0 the Hamiltonian of the particle before t, H_M the term due to its interaction with the measuring device and $H=H_0+H_M$ the total Hamiltonian acting on the particle in the time interval $(t, t+\Delta t)$. If φ is the state of the particle at t, assuming that during the time interval $(t, t+\Delta t)$ the Schrödinger equation rules the process, the state of the particle at $t+\Delta t$ should be $\psi=\varphi+\Delta\varphi$, where

$$\Delta\varphi = \left(\Delta t\, H / i\hbar\right)\varphi \tag{3}$$

and

$$\psi = \left\{1+\left[\Delta t\left(H_0+H_M\right)/ i\hbar\right]\right\}\varphi \tag{4}$$

($\hbar$being Planck's constant). Now, on the one hand the several possible states of the particle immediately after the measurement has been completed should be one of the ψ_j; the uncontrollable character of the measurement process implies that it is not possible to predict which one of them will result. But, on the other hand, ψ given by (4) is just one function, whatever the specific form of H_M may be.

Margenau's suggestion to abandon the Projection Postulate and the arguments which support this idea were included in a manuscript he sent to Einstein on November 13, 1935. Einstein, however, replied: "the formalism of quantum mechanics requires inevitable the following postulate: 'If a measurement performed upon a system yields a value *m*, then the same measurement performed immediately afterwards yields again the value *m* with certainty.' He illustrated this postulate by the example of a quantum of light which, if it has passed a polarizer P_1, is known to pass with certainty a second polarizer P_2 with orientation parallel to P_1." [2]

According to Einstein a particle should always be considered as possessing a definite though perhaps unknown position, even when no such definite position is described by the wave function ψ. [12] In his "Reply to Criticism," he asserts: "One arrives at very implausible theoretical conceptions if one attempts to maintain the thesis that the statistical quantum theory is in principle capable of producing a complete description of an individual physical system. On the other hand, those difficulties of theoretical interpretation disappear if one views the quantum-mechanical description as the description of ensembles of systems." [13] It is not surprising that Einstein, as Margenau, uphold the Statistical Interpretation of Quantum Mechanics (SIQM). According to this approach, "*a pure state* Φ (and hence also a general state) provides a description of certain statistical properties of an *ensemble* of similarly prepared systems, but need not provide a complete description of an individual system." [14] By contrast, Postulate A of OQM explicitly establishes that a pure state Φ completely describes the state of an individual system. *We have then two versions of quantum mechanics which do not deal with the same referent: OQM refers to individual systems and SIQM to ensembles of similarly prepared systems.* In addition: As already mentioned, in classical theories measurement processes are supposed to be ruled by the same laws which govern spontaneous processes. By contrast, in OQM spontaneous processes follow Schrödinger evolutions (given by Postulate D) and measurement processes are ruled by the Projection Postulate (Postulate E). *This is a very important difference between OQM and classical theories.* SIQM avoids this difference by adopting formalism where no mention to measurement is made. In particular, instead of Postulate C (see Section 3), SIQM states: "The only values which an observable [physical quantity or dynamical variable represented by a self-adjoint operator] *may take on* are its eigenvalues..." [14; emphasis added] *So, where OQM talks about the possible results of a measurement, SIQM talks about the values which an observable may take on. SIQM is a theory* which has a referent and postulates different from OQM; in particular, SIQM and OQM *include different generalizations of Born's probabilistic interpretation of the wave function.* By contrast, *Born's probabilistic interpretation of the wave function is not a theory.* This is a difference between Born's probabilistic interpretation of the wave function and SIQM worth to be stressed.

D. Bes points out: "rather than dwell on philosophical interpretations of equations, most physicists proceed to carry out many exciting applications of quantum mechanics." [10] On their side, M. Tegmar and J. A. Wheeler argue: "This approach proved stunningly successful. Quantum mechanics [we would say OQM or similar versions of quantum mechanics] was instrumental in predicting antimatter, understanding radioactivity (leading to nuclear power), accounting for the behavior of materials such as semiconductors, explaining superconductivity and describing interactions such as those between light and matter (leading to the invention of the laser) and of radio waves and nuclei (leading to magnetic resonance imaging)." [15] By contrast, SIQM has not been so successful. This is, we think, the main reason why *most physicists and chemists prefer versions of quantum mechanics which refer to individual systems and, in particular, OQM.*

5. Von Neumann's theory of measurement

Von Neumann's formalism is generally based on the so-called Copenhagen Interpretation whose founding father was mainly N. Bohr; even if E. Schrödinger, W. Heisenberg, M. Born and other physicists also made very important contributions to this interpretation. But Bohr did not care about the absence of quantum mechanics formalism, let alone a theory of measurement. In his view, any formalism would become meaningful only if it was possible to interpret it in terms of classical concepts. So he assigned a double nature to the measuring apparatus: on the one hand it should behave as a classical object; on the other hand, it should follow quantum mechanical laws. As a result, his position remained a somewhat questionable or, at least, obscure. [2]

Contrary to Bohr, von Neumann treated the measuring apparatus as a purely quantum system. His starting point was the assumption that "there are two kinds of changes of quantum mechanical states: (1) 'the discontinuous, non-causal and instantaneously acting experiments or measurements,' which he called 'arbitrary changes by measurements'; and (2) 'continuous and causal changes in the course of time,' which evolve in accordance with the equations of motion and which he called 'automatic changes.' The former, or briefly, 'processes of the first kind' are irreversible whereas the latter, the 'processes of the second kind,' are reversible." [2]

In the first place von Neumann showed how the formalism of quantum mechanics is capable of accounting consistently for the operation of the measuring apparatus. Then he continued his analysis by regarding the measurement process as consisting of two stages: (I) the interaction between the object and the apparatus, and (II) the act of observation. In the following we summarize his arguments.

Stage I does not present any conceptual difficulty. Let S be the operator representing the physical quantity S to be measured. We assume that it has the eigenvalues $s_j(j=1, 2, \cdots)$ and the corresponding eigenvectors are σ_j (for simplicity we refer to the discrete non-degenerate case). If before the interaction of the system (object) with the apparatus the pure state of the system is

$$\sigma = \sum_n c_n \sigma_n \tag{5}$$

during the interaction the object is coupled to the measuring apparatus designed to measure S and, once the interaction ceased, the system+apparatus is in the state

$$\psi = \sum_n c_n \sigma_n \alpha_n \tag{6}$$

where α_n would be the apparatus state if the system state before the interaction were σ_n. Let us stress that ψ, being causally determined, is a pure state as long as the combined system +apparatus remains isolated.

Stage II. The conceptual difficulties concerning the measurement problem become apparent at this stage. "Von Neumann was fully aware that the knowledge of the state of the combined system does not suffice to infer the state of the object or the value of S. If it could be ascertained that after the interaction the apparatus is in the state α_j, it would be known that the object is in the state α_j and S has the value s_j. But how can we find out whether the apparatus is in the state α_j ? It may be suggested that one couple the apparatus to a second measuring device. This proposal, however, would lead to an infinite regress... But clearly, von Neumann reasoned, a measurement must be a finite operation; usually it is completed by an act of observing the pointer position of [the apparatus]. The process leading to this result, von Neumann concluded, can therefore no longer be of the second kind but has to be a discontinuous, non-causal, and instantaneous act." [2]

Where and how does this act take place? In our view, von Neumann's answer to this question is not satisfactory. In his Mathematical Foundations of Quantum Mechanics he asserts [16]: "We must always divide the world into two parts, the one being the observed system, the other the observer. In the former, we can follow up all physical processes (in principle at least) arbitrarily precisely. In the latter, this is meaningless. The boundary between the two is arbitrary to a very large extent... That this boundary can be pushed arbitrarily deeply into the interior of the body of the actual observer is the content of the principle of the psycho-physical parallelism – but this does not change the fact that in each method of description the boundary must be put somewhere... Now quantum mechanics describes the events which occur in the observed portion of the world, so long as they do not interact with the observing portion, with the aid of the process [of the second kind], but as soon as such interaction occurs, *i.e.* a measurement, it requires the application of [a] process [of the first kind]."

M. Jammer points out that "this argument for the indispensability of processes of the first kind also seems to suggest that these processes do not occur in the observed portions of the world, however deeply in the observer's body the boundary is drawn. They can thus occur only in his consciousness. *A complete measurement, according to von Neumann's theory, involves therefore the consciousness of the observer.*" [2; emphases added]

6. Entangled states: The Schrödinger cat and the EPR paradox

The state ψ given by (6) is an *entangled* state where each term in the sum is the product of a possible state σ_n of a microsystem, the corresponding final state α_n of the apparatus, and the numberc_n. So, as long as the total system+apparatus remains isolated, we have a linear superposition of different states of the apparatus, the coefficients being$c_n\sigma_n$.

Entangled states do not have a classical equivalent and are an unavoidable consequence of the superposition principle, considered by some physicists the fundamental principle of quantum mechanics. Schrödinger showed how strange some entangled states are with his well-known example of the cat: Imagine that the microsystem is a radioactive element with two possible states: σ_1 (atom non-decayed), and σ_2 (atom decayed). If the atom decays, a mechanism is activated and kills the cat, a macrosystem with the possible states α_1 (cat alive, if the atom has not decayed) and α_2 (cat dead, if the atom has decayed). Then, if at a given instant the probability the radioactive element has of being decayed is ½, the coefficients take on the value $c_2 = c_1 = \sqrt{(½)}$ and

$$\psi = \sqrt{(½)}\sigma_1\alpha_1 + \sqrt{(½)}\sigma_2\alpha_2 \tag{7}$$

This entangled state is a superposition in which the two states "cat alive" and "cat dead" are mixed or smeared together by equal amounts. Following von Neumann, one should say that only through the act of observation, that is, looking at the cat, the system is thrown into a definite state. On his hand, Schrödinger asserts [2]: "*states of a macroscopic system which could be told apart by a macroscopic observation are distinct from each other whether observed or not.*" So, in his view, "it would be naïve to consider the ψ-function in (7) as depicting the reality." [2]

In [11] Einstein, Podolsky and Rosen demonstrate that the idea that "the wave function *does* contain a complete description of the physical reality of the system in the state to which it corresponds... together with the criterion of reality [see below] leads to a contradiction." Referring to this paper, frequently people speak of EPR paradox, but in fact one should talk about the EPR theorem. It states: "*if the predictions of quantum mechanics are correct (even for systems made of remote correlated particles) and if physical reality can be described in a local (or separable) way, then quantum mechanics is necessarily incomplete: some elements of reality exists in Nature that are ignored by this theory.*" [17]

At a first glance the Schrödinger cat and the EPR paradox look very different. Nevertheless, they share the conceptual problem implied in entangled states and, in this sense, it could be said that they are variations on the same theme. To pin point what we mean, let us suppose that instead of having a particle and a measuring apparatus (as in the previous section), or a radioactive element and a cat (as in the previous example) we have two spin ½ particles which propagate in opposite directions after leaving the source where they have been emitted in a singlet spin state

$$|\Psi\rangle = \sqrt{(½)}|1:+,\ 2:-\rangle - \sqrt{(½)}|1:-,\ 2:+\rangle \tag{8}$$

In this entangled state "spin up" and "spin down" of particle 1 are mixed or smeared together by equal amounts. The same is valid for particle 2.

Let us suppose that after leaving the source every interaction between both particles ceases. Then, if one of them is submitted to a measurement of spin in a direction orthogonal to that of propagation, quantum mechanics tells us that a measurement of spin in the same direction (orthogonal to the direction of propagation) upon the other particle will yield the opposite value to that obtained in the first measurement: for instance, if the first result is $+\hbar/2$, the other will be $-\hbar/2$ *with certainty*, and this must happen independently of the distance between both

http://dx.doi.org/10.5772/59209

particles. So, by measuring the spin of the particle going to one side, *e.g.* particle 1, it is possible to know the spin of the particle going to the other side, *i.e.* particle 2, without performing any measurement upon it or disturbing it in any way.

Now, on the one hand the EPR criterion of reality states [11]: "If, without in any way disturbing a system, we can predict with certainty (i.e., with probability equal to unity) the value of a physical quantity, then there exists an element of physical reality corresponding to this physical quantity." And, on the other hand, according to the condition of completeness formulated by EPR [11]: "every element of the physical reality must have a counterpart in the physical theory." So, applying the EPR criterion of reality we can conclude that the spin of the non-disturbed particle (particle 2), which has for instance the value $-\hbar/2$, is an element of reality. But this supposed element of reality is, however, absent from the state (8) where "spin up" and "spin down" of particle 2 are mixed or smeared together by equal amounts.

Einstein, Podolsky and Rosen end their article with the assertion [11]: "While we have thus shown that the wave function does not provide a complete description of the physical reality, we left open the question of whether or not such a description exists. We believe, however, that such a theory is possible."

7. Controversies about the projection postulate and the theory of measurement

Most authors agree on the following point: neither the primitive notions nor the first four postulates of OQM are controversial. But this is the case neither of the Projection Postulate nor of the Theory of Measurement. The list of authors who have tried to solve the measurement problem in quantum mechanics is very long. In the following we shall sum up and comment the points of view of a few of them.

7.1. Dirac's notion of observation

Referring to an experiment with a single obliquely polarized photon incident on a crystal of tourmaline, Dirac says: "When we make the photon meet a tourmaline crystal, we are subjecting it to an observation. We are observing whether it is polarized parallel or perpendicular to the optic axis. *The effect of making this observation is to force the photon entirely into the state of parallel or entirely into the state of perpendicular polarization*. It has to make a sudden jump from being partly in each of these two states to being entirely in one or the other of them. Which of the two states it will jump into cannot be predicted, but is governed only by probability laws. If it jumps into the parallel state it gets absorbed and if it jumps into the perpendicular state it passes through the crystal and appears on the other side preserving this state of polarization." [6; emphases added]

Our comments: Dirac seems to suggest that these jumps (or projections), even if induced by observations, happen in the real, material world. We ask: what would happen if a photon polarized by nature (i.e. without the intervention of humans) meets a tourmaline crystal?

Would it jump from being partly in each of these two states to being entirely in one or the other? Or would it remain in an entangled state with the tourmaline crystal like that of the system+apparatus given by (6)?

7.2. Landau and Lifshitz' point of view

Following Bohr, L. Landau and E. Lifshitz deal with the measurement problem in the following terms [18]: "The possibility of a quantitative description of the motion of an electron requires the presence also of physical objects which obey classical mechanics to a sufficient degree of accuracy [for brevity the authors speak here of 'an electron,' meaning in general any object of a quantum nature, i.e. a particle or system of particles obeying quantum mechanics and not classical mechanics]. If an electron interacts with such a 'classical object', the state of the latter is, generally speaking, altered... In this connection the 'classical object' is usually called *apparatus*, and its interaction with the electron is spoken of as *measurement*. However, it must be emphasized that we are not discussing a process of measurement in which the physicist-observer takes part. *By measurement*, in quantum mechanics, we understand any process of interaction between classical and quantum objects, occurring apart from and independently of any observer."

They further add [18]: "[Let us] consider a system consisting of two parts: a classical apparatus and an electron (regarded as a quantum object). The process of measurement consists in these two parts coming in interaction with each other, as a result of which the apparatus passes from its initial state into some other; from this change of state we draw conclusions concerning the state of the electron. The states of the apparatus are distinguished by the values of some physical quantity (or quantities) characterizing it – the 'readings of the apparatus'. We conventionally denote this quantity by g, and its eigenvalues by g_n ··· [we shall] suppose the spectrum discrete. The states of the apparatus are described by means of quasi-classical wave functions which we shall denote by $\Phi_n(\xi)$, where the suffix n corresponds to the 'reading' g_n of the apparatus, and ξ denotes the set of its coordinates. The classical nature of the apparatus appears in the fact that, at any given instant, we can say with certainty that it is in one of the known states Φ_n with some definite value of the quantity g; for a quantum system such an assertion would, of course, be unjustified."

It follows the description of an analogous process to that mentioned in Stage I of von Neumann's theory of measurement: If $\Phi_0(\xi)$ is the wave function of the initial state of the apparatus (before the measurement), and $\Psi(q)$ some arbitrary normalized initial wave function of the electron (q denoting its coordinates), the initial wave function of the whole system is the product

$$\Psi(q)\,\Phi_0(\xi) \tag{9}$$

Then, applying the equations of quantum mechanics, we can in principle follow the change of the total system wave function with time. The measurement process finished, we can expand this wave function in terms of the Φ_n and obtain the sum

$$\sum_n A_n(q)\Phi_n(\xi) \tag{10}$$

where the $A_n(q)$ are some functions of q.

At this point Landau and Lifshitz assert [18]: "The classical nature of the apparatus, and the double role of classical mechanics as both the limiting case and the foundation of quantum mechanics, now make their appearance. As mentioned above, the classical nature of the apparatus means that, at any instant, the quantity g (the 'reading of the apparatus') has some definite value. This enables us to say that the state of the system apparatus+electron after the measurement will in actual fact be described, not by the entire sum (10), but by only the one term which corresponds to the 'reading' g_n of the apparatus,

$$A_n(q)\ \Phi_n(\xi) \tag{11}$$

It follows from this that $A_n(q)$ is proportional to the wave function of the electron after the measurement."

Our comments: We have already pointed out that Stage I of the measurement process does not involve any conceptual difficulty. In addition, there is no substantial difference between the analysis due to Landau and Lifshitz, which leads to sum (10), and that due to von Neumann, which leads to sum (6).The problem arises in Stage II, where the reduction to one and only one term of sum (6) or of sum (10) must be achieved. This problem is faced in different ways by different authors: von Neumann, for whom the apparatus is a purely quantum system, makes appeal to observer's consciousness; Landau and Lifshitz, for whom quantum measurements occur apart from and independently of any observer, make appeal to the classical character of the apparatus.

7.3. Bunge's epistemological realism

In [19] M. Bunge asserts: "The main epistemological problem about quantum theory is whether it is compatible with epistemological realism. (The latter is a family of epistemologies which assume that (a) the world exists independently of the knowing subject, and (b) the task of science is to produce maximally true conceptual models of reality…)"

On the one hand, in Bunge's view the question of reality has nothing to do with scientific problems such as whether all properties have sharp values, and whether all behavior is causal. On the other hand, he thinks the Schrödinger equation rules every quantum process. Then, when referring to projections, he says [19]: "we would like to see a rigorous *proof* that the projection, or something close to it, occurs partly as a consequence of the Schrödinger equation, not as a result of an arbitrary decision of an omnipotent Observer placed above the laws of nature. More precisely, we should like to derive a projection (or semi-projection) *theorem* from physical (quantum and classical) first principles. And we should like to have a proof that the projection (or semi-projection) is *a swift but not instantaneous process caused by certain interactions,* in particular those between quanton and apparatus." And in [20]: "one should attempt to *deduce* the reduction of the state function instead of *postulating* it."

Our comments: In [21] and [22] we have asserted that quantum theory is compatible with realism. And we also think that the question of reality has nothing to do with scientific problems such as whether all properties have sharp values or not and whether all behavior is causal or not. We fully agree with Bunge on these points. Nevertheless, we would add to the list of scientific problems which have nothing to do with the question of reality: the issue of action-at-a-distance and the validity of conservation laws, in particular conservation of energy. Concerning this last point, H. Poincaré declares: *"[cette loi] ne peut avoir qu'une signification, c'est qu'il y a une propriété commune à tous les possibles; mais dans l'hypothèse déterministe il n'y a qu'un seul possible et alors la loi n'a plus de sens. Dans l'hypothèse indéterministe, au contraire, elle en prendrait un, même si on voulait l'entendre dans un sens absolue… "* [23] This remark seems to us pertinent for, if there are quantum processes not ruled by deterministic laws, one could suspect that conservation laws are not valid in these kinds of processes.

Now, concerning Bunge's suggestion: it would be grateful to see the Projection Postulate deduced from the Schrödinger equation; the problem is to know whether achieving this task is possible or not. In 1935 Margenau showed that the Projection Postulate contradicts the more fundamental Schrödinger equation of motion; see Section 4. And according to Bes, "because of the linearity of the Schrödinger evolution, there is no mechanism to stop the evolution and yield a single result for the measurement: *the state reduction is beyond the scope of the Schrödinger evolution."* [10; emphases added] So, as long as these assertions have not proven wrong, we do not see in which way somebody could be inspired to face the task Bunge proposes us.

7.4. The many worlds interpretation

In [24] H. Everett proposes an alternative to observation-triggered wave. He assumes that the equations of physics that model the time evolution of systems without observers are sufficient for modeling systems which do contain observers. As a result, the universe which includes the system, the measuring apparatus and the observer, always evolves in agreement with the Schrödinger equation, even when the observer performs a measurement. In this approach the system+apparatus+observer+environment splits into as many branches as results of the measurement are possible. All possibilities are realized at the same time and these branches coexist without interfering, so the component of the observer in one branch is unaware of the others, and he/she perceives what happens as if the system state has been projected. But this is a delusion of the mind of the observer for "there does not exist anything like a single state for one subsystem…" [24]

Everett originally called his approach the "Correlation Interpretation," where correlation refers to entanglement, as that obtained at the end of Stage I of von Neumann's theory of measurement; see sum (6) in Section 5 and sum (10) in Section 7.2. The phrase "many-worlds" is due to B. DeWitt, who was responsible for the wider popularization of Everett's theory. [25]

Our comments: Since each component of the observer is condemned to remain in his/her branch there is no way he/she could know what is happening in the others. As a consequence, Bunge says, "this solution to the contradictions generated by the orthodox version of the projection hypothesis is unscientific because the splitting is unobservable, so the conjecture is untestable." [19] We fully share this assertion.

7.5. Decoherence

Decoherence is a process which prevents different elements in the quantum superposition of the total system's wave function from interfering with each other. So, it has been said, "it looks and smells as a collapse." [15]

W. Zurek, one of its conspicuous defenders, introduces the concept of decoherence in the following way [26-27]: Let $|\uparrow\rangle$ and $|\downarrow\rangle$ be the orthonormal states of a particle of spin ½ in interaction with a detector whose orthonormal states are $|d_\uparrow\rangle$ and $|d_\downarrow\rangle$. If the detector begins in the $|d_\downarrow\rangle$ state and "clicks," $|\uparrow\rangle|d_\downarrow\rangle \rightarrow |\uparrow\rangle|d_\uparrow\rangle$ when the spins are in the state $|\uparrow\rangle$ but remains unchanged otherwise.

If before the interaction the particle is in the pure state $|\psi_S\rangle = \alpha|\uparrow\rangle + \beta|\downarrow\rangle$, the composite system SD (system+detector) starts as $|\Phi^i\rangle = |\psi_S\rangle|d_\downarrow\rangle$ and the interaction results in the evolution of $|\Phi^i\rangle$ into the correlated state $|\Phi^c\rangle$

$$|\Phi^i\rangle = (\alpha|\uparrow\rangle + \beta|\downarrow\rangle)|d_\downarrow\rangle \rightarrow \alpha|\uparrow\rangle|d_\uparrow\rangle + \beta|\downarrow\rangle|d_\uparrow\rangle = |\Phi^c\rangle \quad (12)$$

The corresponding pure state density matrix is

$$\begin{aligned}\rho^c &= |\Phi^c\rangle\langle\Phi^c| \\ &= |\alpha|^2 |\uparrow\rangle\langle\uparrow||\mathbf{d}_\uparrow\rangle\langle\mathbf{d}_\uparrow| + \alpha\beta^* |\uparrow\rangle\langle\downarrow||\mathbf{d}_\uparrow\rangle\langle\mathbf{d}_\downarrow| \\ &+ \alpha^*\beta|\downarrow\rangle\langle\uparrow||\mathbf{d}_\downarrow\rangle\langle\mathbf{d}_\uparrow| + |\beta|^2|\downarrow\rangle\langle\downarrow||\mathbf{d}_\downarrow\rangle\langle\mathbf{d}_\downarrow|\end{aligned} \quad (13)$$

As happened in the analyses performed by von Neumann and by Landau and Lifshitz, this first stage of the detection process is a Schrödinger evolution which does not involve any conceptual difficulty.

Now, there are two branches of the detector state in this correlated state $|\Phi^c\rangle$, but we know the alternatives are distinct outcomes rather than a mere superposition of states. Nevertheless, cancelling the off-diagonal terms, which express quantum correlations, the reduced density matrix results:

$$\rho^{r} = |\alpha|^{2}|\uparrow\rangle\langle\uparrow||d_{\uparrow}\rangle\langle d_{\uparrow}| + |\beta|^{2}|\downarrow\rangle\langle\downarrow||d_{\downarrow}\rangle\langle d_{\downarrow}| \tag{14}$$

In Zurek's view, the key advantage of ρ^{r} over ρ^{c} is that its coefficients $|\alpha|^{2}$ and $|\beta|^{2}$ may be interpreted as classical probabilities. Unitary evolution condemns every closed quantum system to 'purity.' Yet if the outcomes of a measurement are to become independent, with consequences that can be explored separately, a way must be found to dispose of the excess of information (contained in the off-diagonal terms). This disposal can be caused by interaction with the degrees of freedom external to the system, which we shall summarily refer to as the 'environment'… [26]

Following the first step of the measurement process –establishment of the correlation as shown in (12)– the environment E initially in the state $|\varepsilon_0\rangle$, becomes correlated with SD (system +detector):

$$\begin{aligned} &|\Phi^{c}\rangle|\varepsilon_{0}\rangle = (\alpha|\uparrow\rangle|d_{\uparrow}\rangle + \beta|\downarrow\rangle|d_{\downarrow}\rangle)|\varepsilon_{0}\rangle \\ &\rightarrow \alpha|\uparrow\rangle|d_{\uparrow}\rangle|\varepsilon_{\uparrow}\rangle + \beta|\downarrow\rangle|d_{\downarrow}\rangle|\varepsilon_{\downarrow}\rangle = |\Phi\rangle \end{aligned} \tag{15}$$

with obvious notation. "This final state extends the correlation beyond the system-detector pair. When the states of the environment corresponding to spin up and spin down states of the detector are orthogonal, we can take the trace over the uncontrolled degrees of freedom to get the same results as the reduced matrix." [26] The density matrix that describes the detector-system combination obtained by ignoring (tracing over) the uncontrolled (and unmeasured) degrees of freedom is

$$\rho_{SD} \equiv Tr_{E}|\Phi\rangle\langle\Phi| = \sum_{i}\langle\varepsilon_{i}|\Phi\rangle\langle\Phi|\varepsilon_{i}\rangle = \rho^{r} \tag{16}$$

which coincides with the reduced matrix given by (14).

It has been proven that for large classical objects decoherence would be virtually instantaneous because of the high probability of interaction of such systems with some environmental quantum. A quantitative model due to Zurek [26] illustrates the gradual cancellation of the off-diagonal elements with decoherence over time.

Our comments: If we want to describe processes ruled by the Schrödinger equation, disposing of terms which give an account for something that is happening is not a good idea. When SD is coupled to the environment E, the Schrödinger evolution leads the total system SDE to the pure state $|\Phi\rangle$, *it does not lead the SD system to the mixture* ρ^{r}. In addition, the *mixture* ϱ^{r} is *unique* and completely different from the SD *pure states* $|\psi_{\uparrow}\rangle = |\uparrow\rangle|d_{\uparrow}\rangle$ and $|\psi_{\downarrow}\rangle = |\downarrow\rangle|d_{\downarrow}\rangle$ which are, according to the Projection Postulate, the *only two* possible final states of SD. So, in our view decoherence does not provide a solution to the measurement problem. In [28] we have advanced similar arguments to object the contributions of Griffith, Gell-mann, Hartle and Omnès.

Other authors have criticized the solution to the measurement problem which involves decoherence. In particular, in [29] it is asserted that to obtain ρ^r "... an appeal has been made that goes beyond the ordinary Schrödinger equation, to a prior split of [the total] physical system into microscopic system S, detector D and environment E. But no rules have ever been given for making such a split, and certainly a physical system does not come with a subsystem containing a little sign reading, 'I am the environment: Trace over me.' Without such rules one cannot, in the general case, apply the environment-trace prescription..." And in [17] F. Laloë points out: "Indeed, in common life as well as in laboratories, one never observes superposition of results; we observe that Nature seems to operate in such a way that a single result always emerges from a single experiment; this will never be explained by the Schrödinger equation, since all that it can do is to endlessly extend its ramifications into the environment, without ever selecting one of them only."

7.6. Complementing Schrödinger dynamics

In order to find a solution to the measurement problem keeping as valid the individual interpretation of the state vector, other theories close to, but different from, quantum mechanics have been proposed. In these theories, the Schrödinger equation is complemented in a way that leads to spontaneous collapses. This is the case of those developed by D. Bohm [30-31], G. Ghiradi, A. Rimini and T. Weber [32], L. Diosi [33], and E. Joos and H. D. Zeh [34]. Ballentine [35] has demonstrated that these theories violate energy conservation and are incompatible with the existence of stationary states. Let us summarize two of these contributions and reproduce some additional comments on them.

i. In the theory of measurement proposed by Bohm [27-28] the state function $|\psi\rangle$ refers to an ensemble and every particle of the ensemble has a position x, which is a hidden variable. In addition to the usual potential V(x), a quantum potential U(x) is introduced. This allows Bohm to explain in an elegant way the double-slit experiment. In the EPR experiment disturbances from one particle to the other are transmitted instantaneously by the potential U(x).

ii. In CSL (Continuous Spontaneous Localization) theory [32], particles can undergo spontaneous wave-function collapses. For individual particles, these collapses happen probabilistically and will occur at a given rate with high probability but not with certainty; groups of particles behave in a statistically regular way, however. Since experimental physics has not already detected an unexpected spontaneous collapse, it can be argued that CSL collapses happen extremely rarely. The authors suggest that the rate of spontaneous collapse for an individual particle is of the order of once every hundred million years.

In two interesting comments F. Laloë [17] emphasizes that in those theories which modify the Schrödinger equation (i) "new constants appear which may in a sense look like ad hoc constants, but actually have an important conceptual role: They define the limit between the microscopic and macroscopic world (or between reversible and irreversible evolution); the corresponding border is no longer ill-defined, as opposed to the situation, for instance, in the Copenhagen Interpretation." And (ii) "in the initial Bohm-Bub theory, a complete collapse of the wave function is never obtained in any finite time. The same feature actually exists in CSL: *There is always what is called a 'tail' and even when most of the wave function goes to the component corresponding to one single outcome of an experiment, there always remains a tiny component on the others* (extremely small and continuously going down in size)." [17, emphases added]

8. Measurement processes versus spontaneous processes

In "Against 'measurement'," J. Bell complains about quantum mechanics formulations in the following terms [1]: "Surely, after 62 years, we should have an exact formulation of some serious part of quantum mechanics? By 'exact' I do not of course mean 'exactly true'. I mean only that the theory should be fully formulated in mathematical terms, with nothing left to the discretion of the theoretical physicist... until workable approximations are needed in applications. By 'serious' I mean that some substantial fragment of physics should be covered. Nonrelativistic 'particle' quantum mechanics, perhaps with the inclusion of the electromagnetic field and a cut-off interaction, is serious enough. For it covers 'a large part of physics and the whole of chemistry'; see [36]. I mean too, by 'serious', that 'apparatus' should not be separated off from the rest of the world into black boxes, as if it were not made of atoms and not ruled by quantum mechanics."

In the following table the most significant differences between measurement processes and spontaneous processes are reported.

Spontaneous processes	Measurement processes
The observer plays no role	The observer plays a paramount role
The state vector $\vert\psi(t)\rangle$ is necessarily continuous	In general the state vector $\vert\psi(t)\rangle$ is projected
The superposition principle is valid: there is interference	Superposition breaks down: interference is lost
The process is ruled by deterministic laws	The process is ruled by probability laws
Every action is localized	There is a kind of action-at-a-distance
Conservation laws are strictly valid	They have only a statistical sense

The mere comparison of the characteristics of both kinds of processes facilitates the understanding of why so many scientists are disappointed with quantum mechanics formalism.

At this stage it seems superfluous to comment on the first three lines of the previous table. Concerning determinism (fourth line of the table), let us recall that during the Fifth Solvay Congress, *i.e.* less than a century ago, H. Lorentz expressed his dissatisfaction with the rejection of determinism in atomic physic. Nowadays the notion of indeterminism is normally accepted, despite many scientists' aspirations for a version of quantum theory based on deterministic laws, and the "Old One" not playing at dice.

Something similar happened with the idea of action-at-a-distance (fifth line of the table) pointed out by Einstein in the Fifth Solvay Congress. First this notion was rejected by the majority of scientists. Then, in 1964 J. Bell proved a theorem stating that a local hidden variable theory cannot reproduce all statistical predictions of quantum mechanics [37]: More precisely, he showed that in the framework of any deterministic and local theory the correlations between some properties of two particles should satisfy an inequality (Bell's inequality) and that this inequality could be violated if the two particles were in an entangled state like that given by (8). In the following years many experiments yielded results which are compatible with the predictions of quantum mechanics and violate Bell's inequality. [38-41]

Now the door was opened to explore an even more strange and fascinating phenomenon: quantum teleportation. [44] This is a process by which quantum information (*e.g.* the exact state of an atom or photon) can be transmitted from one location to another, with the help of classical communication and previously shared quantum entanglement between the sending and receiving location. Because it depends on classical communication, which cannot proceed faster than the speed of light, it cannot be used for superluminal transport or communication. The seminal paper first expounding the idea was published in 1993. Since then, quantum teleportation has been realized in various physical systems. At present the record distance for quantum teleportation is 143 km (89 mi) with photons, and 21 m with material systems. In August 2013, the achievement of "fully deterministic" quantum teleportation, using a hybrid technique, was reported. On 29 May 2014, scientists announced a reliable way of transferring data by quantum teleportation. Quantum teleportation of data had been done earlier but with highly unreliable methods. The important point in what concerns the measurement problem is that, thanks to these astonishing results, the idea that projections imply a peculiar action-at-a-distance is nowadays frequently accepted.

On the last line of the previous table one reads: Conservation laws are strictly valid in spontaneous processes and have only a statistical sense in measurement processes. We have dealt with this subject a few years ago, but surely our results are not known by everybody. So in the next section we shall reproduce the essential of the paper where this problem is discussed; see [45]

9. Validity of conservation laws in spontaneous processes and in measurement processes

In the framework of OQM, in general, physical quantities are not sharp. "A popular working rule of pragmatic quantum mechanics says that a physical quantity has no value before a measurement." [46] Now, if the operator A_S represents the physical quantity A_S referred to the individual system S, when the system state is $|\Phi_S\rangle$ the mean value of A_S can be defined as

$$\langle A_S \rangle = \langle \Phi_S | A_S | \Phi_S \rangle \tag{17}$$

see for instance [7]. Hence, even if the physical quantity A_S has in general no value, it has a mean value $\langle A_S \rangle$ which is perfectly sharp.

A necessary condition for the physical quantity A_S to be conserved is that $\langle A_S \rangle$ be a constant. If H_S is the Hamiltonian of S, the validity of conditions

$$\partial A_S / \partial t = 0 \tag{18}$$

and

$$[A_S, H_S] = 0 \tag{19}$$

ensure that in those processes that are governed by the Schrödinger equation $\langle A_S \rangle$ remains a constant in time for every state $| \Phi_S \rangle$. As a consequence, according to OQM there is no inconvenience in saying that if conditions (18) and (19) are fulfilled, A_S is conserved in spontaneous processes.

We shall now address the problem of the validity of conservation laws when a measurement of A_S is performed; for simplicity we shall deal with the discrete case. Let $a_k (k=1, 2, \cdots)$ be an eigenvalue of the operator A_S, g_k its degree of degeneracy and $| a_k^v \rangle (v=1, 2, \cdots g_k)$ an eigenvector corresponding to the eigenvalue a_k. We shall assume that $| m_0$ represents the initial state of a measuring device M of A_S, and $| \psi_k^v \rangle$ the orthonormal states of S+M when the measurement process is over. To ensure that measurements of A_S can be performed according to the ideal measurement scheme, we shall suppose that A_S commutes with every operator representing another conserved quantity referred to S+M. [47-51]

According to the ideal scheme the transition

$$| a_k^v \rangle | m_0 \rangle \rightarrow | \psi_k^v \rangle \tag{20}$$

has a probability of one, hence it can be assumed that *it is a result of the Schrödinger evolution.*

Let A be the operator representing a physical quantity A referred to S+M, and H be its Hamiltonian. We can then write

$$H = H_S + H_M + H_{int} \tag{21}$$

where H_M refers to M, and H_{int} is due to the interaction between S and M. We assume that the conditions

$$\partial A / \partial t = 0 \tag{22}$$

and

$$\left[A, H\right] = 0 \tag{23}$$

are fulfilled. If at t_0 (when the interaction between S and M starts) it is possible to write

$$A = A_S + A_M \tag{24}$$

(where A_M refers to M), we have

$$\langle A\rangle_k^v(t_0) = \langle a_k^v | A_S | a_k^v\rangle + \langle m_0 | A_M | m_0\rangle = a_k + \langle m_0 | A_M | m_0\rangle \tag{25}$$

And, since at t_f (when the interaction between S and M is over)

$$\langle A\rangle_k^v(t_f) = \langle \psi_k^v | A | \psi_k^v\rangle \tag{26}$$

the validity of (22) and (23) implies that $\langle A\rangle_k^v(t_f) = \langle A\rangle_k^v(t_0)$, and hence

$$\langle \psi_k^v | A | \psi_k^v\rangle = a_k + \langle m_0 | A_M | m_0\rangle \tag{27}$$

for every ν. As $\langle \psi_k^v \mid A \mid \psi_k^v\rangle$ does not depend on ν, it can be written

$$\langle A\rangle_k(t_f) = \langle A\rangle_k^v(t_f) = a_k + \langle m_0 | A_M | m_0\rangle \tag{28}$$

This relation must necessarily be fulfilled in the ideal measurement scheme. As a consequence, it can be said that *in those cases where the initial state of* S *is an eigenstate of the operator* A_S *representing the physical quantity* A_S *to be measured, the corresponding conservation law of A is valid.* This result can also be seen as a natural consequence of the hypothesis that the process described by (20) is governed by the Schrödinger equation.

$$\left|\Phi_S(t_0)\right\rangle = \sum_{l,\mu} c_l^\mu \left|a_l^\mu\right\rangle \tag{29}$$

(where at least two coefficients c_l^μ and $c_{l'}^{\mu'}$ with $l \neq l'$ are non-null) and the Schrödinger equation rules the measurement process, then the Hamiltonian H, referred to S+M, induces the evolution

$$\sum_{l,\mu} c_l^\mu \left|a_l^\mu\right\rangle \left|m_0\right\rangle \to \sum_{l,\mu} c_l^\mu \left|\psi_l^\mu\right\rangle \tag{30}$$

Making

$$\langle A\rangle(t_0) = \langle \Phi_S(t_0)|\langle m_0|A|\Phi_S(t_0)\rangle|m_0\rangle = \sum_{l,\mu} \left|c_l^\mu\right|^2 a_l + \langle m_0|A_M|m_0\rangle \tag{31}$$

and

$$\langle A\rangle(t_f) = \left(\sum_{l,\mu} c_l^{\mu *}\left\langle \psi_l^\mu\right|\right) A \left(\sum_{l',\mu'} c_{l'}^{\mu'}\left|\psi_{l'}^{\mu'}\right\rangle\right) \tag{32}$$

the validity of (22) and (23) allow us to ensure that $\langle A\rangle(t_0)=\langle A\rangle(t_f)$: *As long as the Schrödinger equation rules the process, the mean value of the physical quantity A, referred to the total system S+M, remains a constant* and the state of S+M continues to be the superposition which appears in (30).

But in view of the Projection Postulate such a superposition is broken down. Hence, the change of S+M is not given by (30) and the transition

$$\sum_{l,\mu} c_l^\mu \left|a_l^\mu\right\rangle \left|m_0\right\rangle \to \sum_{\mu} c_k^\mu \left|\psi_k^\mu\right\rangle \tag{33}$$

has probability $\sum_\mu |c_k^\mu|^2$ to happen. In this last case,

$$\langle A\rangle_k(t_f) = a_k + \langle m_0|A_M|m_0\rangle \tag{34}$$

as stated in (28). As a consequence, it results

$$\langle \mathbf{A}\rangle_k(\mathbf{t}_f) \neq \langle \mathbf{A}\rangle(\mathbf{t}_0) \tag{35}$$

for every k, even though conditions (22) and (23) are fulfilled.

It is worth noticing that inequalities (35) are obtained under the assumptions that the individual interpretation of the state vector and the Projection Postulate are valid. In this case the condition that $\langle A\rangle$ be a constant, a necessary condition for A to be conserved, is not satisfied. We are thus forced to conclude that *if the initial state of S is not an eigenvector of* A_S, *the physical quantity A is not conserved in processes of measurement of* A_S. In [52-55] we give examples of processes of measurement of the type analyzed in this section; and in [45] we deal with the continuous case. The same result is obtained.

A similar conclusion resulting from a different analysis has been obtained by P. Pearle. [56] He says that "it should first be noted that quantum theory itself, with the reduction postulate indiscriminately applied, does not necessarily satisfy the conservation laws..." In his view, "this is a serious problem for quantum theory with a reduction postulate."

Our next step is to calculate the *average* of $\langle A\rangle_k(t_f)$ when the process of measurement of A_S is repeated many times. Let f_k be the frequency corresponding to the possible results $a_k\,(k=1,\,2,\,\cdots)$ and to the mean value $\langle A\rangle_k(t_f)$. If the process is repeated N times, the resulting average is

$$\overline{A} = \sum_k f_k \langle A\rangle_k(t_f) = \sum_k f_k a_k + \langle m_0|A_M|m_0\rangle \tag{36}$$

where (34) has been taken into account. Now, if N is big enough, we can assert that $f_k \approx \sum_\mu |c_k^\mu|^2$ and, in view of (31) we obtain

$$\overline{A} \approx \langle A\rangle(t_0) \tag{37}$$

To sum up, in individual processes of measurement of A_S *the conservation law of A is in general not valid; but this law still has a statistical sense.*

10. Ad-hoc use of the projection postulate

We shall start this section with some remarks concerning the concept of probabilities. Following tradition, we are going to adopt the expression *subjective probabilities* for probabilities related to the lack of knowledge in processes governed by deterministic laws; and *objective probabilities* for probabilities where the process is not ruled by deterministic laws. Accordingly, there is no room for objective probabilities in classical mechanics, electromagnetism and relativity. Moreover, as in the framework of OQM every spontaneous process is ruled by the

Schrödinger equation, which is a deterministic equation, objective probabilities have nothing to do with these kinds of processes.

Then, we explicitly state that *a system cannot be in two different states at the same time*. Hence, if the system is in the state $|\psi(t_0)\rangle$ at time t_0 and the process is ruled by the Schrödinger equation, there is no more than one possibility: *at time t its state must certainly be*

$$|\psi(t)\rangle = U(t,t_0)|\psi(t_0)\rangle \tag{38}$$

where $U(t, t_0)$ is the evolution operator. As a consequence, if at time t_0 the system is in the state $|\psi(t_0)\rangle$ and the process is spontaneous, the *objective probability* the system has of being in $|\psi(t)\rangle = U(t, t_0)|\psi(t_0)\rangle$ at time t is P=1 and the *objective probability* the system has of being in another, different state from $|\psi(t)\rangle = U(t, t_0)|\psi(t_0)\rangle$, is P=0.

There is no doubt that quantum mechanics has been extremely successful in explaining radioactivity, electron-phonon scattering, interactions between light and matter and many other phenomena which involve, supposedly, only *spontaneous processes*. So, in principle one could expect that the analysis of these processes does not involve projections (for they play a role just in cases measurements are performed). Nevertheless, reading quantum mechanics textbooks one is forced to conclude the opposite; see for instance any book of the following list: [6-10, 18, 57].

To deal with spontaneous processes involved in phenomena such as those previously mentioned, in most cases time-dependent perturbation theory is necessary. In Dirac's view, "[time-dependent perturbation theory] must be used for solving all problems involving a consideration of time." [6] And W. Heitler states: "for all problems of physical interest the application of [time-dependent] perturbation theory is beyond doubt." [57] So let us examine in which way this theory is needed to confront these kinds of problems. According to Dirac, "with [time-dependent perturbation theory] one takes a stationary state of the unperturbed system and sees how it varies with time under the influence of the perturbation." [6] And "the perturbation causes the state to change." [6] We are going to analyze this point in detail.

Consider a system with Hamiltonian H_0 which does not depend explicitly on time. It is assumed that the eigenvalues equations of H_0 have previously been solved. We shall denote by ε_k and $|\varphi_k\rangle (k=1, 2, \cdots)$ its eigenvalues and eigenvectors, respectively; for simplicity we shall deal with the discrete non-degenerate case. Then, if at t_0 a perturbation W(t) depending explicitly on time is added to H_0, the Hamiltonian of the system for $t > t_0$ becomes

$$H(t) = H_0 + W(t) \tag{39}$$

and the system evolves according to the Schrödinger equation

$$i\hbar d|\psi(t)\rangle / dt = \left[H_0 + W(t)\right]|\psi(t)\rangle \tag{40}$$

"The solution $|\psi(t)\rangle$ of this first-order differential equation which corresponds to the initial condition $|\psi(t_0)\rangle = |\varphi_i\rangle$ is unique." [9] Then it is said that at time t_f the probability $P_{if}(t_f)$ of *finding* the system in another eigenstate $|\varphi_f\rangle$ of H_0 is

$$P_{if}(t_f) = \left|\langle\varphi_f|\psi(t_f)\rangle\right|^2 \tag{41}$$

Taking into account what has been previously said, we shall write $|\psi(t_f)\rangle = U(t_f, t_0)|\psi(t_0)\rangle$. Now, to find *a system which at t_f certainly is in the state* $|\psi(t_f)\rangle = U(t_f, t_0)|\psi(t_0)\rangle$ in another, different state like $|\varphi_f\rangle$ is a task impossible to achieve. By contrast, to find such a system *immediately after* t_f in $|\varphi_f\rangle$ is a task possible to achieve but it requires a measurement to be performed at time t_f. And which one should be the physical quantity to be measured? The answer is not obvious for, in particular, if this physical quantity were the energy, the system should not be projected to $|\varphi_f\rangle$, an eigenstate of the operator H_0 which does not represent the energy of the system at time$t > t_0$. This last remark, however, does not apply in cases where W(t) is a perturbing interaction limited in time and it can be considered that $W(t_f)=0$. [8]

But let us come back to the declared aim of time-dependent perturbation theory. Conspicuous authors make statements such as "Our objective is to calculate transition amplitudes between the relevant unperturbed eigenstates, owing to the presence of the perturbation…" [8]; "we want to study the transitions which can be induced by the perturbation…" [9]; "the transition probability between the initial state $|\varphi_i\rangle$ and the final state $|\varphi_f\rangle$ [is] induced by the perturbation…" [10] As the perturbation W(t) modifies the Hamiltonian, it is evident that the state $|\psi(t_f)\rangle = U(t_f, t_0)|\psi(t_0)\rangle$ resulting when W(t) is applied will be different from the state $|\psi(t_f)\rangle$ resulting when W(t) is absent. *But perturbations do not induce transitions.* In this sense Messiah is very clear. Referring to the objective of time-dependent perturbation theory, he asserts: *"Supposons qu'à l'instant initial t_0, le système se trouve dans l'un des états propres de H_0, l'état a par exemple. Nous nous proposons de calculer la probabilité de le trouver à l'instant t dans un autre état propre de H_0, l'état b par exemple, dans l'éventualité d'une mesure à cet instant"* [7]; we emphasize: *dans l'éventualité d'une mesure à cet instant.* On the contrary, other authors seem to have forgotten that in the framework of OQM measurements are absolutely necessary in order to obtain the transition probability$P_{if}(t_f)$.

11. Who is afraid of the projection postulate?

C. M. Caves asserts [58]: "Mention collapse of the wave function and you are likely to encounter vague uneasiness or, in extreme cases, real discomfort. This uneasiness can usually be traced to a feeling that a wave-function collapse lies 'outside' quantum mechanics. The real quantum mechanics is said to be the unitary Schrödinger evolution; wave-function collapse is regarded as an ugly duckling of questionable status, dragged in to interrupt the beautiful flow of Schrödinger evolution."

Projections are disliked for many reasons; one of them is that they imply discontinuities. But is there a way of give an account for processes of emission and absorption of light without invoking discontinuities? We think there is not, as shown in the following.

To start with, let us face this question in an intuitive and nearly classic way. Consider the absorption of one photon by one atom. We shall assume that (i) initially the atom and the photon exist as separated things; (ii) the photon can only travel with a speed c and has an energy $\hbar\omega$ at each instant; and (iii) the energy of the system atom+photon is conserved in the process of absorption. Note that the photon cannot be absorbed through a swift, not instantaneous change: *either it is, travels with speed c and carries the energy $\hbar\omega$ or it is not. This implies that the photon must be absorbed by the atom at once and that the energy of the atom must be increased in an instantaneous way.* [59] In more elaborated treatments of the subject, probabilities of projections and hence, indirectly, state function discontinuities are mentioned frequently. For instance, in the classical textbook of Heitler one reads: " $| c_{\cdots n_\lambda}(t) |^2$is the probability for *finding* n_1 photons of type 1, n_λ photons of type λ, etc."; "The probability for *finding* the system at time t in the state n when it was in state 0 at t=0 is thus..."; "We now calculate the probabilities $| b_n(t) |^2$ for *finding* the system in a state n at the time t." [57, emphases added]

On his side, Jammer points out a serious problem which becomes apparent when the notion of projections is rejected: "As long as a quantum mechanical one-body or many-body system does not interact with macroscopic objects, as long as its motion is described by the deterministic Schrödinger time-dependent equation, no events could be considered to take place in the system. Even such elementary process as the scattering of a particle in a definite direction could not be assumed to occur (since this would require a 'reduction of the wave packet' without an interaction with a macroscopic body). In other words, if the whole physical universe were composed only of microphysical entities, as it should be according to the atomic theory, it would be a universe of evolving potentialities (time-dependentψ-functions) but not of real events." [2]

A few authors have considered the possibility that projections may happen at the microscopic level. One of them is H. Primas, for whom "the reality of the breakdown of the superposition principle of traditional quantum mechanics on the molecular level is dramatically demonstrated by the terrible Contergan tragedy which caused many severe birth defects." [46] And Bell complains: "during 'measurement' the linear Schrödinger evolution is suspended and an ill-defined 'wave-function collapse takes over. There is nothing in the mathematics to tell what is 'system' and what is 'apparatus' nothing to tell which natural processes have the special status of 'measurements'. Discretion and good taste, born from experience, allow us to use quantum theory with marvelous success, despite the ambiguity of the concepts named above in quotation marks." [60] In [28] we have given an answer to the question "which natural

http://dx.doi.org/10.5772/59209

processes have the special status of measurements?" In the next section we shall summarize the most important points of our approach.

12. The spontaneous projections approach

In the Spontaneous Projection Approach (SPA) it is assumed that two kinds of processes, irreducible to one another, occur in nature: (i) the strictly continuous and causal ones, which are governed by the Schrödinger equation and (ii) those implying discontinuities, which are ruled by probability laws. A postulate ensuring the statistical sense of conservation laws is adopted. Taking into account this postulate the concept of preferential states is introduced. If the system does not have preferential states, the Schrödinger evolution follows. By contrast, if the system has preferential states projections may happen. Spontaneous and measurement processes are treated on the same footing.

SPA is compatible with epistemological realism: we assume that the world exists independently of the knowing subject and that it is possible to know it, at least in a partial way. So our discourse will be about what happens, not about what is measured or observed. (This does not mean, obviously, that it has to be right; it could happen that it be completely wrong.) We share Bunge's assertion [19]: "the question of reality has nothing to do with scientific problems such as whether all properties have sharp values and whether all behavior is causal." And, as we have already said, we would add to the list of scientific problems which have nothing to do with the question of reality the issue of action-at-a-distance and the validity of conservation laws in individual processes.

The primitive (undefined) notions of SPA are: *system, physical quantity, state system and probability*; the term probability will be used as a synonym of *objective probability*; see Section 10. Nevertheless, we have taken into account Bell's remark [1]: "The concepts 'system', 'apparatus' 'environment' immediately imply an artificial division of the world, and an intention to neglect, or take only schematic account of, the interaction across the split." We do not make such an artificial division for *apparatus and environment are absent of SPA postulates.* And systems mean either objects or collections of objects.

The two first postulates of SPA coincide with those of OQM. They state:

Postulate I: To every system corresponds a Hilbert space H whose vectors (state vectors, wave functions) completely describe the state of the system.

Postulate II: To every physical quantity A corresponds uniquely a self-adjoint operator A acting in H. It has associated the eigenvalue equations

$$A\left|a_j^v\right\rangle = a_j\left|a_j^v\right\rangle \tag{42}$$

(v is introduced in order to distinguish between the different eigenvectors that may correspond to one eigenvalue a_j), and the closure relation

$$\sum_{j,v} \left| a_j^v \right\rangle \left\langle a_j^v \right| = I \tag{43}$$

is fulfilled (here I is the identity operator). If j or v is continuous, the respective sum has to be replaced by an integral.

Postulate III: If the conditions

$$\partial A / \partial t = 0 \tag{44}$$

and

$$\left[H,\ A \right] = 0 \tag{45}$$

are fulfilled (here H is the Hamiltonian of the system), and there is a *generic* orthonormal set $\{|u_k\rangle\}(k=1, 2, \cdots)$ such that the normalized state $|\Phi\rangle$ of the system can be written

$$\left| \varphi \right\rangle = \sum_k c_k \left| u_k \right\rangle \tag{46}$$

the validity of

$$\left| \Phi \right\rangle A \left| \Phi \right\rangle = \sum_k \left| c_k \right|^2 \left\langle u_k \right| A \left| u_k \right\rangle \tag{47}$$

is a necessary condition for projections of the state $|\Phi\rangle$, given by (46), to the vectors of the set $\{|u_k\rangle\}$ to happen, *i.e.* for jumps like $|\Phi\rangle \to |u_1\rangle$, or $|\Phi\rangle \to |u_2\rangle$ etc., to occur.

Comments: (i) By definition, A is a constant of the motion if it satisfies conditions (44) and (45). (ii) Postulate III ensures the statistical sense of the conservation of the physical quantity A. [28]

Hypothesis: A system in the state $|\Phi\rangle$ has tendency to jump to the eigenstates of its constants of the motion.

Comments: (iii) This tendency should not become actualized if the projections it induces results in a violation of Postulate III or lead the state vector outside the Hilbert space. (iv) Taking into account this Hypothesis and Postulate III, the concept of *preferential states* is introduced; see [28, 61]. For simplicity, instead of dealing with the general case, here we shall refer to the following one: Let H, A and B be three operators representing respectively the energy, the physical quantity A and the physical quantity B of the system. It will be assumed that they have discrete spectra and satisfy (44), and that $\{H,\ A,\ B\}$ is the *unique* complete set of compatible operators of the system. The vectors of its common basis will be denoted by $|E_p, a_q, b_r\rangle$, where E_p, a_q and b_r are respectively the eigenvalues of H, A and B. (v) On the one hand, taking into account the previous hypothesis, we can say that the system's state $|\Phi$ has tendency to be projected to the eigenvectors of H, to the eigenvectors of A and to the eigenvectors of B. On the other hand, as the relations

$$\langle\Phi|H|\Phi\rangle = \sum_{p,q,r} |c_{p,q,r}|^2 \langle E_p, a_q, b_r | H | E_p, a_q, b_r\rangle \tag{48}$$

$$\langle\Phi|A|\Phi\rangle = \sum_{p,q,r} |c_{p,q,r}|^2 \langle E_p, a_q, b_r | A | E_p, a_q, b_r\rangle \tag{49}$$

and

$$\langle\Phi|B|\Phi\rangle = \sum_{p,q,r} |c_{p,q,r}|^2 \langle E_p, a_q, b_r | B | E_p, a_q, b_r\rangle \tag{50}$$

are satisfied for the state $|\Phi\rangle = \sum_{p,q,r} c_{p,q,r} | E_p, a_q, b_r\rangle$, Postulate III does not prohibit projections like $|\Phi\rangle \to | E_p, a_q, b_r\rangle$. Then we state:

Definition: The preferential states of the system are the common eigenstates of H, A and B.

Comment: (vi) The previous definition is valid in cases restrictions established in Comment (iii) are fulfilled; in this particular case the preferential states do not depend on $|\Phi\rangle$. Cases where $[A, B] \neq 0$ or where the spectrum of H is partially continuous, have been analyzed in [28, 61]. In these last cases the preferential states depend on $|\Phi\rangle$. (vii) As we have assumed that $\{H, A, B\}$ is the unique complete set of compatible operators of the system, the set of preferential states $\{| E_p, a_q, b_r\rangle\}$ is necessarily unique. *The condition of uniqueness of the set of preferential states remains valid in the general case.* [28, 61]

Postulate IV: The system's state $|\Phi\rangle$ can be projected to the state $|u_j\rangle$ if and only if $|u_j\rangle$ is a preferential state. If the system in the state $|\Phi\rangle$ does not have preferential states, *the Schrödinger evolution must follow.*

Postulate V: Let $|u_k\rangle (k = 1, 2, \cdots)$ be the preferential states of the system in the state

$$|\Phi(t)\rangle = \sum_k c_k(t) |u_k(t)\rangle \tag{51}$$

where$c_k(t) = \langle u_k(t) | \Phi(t)\rangle$. In the small interval $(t, t + dt)$ the system's state can undergo the following changes:

$$\left|\Phi(t)\right\rangle \rightarrow \left|\Phi(t+dt)\right\rangle = \left|u_j(t)\right\rangle \tag{52}$$

with probability $dP_j(t) = |c_j(t)|^2(dt/\tau)$; or

$$\left|\Phi(t)\right\rangle \rightarrow \left|\Phi_{Sch}(t+dt)\right\rangle = U(t+dt,t)\left|\Phi(t)\right\rangle \tag{53}$$

with probability$dP_{Sch}(t) = 1 - dt/\tau$, where $U(t + dt, t)$ is the evolution operator,

$$\tau\Delta H = \hbar/2 \tag{54}$$

and

$$(\Delta H)^2 = \left\langle\Phi(t)\right|H^2\left|\Phi(t)\right\rangle - \left\langle\Phi(t)\right|H\left|\Phi(t)\right\rangle \tag{55}$$

Comments: (viii) The change given by (53) is a Schrödinger evolution and those given by (52) are projections to the preferential states of the system in the state $|\Phi(t)\rangle$ (ix) Since preferential states are members of an orthonormal set of vectors, a system's state projected to a preferential state remains there evolving in agreement with the Schrödinger equation. (x) The state $|\Phi_{Sch}(t)\rangle$ may be considered as un unstable state that can decay to one of the preferential states $|u_j(t)\rangle$, the relaxation time being τ. Calling $P_{Sch}(t)$ to the probability that the system's state has not been projected to any preferential state in the interval (0, t) the well-known exponential decay law is obtained; see [28, 61].

13. The ideal measurement scheme in the framework of SPA

Let us start this section with the question: "what can be observed?" In his answer Bell quotes Einstein saying "it is theory which decides what is 'observable'." He adds: "I think he was right – 'observation' is a complicated and theory-laden business." [1] We agree with these assertions.

Consider, for instance, the determination of the energy levels of the Hg atom in the Franck-Hertz experiment, where a curve of electrical current versus the applied voltage is obtained; this curve presents peaks of the current at regular intervals of voltage. [62] Relating the values of the voltage where the peaks are located to the first excited energy level of the atom requires a quite elaborate theory of what is happening inside the tube. But once the way the device works has been understood, the Franck-Hertz experiment provides a *direct* measurement of the energy difference between the quantum states of the atom: it appears on the dial of a voltmeter! It is worth stressing that no entanglement is invoked in the analysis of this experi-

http://dx.doi.org/10.5772/59209

ment and the same is true of many others related, *e.g.*, to blackbody radiation, photoelectric effect and Compton shift. By contrast, *in the ideal measurement scheme entanglements are unavoidable.* This is for instance the case of the system *photon* meeting the device tourmaline crystal mentioned by Dirac; see Section 7.1.

In the following we shall address the conceptual problem of the ideal measurement scheme in the framework of SPA. We are going to analyze the measurement of the physical quantity A_S pertaining to the system S; for simplicity we shall deal with the discrete non-degenerate case. Let $a_k (k=1, 2, \cdots N)$ be an eigenvalue of the operator A_S representing A_S and $|a_k\rangle$ the corresponding eigenvector. The operator A_S acts in the Hilbert space H_S of S and its extension $A=A_S \otimes I_M$ (here I_M is the identity operator in the Hilbert space H_M of M) acts in the Hilbert space H_{S+M} of S+M. The Hamiltonian of the total system S+M will be denoted by H, the operator B will represent a physical quantity B referred to S+M, the initial state of the measuring device M of A_S will be denoted by $|m_0\rangle$, and the state of the total system S+M at time t by $|\Phi(t)\rangle$.

In a first step we shall suppose that at t_0, when the interaction between S and M starts, the state of S is $|a_k\rangle$ and that of S+M is

$$\left|\Phi_k(t_0)\right\rangle = \left|a_k\right\rangle\left|m_0\right\rangle \quad (56)$$

It is easily verified that the state $|\Phi_k(t_0)\rangle$ is an eigenstate of A corresponding to the non-degenerate eigenvalue a_k. In addition, if [A, B]=0 and H, A, and B are constants of the motion, the state $|\Phi_k(t_0)\rangle$ will be a common eigenstate of these three operators. Hence, if $\{H, A, B\}$ is the unique complete set of compatible operators of the system, according to SPA the state $|\Phi_k(t_0)\rangle$ will be a preferential state of the system S+M. As we have already pointed out, it must remain evolving in agreement with the Schrödinger equation (see previous section). So at time t the state of S+M will be

$$\left|\Phi(t)\right\rangle = \left|\Phi_k(t)\right\rangle = U(t,t_0)\left|\Phi_k(t_0)\right\rangle \quad (57)$$

as it happens in the traditional treatment of the ideal measurement scheme.

Note that if H, A, and B are constants of the motion but $[A, B]\neq 0$, the operators A and B do not have a common basis. In this case it has been shown that collapses to the basis of the eigenvectors common to H and A violate the statistical sense of the conservation of B, hence Postulate III of SPA prevents these projections; in the same way it is concluded that collapses to the eigenvectors common to H and B are forbidden [28]: as stated in the traditional treatment, for the ideal measurement scheme to be valid, the measured physical quantity must be compatible with every conserved quantity referred to S+M.

Now we shall consider the case where the initial state of S is $\sum_{k=1}^{N} c_k \mid a_k\rangle$. The initial state of S+M will be

$$\left|\Phi\left(t_0\right)\right\rangle = \sum_{k=1}^{N} c_k \left|a_k\right\rangle \left|m_0\right\rangle = \sum_{k=1}^{N} c_k \left|\Phi_k\left(t_0\right)\right\rangle \tag{58}$$

Postulate V of SPA tells us that at $t > t_0$ the state of S+M can be

one of its N preferential states $\mid \Phi_k(t)\rangle$, in case in the interval (t_0, t) the state of S+M has been projected; or
$\mid \Phi_{Sch}(t)\rangle = U(t, t_0) \mid \Phi(t_0)\rangle$, in case in the interval (t_0, t) the state of S+M has not been projected and hence its behavior has been ruled by the Schrödinger equation.

Which one of these (N+1) states will result at time t cannot be predicted, but each one of them has an associated probability given by Postulate V. In case $t \gg \tau$, the relaxation time given by (54), the probability the system has to remain in the state $\mid \Phi_{Sch}(t)\rangle$ goes to zero and all we can "observe" is the result corresponding to one of the preferential states onto which the system can decay.

14. Conclusions

OQM formalism includes two different laws: a strictly continuous and causal Schrödinger evolution which governs spontaneous processes and the Projection Postulate, a rule implying discontinuities and changes of the state vector in agreement with probability laws. On the one hand, the inclusion in the formalism of two laws irreducible to one another has been a source of dissatisfaction from quantum mechanics birth. On the other hand, OQM (which includes the Projection Postulate) has been extremely successful in the area of experimental predictions; and *even if the Projection Postulate should be applied only in cases where measurements are performed, in the present work we have shown that it is also used ad-hoc, when needed to explain processes which supposedly are spontaneous.*

Some authors have suggested that measurement processes could be a particular kind of natural processes. But, then, we confront the problem pointed out by Bell [60]: there is nothing in OQM formalism to tell which natural processes have the special status of measurements, *i.e.* to decide whether one or the other law rules the process.

Looking for a solution to this problem, we have proposed a Spontaneous Projection Approach (SPA) to quantum mechanics, a theory where *spontaneous and measurement processes are treated on the same footing* and the behavior of macroscopic and microscopic objects are ruled by the same laws. The first step to achieve this objective is to admit that *projections can occur spontaneously in nature, even in closed systems, without being acted by any external perturbation.* But, then, the theory must say in which situations and to which vectors the state vector can collapse, and

http://dx.doi.org/10.5772/59209

which are the corresponding probabilities. These goals have been achieved in the framework of SPA.

It is worth stressing that our approach does not modify OQM in a substantial way: it does not change the Schrödinger equation and it recovers a version of Born postulate where no reference to measurements is made. So, in general its predictions coincide with those of OQM.

Concerning the treatment of the ideal measurement scheme in the framework of SPA, we are aware of its limitations derived, among other reasons, from the hypotheses introduced "for simplicity." For instance, we have considered that there are only three relevant physical quantities referred to the total system (which includes the measuring apparatus), that the operators which represent them are constants of the motion, and that the physical quantity to be measured is represented by an operator having discrete non-degenerate spectrum. Our treatment, however, has the merit of predicting results which completely agree with those obtained in the framework of OQM, without having recourse to the observer consciousness, to the macroscopic character of the measurement device, or to interactions with the environment producing decoherence, something that in the long term looks and smells like a collapse. We should also stress that in other theories such us that due to Ghirardi, Rimini and Weber, even when most of the wave function goes to the component corresponding to one single outcome of an experiment, there always remains a 'tail', *i.e.* a tiny component of the system's state on the others. By contrast, in SPA the system's state either evolves according to the Schrödinger equation, or is at once entirely projected into one of its preferential states.

To end this chapter let us highlight the most important differences between SPA and OQM:

i. SPA is compatible with epistemological realism.

ii. In SPA projections occurring in spontaneous processes such as those involved in radioactivity, interactions between light and matter, etc., are not surreptitiously but explicitly included. In this sense it could be said that SPA enjoys of a coherence which is absent from OQM.

iii. Differing from OQM, SPA yields an expression for the probability of transitions to the continuum which is valid for every time and, except for some minimal restrictions, for every added potential. We have pointed out in [61] that these predictions could be experimentally tested.

Theories which include only deterministic laws in their formalism can give an account for nothing but "automatic changes." On the contrary, by including probabilistic laws in its formalism, SPA opens the door to novelty.

Acknowledgements

I am grateful to Professors D. R. Bes, J. C. Centeno and F. G. Criscuolo for fruitful discussions.

Author details

M. E. Burgos*

Address all correspondence to: mburgos25@gmail.com

Departamento de Física, Facultad de Ciencias, Universidad de Los Andes. Mérida, Venezuela.

References

[1] Bell J. Against 'measurement'. Physics World August 1990; 33-40.

[2] Jammer M. The Philosophy of Quantum Mechanics. New York: John Wiley & Sons; 1974.p.5, p.7, p. 27, p. 114-116, 44, p.226-227, p.228, p.474, p. 475, p.479,, pp.481-482, p. 216,, p.217, p.474

[3] Born M. Experiment and Theory in Physics. London: Cambridge University Press; 1943; p. 23

[4] The Born Einstein Letters, Letters 50, 53 and 81

[5] von Neumann J. Mathematische Grundlagen der Quantenmechanik. Berlin: Springer; 1932.

[6] Dirac PAM. The Principles of Quantum Mechanics. Oxford: Clarendon Press; 1930; Chapter 7; p.7, p.168, p.172

[7] Messiah A. Mécanique Quantique. Paris: Dunod; 1965; Chapter 17; pp.250-251 and pp.261-263; p.140;

[8] Merzbacher E. Quantum Mechanics. New York: John Wiley & Sons; 1998; Chapter 19; p.25, p.59 and p.180; p.483

[9] Cohen-Tannoudji C, Diu B, Laloë F. Quantum Mechanics. New York: John Wiley & Sons; 1977; Chapter 13 and pp.215-218 and p.222; p.1285

[10] Bes DR. Quantum Mechanics. Berlin: Springer-Verlag; 2004; Chapter 9; pp.9-11and p. 137, pp.183-184, p.168, p.142

[11] Einstein A, Podolsky B, Rosen N. Can Quantum-Mechanical Description of Physical Reality Be Considered Complete? Physical Review 1935; 47, 777-780.

[12] Ballentine LE. Einstein Interpretation of Quantum Mechanics. American Journal of Physics 1972; 40, 1763-1771.

[13] Schilpp PA., editor. A. Einstein: Philosopher-Scientist. New York: Library of the Living Philosophers; reprinted by Harper and Row; 1959.

[14] Ballentine LE. The Statistical Interpretation of Quantum Mechanics. Reviews of Modern Physics 1970; 42(4), 358-381.

[15] Tegmar M, Wheeler JA. 100 Years of Quantum Mysteries. Scientific American 2001; 284(2), 68-75.

[16] von Neumann J. Mathematical Foundations of Quantum Mechanics. New Jersey: Princeton University Press; 1955; p.420

[17] Laloë F. Do We Really Understand Quantum Mechanics? Strange Correlations, Paradoxes, and Theorems. American Journal of Physics 2001; 69, 655-701.

[18] Landau LD, Lifshitz EM. Quantum Mechanics. London: Pergamon Press; 1959; Chapter 6 and pp.2-22.

[19] Bunge M. Philosophy of Science and Technology Part I, Vol 7 of Treatise on Basic Philosophy. Dordrecht: D. Reidel Publishing Company; 1985; pp.191-202.

[20] Bunge M. Reply to Burgos on the projection (reduction) of the state function in Studies on Mario Bunge's Treatise. Amsterdam: Rodopi; 1990

[21] Burgos ME. Can the EPR Criterion of Reality Be Considered Acceptable? Kinam 1983; 5, 277-284.

[22] Burgos ME. Quantum Mechanics Is Compatible with Realism. Foundations of Physics 1987; 17, 809-812.

[23] Poincaré H. La science et l'hypothèse. Paris: Flammarion; 1906; p.161

[24] Everett H. Relative State Formulation of Quantum Mechanics. Reviews of Modern Physics 1957; 29, 454-462.

[25] Wikipedia, The Free Encyclopedia: Many-Worlds Interpretation.

[26] Zurek W. Decoherence and the Transition from Quantum to Classical. Physics Today October 1991; 44, 36-44.

[27] Zurek W. Decoherence, einselection, and the quantum origins of the classical. Reviews of Modern Physics 2003; 75(3), 715-775.

[28] Burgos ME. Which Natural Processes Have the Special Status of Measurements? Foundations of Physics 1998; 28(8), 1323-1346.

[29] Ghirardi GC, Grassi R, Pearle P. Physics Today, Section Letters, 1993; 44, 13

[30] Bohm D. A suggested Interpretation of the Quantum Theory in Terms of 'Hidden' Variables. I. Physical Review 1952; 85(2), 166-179.

[31] Bohm D. A suggested Interpretation of the Quantum Theory in Terms of 'Hidden' Variables. II. Physical Review 1952; 85(2), 180-193.

[32] Wikipedia, The Free Encyclopedia: Ghirardi-Rimini-Weber Theory.

[33] Diosi L. Models for Universal Reduction of Macroscopic Quantum Fluctuations. Physical Review A 1989; 40(3), 1165-1174.

[34] Joos E, Zeh HD. The Emergence of Classical Properties through Interaction with the Enviroment. Zeitschrift für Physik B 1985; 59, 223.

[35] Ballentine LE. Failure of some theories of state reduction. Physical Review A 1991; 43(1), 9-12.

[36] Dirac PAM. Algebra, Mathematical Analysis, Geometry. Proceedings of the Royal Society A 123(192) 1929, 714-733

[37] Bell JS."On the Einstein Podolsky Rosen Paradox. Physics 1964; 1, 195-200.

[38] Clauser JF, Horne MA, Shimony A, Holt RA. Proposed experiment to test local hidden-variable theories. Physical Review Letters 1969; 23(15), 880-884.

[39] Freedman SJ, Clauser JF. Experimental Test of Local Hidden-Variable Theories. Physical Review Letters 1972; 28, 938-941.

[40] Clauser JF, Horne MA. Experimental consequences of objective local theories. Physical Review D 1974; 10(2), 526-535

[41] Aspect A, Grangier P, Roger G. Experimental Tests of Realistic Local Theories via Bell's Theorem. Physical Review Letters 1981; 47(7), 460-463.

[42] Aspect A, Grangier P, Roger G. Experimental Realization of Einstein-Podolsky-Rosen-Bohm Gedanken experiment: A New Violation of Bell's Inequalities. Physical Review Letters 1982; 49(2), 91-94.

[43] Tittel W, Brendel J, Zbinden H, Gisin N. Violation of Bell inequalities by photons more than 10 km apart. Physical Review Letters 1998; 81, 3563-3566.

[44] Wikipedia, The Free Encyclopedia: Quantum Teleportation.

[45] Burgos ME. Contradictions between Conservation Laws and Orthodox Quantum Mechanics. Journal of Modern Physics 2010; 1, 137-142; DOI 10.4236.

[46] Primas H. Realism and Quantum Mechanics. Proceedings of the 9th International Congress of Logic, Methodology and Philosophy of Science 134. Upsala 1991, 609-631.

[47] Wigner EP. Die Messung Quantenmechanischer Operatoren. Zeitschrift für Physik 1952; 131, 101-108.

[48] Araki H, Yanase MM. Measurement of Quantum Mechanical Operators. Physical Review 1960; 120(2), 622-626.

[49] d'Espagnat B, editor. Stein H, Shimony A in Foundations of Quantum Mechanics. New York: Academic; 1971.

[50] Ghirardi GC, Miglietta F, Rimini A, Weber T. Limitation on Quantum Measurements. Physical Review D 1981; 24(2), 353-358.

[51] Ozawa M. Does a Conservation Law Limit Position Measurements? Physical Review Letters 1991; 67(15), 1956-1159.

[52] Burgos ME. Conservation Laws and Deterministic Evolutions. Physics Essays 1994; 7(1), 69-71.

[53] Burgos ME. Does Conservation of Energy Apply in Processes Ruled by Quantum Mechanical Laws? Speculations in Science and Technology 1997; 20, 183-187.

[54] Burgos ME, Criscuolo FG, Etter TL. Conservation Laws, Machines of The First Type and Superluminal Communication. Speculations in Science and Technology 1999; 21(4), 227-233.

[55] Criscuolo FG, Burgos ME. Conservation Laws in Spontaneous and Mesurement-Like Individual Processes. Physics Essays 2000; 13(1), 80-84.

[56] Pearle P. Suppose the State Vector is Real: The Description and Consequences of Dynamical Reduction. Annals of the New York Academy of Sciences 1986; 480, 539-551.

[57] Heitler W. The Quantum Theory of Radiation. New York: Dover Publications, Inc.; 1984; Chapter 6, p.138, p.61, p.139, p.168

[58] Caves CM. Quantum Mechanics of Measurements distributed in Time. A path-integral Formulation. Physical Review D 1986; 33; 1643–1665.

[59] Weingartner P, Dorn GJW, editors. Burgos ME in Studies on Mario Bunge's Treatise. Amsterdam: Rodopi; 1990; 365-376.

[60] Bell JS. Beables for Quantum Field Theory. CERN-TH: 1984; 4035; 1-10.

[61] Burgos ME. Transitions to the Continuum: Three Different Approaches. Foundations of Physics 2008; 38(10); 883-907.

[62] Eisberg R, Resnick R. Quantum Physics of Atoms, Molecules, Solids, Nuclei and Particles. New York: John Wiley & Sons; 1974; pp. 118-120.

Chapter 6

The Computational Unified Field Theory (CUFT) – Revising Quantum & Relativistic Models

Jehonathan Bentwich

Additional information is available at the end of the chapter

http://dx.doi.org/10.5772/59175

1. Introduction

1.1. The Computational Unified Field Theory (CUFT)

Over the past three years, a new hypothetical 'Computational Unified Field Theory' (CUFT) has been discovered which sets to unify between Quantum Mechanics and Relativity Theory (e.g., whose current theoretical contradiction is considered to be most likely the greatest unresolved enigma in modern Science). Indeed, several previous articles have demonstrated that this new hypothetical CUFT is capable of resolving the principle quantum-relativistic theoretical inconsistencies, replicating all of their key empirical phenomena, and was able to identify (three) "differential-critical" predictions differentiating it from both quantum and relativistic models of physical reality; Indeed, before proceeding to describe a (recent) empirical validation of one of these three CUFT 'differential-critical' predictions it may be helpful to delineate the key theoretical postulates underlying the CUFT as well as its associated "'Cinematic-Film Metaphor";

1.1.1. The 'Duality Principle'

The first theoretical postulate underlying the CUFT is the computational **'Duality Principle'** [4] which identified a basic "computational flaw" associated with both Quantum and Relativistic computational systems; The Duality Principle demonstrates that both quantum and relativistic computational systems comprise a 'Self-Referential Ontological Computational System' (SROCS) which assumes that it is possible to determine the value of any given 'y' (e.g., subatomic 'target' or relativistic 'space-time' or 'energy-mass' entity) strictly based on its direct (or indirect) physical interaction with another (exhaustive) 'x' factor/s (e.g., subatomic 'probe' or relativistic observer). It proves that such SROCS computational structure inevitably leads

to both "logical inconsistency" and "computational indeterminacy" which are contradicted by robust empirical evidence indicating the capacity of both quantum and relativistic to determine the particular value of any given subatomic 'target' element or relativistic 'space-time' or 'energy-mass' phenomenon. Hence, the 'Duality Principle' negates the assumed SROCS computational structure underlying both quantum and relativistic computational systems, instead pointing at the existence of a (singular) higher-ordered 'Universal Computational Principle' (UCP) which computes the "simultaneous co-occurrence" of all exhaustive subatomic 'probe-target' or relativistic 'observer- (space-time or energy-mass) phenomenon' physical interactions (at any given point in time).

Indeed, the identification of such a singular **'Universal Computational Principle' (UCP)** responsible for the computation of all quantum and relativistic (exhaustive) physical interactions which is also postulated to possess three **'Computational Dimensions'** constitutes the second theoretical postulate of the CUFT; In order to perhaps better sense these three *'Computational Dimensions'* of the UCP let us examine a closely related **"Cinematic-Film Metaphor"** which may be used to explain these three 'Computational Dimensions' (as well as some other features and aspects of the CUFT):

Imagine yourself sitting in a cinema film presentation (e.g., seeing a film for the first time – unaware of the 'mechanics' of a film being presented to you)... In this case you could measure (for instance) the "velocity" (or energy) of a jet-plane zooming through the screen, the "time" it took this jet-plane to get from point 'A' to point 'B' (on the screen), the "spatial" length of the plane etc. – being unaware that (in truth) all of these 'spatial', 'temporal', 'energy' (and 'mass') "physical" features are produced based on the 'higher-ordered' computation of the degree of "displacement" or "lack of displacement" occurring across the series of cinematic-film frames!? Thus, for instance, the plane's "energy" (or velocity) is computed based on the number of 'pixels' that plane has been displaced across a given series of frames... Conversely, the plane's "spatial" measure is give based on the computation of the number of 'spatial pixels' that remain constant across a series of cinematic film frames (e.g., resulting in the fact that the plane's length doesn't "increase" or "decrease" across these frames)... Likewise, the "temporal" length of the plane's flight is computed based on the number of changes that occur in- or around- the plane (across a given number of film frames): imagine for instance what would happen to that plane's flight temporal value if the frames were projected more slowly (e.g., in "slow-motion" where there is a smaller number of changes taking place in the plane's flight, giving rise to a "dilated time" measure) or in a case in which precisely the same frame was presented over and over again for say one minute – time would "stand-still"... Similarly, we can devise a special 'cinematic-film' operation in which any given object is projected at "below-threshold" intensity at any given single frame such that only the presentation of the same object (in the same spatial configuration) across multiple number of frames may produce a visible object and that its apparent "mass" value will be computed as a function of the number of frames in which that object appeared 'spatially-consistent'... So, we can see that at least in the "cinematic-film metaphor", 'energy', 'space'; 'time' or 'mass' – are all produced as secondary computational measures being computed by a higher-ordered (singular) computation relating to the degree of 'changes'- or 'lack of changes'- of a given object across the frame, or as measured in the object itself (across a given series of cinematic film-frames)...

Quite similarly, the CUFT posits that the four basic physical features of 'space', 'time', 'energy' and 'mass' are produced through the computation of a singular (higher-ordered) 'Universal Computational Principle' (represented by the Hebrew letter "yud") – of the degree of 'consistency' or 'inconsistency' across a series of extremely rapid (c^2/h) 'Universal Simultaneous Computational Frames' (USCF's): According to the CUFT, this Universal Computational Principle (UCP) employs two 'Computational Dimensions' to compute these four (secondary computational) physical features of 'space', 'time', 'energy' and 'mass' which are: 'Consistency' ('consistent' vs. 'inconsistent') and 'Framework' ('frame' vs. 'object'), and an additional Computational Dimension of 'Locus' ('global' vs. 'local') which accounts for relativistic phenomena.

1.1.2. The UCP's computational dimensions

Hence, the CUFT hypothesizes that the above mentioned 'Universal Computational Principle' (UCP) possesses three 'Computational Dimensions': The *'Framework'* Dimension relates to certain 'computational features' that are computed at the 'object' level, or at the 'frame' (USCF's) level; The 'Consistency' Dimension relates to the UCP's computation of the degree of 'consistency' or 'inconsistency' of an object across a series of USCF's frames (e.g., regarding its above mentioned 'object' or 'frame' measures and also relating to the below mentioned 'Locus' Dimension computation); and the 'Locus Dimension' relates to the UCP's computation of any 'Framework-Consistency' combination from computational perspective of the 'frame' (termed: 'global') or from the 'object's' computational perspective (termed: 'local'); The fascinating facet of these UCP's three Computational Dimensions is that they produce the four physical features of 'space', 'energy', 'mass' and 'time' – i.e., as secondary computational combinations of the 'Framework' and 'Consistency' Computational Dimensions: The CUFT posits that 'space' and 'energy' emerge as a result of the UCP's computation of the degree of 'consistent' or 'inconsistent' measure of an 'object' (e.g., comprising one of the computational levels of the 'Framework' Dimension) the 'Framework' Dimension; Likewise, the basic physical features of 'mass' and 'time' arise as secondary computational features associated with the degree of 'consistent' or 'inconsistent' measure of an object relative to the 'frame' (also comprising the 'Framework' Dimension)!

Hence, the (new) computational definitions of 'space', 'energy', 'mass' and 'energy' are given by:

S:(fi{x,y,z}[USCF(*i*)]+… f*j*{x,y,z}[USCF(*n*)]) / h x n{USCF's}

such that:

fj{x,y,z}[USCF(i)]) ≤ fi{x+(hxn),y+(hxn),z+(hxn)}[USCF(i…n)]

where the 'space' measure of a given object (or event) is computed based on a *frame consistent* computation that adds the specific USCF's (x,y,z) localization across a series of USCF's [1...n] – which nevertheless do *not exceed* the threshold of Planck's constant per each ('n') number of frames (e.g., thereby providing the CUFT's definition of "space" as 'frame-consistent' USCF's measure).

Conversely, the 'energy' of an object (e.g., whether it is the spatial dimensions of an object or event or whether it relates to the spatial location of an object) is computed based on the *frame's differences* of a given object's location/s or size/s across a series of USCF's, divided by the speed of light 'c' multiplied by the number of USCF's across which the object's energy value has been measured:

E:(f*j*{x,y,z}[USCF(n)])–f*i*{(x+n),(y+n),(z+n)} [USCF(*i*...*n*)]) /c x n{USCF's}

such that:

fj{x,y,z}[USCF(n)])>fi{x+(hxn),y+(hxn),z+(hxn) [USCF(i…n)])

wherein the energetic value of a given object, event etc. is computed based on the subtraction of that object's "universal pixels" location/s across a series of USCF's, divided by the speed of light multiplied by the number of USCF's.

In contrast, the of 'mass' of an object is computed based on a measure of the number of times an *'object'* is presented *'consistently'* across a series of USCF's, divided by Planck's constant (e.g., representing the minimal degree of inter-frame's changes):

M: Σ[o*j*{x,y,z} [USCF(*n*)] = o(*i*…*j*-*1*) {(x),(y),(z)} {*USCF*(*i*...*n*)} / h x n{USCF's} {*USCF*(*1*...*n*)} / h x n{USCF's}

where the measure of 'mass' is computed based on a comparison of the number of instances in which an object's (or event's) 'universal-pixels' measures (e.g., along the three axes 'x', y' and 'z') is identical across a series of USCF's (e.g., Σoi{x,y,z} [USCF(n)] = oj{(x+m),(y+m),(z+m)} [USCF(1...n)]), divided by Planck's constant.

Again, the measure of 'mass' represents an *object-consistent* computational measure – e.g., regardless of any changes in that object's spatial (frame) position across these frames.

Finally, the *'time'* measure is computed based on an *'object-inconsistent'* computation of the number of instances in which an 'object' (i.e., corresponding to only a particular segment of the entire USCF) changes across two subsequent USCF's (e.g., Σ o*i*{x,y,z} [USCF(*n*)] ≠ o*j*{(x+m), (y+m),(z+m)}[USCF(*1*...*n*)]), ivided by 'c':

T : Σ o*j*{x,y,z} [USCF(*n*)] ≠ o(*i*…*j*-*1*){(x),(y),(z)} [USCF(*1*...*n*)] /c x n{USCF's}

such that:

T:Σo*i*{x,y,z}[USCF(*n*)]-*j*{(x+m),(y+m),(z+m)} [USCF(*1*...*n*)] ≤ c x n{USCF's}

Hence, the measure of 'time' represents a computational measure of the number of *'object-inconsistent'* presentations any given object (or event) possesses across subsequent USCF' (e.g., once again- regardless of any changes in that object's 'frame's' spatial position across this series of USCF's). Finally, the combination of the 'Locus' Dimension together with the 'Framework-Consistency' Dimensions, e.g., producing the four physical features of 'space', 'energy', 'mass', and 'time' – produces all known relativistic effects and phenomenon, e.g., such as 'time-dilation', 'energy-mass' equivalence and even the curvature of 'space-time'!

http://dx.doi.org/10.5772/59175

1.1.3. The Computational Invariance Principle

Another key theoretical postulate comprising the CUFT is the 'Computational Invariance Principle' which identifies this 'Universal Computational Principle' as the sole 'computationally invariant' element which both produces all four 'computationally variant' physical features of 'space', 'time', 'energy' and 'mass' and also exists independently of these physical features "in-between" any two subsequent 'USCF's frames; As such, the 'Computational Invariance Principle' recognizes the Universal Computational Principle as the sole (and singular) 'invariant' reality underlying the production of the four secondary computational 'variant' physical properties of 'space', 'time', 'energy' and 'mass' (based in part on a well-known scientific principle: "Ockham's Razor" which prefers the simplest most parsimonious theoretical account for complex phenomena) [1]...

1.1.4. The Universal Computational Formula

Finally, this recognition of the Universal Computational Principle as the sole and singular reality producing and sustaining all four (secondary computational) physical properties of space, time, energy and mass has also lead to the formulation of a singular 'Universal Computational Formula' which completely integrates these four secondary computational physical properties, as well as all known quantum and relativistic properties (e.g., as embedded within the higher-ordered Universal Computational Formula):

$$\text{Universal Computational Formula}: \left\{ \frac{`c^2}{h} \right\} = \frac{s}{t} \times \frac{e}{m}$$

2. The CUFT: Quantum-relativistic harmonization- embedding- & transcendence

Hence, the next necessary step in validating the CUFT as a satisfactory TOE is to demonstrate that it's capable of harmonizing between quantum and relativistic models, embed both models within the CUFT's Universal Computational Formula, and providing certain "differential critical predictions" which transcend these quantum and relativistic models (e.g., and if validated empirically validate the CUFT as an satisfactory TOE!) First, we set to demonstrate that the CUFT is able to bridge the (above mentioned) key theoretical inconsistencies that seem to exist between quantum and relativistic models based on its reformulation- and embedding- of quantum and relativistic computation within the singular (higher-ordered) Universal Computational Principle (e.g., due to the Duality Principle's identification of a mutual computational flaw underlying both models, as shown above); Interestingly, based on the singularity of the Universal Computational Principle' computation of both quantum and relativistic relationships (e.g., as embedded within an exhaustive USCF's frames' series) the CUFT is able to embrace both quantum's probabilistic and positivistic relativistic modeling. This is because the Universal Computational Principle's (rapid) production of all exhaustive spatial-pixels in the physical universe comprising each USCF frame – allows it to embed "single spatial-temporal" relativistic objects' (or subatomic 'particles') measurements as well as "multi spatial-temporal" subatomic 'wave' measures! In fact, one of the elegant features of the CUFT is precisely the fact that it conceptualizes such 'single spatial-temporal' relativistic 'objects' or subatomic quantum 'particles' and 'multi spatial-temporal' quantum subatomic 'wave' measurements – within the exhaustive computational framework of the Universal Computational Principle's rapid production of the series of USCF's frames (e.g., comprising all such 'single spatial-temporal' relativistic object or subatomic particle and 'multi spatial-temporal' quantum wave measurements...) Moreover, this exhaustive computational framework of the CUFT allows it to reconceptualize quantum's Uncertainty Principle's 'complimentary pairs' of 'space and energy' or 'time and mass' merely representing a computational constraint intrinsically embedded within the Universal Computational Principle's computation of the two 'Framework' and 'Consistency' Computational Dimensions – i.e., based on the fact that 'space' and 'energy' exhaustively comprise the Framework's Dimensions' 'frame' level, and likewise 'mass' and 'time' exhaustively comprising Framework's 'object' computational level... Hence, the CUFT is capable of embedding both 'single spatial-temporal' relativistic objects (and quantum 'particles'), and (apparently) 'probabilistic' 'multi spatial-temporal' quantum wave measures within the broader and more exhaustive Universal Computational Principle's rapid computation of the series of USCF's frames (e.g., thereby also resolving the 'particle-wave duality' postulate of Quantum Mechanics!)

The CUFT's resolution of the second key quantum-relativistic theoretical inconsistency relating to quantum's instantaneous 'entanglement' phenomenon as opposed to Relativity's speed of light constraint set on the transmission of any signal across space is also anchored in the above mentioned Universal Computational Principle's rapid computation of these USCF's frames ; Since the CUFT posits that the Universal Computational Principle's (rapid) computation of each of the Universal Simultaneous Computational Frame (USCF) simultaneously computes all of the spatial-pixels in the physical universe at a minimal time-point (e.g., c^2/h), then this computation extends the phenomenon of 'quantum entanglement' to all exhaustive spatial points in the universe (e.g., at any such minimal time-point! On the other hand, based on the above embedding of all 'single spatial-temporal' relativistic objects (or subatomic particles) as well as 'multi spatial-temporal' subatomic wave measures within the Universal Computational Principle's exhaustive USCF's computation – it allows for Relativity's apparent speed of light constraint imposed on any such 'single spatial-temporal' relativistic object (or subatomic 'particle') transmission!

The next step towards the validation of the CUFT as a satisfactory TOE involves an articulation of the embedding of quantum and relativistic models within the singular higher-ordered CUFT's Universal Computational Formula – i.e., which is shown to both maintain- and transcend- the (currently) known quantum and relativistic relationships! As can be seen from the two 'quantum' and 'relativistic' formats of the Universal Computational Principle (below), the highlighted portions of these formats conforms to the known mathematical relationships

http://dx.doi.org/10.5772/59175

found in quantum and relativistic models, e.g., Relativity's energy and mass equivalence, and Quantum's 'complimentary pairs' of 'space and energy', 'mass and time' as constrained by the Uncertainty Principle's 'h' Planck's constant simultaneous measurement accuracy constraint:

I. Relativistic Format: $e \times \frac{s}{t} = m \times \frac{c^2}{h}$

II. Quantum Format: $t \times m \times \frac{c^2}{h} = s \times e$

3. The CUFT's "differential-critical predictions"

However, it also becomes clear that the CUFT's Universal Computational Principle's embedding of those empirically validated quantum and relativistic relationships – also transcends and critically differs from these relationships! Indeed, these computational differences between the Universal Computational Formula's 'quantum' and 'relativistic' formats and the 'standard' relativistic '$E=Mc^2$' and quantum 'complimentary pairs' constitutes one (of three) "differentia-critical predictions" that differentiate the CUFT model from both quantum and relativistic predictions, e.g., thereby providing an empirically testable means for validating the CUFT as a satisfactory TOE...

Another key "differential-critical prediction" that differentiates the CUFT from both relativistic and quantum models' predictions are: the CUFT's prediction regarding the more consistent (spatial) presentation of more massive particles (or elements) – across a given series of USCF's frames, relative to less massive particles' appearance across the same series of USCF's frames. In fact, this 'differential critical prediction' regarding the more consistent spatial presentation of more massive particles (or elements) across a series of USCF's frames, relative to the spatial presentation of less massive particles (or elements) precisely replicates the empirical findings of the recently discovered 'Proton-Radius Puzzle', thereby providing a first empirical validation for the CUFT as a satisfactory TOE!

The third 'differential-critical prediction' differentiating the CUFT from both quantum and relativistic models involves a possible "reversal of the space-time spatial-electromagnetic pixels sequence" across a series of USCF's frames electromagnetic spatial-pixels' sequence of a given object or phenomenon; This may be achieved through a precise recording of that object (or phenomenon's) spatial-electromagnetic pixels values (across a given series of USCF's), and a manipulation of these electromagnetic-spatial pixels values (through precise electromagnetic stimulation) so as to produce the reverse sequence of the recorded spatial-electromagnetic values sequence! Interestingly, due to the fact that quantum theory precludes the possibility of the "un-collapse" of the probability wave function following a certain interaction between the any such probe particle and the target particle's wave function – this 'differential critical prediction' is ruled out as a possible prediction of Quantum Mechanics; Likewise, since Relativity sets the speed of light as a clear "unsurpassable" limit for the transference of any signals it also precludes the possibility of "reversing time"; In contrast, since the CUFT defines 'time' (e.g., alongside the other three physical features of 'space', 'energy' and 'mass') merely as a secondary computational property produced by the Universal Computational Principle's three Computational Dimensions' computation of the degree of an "object's-inconsistency" across a series of USCF's frames – then it should allow for the "reversal" of the 'space-time' sequence (e.g., of the particular spatial-electromagnetic pixels' values) across a series of USCF's frames!

4. Empirical validation of the CUFT as satisfactory 'TOE': The 'proton radius puzzle'

Fortunately, the second (abovementioned) 'differential-critical prediction' of the CUFT regarding the more consistent spatial presentations of a more massive particle (or element), relative to the spatial-consistency of a less massive particle (or element) across a given series of USCF's frames – has now received initial empirical validation through the findings associated with the 'Proton-Radius Puzzle'! This is because the 'Proton-Radius Puzzle' empirical findings indicate that the more massive 'Moun Hydrogen Proton' is measured (approximately) 200 times – smaller and more accurate than the standard Hydrogen (e.g., with the 200 times lighter electron particle instead of the Muon)... In order to fully understand how these 'Proton-Radius Puzzle' findings (Bernauer & Pohl, 2014) empirically confirm the differential-critical prediction of the CUFT, lets us return to the CUFT's computational definitions of "mass"; Mass is defined by the CUFT as a measure of the degree of "spatial-consistency" of a particle across a given series of USCF's frames! In mathematical terms, it is measured as the number of times that this particle was presented across the same spatial pixels (measured from within the object's frame of reference) across a series of USCF's frames... This computational definition of 'mass' implies at least two empirically measurable predictions:

a. That the more massive 'Muon' particle should be measured as more accurate- and as smaller- than the less massive electron particle; this is due to the fact that the more massive a particle is the greater its spatial-consistency across USCF's frames! And/or:

b. That more massive particles (e.g., such as the Muon) should be measured across a greater number of USCF's frames, relative to less massive particles (such as the electron); In other words, we could expect to measure the (more massive) Muon across a greater number of USCF's frames than the (lighter) electron!

Interestingly, the 'Proton-Radius Puzzle' precisely confirms the first of these two CUFT 'differential critical' predictions – i.e., indicating that the (200 times) more massive Muon particle (e.g., when embedded within the Hydrogen Proton) is measured as (200 times) 'smaller' and 'more accurate' than the (200 times) less massive electron (associated) Hydrogen Proton! Hence, these findings provide an initial empirical confirmation of the CUFT – as differing from the predictions of both quantum and relativistic models' predictions (e.g., which cannot account for these "Proton-Radius Puzzle" findings)!

http://dx.doi.org/10.5772/59175

Efforts should be made to empirically validate the second (abovementioned) aspect of the CUFT's differential-critical prediction regarding the appearance of 'more massive' particles such as the Muon across a greater number of USCF's frames than the appearance of less massive particles (such as the electron)!

5. The CUFT: Challenging quantum & relativistic "materialistic-reductionistic" assumption

Thus far, we've been able to demonstrate that the CUFT may be considered a satisfactory 'TOE' capable of resolving all major quantum-relativistic theoretical inconsistencies, replicating their primary empirical phenomena, identifying and empirically validating one of the CUFT's 'differential-critical' predictions differentiating it from both quantum and relativistic predictions... The primary aim of the current manuscript is to utilize this recognition of the CUFT as a satisfactory TOE, e.g., which also embeds both quantum and relativistic models within its broader more comprehensive (singular) 'Universal Computational Principle' theoretical framework – *towards recognizing the need to revise certain key theoretical aspects of both quantum and relativistic fields, i.e., based on the CUFT's singularity of the UCP sole production of the (extremely rapid:* c^2/h) *series of the 'Universal Simultaneous Computational Frames' (USCF's); Specifically, the CUFT's emphasis on the singularity of the UCP (rapid) production of the USCF's series – forces us to revise both Quantum and Relativistic "materialistic-reductionistic" basic assumption whereby any (quantum or relativistic) physical relationship (or entities, value/s, phenomenon) can be determined solely based on an exhaustive probe-target (subatomic) interaction or observer-phenomenon (e.g., space-time or energy-mass) relationship in such a manner as to point at the sole and singular production of all such quantum and relativistic entities, phenomena, relationship/s by the UCP's USCF's production...*

Indeed, if we revert back to the CUFT's (first) 'Duality Principle' theoretical postulate, we can see that both Quantum and Relativistic computational systems comprise a *'Self-Referential Ontological Computational Systems' (SROCS)* structure; this quantum and relativistic SROCS computational structure is synonymous with a **"materialistic-reductionistic" assumption**, wherein it is assumed that the determination of the "existence"/"non-existence" of any given 'y' entity (or value) is determined solely based on its direct (or indirect) physical interaction with another (exhaustive) 'x' factor/s... As we've seen, the Duality Principle in fact negates the validity of such assumed (quantum or relativistic) SROCS systems – instead, pointing at the existence of the singular higher-ordered (D2) 'Universal Computational Principle' (UCP) which alone computes the "simultaneous co-occurrence" of all (exhaustive) quantum and relativistic 'probe-target' and 'observer-phenomenon' pairs series (e.g., subsequently shown by the CUFT to comprise any minimal time-point 'Universal Simultaneous Computational Frame'). The CUFT further developed this 'Duality Principle' and 'UCP' (alongside its three 'Computational Dimensions') postulates towards the recognition of the *'Computational Invariance Principle'*: i.e., recognizing the fact that since only the UCP "exists" both 'during' each of the USCF's frames (in fact producing all of its exhaustive universal spatial pixels simultaneously at any such minimal time-point) as well as 'solely existing' "in-between" any two subsequent USCF frames (whereas the four secondary computational 'physical' features of 'space', 'time', 'energy' and 'mass' only exist "during" the UCP's production of the USCF's and its computation of these four secondary computational physical features), then we must conclude that only this singular UCP comprises an invariant "reality" (whereas these four secondary-computational 'physical' features may only be considered *'phenomenally' variant*)... Hence, the CUFT's 'Computational Invariance Principle' in fact points at the sole reality of the UCP (e.g., computationally invariant), as opposed to the "phenomenal" nature of the four secondary computational 'physical' features of 'space', 'time', 'energy' and 'mass' (e.g., computationally variant).

It is hereby suggested that a deeper analysis of these three particular theoretical postulates of the CUFT (e.g., the 'Duality Principle', 'Universal Computational Principle' and 'Computational Invariance Principle') may negate the current (quantum and relativistic) *"materialistic-reductionistic assumption"* (e.g., represented by the SROCS computational structure) based on the sole and singular reality of the 'Universal Computational Principle'; This is made particularly clear based on (above mentioned) 'Computational Invariance' Principle's proof for the singular reality of the Universal Computational Principle – which is the only 'computationally invariant' element which "exists" both during its sole production the (rapid series of) USCF's frames, and "in-between" any two such (subsequent) USCF's frames! This is because based on this 'Computational Invariance Principle', the four physical features of 'space', 'time', 'energy' and 'mass' constitute computationally 'variant' properties and are therefore 'transient' (i.e., exist only "during" the Universal Computational Principle's production of the USCF's frames but ceases to exist "in-between" any two such USCF's frames)... Indeed, their "computational variant" composition makes them possess only "phenomenal" validity as opposed to the singular reality of the 'Universal Computational Principle' which exists permanently (and solely) – both as producing these 'phenomenal' (computationally variant) four 'physical' features "during" the USCF's frames and also "in-between" these USCF's frames... Therefore, the sole production- sustenance- and "transference" of any of these four "physical" features – during or across – any USCF frame/s is only made possible through the singular existence (and operation) of the 'Universal Computational Principle'! In other words, since these four 'phenomenal-physical' features "exist" only during each USCF frame, e.g., as produced by the singular 'Universal Computational Principle', but not "in-between" any two such subsequent USCF's frames – as opposed to the singularity of the UCP which solely produces these four physical features "during" the USCF's frames and also exists "in-between" the USCF's frames; then, we must conclude that the only means for the "production"- "sustenance"- or "transference" of any given 'physical' feature across any two (subsequent) USCF's frames may only be done based on the UCP!

Hence, we must conclude that the basic assumption of "materialistic-reductionism", e.g., whereby it is possible to determine the "existence" or "non-existence" of any given 'physical' 'y' feature solely based on its direct or indirect physical interaction with another (exhaustive) 'x' factor/s – is negated (not only by the above mentioned Duality Principle) but even more explicitly through the recognition that it is not possible for any of these four (phenomenal) 'physical' features to be "transferred" across any two subsequent USCF's frames – except through the computation of the Universal Computational Principle, which constitutes the sole (computationally invariant) "reality" (which exists both "during" the USCF frames producing these four phenomena 'physical' features and solely exists "in-between" any two USCF's frames...) What this means is that the basic "materialistic-reductionistic" assumption underlying both quantum and relativistic SROCS computational systems, i.e., which assumes that the "existence" (or "non-existence") of any given subatomic 'target' or relativistic (space-time or energy-mass) -'phenomenon' is determined solely based on their direct or indirect physical interaction/s with another (exhaustive) subatomic 'probe' element or relativistic 'observer' – is negated! Instead, the 'production'- 'sustenance'- or 'development'- of any 'physical' feature (relationship or phenomenon) – at the quantum or relativistic frameworks can only be computed through the singularity of the Universal Computational Principle!

6. Revising physics: UCP a-causal computation

Hence, there seems to arise a necessity to revise both quantum and relativistic computational systems such that the 'existence' of any of the four (computationally variant phenomenal) physical features (of 'space', 'time', 'energy' or 'mass') in either quantum or relativistic theoretical frameworks be solely produced- sustained- or developed- solely based on the Universal Computational Principle's singular production of all spatial pixels in the universe at any minimal USCF frame/s time-point; Note, however, that this revision does not represent merely a 'philosophical' concept – i.e., in fact, it is suggested that this revision signifies a fundamental shift in Physics as it relies on the **UCP** singular (higher-ordered) **"A-Causal Computation"!**

In order to fully grasp the potential significance of this (novel) UCP 'A-Causal Computation', it may be helpful to specify the theoretical ramifications of recognizing the fact that in both Quantum Mechanics and Relativity Theory the sole production- sustenance- and development- of any of the four 'physical' (phenomenal) features can only be computed by the UCP; Given the fact that the UCP is postulated to compute *"simultaneous co-occurrence"* of all (exhaustive) quantum 'probe-target' and relativistic 'observer-phenomenon' relationships, this means that both within a *single USCF* frame and across a *series* of such *USCF's* frames – we cannot (any longer) rely on any 'materialistic-reductionistic' (SROCS) subatomic 'probe-target' or relativistic 'observer-phenomenon' interactions for determining any of the four (quantum or relativistic) 'physical' features... Instead, we must revise any such quantum or relativistic 'physical' feature based on the UCP computation of the "simultaneous co-occurrence" of all (exhaustive) quantum and relativistic relationships comprising any (single or multiple) USCF frame/s! Indeed, this fundamental shift from the current "materialistic-reductionistic" quantum or relativistic (SROCS) assumption towards a recognition of the sole computation of the UCP of all 'simultaneously co-occurring' quantum and relativistic ('probe-target' and 'observer-phenomenon') interactions is termed: the 'UCP A-Causal Computation' (i.e., of all 'simultaneously co-occurring quantum and relativistic relationships comprising any single or multiple USCF's)...

Now, in order to understand the far reaching theoretical implications of recognizing this UCP (singular) higher-ordered 'A-Causal Computation' let us turn our attention to the two (overarching) conceptual models of the "probabilistic interpretation of Quantum Mechanics" (i.e., represented by the 'probability wave function' and its "collapse" following any given subatomic probe measurement) and Einstein's (famous) General Relativity Einstein field equations (EFE): $R_{\mu\nu} - \frac{1}{2} g_{\mu\nu} R + g_{\mu\nu} \Lambda = \frac{8\pi G}{c^4} T_{\mu\nu}$ (describing the dynamic interaction that exists between massive object's curvature of the fabric of 'space-time' which in return determines their travelling pathway, and vice versa...) It is suggested that in both of these cases, their current theoretical formulation represents the above mentioned "materialistic-reductionistic" assumption – i.e., whereby it is the direct or indirect physical interaction/s between a given subatomic 'probe' and 'probability wave function' target element which determines the "collapse" of that wave function and hence the value of the measured 'target particle; or it is the direct physical interaction of a given (massive) object with the 'space-time' which determines its curvature – and this curvature of 'space-time' (in return) interacts with this given (massive) object thereby determining its pathway movement...

Indeed, the 'materialistic-reductionistic' structure of any such (hypothetical) quantum 'probe-target' or relativistic 'observer-phenomenon' relationships was analyzed earlier, and proven to comprise a SROCS computational structure, e.g., being negated by the CUFT 'Duality Principle' – pointing at the necessity to reformulate both quantum and relativistic computational systems based on the singularity of the Universal Computational Principle...

But, what becomes apparent here, e.g., based on the recognition of the UCP "A-Causal Computation" is that the basic (overarching) theoretical model of both Quantum Mechanics and (General) Relativity Theory must be revised based on this 'UCP A-Causal Computation'! This is because once we accept the CUFT assertion that there exists only one singular (computationally invariant) UCP "reality" and that this UCP singular 'reality' solely computes the "simultaneous co-occurrence" of all exhaustive (quantum and relativistic) relationships comprising a (minimal time-point) USCF frame/s (e.g., termed: 'UCP A-Causal Computation'), then we cannot (any longer) retain either QM's current model regarding the 'materisalistic-reductionistic collapse of the probability wave function' or General Relativity's EFE. This is because both QM's assumed (SROCS) collapse of the probability wave function, as well as RT's assumed SROCS (massive) object –space-time curvature computational structure – are based on this "materialistic-reductionistic SROCS' assumption; In the case of QM this 'materialistic-reductionistic' SROCS computational structure is represented in the assumption wherein the determination of the values of any subatomic 'target' (probability wave function) element is contingent upon its direct (or indirect) physical interaction with another subatomic 'probe' element – i.e., which "causes" the "collapse" of the probability wave function (see earlier description of the Quantum 'probe-target' SROCS structure and its violation of the Duality Principle); In the case of (General) Relativity Theory, this 'materialistic-reductionistic SROCS assumption' is represented by through Relativity's EFE which determines the 'curvature of space-time' based on its direct physical interaction with 'massive objects' and vice versa determines the movement of these 'massive objects' based on their interaction with the curvature of 'space-time' (also see the earlier Duality Principle's analysis of relativistic 'observer-phenomenon' SROCS computational structure);

http://dx.doi.org/10.5772/59175

However, based on the above recognition of the singular reality of the UCP which solely computes the "simultaneous co-occurrence" of all (exhaustive) quantum and relativistic interactions, i.e., comprising any single or multiple USCF's frames (e.g., termed: the UCP 'A-Causal Computation), we must revise this basic 'materialistic-reductionistic SROCS assumption' based on this higher-ordered singular UCP A Causal Computation! In other words, since the sole production- sustenance- and development- of any of the four (phenomenal) 'physical' features (e.g., say in the quantum domain) is based on the UCP 'A-Causal Computation', i.e., of all exhaustive 'probe-probability wave function target' interactions comprising a single (or multiple) USCF frame/s, then we must also conclude that the apparent "collapse" of the probability wave function – cannot be "caused" by the direct interaction between any given probe element and given probability wave function! This is simply due to the fact that according to the above 'UCP A-Causal Computation' all exhaustive values of all quantum subatomic 'probe-probability wave function target' interactions are *computed simultaneously by the UCP – comprising all spatial pixels comprising any single or multiple USCF frame/s...* And since the only "transference" of any of the four (phenomenal) 'physical' features from one USCF frame to another – can only be carried out through the singular operation of the UCP's production- sustenance- and development- of any spatial pixel in te universe (across USCF's frames), then we cannot attribute the "collapse of the wave function" to any physical interaction taking place between any 'probe' and 'probability wave function target' entities (e.g., at any particular USCF frame/s)... Hence, the UCP's A-Causal Computation which produces simultaneously all exhaustive quantum 'probe-target' interactions – at any single or multiple USCF frame/s negates the validity of the "materialistic-reductionistic SROCS" assumption of the "collapse of the probability wave function" (target element) as a result of its direct interaction with another subatomic 'probe' element.

Likewise, based on the recognition of the UCP's singular 'A-Causal Computation' which is solely responsible for the production- sustenance- and development- of all exhaustive 'observer – (space-time, energy-mass) phenomenon' interactions comprising any (single or multiple) USCF frame/s, we must revise the current EFE representing a "materialistic-reductionistic SROCS assumption"; Once again, this is due to the fact that contrary to this 'materialistic-reductionistic SROCS' assumption represented by Relativity Theory EFE according to the UCP's singular 'A-Causal Computation' the UCP computes the "simultaneous co-occurrence" of all exhaustive relativistic 'observer – (space-time or energy-mass) phenomenon' relationships comprising any (single or multiple) USCF frame/s! Therefore, contrary to (General) Relativity Theory's currently assumed 'materialistic-reductionistic SROCS' assumption, wherein the 'curvature of space-time' is determined through its direct physical interaction with 'massive objects' (and vice versa, the movement of these 'massive objects' is determined strictly based on the 'curvature of space-time') – the UCP's A-Causal Computation asserts that it is solely the singularity of the UCP which computes the "simultaneous co-occurrences" of all exhaustive relativistic 'observer – (space-time or energy-mass) phenomenon' interactions (e.g., comprising any single or multiple USCF/s)...

Hence, the fundamental necessary revision of both (contemporary) probabilistic interpretation of QM and of (General) Relativity Theory involves a shift from the current 'materialistic-reductionistic SROCS' assumption underlying both QM and RT – towards the UCP's singular 'A-Causal Computation'! Essentially, this revision implies that instead of the currently assumed (quantum or relativistic) 'materialistic-reductionistic SROCS' assumption wherein the "collapse of the target probability wave function' is "caused" by its direct physical interaction with the subatomic 'probe' element, and the 'curvature of space-time' is "caused" by its direct physical interaction with 'massive object/s' (and vice versa the movement of these 'massive object/s' is "caused" by the 'curvature of 'space-time'; the UCP 'A-Causal Computation' negates any such 'materialistic-reductionistic' "causal" relationships, instead pointing at the fact that it is only the singularity of the UCP which computes the "simultaneous co-occurrence" of all (exhaustive) quantum 'probe-target' and relativistic 'observer-phenomenon' interactions comprising any (single or multiple) USCF frame/s... Perhaps another (lucid) manner of demonstrating the UCP's negation of contemporary Quantum and Relativistic 'materialistic-reductionistic SROCS' computational structure can be given through an analysis of the minimal-temporal, i.e., USCF's frames dynamics representing Quantum Mechanics' currently assumed "collapse of the target probability wave function", as well of the USCF's frames' dynamics representing Relativity's EFE (e.g., describing the interactive effect of 'massive objects' on the 'curvature of space-time' and vice versa as explained above); According to the contemporary 'probabilistic interpretation of QM' the target's probability wave function "collapses" as a result of its direct physical interaction with another subatomic probe element: this means that at a particular minimal-time USCF frame there occurs a direct physical interaction between the 'target's probability wave function' and the 'probe element' – and that based on this direct 'probe-target probability wave function' physical interaction this 'target probability wave function' "collapses"; But since each (single) USCF frame comprises the "minimal time-point" (possible) at which the (singular) UCP produces all exhaustive (quantum and relativistic) spatial-pixels in the universe, then necessarily the initial direct physical interaction between the 'probe' element and the 'target's probability wave function' – takes place at a given USCF frame, whereas the (assumed) "resulting collapse" of this 'target's probability wave function' must occur at a subsequent USCF frame! But, since according to the above mentioned 'Computational Invariance Principle' the sole and singular (computationally invariant) principle which exists both "during" and "in-between" any two subsequent USCF frames is the UCP (whereas the four phenomenal 'physical' features of 'space', 'time', 'energy' and 'mass' exist only "during" any given USCF frame), then the sole 'production'-sustenance- and 'development' of any quantum (or relativistic) (phenomenal) 'physical' feature may only be carried out by the singularity of the UCP – which indeed computes the "simultaneous co-occurrence" of all exhaustive quantum (or relativistic) 'probe-target' interactions comprising any single or multiple USCF's... Hence, since it is not possible for any physical interaction (say) between the 'subatomic probe' and (assumed) 'target's probability wave function' at a given USCF frame (i) to have any effect on their (phenomenal) 'physical' features at a subsequent USCF frame (i+1) – but rather it is the sole (and singular) computation of the UCP of the "simultaneous co-occurrences" of any exhaustive 'probe-target' subatomic interaction/s (at any single or multiple USCF/s frames) which produces- sustains- and develops- any exhaustive subatomic probe-target relationships! Likewise, we can show that based on such 'minimal time-point' USCF's frames analysis, that the currently assumed 'materialistic-reductionistic SROCS' General Relativity Theory model's interactive 'curvature of space-time' based on its direct physical interaction with 'massive objects' (and vice versa) is negated – instead, pointing at the sole production- sustenance- and development- of any of the four phenomenal 'physical' features (of 'space', 'time', 'energy' and 'mass') including the phenomena of the curvature of 'space-time' or of the (apparent) movement of massive objects solely based on the singular UCP 'A-Causal Computation'; In order to demonstrate the impossibility of the currently assumed Relativistic 'materialistic-reductionistic SROCS' assumption, let us (once again) imagine the 'minimal time-point dynamics' of the currently assumed (General) Relativity 'materialistic-reductionistic SROCS' direct physical interaction between certain massive object/s and the curvature of space time (and vice versa, as outlined above): According to this 'materialistic-reductionistic SROCS' Relativistic assumption, it is the direct physical interaction that exists between (one or more) 'massive object/s' and 'space-time' which "causes" this 'space-time' (fabric) to curve, and vice versa, this 'curvature of space-time' "causes" any given 'massive object/s' to travel in a particular (curved) space-time pathway... But, when analyzed from the perspective of a 'minimal time-point' USCF frame/s, then we see that in the first (hypothetical) USCF frame, there is a direct physical interaction between the given 'massive object' and the (fabric of) 'space-time', whereas this 'curving of space-time' may affect the space-time movement pathway of this 'massive object' – only in a subsequent USCF frame/s... However, such 'materialistic-reductionistic' (SROCS) relativistic assumption – not only violates the Duality Principle (as shown earlier and previously), but in the context of this 'minimal time-point' USCF's frames analysis seems to negate the (proven) singularity of the UCP's 'A-Causal Computation': This is because, according to the above 'materialistic-reductionisitc' relativistic assumption it is the direct physical interaction between the given 'massive object' and the (fabric of) 'space-time' – in the first USCF frame/s which "causes" this (fabric of 'space-time') to curve in subsequent USCF frame/s, and this 'curvature of space-time' in turn "causes" this given "massive object" to travel in a "curved space-time pathway" in still later USCF frame/s... However, based on the UCP's 'A-Causal Computation', which computes the "simultaneous co-occurrences" of all exhaustive relativistic 'observer-phenomenon' relationships, we have to negate this relativistic 'materialistic-reductionistic SROCS' assumption – since there cannot exist any "cause and effect" relationship between any direct (or indirect) physical interactions (e.g., such as 'massive object/s' which 'curve the fabric of space-time') at an initial USCF frame/s, and its effect on the 'space-time movement pathway of that massive object' at a subsequent USCF/s... Instead, the singularity of UCP's 'A-Causal Computation' forces us to recognize its computation of the 'simultaneous co-occurrences' of all (four) phenomenal 'physical' features – i.e., including the simultaneous computation of any 'massive object' and any 'curvature of space-time'!

Hence, we reach the inevitable conclusion whereby both Quantum and Relativistic models have to be revised in terms of their basic 'materialistic-reductionistic SROCS' assumption, i.e., recognizing the fact that either Quantum Mechanics' assumed 'collapse of target's probability wave function' as 'caused' by its direct physical interaction with another subatomic 'probe' element; or Relativity's assumed 'curvature of space-time' as "caused" by its direct physical interaction with 'massive objects' – is negated by the CUFT's recognition of the singularity of the Universal Computational Principle's (UCP) (higher-ordered) 'A-Causal Computation', which computes the 'simultaneous co-occurrences' of all quantum and relativistic (exhaustive) 'probe-target' and 'observer-phenomenon' interactions. The key revision brought about by the UCP (higher-ordered) 'A-Causal Computation' is that it negates, i.e., in principle, the existence of any 'materialistic-reductionistic SROCS "causal" relationships in QM or Relativity Theory; this is because once we accept the UCP's (higher-ordered) 'A-Causal Computation' as the sole and singular source for producing- sustaining- and evolving- any of the four phenomenal 'physical' features (of 'space', 'time', 'energy' and 'mass'), e.g., across all exhaustive spatial-pixels (in the universe) comprising any single or multiple USCF frame/s (at the minimal USCF time-point), then we must reject any 'materialistic-reductionistic SROCS' physical relationship/s between any hypothetical quantum 'probe' and 'target' or between any relativistic 'observer' and 'phenomenon' entities as "causing" any hypothetical change or effect in that subatomic target (e.g., such as the assumed "collapse of the target's probability wave function" as "caused" by its interaction with the subatomic probe element) or as "causing" any effect in the given relativistic 'phenomenon' (e.g., such as in Relativity's assumed "curvature of space-time" as "caused" by its interaction with 'massive objects'). Thus, the necessary revision in both QM and RT brought about by the UCP's 'A-Causal Computation' is to base all 'phenomenal' (quantum or relativistic) physical features of 'space', 'time', 'energy' and 'mass' on the singularity of the UCP which solely produces- sustains- or evolves- all of these physical features at all spatial-pixels in the universe comprising any (single or multiple) exhaustive USCF's frame/s...

7. The CUFT's embedding & transcendence of QM and RT models

It should, nevertheless, be made clear that this necessary revision of both Quantum Mechanical and Relativistic Models – does not negate any of the validated empirical phenomena or known quantum or relativistic laws and relationships, but rather broadens our theoretical understanding of these quantum and relativistic phenomena, as embedded in- and (indeed) transcended by- the CUFT theoretical framework; This is due to the fact that whereas Relativity Theory may represent the characterization of single spatial-temporal (relativistic) objects and phenomena, and Quantum Mechanics represents multi spatial-temporal 'probability wave function' (subatomic) entities (e.g., which also embeds 'single' multi spatial-temporal 'particle' elements) – the CUFT expands the theoretical framework to include all single- multiple- and indeed *exhaustive*- spatial-pixels comprising any minimal time-point USCF frame/s... By doing so, and based on the CUFT's identification of this minimal time-point (extremely rapid-series: 'c^2/h') series of USCF's produced solely by the singular 'Universal Computational Principle', the CUFT is capable of fully integrating between quantum and relativistic components and phenomena – which is made most apparent in the CUFT's Universal Computational Formula (e.g. that fully integrates between quantum and relativistic relationships, as well as between the four basic 'physical' features of 'space', 'time', 'energy' and 'mass'):

$$`\frac{c^2}{h} = \frac{s}{t} \times \frac{e}{m}$$

This embedding- and transcendence- of both quantum and relativistic phenomena and relationships within the broader (higher-ordered) CUFT is made most apparent in the two (above mentioned) Relativistic and Quantum formats, which include the known relativistic 'energy and mass equivalence' ($E = Mc^2$) and the quantum 'complimentary pairs';

I. Relativistic Format: $e \times \frac{s}{t} = m \times \frac{c^2}{h}$

II. Quantum Format: $t \times m \times \frac{c^2}{h} = s \times e$

Note, however, that in both quantum and relativistic formats the Universal Computational Formula transcends these (known) quantum and relativistic relationships based on the incorporation of these known relationships within the broader (quantum and relativistic) formats computational structure: Specifically, it becomes apparent that these known quantum and relativistic relationships may represent "special cases" – within the broader Relativistic or Quantum Formats, which in fact represent the complete integration of both "quantum" and "relativistic" computational components within the singular higher-ordered (fully integrated) CUFT's Universal Computational Formula; *Indeed, a more comprehensive mathematical (and empirical) validation of these two Quantum and Relativistic Formats – as different from the (above-mentioned) known quantum and relativistic relationships constituted one of the three (abovementioned) "differential-critical predictions" differentiating the CUFT from both QM and RT* [2], *and should be further investigated and validated – i.e., both mathematically and empirically.*

On the principle theoretical level, it can be pointed out that whereas the current 'Quantum Mechanical' and 'Relativistic' models represent particular phenomena and relationships – corresponding to 'single' spatial-temporal relativistic objects (and phenomena) or subatomic 'multi spatial-temporal' "wave" (and embedded 'single spatial-temporal' "particle") elements and phenomena, the CUFT fully integrates these apparently distinct, particular phenomena within the higher-ordered series of exhaustive Universal Simultaneous Computational Frames (USCF's) which is produced by the Universal Computational Principle based on its singular 'A-Causal Computation' – which produces- sustains- and evolves- all four (phenomenal) 'physical' features of 'space', 'time', 'energy' and 'mass'; Indeed, as we've seen, this singular (higher-ordered) UCP 'A-Causal Computation' negates any "materialistic-reductionistic" (SROCS) quantum or relativistic physical relationships (e.g., such as RT's SROCS assumption regarding the 'curvature of space-time' as "caused" by 'mass', or as QM's SROCS assumed 'collapse of the target's probability wave function' as "caused" by its direct physical interaction with another 'subatomic probe element'); Instead, the singularity of this higher-ordered UCP brings about the complete integration- and indeed transcendence- of the four (quantum and relativistic) phenomenal 'physical' features of 'space', 'time', 'energy' and 'mass'; A such, the CUFT goes beyond RT's integration of 'space-time' and 'energy-mass', and its curvature of space-time by mass (and vice versa) – by fully integrating 'space', 'time', 'energy' and 'mass' as four secondary computational (phenomenal) 'physical' features produced- sustained- and evolved- by the singular (computationally invariant) UCP... Likewise, the CUFT goes beyond QM's complimentary pairs of 'space and energy', 'time and mass' as constrained by Planck's constant ('h' simultaneous accuracy measurement) – as representing the exhaustive (complimentary) computational levels of the two UCP's Computational Dimensions of (Framework and Consistency) (as explained in [1]. Ultimately, the CUFT completely integrates the apparently "distinct" aspects of Quantum and Relativistic models as comprising integral computational aspects of the same singular higher-ordered UCP 'A-Causal Computation' – i.e., such as the complete integration of Relativity's 'c^2' (associated with the speed of light constraint imposed on the transmission of any signal) with Quantum's Planck's constant ('h', associated with subatomic complimentary pairs' simultaneous measurement accuracy constraint) to signify the CUFT's identified rate of UCP rapid production of the series of USCF's frames. Finally, as noted above, the CUFT's unique recognition of the singularity of the UCP's (e.g., computationally invariant) higher-ordered 'A-Causal Computation' which solely produces-sustains- and evolves- all four (e.g., computationally variant) phenomenal 'physical' features of 'space', 'time', 'energy' and 'mass' negates the current basic "materialistic-reductionistic" (SROCS) assumption underlying QM and RT – and forces Physics to recognize this singular UCP 'A-Causal Computation' as the sole reality giving rise to the phenomenology of all 'physical' features, including all quantum and relativistic phenomena and relationships.

8. The CUFT revision of 'dark energy/matter' & 'second law of thermodynamics'

One of the initial theoretical implications of the acceptance of the CUFT as a satisfactory TOE and the acceptance of its singular (higher-ordered) UCP 'A-Causal Computation' as revising contemporary Physics' Quantum and Relativistic "materialistic-reductionistic" SROCS assumption – is its capacity to explain the unresolved "enigma" of 'Dark Matter' and 'Dark Energy' and its potential revision of the 'Second Law of Thermodynamics' (and its associated 'Arrow of Time' enigma); Essentially, the 'Dark Matter, Dark Energy' enigma constitutes the inability of contemporary Physics to account for the acceleration in the rate of expansion of the physical universe – solely based on the observed (and calculated) total mass and energy associated with all planetary object comprising this physical universe... According to these calculations roughly 70-90% of all the mass and energy in the universe is "missing", i.e., cannot be observed! Hence, the working assumption (of Contemporary Physics) is that this (70-90%) of the "missing" mass and energy in the universe is "dark", i.e., it cannot be observed empirically (for some unexplained reason)...

Interestingly, this 'Dark Matter, Dark Energy' enigma is closely connected with the above mentioned UCP 'A-Causal Computation' constraining Relativity's SROCS (interactive) determination of 'massive objects' "causing" the 'curvature of space-time' and vice versa: 'curved space-time' "causing" these 'massive objects' to travel along curved space-time pathways... As delineated above, both the CUFT's 'Duality Principle' theoretical postulate and the discovery of the UCP's singular 'A-Causal Computation' prove (unequivocally) the impossibility of any such Relativity's SROCS "materialistic-reductionistic" assumption: e.g., due to such Relativistic SROCS inevitably leading to both 'logical inconsistency' and ensuing 'computational indeterminacy' which are contradicted by Relativistic Systems empirical capacity to determine both the curvature of space-time and the movement pathways of massive objects; as well as due to the UCP's 'A-Causal Computation' "minimal time-point" USCF's analysis which indicates that based on the 'Computational Invariance Principle' proof that only the 'computationally invariant' UCP exists constantly both "during" the USCF frames and also "in-between" USCF frames, whereas the four phenomenal 'physical' features of 'space', 'time', 'energy' and 'mass' only exist "during" the USCF frames as produced by the singular UCP but cease to exist "in-between" these USCF frames – we must conclude that it is not possible for any of these four phenomenal 'physical' features to "cause" any change across USCF's frames... In other words, the discovery of the UCP singular 'A-Causal Computation' (alongside the Duality Principle) negates the basic "materialistic-reductionistic" SROCS (Relativistic) current assumption, wherein it may be possible for any direct (phenomenal) 'physical' interaction, e.g., of any of these four phenomenal 'physical' features (of 'space', 'time', 'energy' and 'mass') to "cause" any change (or effect) upon another physical attribute (e.g., such as the abovementioned 'curvature of space-time by massive objects' or its vice versa: 'curved space-time' "causing" 'massive object to move along these curved space-time pathways...) Instead, the UCP asserts that the only singular 'A-Causal Computation' solely responsible for the production- sustenance- and development- of any of these four 'phenomenal physical' features (e.g., at any hypothetical 'spatial pixels' comprising any single or multiple USCF's) is singularly conducted by the UCP's computation of the "simultaneous co-occurrence" of all spatial pixels comprising any such USCF.

Hence, the UCP's singular (proven) computation of the 'simultaneous co-occurrence' of all (four) phenomenal 'physical' features of 'space', 'time', 'energy' and 'mass' comprising all spatial pixels in the universe at any minimal time-point (single or multiple) USCF's necessarily negates the possibility of any 'materialistic-reductionistic' relationship existing between any of these four (secondary computational) phenomenal 'physical' features – i.e., including both the curvature of space time by massive objects (or vice versa) as well as the "expansion of the physical universe" – as "caused" by the phenomenal features of the amount of "mass" or "energy" comprising any single or multiple 'USCF' frame/s... In other words, based on the CUFT's proven singularity of the UCP – in producing- sustaining- and developing- all (four) phenomenal 'physical' features of 'space', 'time', energy' and 'mass' (across all exhaustive spatial pixels comprising the totality of the physical universe at any minimal time-point single or multiple USCF frame/s), the UCP's 'A-Causal Computation' is seen as solely responsible for all quantum, relativistic and CUFT known (or predicted) phenomena: This includes also the observed accelerated expansion of the physical universe – i.e., which indeed cannot be accounted for through any 'materialistic-reductionistic ' interactions between any of these secondary computational phenomenal 'physical' features. Indeed, viewed from this singular (higher-ordered) perspective of the UCP'S sole production- sustenance- and evolution- of all spatial pixels in the universe (comprising any minimal time-point single or multiple USCF frame/s) all quantum and relativistic phenomena, e.g., including the accelerated expansion of the physical universe must be accounted for solely through the UCP Causal Computation; Hence, according to the CUFT the relativistic phenomenon of the accelerated expansion of the physical universe cannot be accounted for by the currently assumed "Dark Energy and Dar Matter" – which represent a "materialistic-reductionistic" assumption (as explained in detail above), but instead must be explained as arising from the sole and singular production- sustenance- and evolution- of the physical universe by the UCP...

Another interesting potential theoretical ramification of the adoption of the CUFT as a satisfactory 'TOE' – including its discovery of the UCP singular 'A-Causal Computation', may be its potential revision of the 'Second Law of Thermodynamic' (and associated 'Arrow of Time' phenomenon). The 'Arrow of Time' enigma refers to the observation that the laws of Physics are "biased" in such a manner that events (and phenomena) always occur in a uni-directional temporal direction: thus, for instance, a glass may break into a hundred pieces – but those hundred pieces will not (of themselves) revert back to form a single unitary glass... Indeed, closely associated with this 'Arrow of Time' unidirectional temporal characteristic of physical phenomena is the (famous) 'Second Law of Thermodynamics' which states that in any given physical system the degree of entropy always increases with time... However, based on the CUFT's discovery of the singular UCP 'A-Causal Computation' – which was shown (above) to negate the basic "materialistic-reductionistic" assumption (underlying both Quantum and Relativistic models of physical reality), and one of the CUFT's (previous: [2] 'differential-critical' predictions regarding the possibility of "reversing the sequence of spatial-electromagnetic pixels" based on the application of certain electromagnetic effects, we may need to revise this Second Law of Thermodynamics (and associated 'Arrow of Time' enigma); This is because as explained earlier, none of the quantum or relativistic physical phenomena, relationships (or even laws) can continue to be based on any "materialistic-reductionistic" assumption/s. Hence, as shown above, neither the curvature of space-time by massive objects or (vice versa) the determination of the movement of massive objects based on the curvature of space-time, nor the observed accelerated expansion of the physical universe – can be explained by the current 'materialistic-reductionistic' relativistic (or quantum) assumption, but must be based on the singularity of the UCP 'A-Causal Computation'; Indeed, as we've shown (above), a fine temporal analysis of the dynamics of this singular UCP's production-sustenance- and evolution- of every (exhaustive) spatial pixel in the universe (comprising any single or multiple USCF frame/s) indicates that there cannot exist any 'materialistic-reductionistic' effect of any of the four phenomenal 'physical' features (of 'space', 'time', 'energy' or 'mass') between any two (or more) spatial pixels, i.e., either within the same USCF frames or across different USCF frames. This is due to the UCP's singular asserted computation of the "simultaneous co-occurrence" of all spatial pixels in a given USCF frame (which prohibits any "causal" materialistic-reductionistic" effects existing between any two or more spatial pixels in the same USCF frame), as well as the UCP 's 'A-Causal Computation' associated 'Computational Invariance Principle' which indicates that the sole and singular reality existing invariantly "during" the USCF's frame/s (e.g., as producing, sustaining and evolving any of the four phenomenal 'physical' features of all of its spatial pixels) and "in-between" these USCF's frames is the UCP. Hence, the only source for producing- sustaining- and evolving- any spatial pixel in the physical universe (e.g., at any given USCF frame/s) is the singular UCP, but not any of its (computationally variant) phenomenally produced 'physical' features...

Therefore, also the 'Second Law of Thermodynamics' which asserts the increase in entropy of any physical system with the progression of time – must be revised based on this new higher-ordered recognition of the singularity of the UCP 'A-Causal Computation': Hence, instead of the currently assumed 'materialsitic-reductionistic' basis for this Second Law of Thermodynamics, i.e., wherein it is the physical relationships that exist between a given physical system's material components which "causes" the degree of entropy in that system to necessarily increase with time, the UCP's 'A-Causal Computation' points unequivocally at the singularity of the UCP as producing- sustaining- and evolving- all spatial pixels in the universe and all associated physical phenomena and laws... Moreover, since all four phenomenal 'physical' features of 'space', 'time', 'energy' and 'mass' – are shown to comprise only 'computationally variant' features singularly produced by this UCP then we can foresee a condition in which the spatial-temporal sequence of a given USCF's frames can be reversed (i.e., at least when it is limited to a particular physical phenomenon); This was indeed predicted as one of the CUFT's 'differential-critical' predictions [2] – i.e., regarding the possibility of reversing the 'spatial-electromgnetic' sequence of a given phenomenon such as the growth and decay of a given amoeba. Essentially, this 'critical-differential' prediction of the CUFT states that it should be possible at least in principle) to reverse any given physical phenomenon by recording its precise USCF's spatial-electromagnetic values (e.g., of each of its constituting spatial pixels across a give number of USCF's frames), and then applying a specific electromagnetic stimulation (to each of this phenomenon or physical object's spatial pixels) in such a manner as to produce the "reversed spatial electromagnetic sequence" across the same number of given USCF's frames! Therefore, this CUFT's 'differential-critical' prediction predicted that it should be possible (at least in principle) to "cause" an 'amoeba' to "go back in time" – reversing its spatial-electromagnetic spatial pixels' sequence (by applying the particular electromagnetic stimulation to each of its spatial-pixels across a given number of USCF's frames... More generally then, the CUFT asserts the possibility of reversing the sequence of temporal events comprising any physical phenomenon! Therefore, it should be possible to increase the degree of entropy in any given physical system – contrary to the (currently accepted) 'Second Law of Thermodynamics'!

Although apparently "radical" this 'differential-critical' prediction of the CUFT does not aim to "topple down" the foundations of theoretical Physics, but rather expand our understanding of the physical reality by incorporating both Quantum Mechanics and Relativity Theory within a broader (higher-ordered) theoretical framework based on the discovery (and initial empirical verification) of the CUFT and its associated singularity of the UCP 'A-Causal Computation'; This is simply because in light of contemporary Physics basic contradiction between its two primary theoretical pillars (e.g., Quantum Mechanics and Relativity Theory) which has been shown to be resolved by the CUFT, the (initial) empirical validation of the CUFT's 'differential-critical' prediction associated with the 'Proton-Radius Puzzle', and its discovery of the singularity of the UCP's 'A-Causal Computation' – the fundamental concepts of 'space', 'time', 'energy' and 'mass' as representing merely secondary ('computationally variant') 'phenomenal' features produced by the sole reality of the ('computationally invariant') UCP have to be revised: Specifically, since "time" (alongside all three other 'phenomenal' physical features) is conceptualized as being singularly produced by the UCP – e.g., representing the degree of change of any given object or phenomena across a series of USCF's frames, then it should be possible (at least in principle) to reverse the sequence of spatial-change across frames (through the application of specific electromagnetic stimulation to the relevant spatial pixels comprising this physical phenomenon), thereby reversing the temporal events comprising this physical phenomenon... Hence, it may be said that the Second Law of Thermodynamics accurately represents the "natural progression" or temporal phenomena – but must be revised to include the possibility of reversing these 'natural phenomena' (thereby increasing their measured degree of entropy) across a series of USCF's frames. In a broader theoretical sense, the discovery of the CUFT's 'A-Causal Computation' necessitates us to revise our basic 'materialistic-reductionistic' assumptions underlying contemporary Physics, in such a manner that Quantum Mechanics and Relativity Theory will be anchored and based on the singular higher-ordered operation of the UCP's A-Causal Computation...

Therefore, we see that there is an urgent need to revise both quantum and relativistic models (laws and phenomena) based on the CUFT's discovery of the singularity of the UCP's production- sustenance- and evolution- of the physical universe; This important task involves several future steps, including: an empirical and mathematical verification of all of the CUFT's "differential-critical" predictions (e.g., beyond the initial empirical validation of one of its 'differential-critical' predictions associated with the 'Proton-Radius Puzzle' findings, mentioned earlier), a revision of the laws of Physics based on the CUFT's 'Universal Computational Formula' (which in fact fully embeds and integrates the key quantum and relativistic components) and further explication and exploration of the new theoretical vistas offered by the CUFT higher-ordered and broader theoretical framework (including the potential connection between this singular Universal Computational/Consciousness Principle and individual human Consciousness).

Author details

Jehonathan Bentwich

Address all correspondence to: drbentwich@gmail.com

'BLIS' LTD, Israel

References

[1] Bentwich, J. (2012a) "Harmonizing Quantum Mechanics and Relativity Theory". *Theoretical Concepts of Quantum Mechanics*, Intech (ISBN 979-953-307-377-3), Chapter 22, pp. 515-550.

[2] Bentwich, J. (2012b) "Theoretical Validation of the Computational Unified Field Theory". Theoretical Concepts of Quantum Mechanics, (ISBN 979-953-307-3773), Chapter 23, pp. 551-598.

[3] Bentwich, J. (2013a). "The Theoretical Ramifications of the Computational Unified Field Theory". Advances in Quantum Mechanics (ISBN 978-953-51-1089-7), Chapter 28, pp. 671-882.

[4] Bentwich, J. (2013b). The Computational Unified Field Theory (CUFT): A Candidate Theory of Everything. Advances in Quantum Mechanics (ISBN 978-953-51-1089-7) Chapter 18, pp. 395-436.

[5] Bentwich, J. (2014a) What if Einstein was Right? *Amazon Kindle Book Store.*

[6] Bentwich, J. (2014b) The Next Scientific Shift: The Computational Unified Field Theory. Intech Publication (In Press).

[7] Bernauer & Pohl (2014). The Proton Radius Puzzle. *Scientific American,* 310 (2), p. 20-24.

[8] Bentwich, J. The 'Duality Principle': Irreducibility of sub-threshold psychophysical computation to neuronal brain activation. *Synthese,* Vol. 153, No. 3, pp. (451-455) (2006a).

[9] Bentwich, J. *Universal Consciousness: From Materialistic Science to the Mental Projection Unified Theory,* iUniverse Publication (2006).

[10] 10. Born, M. The statistical interpretation of quantum mechanics, *Nobel Lecture, December* 11, *1954.* Brumfiel, G. Our Universe: Outrageous fortune. *Nature,* Vol. 439, pp. (10-12) (2006).

[11] Ellis, J. The Superstring: Theory of Everything, or of Nothing? *Nature,* Vol. 323, No. 6089, pp. (595–598) (1986).

[12] Greene, B. *The Elegant Universe,* Vintage Books, New York (2003).

[13] Heisenberg, WÜber den anschaulichen Inhalt der quantentheoretischen Kinematik und Mechanik. *Zeitschrift für Physik,* Vol. 43 No. 3–4, pp. (172–198) (1927).

[14] D. Hilbert. 'Die Grundlagen Der Elementaren Zahlentheorie'. *Mathematische Annalen* 104:485–94. Translated by W. Ewald as 'The Grounding of Elementary Number Theory', pp. 266–273 in Mancosu (ed., 1998) From Brouwer to Hilbert: The debate on the foundations of mathematics in the 1920's, Oxford University Press. New York.

[15] Horodecki, R.; Horodecki, P.; Horodecki, M. & Horodecki, K. Quantum entanglement. *Rev. Mod. Phys,* Vol. 81, No. 2, pp. (865–942) (2007).

[16] "Ockham's razor". Encyclopædia Britannica. Encyclopædia Britannica Online. http://www.britannica.com/EBchecked/topic/424706/Ockhams-razor. Retrieved 12 June 2010.

[17] Polchinski, J. All Strung Out? *American Scientist,* January-February 2007 Volume 95, Number 1 (2007).

[18] Stephen W. Hawking Godel and the end of Physics. (Public lecture on March 8 at Texas A&M University) 2003.

[19] On the right track. Interview with Professor Edward Witten, *Frontline,* Vol. 18, No. 3, February 2001.

Chapter 7

A Lie-QED-Algebra and their Fermionic Fock Space in the Superconducting Phenomena

Francisco Bulnes

Additional information is available at the end of the chapter

http://dx.doi.org/10.5772/59078

1. Introduction

It's created a canonical Lie algebra in electrodynamics with all the "nice" algebraic and geometrical properties of an universal enveloping algebra with the goal of can to obtain generalizations in electrodynamics theory of the TQFT (*Topological Quantum Field Theory*) and the Universe based in *lines and twistor bundles* to the obtaining of *orbital spaces* [1] that will be useful in the study of superconducting phenomena. The obtained object haves the advantages to be an algebraic or geometrical space at the same time. This same space of certain $\pounds$- modules can explain and model different electromagnetic phenomena as superconductor and quantum processes where is necessary an organized transformation of the electromagnetic nature of the space-time and obtain nanotechnology of the space-time and their elements when this is affected by the superconducting fields created by the different *electrodynamic Majorana states* in the matter and space [2]. Then using the second quantizing formalism to the description of the behavior of the particles of the affected space for superconducting fields created by the BSC-theory (*Bardeen-Cooper-Schrieffer-theory*) in condensation of the matter where is conformed a fermionic space called set of *pairs of Cooper*, which are comported as bosons will be the domain of the QED (*Quantum Electro-Dynamics*) transformations of the space to different actions as magnetic levitation, electromagnetic impulse, etc. Then considering the Hilbert space of their multiple fermionic modes we obtain the *fermionic Fock space* which describes the *quantum organized transformation* of the particles to a *Fock space* of an arbitrary number of identical fermions explaining the superconductivity as a Bose-Einstein condensation in this process. The fluidity obtained in the Bose-Einstein condensation obeys to a Bogoliubov transformation which in this case, is our organized transformation required to produce the micro-electromagnetic effects from the actions of the operators of Lie-QED-algebra whose fermionic Fock space is the given for the affecting of the energy space where exist the fermionic modes as a Plasmon resonances.

From this study are mentioned some applications to photonics using the *boson link-wave* to hyper-telecommunications and superconducting materials.

2. A Lie-quantum electrodynamical-algebra and their corollaries

2.1. A Lie Algebra in Electro-dynamics

We consider the electromagnetic field or Maxwell field defined as the differential $2-$ form of the forms space $\Omega^2(\mathbb{R}^4)$;

$$F = F_{ab} dx^a \wedge dx^b, \tag{1}$$

which can be described in the endomorphism space of M, by the matrix (where ^{a}, and ^{b}, are equal to $1,2,3$):

$$F_{ab} = \begin{pmatrix} 0 & B^3 & -B^1 & E^1 \\ -B^3 & 0 & B^1 & E^2 \\ B^2 & -B^1 & 0 & E^3 \\ -E^1 & -E^2 & -E^3 & 0 \end{pmatrix}, \tag{2}$$

where E (respectively B) the corresponding forms of electric field (respectively magnetic field).

We want to obtain a useful form to define the actions of the group $\pounds$, on the space of electromagnetic fields F, which is resulted of generalize to the space $\Omega^2(M)$, as an *anti-symmetric tensor algebra* through from induce to the product in the product , shape that will be useful to the localizing and description of the irreducible unitary representations of the groups $SO(4), O(1,3)$, and *representations of spinor fields* in the space-time furthermore of their characterizing as principal $G-$ bundle of M. In the context of the gauge theories (that is to say, in the context of *bundles with connection* as the principal $G-$bundles) we first observe that F, is an *exact form* and thus there exists a $1-$form A^b (*electromagnetic potential*) that defines a connection in a $U(1)-$ bundle on M, and such that[1]

---

1 The anti-symmetric nature of this form results obvius:
$F_{ab} = \partial_a A_b - \partial_b A_a = -F_{ba} = -(\partial_b A_a - \partial_a A_b)$.
Likewise, the electromagnetic field is the 2-form given by (6) with the property of the transformation
$F'_{ab} = \partial_a^{'} A_b^{'} - \partial_b^{'} A_a^{'} = a_{ac} a_{bd} (\partial_c A_d - \partial_d A_c)$
$= a_{ac} a_{bd} F_{cd}$.

In $\mathbb{R}^3$, said 2-form match with the 3×3 - matrix to B_{ab} . Remember that $\mathbf{B} = \nabla \times \mathbf{A}$.

$$F_{ab} = \partial_a A_b - \partial_b A_a, \tag{3}$$

Consider the $K-$invariant $G-$structure $S_G(M)$, of the differentiable manifold $M \cong \mathbb{R}^4$, with Lorentzian metric (and thus pseudo-Riemannian) g, on $\mathbb{R}^4$, with $Diag(g) = (1,1,1,-1)$, in the system of canonical coordinates

$$\varphi\{U, x, y, z, t\}, \tag{4}$$

and let the spaces E, H, two free $\mathbb{R}-$ modules (modules belonging to a commutative ring with unit $\mathbb{R}$) such that

$$\mathrm{E} = \{E \in \mathcal{X}(\mathbb{R}^4) \big| E^b = \iota\frac{\partial}{\partial t}F\}, \tag{5}$$

and

$$\mathrm{H} = \{B \in \mathcal{X}(\mathbb{R}^4) \big| B^b = -\iota\frac{\partial}{\partial t}F\}, \tag{6}$$

where ^{b}, is Euclidean in $\mathbb{R}^3$, and $*$, is Loretzian in $\mathbb{R}^4$. Such $\mathbb{R}-$ modules are $\pounds-$ modules where $\pounds(M) \cong O(1,3)$, is the orthogonal group of range 4. The two modules in (5) and (6) intrinsically define all electric and magnetic fields E, and B, in terms of F. Thus also their tensor, exterior, and scalar products between elements must be expressed in terms of F. To it we consider the tensor product of (5) and (6) as free $\mathbb{R}-$ modules elements, to know[2],

$$E^b \otimes B^b = \frac{1}{c} F \otimes \overline{F}, \tag{7}$$

where c, is light speed and $\overline{F}$ [3], is the dual electromagnetic tensor of F. Then, what must be $E \otimes B$?

2This is valid since tensor product of free $\mathbb{R}-$ modules is a free $\mathbb{R}-$ module[5]. Here $E = \sum_b E^b dx_b \wedge dt, \quad B = \sum_b B^b dx_b \wedge dt.$

3The Levi-Civita tensor can be used to construct the dual electromagnetic tensor in which the electric and magnetic components exchange roles (conserving the symmetry, characteristic that can be seen in the matrices of the electromagnetic tensor F, and their dual $\overline{F}$):

Proposition 2. 1 (F. Bulnes) [3]. Said $\mathbb{R}$ – modules are invariant under Euclidean movements of the group $O(1,3)$, and thus are $\pounds$ – modules.

Proof. [3]. □

Now we consider electro-strength field algebra given by the $U(1)$ – *gauge field* coupled to a charged spin 0 scalar field that takes the place of the Dirac fermions in "ordinary" QED.

Let $(\otimes \mathrm{E})$, be the tensor algebra generated by the elements $F_1 \otimes F_2 - F_2 \otimes F_1$. Let J, be the two-seated ideal generated by the elements $F_1 \otimes F_2 - F_2 \otimes F_1 - [F_1, F_2]$. Let $\mathfrak{e}$, be the Lie algebra whose composition rule is $[,]$. Its wanted to construct an associative algebra with unity element corresponding to $\mathfrak{e}$, such that

$$[F_1, F_2]_{\otimes} = F_1 \otimes F_2 - F_2 \otimes F_1 \tag{8}$$

We want describe energy flux in liquid and elastic media in a completely generalized diffusion of electromagnetic energy from the source (*particles of the space-time "infected" for this electromagnetic energy*), which must be very seemed as a *multi-radiative tensor insights space* or a *electromagnetic insights tensor space.* This will permits us to express and model the flux of electromagnetic energy through pure tensor product of Maxwell fields F, which will be useful in the symplectic structure subjacent in the quantum version of this algebra and their actions of group. After and inside of the demonstration of one result where is related the structure of this quantum version algebra with the superconducting phenomena , the quantum macroscopic effects obtained result of this inheritance of structure, having that for the energy conservation and the use of Lagrangians [4, 5]:

"The rate of energy transfer (per unit volume) from a region of space equals the rate of work done on a charge distribution plus the energy flux leaved in that region"

Of fact these are elements $E^b \otimes B^b$, that are constructed from a power space given in by $\mathrm{E} \times \mathrm{B}$, and that conforms the *electromagnetic multi-radiative space which will be the region space with the "transmittance" of the fermions effects obtained one time that let be quantized the space* $\mathrm{E} \otimes \mathrm{H}$,

to obtain our QED-Lie algebra necessary whose operators will act in the wrapped space (of the electromagnetic type) to get the superconducting effects accord to the Bogoliubob transformation [6] required to produce the quantum electromagnetic effects (electro-anti-gravitational effects) from the actions of the operators of Lie-QED-algebra whose fermionic

---

$$\tilde{F}_{\mu\nu} = \frac{1}{2} \varepsilon_{\mu\nu\kappa\lambda} F^{\mu\nu},$$

where $\varepsilon_{\mu\nu\kappa\lambda}$, is the rank-4 Levi-Civita tensor density in Minkowski space.

Fock space is the given for the affecting of the energy space where exist the fermionic modes as a *Plasmon resonances.*

First we demonstrate the nature of Lie algebra of this space of tensors to electro-physics.

Proposition (F. Bulnes) 3. 1. The electrodynamical space $\mathrm{E}\otimes\mathrm{H}$, is a closed algebra under the composition law $[\ ,\ ]$ of the $U(1)-$ connections.

Proof. Let $F_1=\nabla_a^1A_b^1-\nabla_b^1A_a^1$, and $F_2=\nabla_a^2A_b^2-\nabla_b^2A_a^2$, two elements of $\mathrm{E}\otimes\mathrm{H}$, $\forall A$ an $U(1)-$ connection. Then the composition $[F_1,F_2]$, takes the form in function of the $U(1)-$ connections as

$$\begin{aligned}[F_1,F_2]&=(\nabla_a^1A_b^1-\nabla_b^1A_a^1)\otimes(\nabla_a^2A_b^2-\nabla_b^2A_a^2)-\\&\quad(\nabla_a^2A_b^2-\nabla_b^2A_a^2)\otimes(\nabla_a^1A_b^1-\nabla_b^1A_a^1)\\&=\nabla_a^1A_b^1\otimes\nabla_a^2A_b^2-\nabla_b^1A_a^1\otimes\nabla_a^2A_b^2-\nabla_a^1A_b^1\otimes\nabla_b^2A_a^2+\\&\quad\nabla_b^1A_a^1\otimes\nabla_b^2A_a^2-(\nabla_a^2A_b^2\otimes\nabla_a^1A_b^1-\nabla_b^2A_a^2\otimes\nabla_a^1A_b^1-\\&\quad\nabla_b^1A_a^1\otimes\nabla_a^2A_b^2+\nabla_a^2A_b^2\otimes\nabla_b^1A_a^1)=\\&=\nabla_a^1A_b^1\otimes(\nabla_a^2A_b^2-\nabla_b^2A_a^2)-\nabla_b^1A_a^1\otimes(\nabla_a^2A_b^2-\nabla_b^2A_a^2)\\&\quad-\{\nabla_a^2A_b^2\otimes(\nabla_a^1A_b^1-\nabla_b^1A_a^1)-\nabla_b^2A_a^2\otimes(\nabla_a^1A_b^1-\nabla_b^1A_a^1)\}\\&=\nabla_a^1A_b^1\otimes\nabla_a^2A_b^2-\nabla_a^1A_b^1\otimes\nabla_b^2A_a^2-\nabla_b^1A_a^1\otimes\nabla_a^2A_b^2+\\&\quad\nabla_b^1A_a^1\otimes\nabla_b^2A_a^2-\nabla_a^2A_b^2\otimes\nabla_a^1A_b^1+\nabla_b^2A_a^2\otimes\nabla_a^1A_b^1+\\&\quad\nabla_b^1A_a^1\otimes\nabla_a^2A_b^2-\nabla_a^2A_b^2\otimes\nabla_b^1A_a^1\\&=-\nabla_b^1A_a^1\otimes\nabla_a^2A_b^2+\nabla_b^2A_a^2\otimes\nabla_a^1A_b^1,\end{aligned}$$

Since $F_{ab}=-F_{ba}$, in $\mathbb{R}^4$, then

$$\begin{aligned}&-\nabla_b^1A_a^1\otimes\nabla_a^2A_b^2+\nabla_b^2A_a^2\otimes\nabla_a^1A_b^1=\nabla_a^1A_b^1\otimes\\&\nabla_a^2A_b^2-\nabla_b^2A_a^2\otimes\nabla_b^1A_a^1\in\mathrm{E}\otimes\mathrm{H},\end{aligned}$$

Thus $[F_1,F_2]\in\mathrm{E}\otimes\mathrm{H},\forall F_1,F_2\in\Omega^2(M)$. □

Due to that we are using a *torsion-free connection* (e.g. the *Levi Civita connection*), then the partial derivative ∂_a, used to define F, can be replaced with the covariant derivative ∇_a. The Lie derivative of a tensor is another tensor of the same type, i.e. even though the individual terms in the expression depend on the choice of coordinate system, the expression as a whole result in a tensor in $\mathbb{R}^4$.

Proposition (F. Bulnes) 3. 2. The closed algebra $(\mathrm{E}\otimes\mathrm{H},[,])$, is a Lie algebra.

Proof. Consider

$$[F,F]=\nabla_a A_b \otimes \nabla_b A_a - \nabla_b A_a \otimes \nabla_a A_b - [\nabla_a A_b \\ \otimes \nabla_b A_a - \nabla_b A_a \otimes \nabla_a A_b] = 0,$$

Then the other properties of Lie algebra are trivially satisfied. Thus $\mathrm{E} \otimes \mathrm{H}$, haves structure of Lie algebra under the operation $[,]$. □

2.2. Lie-QED-algebra to superconducting phenomena

If first, we consider the Maxwell tensors given by $F = \nabla_a A_b - \nabla_b A_a$, and thus of the Lie-EM-algebra given in the before section, these comply the following variation principle given by the Maxwell Lagrangian.

$$\mathcal{L}_{\mathrm{MAXWELL}} = -\frac{1}{4} F_{ab} F^{ab}, \tag{9}$$

Then the due action to this Maxwell Lagrangian is

$$\begin{aligned} \Im &= -\frac{1}{4}\int dx^4 F_{ab}F^{ab} = -\frac{1}{4}\int dx^4 (\nabla_a A_b - \nabla_b A_a)(\nabla^a A^b - \nabla^b A^a) \\ &= \frac{1}{2}\int d^4k A_a(k)[-k^2 g^{ab} + k^a k^b] A_b(-k), \end{aligned} \tag{10}$$

where is expected that the inner product $\langle A_a(k), A_a(k')\rangle$, must be equivalent to an expression where the inverse of the differential operator defined by $[-k^2 g^{ab} + k^a k^b]$,[4] appears

[4]The general formula for the Gaussian integral of the last integral of (10) takes the form:

$$\int [d\varphi] \exp\left\{-\frac{1}{2}(\varphi k_{ab} \varphi) + (J\varphi)\right\} \approx \frac{1}{\sqrt{\det K}} \exp(JK^{-1}K),$$

However in our case the operator $K = k_{ab}$, comes given by

$$K = k_{ab}(x-y) = [-k^2 g^{ab} - k^a k^b]\delta^4(x-y),$$

Has the property of the projection operator, that is to say,

$$\int d^4 y k_{ab}(x-y) k_\lambda^\nu (y-z) \propto k_{\nu\lambda}(x-y),$$

and has not inverse. This means that the Gaussian integral diverges. The reason that the free-field part of the action integral given in (10) is singular is due to the gauge invariance which projects out the transverse gauge fields. In the path integral for the free-field part given by

$$\int (d^4 A_a) \exp\left\{i \int d^4 x [\mathcal{L}_0 + J_a A^a]\right\},$$

in the product given in the covariant rules of Feynman diagrams. The field equation that must be to solve is

$$[-k^2 g^{ab} + k^a k^b]k_b \alpha(k) = [-k^2 k^a + k^2 k^a]\alpha = 0,$$

which is particular case of the Bulnes's equation in the *curvature context* [4] to $G = U(1)$. Here $k^a = \nabla^a$, $k_a = \nabla_a$, and $-k = \square$.

We consider the model that consists of a complex scalar field $\varphi(x)$, [5]minimally coupled to a gauge field given by $1-$forms ($U(1)-$ gauge field) "coupled to a charged spin 0 scalar field" and that satisfy:

$$\mathcal{L} = \frac{1}{2}(D_a \varphi)^* D^a \varphi - U(\varphi^* \varphi) - \frac{1}{4} F_{ab} F^{ab}, \tag{11}$$

where F_{ab}, has been defined in the section 2. 1. We define to $D_a \varphi = (\partial_a \varphi - ieA_a \varphi)$, as the covariant derivative of the field φ, also e, is the electric charge and $U(\varphi, \varphi^*)$, is the potential for the complex scalar field. This model is invariant under gauge transformations parametrized by $\lambda(x)$, that is to say, are had the following transformations to the fields:

$$\varphi'(x) = e^{ie\lambda(x)}\varphi(x), \qquad A_a^{'}(x) = A_a(x) + \partial_a \lambda(x), \tag{12}$$

If the potential is such that their minimum occurs at non-zero value of $|\varphi|$, this model exhibits the Higgs mechanism. This can be seen studying the fluctuations about the lowest energy configuration, one sees that gauge field behaves as a massive field with their mass proportional to the e, times the minimum value of $|\varphi|$. As shown by Nielsen and Olesen [7], this model, in $2+1$, dimensions, admits time-independent finite energy configurations corresponding to vortices carrying magnetic flux. The magnetic flux carried by these

---

we have summed over all field configurations including "orbits" that are related by gauge transformations. This over counting is the root of the divergent integral. Thus we have to remove this "volume" of the orbit in this quantization. In the case where the quantizing is realized by the scalar field theory to our superconducting phenomena, the "orbits" are considered as part of the interactions spin-orbit, and the Lie-QED structure to the orbits will be conserved.

[5]In a complex scalar field theory, the scalar field takes values in the complex numbers, rather than the real numbers. The action considered normally takes the form

$$\Im = -\frac{1}{4}\int dx^{d-1} xdt\mathcal{L} = \int d^{d-1} xdt[\eta^{ab}\nabla_a \varphi^* \nabla_b \varphi - V(|\varphi|^2)],$$

This has a $U(1)$, or, equivalently $SU(2)$, symmetry, whose action on the space of fields rotates $\varphi \mapsto e^{i\alpha}\varphi$, for some real phase angle α.

vortices is quantized (in units of $\frac{2\pi}{e}$ [6]) and appears as a topological charge associated with the topological current [8]:

$$J^a_{top} = \in^{abc} F_{bc}, \tag{13}$$

These vortices are similar to the vortices appearing in type-II superconductors. These in the superconducting theory are acquaintance as *fluxoids*. There exist some thermo-dynamical conditions established to the existence the superconductor of type II[7].

Theorem. (F. Bulnes) [1, 9, 10] 2. 2. 1. We consider $F = (H_i H_k - \frac{1}{2}H^2\delta_{ik})/4\pi\sigma\mu,$ [8]and $F_S = -H^2_{ext}\vec{n}/8\pi\sigma\mu,$ with the Hamiltonian foreseen in the *Appendix A* given by (A. 12), of the

---

[6]If $\Gamma \subset \mathfrak{S}$, is away from of the borders and rounds to the hollow (see the Figure 1) and suppose that is have applied a magnetic field to this superconductor $\mathfrak{S}$, then

$$\oint_\Gamma \left(\frac{m^*}{\hbar n_s e^*} J + \frac{e^*}{\hbar c} A \right) dl = n2\pi,$$

But $J = 0,$ (inside the superconductor (or ring in the experimental Figure 1) there not are currents) then

$$\oint_\Gamma A dl = n \frac{\hbar c}{e^*},$$

For the Stokes theorem is had that

$$\oint_\Gamma A dl = \iint_{\partial\mathfrak{S}} B dS = \Phi,$$

To it is necessary remember that the superconducting current J_s, haves an unique value in each point, which is equivalent to that the density of superconductor electrons is injective in each point. This bring as consequence that in a close circuit Γ, of length 2π, we have

$$\phi(2\pi) - \phi(0) = n2\pi,$$

For the circulation around a close circuit Γ, and considering that

$$A = -\nabla^a \phi,$$

we have that on the close circuit Γ,

$$\oint_\Gamma \nabla^a \varphi dl = n2\pi,$$

that in our case is

$$\Phi = n\Phi_0.$$

[7]Its considered the superconductor of the type I, in the intermediate state $\boldsymbol{H}_c(1-n) < \boldsymbol{H} < \boldsymbol{H}_c$, (ellipsoidal superconductor) and we calculate the transition to type II, with $\lambda(T) > \xi(T)$[9,10].

[8]Where the term $\frac{1}{2}H^2\delta_{ik}$, is the term of the tensor *F*, corresponding to the free or total energy of the magnetic field of the superconductor, which involves the thermodynamic effects foreseen in (9), for "compression", to which is subject the surface of the object *O* (**see theAppendix A**).

Lemma. A. 1., and their proof, where is satisfied the inequality on magnetic energy necessary to all magnetic process of superconducting

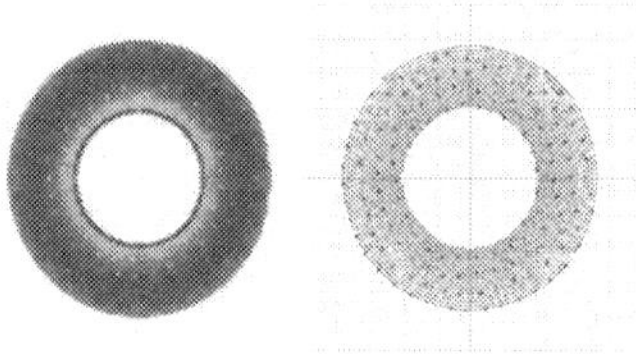

Figure 1. Ring region where is applied a magnetic field in a superconductor $\mathfrak{S}$. This magnetic field could realize iso-rotations and levitation force and impulse. The magnetic flow is intense in this region and the lines of intensity of magnetic field behave like it is pointed out. Inside of hollow the current is zero. The region of *fluxoids* will be generated from the inner of the ring where not exists current but yes magnetic flows doing it in discrete form.

$$8\pi\Phi_0 nH_v \geq \int_O (H(1-h\nabla^2\vec{H}))^2 dV \geq 8\pi\int_O H^2 dV, \tag{14}$$

Then a sensorchip to magnetic flux (pressure on the surface of O) to super-currents is defined by the inequality

$$\int_V j_s(x,y)\delta(z)dV \leq \int_S BdS \leq n\Phi_0, \tag{15}$$

where Φ_0, is a fluxoid ($=\hbar c/e^* = 2.07\times 10^{-7} gauss \bullet cm^2$), being $e^* = 2e$, where e, is the charge of electron.

Proof. [9]. □

Figure 2. One has a view of profile of the magnetic flow of a plate under magnetic field (this simulation was published in the Proceedings of Fluid Flow, Heat Transfer and Thermal Systems of ASME in the paper IMECE2010-37107, British Columbia, Canada with all rights reserved ® [9]).

Developing these topological electromagnetic elements using the tensor $\in^{abc}$, we have to two Maxwell tensors:

$$J_{\alpha\beta} = \nabla_b^1 A_c^1 \otimes \nabla_b^2 A_c^2 - \nabla_c^2 A_b^2 \otimes \nabla_c^1 A_b^1 = (F_1 \otimes F_2 - F_2 \otimes F_1), \tag{16}$$

precisely is our tensor algebra given in the *proposition 3. 1.*, with their conserved Lie structure.

The essential difference between both versions consists in the coupling to a charged $spin0$, scalar field, that in this case is a scalar magnetic field corresponding to a magnetic flow associated to the supercurrent J_s.

Considering the supercurrent J_s in presence of magnetic field of vector potential, this takes the form

$$J_s = \frac{e^* \hbar}{2m^* i}(\psi^* \nabla \psi - \psi \nabla \psi^*) - \frac{(e^*)^2}{m^* c}|\psi|^2 A, \tag{17}$$

where ψ, is a function very general of complex type that are changing spatially and that in an any point this function depends of the order parameter (as coherent length, penetration length, etc parameters that are useful to characterize a superconductor [11]) and $|\psi|^2 = n_s$, is the density of the superconducting electrons.

Considering the action (10) and the proper to the Lagrangian (11), the electromagnetic total action to all electromagnetic phenomena (included the given in ordinary electromagnetism) using the scalar field theory takes the form:

$$\begin{aligned} S_\xi = \int \varphi^* [-\nabla_a \nabla^a - m^2]\varphi + \frac{1}{2}A_a[g^{ab} - (1 - \frac{1}{\xi})\nabla^a \nabla^b]A_b \\ + ieA_a((\nabla^a \varphi^*)\varphi - \varphi^* \nabla^a \varphi) + e^2 A_a A^a \varphi^* \varphi, \end{aligned} \tag{18}$$

in where and under certain physical conditions of symmetries we can establish the following relations between the complex functions ψ, and ψ^*, and their covariant derivatives $\nabla \psi$, $\nabla \psi^*$, and the proper to the scalar fields given by φ, $(\nabla^a \varphi^*)$, φ^*, and $\nabla^a \varphi$ obtaining the pure action to superconductors given by the integral:

$$S_s = \int ieA_a(\psi^* \nabla \psi - \psi \nabla \psi^*) + e^2 A_a A^a \nabla \psi^* \nabla \psi, \tag{19}$$

which appears in the superconductor under energy regime given by their corresponding Hamiltonian defined in (A. 12) of the *appendix A*. The representation spaces that appear are the Fock states spaces and are corresponded in the superconductivity with the photon states spaces where there exist the interaction *photon-phonon-photon* [12] under frame study of the microscopic theory of the superconductivity. In a second affirmation, is necessary consider that electrons in superconductivity are moving in a very special enthrone, that is the crystalline net formed by ions that constitute the solid which we want that be superconductor under the application of field actions as given in (19) on the solid of object O.

We consider a particles system, all identical, that is to say, undistinguishable, that is to say, the interchange of two of them not change the measurable properties of the system. Let e_1, and e_2, two electrons and we suppose that the electron e_1, is in a state that comes represented by the function ψ_n, (*wave function*), for other side, the electron e_2, comes represented by the wave function ψ_m, and we suppose for last that the direct interaction between these two electrons is quasi-vanishing. Then we can describe the system of these two electrons by the wave function $\psi_n(e_1)\psi_m(e_2)$. Remember that the two particles are identical, thus the interchange of the two let us equal the system. Then also is valid as before that $\psi_n(e_2)\psi_m(e_1)$. Then the total wave function with the interchanges realized by the two electrons takes the form:

$$\Psi = \frac{1}{\sqrt{2}}[\psi_n(e_1)\psi_m(e_2) + \psi_n(e_2)\psi_m(e_1)], \tag{20}$$

and their conjugated

$$\Psi^* = \frac{1}{\sqrt{2}}[\psi_n(e_1)\psi_m(e_2) - \psi_n(e_2)\psi_m(e_1)], \tag{21}$$

But if the two states are the same state, then only $\Psi \neq 0$, since $\Psi^* = 0$. Then the system is anti-symmetric in the interchange of two particles, that is to say, the wave function is anti-symmetric under the interchange of electron coordinates. But by the *Pauli Exclusion Principle* the situation described in the total wave function Ψ, is incorrect, being the correct by Ψ^*.

But the before help us to establish the anti-symmetric structure of the interaction between pair of particles in the microscopy superconductivity theory and reflected this anti-symmetry property also in every *spin-orbit interaction* of every wave function to the two pair electrons that satisfy the total wave function Ψ^*.

Considering to an electron field, a representation $\xi : \mathfrak{E} \to V$, where V,, is a Hilbert space and whose correspondence rule is

$$e \mapsto \xi(e), \tag{22}$$

and let J, [9]the two-sided ideal in the tensor algebra defined in the section 2. 1, $(\mathfrak{E}, \otimes)$, generated by the elements of the form $e_1 \otimes e_2 - e_2 \otimes e_1$, where $e_1, e_2 \in \mathfrak{E}$.

Proposition 2. 2. 1. There is a natural one-to-one correspondence between the set of all representations of $\mathfrak{E}$, on V, and the set of all representations of $\mathfrak{E} \otimes \mathfrak{E} / J$, on V.. If ξ, is a representation of $\mathfrak{E}$, on V, and , ξ^*, is a representation of $\mathfrak{E} \otimes \mathfrak{E} / J$, on V, then

$$\xi(e) = \xi^*(e^*), \qquad \forall\, e \in \mathfrak{E}, \tag{23}$$

[9]Remember that J, from a point of view of the superconductors is a topological current associated with the topological charge defined related with the magnetic flux carried by the fluxoinds.

Proof. Let $\xi,$ be a representation of $\mathfrak{E}$, on V. Then there exists a unique representation $\bar{\xi},$ of $(\mathfrak{E},\otimes)$, on V , satisfying that $\bar{\xi}(e)=\xi(e), \quad \forall e\in\mathfrak{E}$. Then mapping $\bar{\xi},$, vanishes on the ideal J, , became

$$\bar{\xi}(e_1\otimes e_2-e_2\otimes e_1-[e_1,e_2])=\xi(e_1)\xi(e_2)-\xi(e_2)\xi(e_1)-\xi([e_1,e_2])=0, \tag{24}$$

Thus we can define a representation ξ^*, of factor algebra $\mathfrak{E}\otimes\mathfrak{E}/J$, on V , by the condition $\xi^*\circ\pi=\bar{\xi}$. [10]Then (23) is satisfied and determines ξ^*, uniquely. Also is unique, indeed suppose other representation, for example σ, $\mathfrak{E}\otimes\mathfrak{E}/J$ of on V. If $e\in\mathfrak{E}$, we put $\xi(e)=\sigma(e^*)$. Then the mapping $e\mapsto\xi(e)$, is linear and is a representation of $\mathfrak{E}$ on V , since

$$\begin{aligned}\xi([e_1,e_2])&=\sigma([e_1,e_2]^*)=\sigma(\pi(e_1\otimes e_2-e_2\otimes e_1))\\&=\sigma(e_1^*e_2^*-e_2^*e_1^*)=\xi(e_1)\xi(e_2)-\xi(e_2)\xi(e_1).\square\end{aligned}$$

Considering in particular the representation given for Ψ^*, [11]we have that Lie-QED-algebra structure to the spin and orbits is conserved. Indeed the spin and orbit parts can be separated in each wave function as for example, from (24) we have that if $\psi(e_1)=\phi_n(r_1)\alpha(s_1)$, and $\psi(e_2)=\phi_m(r_2)\alpha(s_2)$, then (24) defines Ψ^*. Remember that the electron is a *fermion* and the electrons are described for the anti-symmetric wave functions. This property will be fundamental in the section relative to the construction of the fermionic Foch space corresponding to the Lie-QED-algebra.

Def. 2. 2. 1. [10]. A $\mathfrak{E}\otimes\mathfrak{H}$ - field is an element of a bi-sided ideal of the Maxwell fields [1, 6]. Explicitly is the formal space

$$\mathfrak{E}\otimes\mathfrak{H}=\{(F_1,F_2)\in\Omega^2(O)\times\Omega^2(O)\big|F_1\otimes F_2-F_2\otimes F_1-[F_1,F_2],\text{with}\otimes=\otimes_{\mathbb{R}}\}, \tag{25}$$

Before of this, we pass to the fundamental lemma to characterize the algebra $\mathfrak{E}\otimes\mathfrak{H}$, as the fundamental algebra of all movements and electromagnetic phenomena *(for example, magnetic levitation, electromagnetic matter condensation, Eddy currents, etc)* produced to

---

[10] Its realized the following descend map π, from $\bar{\xi}:\mathfrak{E}\to\mathfrak{E}\otimes\mathfrak{E}$, to $\xi^*:\mathfrak{E}\to\mathfrak{E}\otimes\mathfrak{E}/J$.

[11]It is the Slater determinant (that helps to construct wave functions to start of the expressions $\phi_n(r_1)\alpha(s_1)\phi_m(r_2)\beta(s_2)$ where $r_i,s_i(i=1,2)$ are the radius of the orbit and spin respectively) , for example:

$$\Psi^*=\frac{1}{\sqrt{2}}\begin{vmatrix}\phi_n(r_1)\alpha(s_1) & \phi_n(r_2)\alpha(s_2)\\ \phi_m(r_1)\alpha(s_1) & \phi_m(r_2)\alpha(s_2)\end{vmatrix}=\frac{1}{\sqrt{2}}[\phi_n(r_1)\alpha(s_1)\phi_m(r_2)\alpha(s_2)-\phi_m(r_1)\alpha(s_1)\phi_n(r_2)\alpha(s_2)]$$

quantum level by their electromagnetic fields satisfying the variation principle in their field actions.

Lemma (F. Bulnes) [10] 2. 2. 1. All electromagnetic actions and their effects (microscopic and macroscopic) on the superconductor object O, comes from the $\mathfrak{E}\otimes\mathfrak{H}$ - fields.

Proof. [10]. □

One important fact inside the demonstration of the lemma 2. 2. 1, was consider the bi-sided ideal given by the space (25) whose actions are extended to all space from the superconductor O, until the infinite (ambient of O) through the gauge transformations used by the Lie-QED-algebra . Then by the lemma A. 1 (F. Bulnes), given in the *appendix A*, of this work [9], the quantum effects underlying in superconducting phenomena satisfies that

$$H(A,\mathfrak{I}_M) = \mathfrak{I}_M - H(A,B) = \int_O \left[F_i F_k - \frac{1}{2} H^2 \delta_{ik} \right] dV / 4\pi$$
$$= \int_O L_M(x(s))d(x(s)) - \int_O H^2 / 8\pi dV,$$

where $H^2/8\pi$, is the free magnetic energy and the integral of the Lagrangian, of the expulsion by the action , and that is useful to establish the macroscopic wave functions that give place to a microscopic quantum current Js. By the mathematical electrodynamics [3] we can define using the structure of $\mathfrak{E}\otimes\mathfrak{H}$, (that is to say, a module of the exterior algebra which is deduced by the universal map applied to each term of the element $F_1 \otimes F_2 - F_2 \otimes F_1 - [F_1, F_2]$) that:

$$(\psi, J) = \int_{\partial O} (\psi^* \wedge J - J \wedge \psi^*) dS - \int_{\partial O} \langle \psi, \psi^* \rangle dS, \tag{26}$$

where $J = \nabla\psi$,, and $\psi^* = conj(\psi)$,, such that $|\psi|^2 = \langle \psi, \psi^* \rangle$, which is (19) in the supercurrents modality. As the last integral (18) measures effects due to the macroscopic actions to level quantum, this proved the affirmation of the lemma 2. 2. 1, in microscopic theory of superconductivity and also the macroscopic effects due to the Eddy currents must satisfy the magnetic force equation [9] to magnetic levitation.

2.3. Photon spin algebra from $\mathfrak{E}\otimes\mathfrak{H}$

The same Lie structure is conserved to the electromagnetic spin algebra. The Lie structure of the macroscopic level given and demonstrated to the space $\mathfrak{E}\otimes\mathfrak{H}$, in the before subsection

2.1, and using after through the path integral quantization on Lie bracket $[e_1,e_2]$, given in the *subsection* 2. 2 can establish a version of this quantized Lie algebra to quantum spin number (parts $\alpha(s_i)$) associated to the photons that interact in the superconducting phenomena on the quantum macroscopic effects generated by the superconducting currents (for quantized electromagnetic fields). Then the QED- algebra obtained conserves the Lie structure to spin electromagnetic operators.

The photon can be assigned a triplet spin with spin quantum number $s=1$, accord to the classification of particles and their spin. This is similar to, say, the nuclear spin of the N isotope, but with the important difference that the state with $M_s=0$, is zero, only the states with $M_s=\pm 1$, are non-zero. We consider the electro-spin operator as the vector with their Pauli matrices associated[12]:

$$S_k=-i\hbar\varepsilon_{ijk}=\frac{\hbar}{2}\sigma_k,(k=1,2,3), \tag{27}$$

Then we can define the analogous of $\mathrm{E}\otimes\mathrm{H}$ to the quantum spin context as the algebra:

$$\mathfrak{e}\otimes\mathfrak{h}=\left\{S_k\in\left|[S_i,S_j]=-i\hbar\varepsilon_{ijk}\right.\right\}, \tag{28}$$

which is closed under the bracket $[,]_{\otimes}$, operation . Indeed, we consider two elements $S_i,S_j\in\mathfrak{e}\otimes\mathfrak{h}$, given by the relations

$$S_i=-i\hbar(\varepsilon_j\otimes\varepsilon_k-\varepsilon_k\otimes\varepsilon_j),\qquad S_j=-i\hbar(\varepsilon_k\otimes\varepsilon_i-\varepsilon_i\otimes\varepsilon_k), \tag{29}$$

satisfying the cyclically rule $i\to j\to k\to i$. Then their operation under $[,]$, is

$$\begin{aligned}&[S_i,S_j]=-\hbar^2(\varepsilon_j\otimes\varepsilon_k-\varepsilon_k\otimes\varepsilon_j)(\varepsilon_k\otimes\varepsilon_i-\varepsilon_i\otimes\varepsilon_k)+\\&\hbar^2(\varepsilon_k\otimes\varepsilon_i-\varepsilon_i\otimes\varepsilon_k)(\varepsilon_j\otimes\varepsilon_k-\varepsilon_k\otimes\varepsilon_j)=i\hbar\left[-i\hbar(\varepsilon_i\otimes\varepsilon_j-\varepsilon_j\otimes\varepsilon_i)\right]=i\hbar S_k\in\mathfrak{e}\otimes\mathfrak{h}.\end{aligned}$$

For simple inspection it follows that

$$-i\hbar(\varepsilon_i\otimes\varepsilon_j-\varepsilon_j\otimes\varepsilon_i)\bullet\varepsilon^{(\mu)}=\mu\varepsilon^{(\mu)}, \tag{30}$$

---

[12]In the special case of spin $-1/2$ particles, σ_x, σ_y, and σ_z, are the three Pauli matrices given by:

$$\sigma_x=\begin{pmatrix}0&1\\1&0\end{pmatrix},\sigma_y=\begin{pmatrix}0&-i\\i&0\end{pmatrix},\sigma_z=\begin{pmatrix}1&0\\0&-1\end{pmatrix}.$$

http://dx.doi.org/10.5772/59078

with $\mu = +1$ or -1, and therefore labels the photon spin, $S_z|\mathbf{k},\mu\rangle = \mu|\mathbf{k},\mu\rangle$, with $\mu = +1$ or -1. Then due the vector potential is a transverse field the photon has no forward ($\mu = 0$) spin component.

3. Fermionic Fock spaces of the superconductivity

By the BSC-theory we have a Cooper pair is a magnitude whose spin is zero. But spin zero are bosons, then is easy fall in the temptation of treat a Cooper pairs as bosons. Furthermore, we have indicated that few many more Cooper pairs, better energy will be the process of superconductivity. However, the Pauli principle remains in force and the state formed, for example by $(k\uparrow,-k\downarrow)$, cannot be occupied for more than a pair of electrons at the same time (see figure 3).

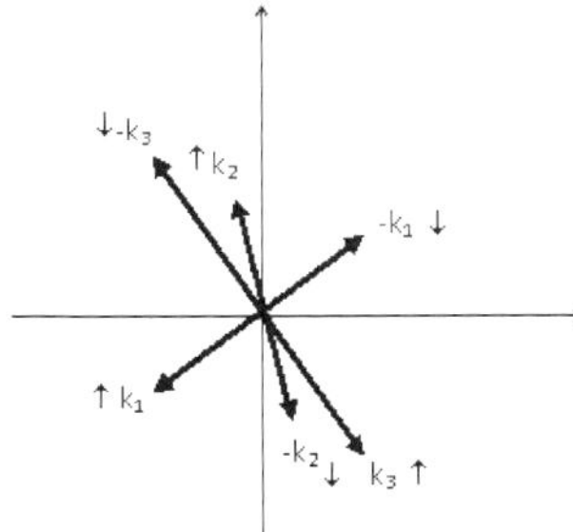

Figure 3. BSC fundamental state with three Cooper pairs.

Also the Hamiltonian in the BSC-theory is constructed by operators (that is to say, their formalism with that is calculated the energy of the *fundamental superconductor state are anti-commutative*) follows anti-commutative rules as was discussed in the introduction of this paper. But involve a boson that is created an interaction between fermions[13]. The wave function such and as is proposed by the BSC-theory to Ξ, electrons (*foreseen in* (21) *to the case of only pairs of electrons*) is the product of wave functions of pair conveniently anti-symmetrized, that is to say:

$$\psi(1,2,\ldots,\Xi) \approx \psi(1,2)\psi(2,3)\cdots\psi(1-\Xi,\Xi), \tag{31}$$

If we not write explicitly the part of spin and only we do with the orbital part we have:

$$\psi(1,2,\ldots,\Xi) \approx \sum_{k_1}\sum_{k_2}\cdots\sum_{k_{\Xi/2}}\zeta_{k_1}\cdots\zeta_{k_{\Xi/2}}e^{i(k_1r_1-k_2r_2+\ldots+k_{\Xi/2}r_{\Xi-1}-k_{\Xi/2}r_{\Xi})}, \tag{32}$$

[13]To difference of the London superconductivity where a charged gas of bosons produces naturally a Meissner effect.

where each term of this wave function describes a configuration where the Ξ, electrons is grouped in $\Xi/2$, pairs that are:

$$(k_1, -k_1)\cdots(k_{\Xi/2}, -k_{\Xi/2}), \tag{33}$$

The spin part is immediate, each electron of each pair haves opposite spines. The wave function is a complicated function that covers all the related pairs between them. This takes the form from excited states are obtained as linear combinations of the ground state excited by some creation operators $\prod_{k=1}^{n} a_{i_k}^{\dagger} \left|0\right\rangle = 0$,to the wave functions as:

$$\psi = \prod_{k} \psi_k, \tag{34}$$

Using the second quantizing formalism to the Fock space in *appendix B*,, we have that the potential energy to said pairs is:

$$V = \sum_{k,k'} V_{k,k'} b_k^{\dagger} b_{k'}, \tag{35}$$

The term of kinetic energy of the corresponding Hamiltonian considering the energy in the Fermi level is:

$$E = \sum_{k,k'} 2\varepsilon_k b_k^{\dagger} b_k + \Delta b_k^{\dagger} b, \tag{36}$$

where to the states of exited electrons (super-electrons) appear the trenches (to break the pairs and get to superconductivity peak state (see the figure 4A) and 4B)).

The fermionic Fock space is (*B*. 1) where for second approximation we have

$$\mathcal{F} = \mathcal{H}_0 \oplus \mathcal{H}_1 \oplus \mathcal{H}_2, \tag{37}$$

whose energies of the electron are ε_k, and ε_{k-q}, with momentums to two electrons in superconducting states hk, and $h(k-q)$ respectively. This interaction is negative since is attractive, Then the potential is:

$$V(k,k',q) =, \frac{A^2 \hbar\omega_q}{(\varepsilon_k - \varepsilon_{k-q})^2 - (\hbar\omega_q)^2}, \tag{38}$$

where A, is a coupling electron-fonon. Then

$$\left|\varepsilon_{k\pm q} - \varepsilon_k\right| < \hbar\omega_q, \tag{39}$$

Then after of realize some calculations in the fermionic Fock space is had that

$$\varepsilon_{\mathcal{F}} = (\Delta^2 + \varepsilon_k{}^2)^{1/2}, \tag{40}$$

where Δ, is the minimum energy of excitation, that is to say, the value of energy trenches that appears in the superconductor state (*see the* Figure 11A, *in the last section*). *These trenches* have a big variation with respect to the absolute temperature of the material. The energy ε_k, is the quasi-particle energy, that is to say, the energy of the holes of the fermions distribution when happen the excited states. The trenches energy as the excited states shape the orthogonal space of infinite dimension separated by the Fermi momentum.

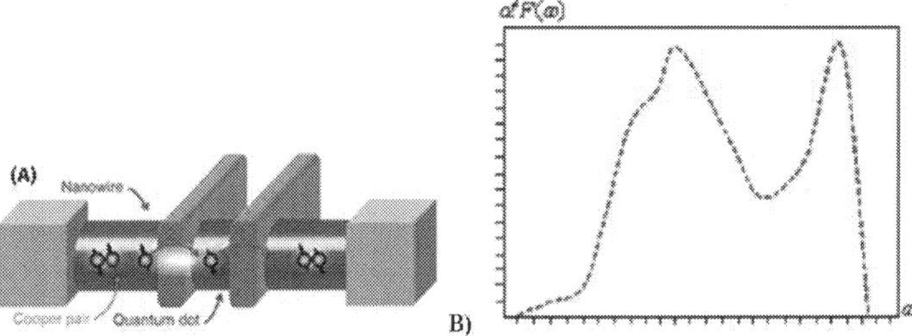

Figure 4. A) Nano-wire device to break Cooper pairs. The Cooper pairs must be break to obtain the maximum superconductor state. The super-electrons are transformed in Fermi liquid which established the required transformation of the immediate region of the space-time where must be executive to transformation due superconductivity [2, 13, 14, 15]. B). Spectral density of electron-phonon.

Indeed, consider a system of fermions with an one-body Hamiltonian of the form (accord to (32) and (34)):

$$\hat{H} = \sum_k \varepsilon_k \hat{a}_k^{\dagger} \hat{a}_k + E_0, \tag{41}$$

When all particle energies ε_k, are positive, the ground state of the system is the vacuum state $\left|\text{vac}\right\rangle$, with all $n_k = 0$. In terms of the creation and annihilation operators said state $\left|\text{vac}\right\rangle$, can be identified as the unique state killed by the the annihilation operators, that is to say, $\hat{a}_k\left|\text{vac}\right\rangle = 0, \forall k$. The excited states of the Hamiltonian (50) are $\Xi -$ particle states which obtain by applying creation operators to the vacuum, $\left|k_1,\ldots,k_\Xi\right\rangle = \hat{a}_{k_\Xi}^{\dagger}\cdots\hat{a}_{k_1}^{\dagger}\left|\text{vac}\right\rangle$, the energy of such a state is $E = E_0 + \varepsilon_{k_1} + \ldots + \varepsilon_{k_\Xi} > E_0$.

Now suppose for a moment that all the particle energies ε_k, are negative instead of positive. In this case, adding particles decreases the energy, so the ground state of the system is not the vacuum but rather the full-to-capacity state

$$\left|\text{full}\right\rangle = \left|\text{all } n_k = 1\right\rangle \prod_{\text{all } k} \hat{a}_k^{\dagger} \left|\text{vac}\right\rangle, \tag{42}$$

with energy

$$E_{\text{full}} = E_0 + \sum_{\text{full } k} \varepsilon_k, \tag{43}$$

Never mind whether the sum here is convergent; if it is not, we may add an infinite constant to the E_0, cancel the divergence. What's important to us here are the energy difference between this ground and the excited states.

The excited states of the system are not completely full but have a few *holes*. If we consider $n_{k_1} = \ldots = n_{k_\Xi} = 0$, for some $\Xi-$ modes $(k_1, \ldots, k_\Xi)$, while all the other $n_l = 1$. The energy of such a state is

$$E = E_0 + \sum_{l \neq k_1, \ldots, k_\Xi} \varepsilon_l = E_{\text{full}} - \sum_{i=1}^{\Xi} \varepsilon_{k_i} > E_{\text{full}}, \tag{44}$$

In other words, an un-filled hole in mode k, carries a positive energy $-\varepsilon_k$.

In terms of the operator algebra, the $\left|\text{full}\right\rangle$, state is the unique killed by all the creation operators, $\hat{a}_k^{\dagger}\left|\text{full}\right\rangle, \forall k$. The holes can be obtained by acting on the $\left|\text{full}\right\rangle$, state with the annihilation operators that remove one particle at a time. Thus,

$$\left|1\text{hole at } k\right\rangle = \left|\hat{n}_k \text{ other } n = 1\right\rangle = \hat{a}_k \left|\text{full}\right\rangle, \tag{45}$$

and likewise

$$\left|N \text{ holes at } k_1, k_2 \ldots, k_\Xi\right\rangle = \hat{a}_{k_\Xi} \cdots \hat{a}_{k_1} \left|\text{full}\right\rangle, \tag{46}$$

Altogether, when the ground state is $\left|\text{full}\right\rangle$, the creation and annihilation operators Exchange their roles. Indeed, the $\hat{a}_k$, make extra holes in the full or almost-full states, while the $\hat{a}_k^{\dagger}$, operators annihilate those holes (by filling them up). Also the algebraic definition of the $\left|\text{full}\right\rangle$, and $\left|\text{vac}\right\rangle$, states are related by the exchange: $\hat{a}_k\left|\text{vac}\right\rangle = 0, \forall k, \text{vs } \hat{a}_k^{\dagger}\left|\text{full}\right\rangle, \forall k.$

To make this exchange manifest, let us define a new family of fermionic creation and annihilation operators, to know,

$$\hat{b}_k = \hat{a}_k^{\dagger}, \qquad \hat{b}_k^{\dagger} = \hat{a}_k, \tag{47}$$

Unlike the bosonic commutation relations, the fermionic anti-commutation are symmetric between $\hat{a}, \hat{a}^{\dagger}$, so the $\hat{b}, \hat{b}^{\dagger}$, satisfy exactly the same anti-commutation relations as the $\hat{a}_k$, and $\hat{a}_k^{\dagger}$,

$$\{\hat{b}_k, \hat{b}_l\} = 0, \{\hat{b}_k^{\dagger}, \hat{b}_l^{\dagger}\} = 0, \{\hat{b}_k, \hat{b}_l^{\dagger}\} = \delta_{k,l}, \tag{48}$$

Physically, the $\hat{b}_k^{\dagger}$, operators *create holes* while the $\hat{b}_k$, operators *annihilate holes* and the holes obey exactly the same Fermi statistics (as given in the Figure 4B) as the original particles. In condensed-matter terminology, the holes are *quasi-particles,* but the only distinction between the quasi-particles and true particles is that the later may exist outside the condensed matter. When viewed from the inside of condensed matter, this distinction becomes irrelevant.

Anyhow, from the hole point of view, the $|\text{full}\rangle$, state is the *hole vacuum* which is the unique state with no holes at all, algebraically defined by $\hat{b}_k|\text{full}\rangle = 0, \forall k$. The excitations are Ξ – hole states obtained by acting with *hole-creation operators* $\hat{b}_k^{\dagger}$, on the hole-vacuum, $|\text{ holes at } k_1, k_2 \ldots, k_{\Xi}\rangle = b_{k_{\Xi}}^{\dagger} \cdots b_{k_1}^{\dagger} |\text{full}\rangle$. Then the Hamiltonian operator (32) of the system becomes[14]

$$\hat{H} = E_0 + \sum_k \varepsilon_k (1 - \hat{b}_k^{\dagger}\hat{b}_k) = E_{\text{full}} + \sum_k (-\varepsilon_k)\hat{b}_k^{\dagger}\hat{b}_k, \tag{49}$$

in accordance with individual holes having positive energies $-\varepsilon_k > 0$.

The k, modes are eigenspaces of some conserved quantum numbers such as momentum or spin (or rather $\hat{S}_z$). When one makes a hole by removing a particle from mode $(\mathbf{p}, s)$, the net momentum of the system changes by $-\mathbf{p}$, while the net S_z, changes by $-s$, so one can say that the hole in that mode has momentum $-\mathbf{p}$, and $S_z = -s$. Consequently the hole operators are usually defined as

$$\hat{b}_{\mathbf{p},s} = \hat{a}^{\dagger}_{-\mathbf{p},-s}, \qquad \hat{b}^{\dagger}_{\mathbf{p},s} = \hat{a}_{-\mathbf{p},-s}, \tag{50}$$

---

[14] $\hat{a}_k^{\dagger}\hat{a}_k = \hat{b}_k\hat{b}_k^{\dagger} = (1 - \hat{b}_k^{\dagger}\hat{b}_k)$.

which leads to

$$\hat{\mathbf{P}}_{\text{Tot}} = \mathbf{P}_{\text{full}} + \sum_{\mathbf{p},s} \mathbf{p} \times \hat{b}^{\dagger}_{\mathbf{p},s} \hat{b}_{\mathbf{p},s}, \tag{51}$$

and likewise

$$\hat{S}^{z}_{\text{Tot}} = S^{z}_{\text{full}} + \sum_{\mathbf{p},s} s \times \hat{b}^{\dagger}_{\mathbf{p},s} \hat{b}_{\mathbf{p},s}, \tag{52}$$

Finally, consider a system where the energies ε_k, take both signs: positive for some modes k, but negative for other modes. For example, a free fermion gas with a positive chemical potential μ, and free-energy operator

$$\hat{H} = \sum_{\mathbf{p},s} (\frac{\mathbf{p}^2}{2m} - \mu) \hat{a}^{\dagger}_{\mathbf{p},s} \hat{a}_{\mathbf{p},s}, \tag{53}$$

has positive ε, for $|\mathbf{p}| > p_f$, where p_f, is the Fermi momentum defined by the threshold $\frac{p_f^2}{2m} - \mu = 0$ (see figure 4). For this system the ground state is the *Fermi sea* where

$$n_{\mathbf{p},s} = \Theta(|\mathbf{p}| < p_f) = \begin{cases} 1 & \text{for } |\mathbf{p}| < p_f \\ 0 & \text{for } |\mathbf{p}| > p_f \end{cases}, \tag{54}$$

In terms of the creation and annihilation operators, the Fermi sea is the state (directly from $\prod_{k=1}^{n} a^{\dagger}_{i_k} |0\rangle = 0$):

$$|\text{Fermi sea}\rangle = \prod_{\mathbf{p},s}^{|\mathbf{p}|<p_f \text{ only}} \hat{a}^{\dagger}_{\mathbf{p},s} |\text{vac}\rangle, \tag{55}$$

which satisfies

$$\hat{a}_{\mathbf{p},s} |\text{Fermi sea}\rangle = 0, \tag{56}$$

for $|\mathbf{p}| > p_f$, and

$$\hat{a}^{\dagger}_{\mathbf{p},s} |\text{Fermi sea}\rangle = 0, \tag{57}$$

for $|\mathbf{p}| < p_f$.

We may treat this state as a quasi-particle vacuum if we re-define all the operators killing the $|\text{Fermi sea}\rangle$, as annihilation operators. Thus we define new (59) for $|\mathbf{p}| < p_f$. only. But keep the original operators $\hat{a}_{\mathbf{p},s}$, and $\hat{a}^{\dagger}_{\mathbf{p},s}$ acting with momentum outside the Fermi surface. Despite the partial exchange the complete set or creation and annihilation operators satisfies the fermionic anticommutation relations having:

$$\begin{aligned} &\text{all}\{\hat{a},\hat{a}\} = \{\hat{b},\hat{b}\} = \{\hat{a},\hat{b}\} = 0, \\ &\text{all}\{\hat{a}^{\dagger},\hat{a}^{\dagger}\} = \{\hat{b}^{\dagger},\hat{b}^{\dagger}\} = \{\hat{a}^{\dagger},\hat{b}^{\dagger}\} = 0, \\ &\text{all}\{\hat{a},\hat{b}^{\dagger}\} = \{\hat{b}^{\dagger},\hat{a}^{\dagger}\} = 0, \end{aligned} \tag{58}$$

and also

$$\{\hat{a}_{\mathbf{p},s},\hat{a}^{\dagger}_{\mathbf{p},s}\} = \delta_{\mathbf{p},\mathbf{p}'}\delta_{s,s'}, \qquad \{\hat{b}_{\mathbf{p},s},\hat{b}^{\dagger}_{\mathbf{p},s}\} = \delta_{\mathbf{p},\mathbf{p}'}\delta_{s,s'}, \tag{59}$$

if we restrict the $\hat{b}_{\mathbf{p},s}$, and $\hat{b}^{\dagger}_{\mathbf{p},s}$, to $|\mathbf{p}| < p_f$, only and the $\hat{a}_{\mathbf{p},s}$, and $\hat{a}^{\dagger}_{\mathbf{p},s}$, to $|\mathbf{p}| > p_f$.

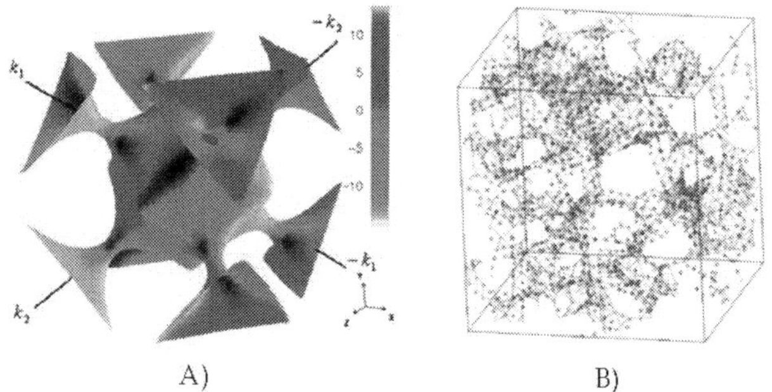

Figure 5. A). Fermi surface to the gold (Au). All Au-quasi-particles must shape this surface with the number of pairs corresponding to the metal to the superconducting state. This surface in the BCS-theory shapes the quantum nucleus of the interaction electron-fonon-electron corresponding to the fermionic Fock superconducting space [16, 17].. B). Fermionic Fock superconducting space conformed with for Cooper pairs: (red particle $k\uparrow$, blue particle $-k\downarrow$). The net is obtained by the adding of quantum Hilbert spaces respectively.

The Fermi sea $|\text{Fermi sea}\rangle$, is the quasi-particle vacuum state of these fermionic operators. The two types of creation operators $\hat{a}^{\dagger}_{\mathbf{p},s}$, and $\hat{b}^{\dagger}_{\mathbf{p},s}$, create two distinct types of quasi-particles (*respectively the extra fermions above the Fermi surface and the holes below the surface*).

Both types of quasi-*particles* have positive energies. Then in terms of our new fermionic operators, the Hamiltonian takes the form:

$$\hat{H} = E_{\text{Fermi Surface}} + \sum_{\mathbf{p},s} (\frac{\mathbf{p}^2}{2m} - \mu > 0) \times \hat{a}^{\dagger}_{\mathbf{p},s} \hat{a}_{\mathbf{p},s} + \sum_{\mathbf{p},s} (\mu - \frac{\mathbf{p}^2}{2m} > 0) \times \hat{b}^{\dagger}_{\mathbf{p},s} \hat{b}_{\mathbf{p},s}, \tag{60}$$

with the domains to every sum $|\mathbf{p}| > p_f,$ only (in the first sum) and $|\mathbf{p}| > p_f,$ only (in the second sum).

Clearly (60) describes elements of the fermionic Fock space given in (37).

4. Fermionic C*-Lie-QED-algebra

Theorem (F. Bulnes) 5. 1. The *electro-anti-gravitational effect*s produced from superconductivity have that to be governed by the actions of the *superconducting Lie-*QED-*algebra* $\mathfrak{C} \otimes \mathfrak{H}$.

To demonstrate the before result is necessary to define the electro-anti-gravity in the formalism of the Lie-QED-algebra and their C*-algebras associated to her. The electro-anti-gravity is obtained through of experiments where a fast rotating superconductor reduces the gravitational effect. Of fact the rotation is fundamental and necessary to the complementing of the anti-gravity effects searched through the magnetic levitation (see the Figure 6. where were realized many experiments with rotating geometrical pin, using the high intense magnetic field).

Figure 6. Magnetic levitation by fast rotating magnet.

The demostration of this theorem has been realized in part, by the lemma 2. 2. 1. However we require other additional lemmas that have that to see with other aspects, as the iso-rotations (condition established and illustrated in the experiments realized in the Figure 6) and the condensation effects of the matter required in the transmission process through a "bosonic cloud" of the plasmons in the quantum transmission of the electro-anti-gravity effect. The concluyent aspect in this last digression is that the rotation movement to very fast velocity and the superconducting phenomena must be togheter and due to the Lie structure of our QED-algebra, this rotations will be inside the $\mathfrak{C} \otimes \mathfrak{H}$ - fields as images of orthogonal transformations of the special orthogonal group $SO(2)$, [15]and their topological essency let to see the inherent geometrical properties to an application of our superconductor electro-fields as was demostrated in the following lemma [9] and mentioned in [18]:

Lemma (F. Bulnes) 5. 1. Let $\mathfrak{S} = G^{\mathbb{C}} / C(T)$, be with $C(T)$, a space of orbits (*hypersurfaces*), generated by the electro-fields on O, by the realization of movements given for $SU(2)$, through of the action of their Maxwell fields F, given by $F = (H_i H_k - \frac{1}{2} H^2 \delta_{ik}) / 4\pi\sigma\mu$, in the superconductor. Then the orbits engendered by the actions $\Im_M$, on M, are magnetic torus engendered by rotations $SO(2)x(s), \forall x \in M$, generates by fluxoids Φ_0, in the vortex zone [8].

Proof. [9]. □

Then an analogous to QED of the fields F, will have that consider in the states generated in a Fock space $\mathcal{F}$, the corresponding transformation of a subgroup of $O(n)$, that is to say, the automorphism of the group must act on fermionic states of the space, where the electro-anti-gravity comes established to change of spin-orbit from $M_s = +1$, to $M_s = -1$, (or viceversa), in a bose-Einstein distribution in the matter condensation phenomena to produce an electro-anti-garavity wrapping of the object O .

One important fact is that there exist orthogonal invariance of the CAR-algebra on a Fock space, that is to say, the Fermionic Fock space is invariant under rotations, that is tosay, $O(n)\psi = \psi, \forall \psi \in \mathcal{F}$, where explicitely the orthogonal group is:

$$O(n, K) = \{Q \in GL(n, K) \mid Q^T Q = QQ^T = I\}, \tag{61}$$

If we consider the subgroup $SO(2)$, of $O(2, K)$, we have that the group $U(1)$, of the 2-forms F^{ab}, satisfy:

$$U(1) = Spin(2) = SO(2), \tag{62}$$

---

[15] $SO(2) = \left\{A \in GL_2 \,\middle|\, A = \begin{pmatrix} \cos\theta & \sin\theta \\ -\sin\theta & \cos\theta \end{pmatrix}, \forall \theta \in [0, 2\pi]\right\}$.

having the considered in the section 2. 3. Then all particle represented for their energy (by their wave function ψ) can change their behavior using a gauge group as $U(1)$, , or $SU(2)$. This last enclose the all electromagnetic phenomena around of the superconductivity that we want cover. Remember that we required to obtain anti-gravity from the $\mathfrak{E}\otimes\mathfrak{H}$ - fields of our superconducting Lie-QED-algebra.

Then considering two elements of the group $SO(2)$, , for example $e_1, e_2 \in \mathfrak{E}\otimes\mathfrak{H}$, the representation fulfils (by proposition 2. 2. 1) is

$$\zeta(e_1)\zeta(e_2) - \zeta(e_2)\zeta(e_1) = \zeta(e_1 \otimes e_2 - e_2 \otimes e_1), \tag{63}$$

and the field is transformed as

$$\Psi \mapsto \Psi', \tag{64}$$

where explicitly the image $\Psi' = \zeta(J_{\alpha\beta})\Psi$. From this always is possible construct a second representation defined by:

$$\zeta^*(J_{\alpha\beta}) = \zeta(({J_{\alpha\beta}}^T)^{-1}), \tag{65}$$

which belongs to the charge-conjugated particle. The anti-particle is obtained of accord to the contragradient $\bar{\zeta}$, representation, which is:

$$\bar{\zeta}(J_{\alpha\beta}) = \zeta(J_{\alpha\beta}^{-1}), \tag{66}$$

There are not charge-conjugated in gravity, since if the gauge group is Lorentz group $SO(3,1)$, then elements $J_{\alpha\beta}^{-1} = J_{\alpha\beta}^T$, , which means that the second representation ζ^*, is equivalent to ζ.

But we need affect the immediate space-time at least locally through of these $\mathfrak{E}\otimes\mathfrak{H}$ - fields, such that we will have the anti-particles given in (75). Also we need a mapping that involves and include in their image the spin connection that is involved in this anti-gravity process from superconductivity.

We define the field Ψ, as a vector field whose application is as given in (64)

$$\Psi' = \Gamma\Psi, \quad \bar{\Psi}' = \bar{\Psi}\Gamma^{-1}, \tag{67}$$

under a general diffeomorphism Γ, that is to say, the mapping belonging to the space $Diff(TM, TM^*)$, where TM, is the dual to T^*M. But we required local transformations at least in the immediate enthrone of object O, such that be anti-gravitational and this local enthrone acts with the space-time to create levitation in O.

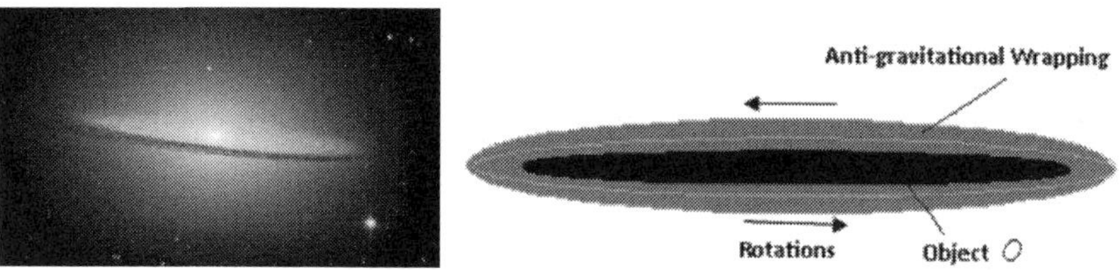

Figure 7. Rotations and anti-gravitational wrapping energy. An example of this idea is the cloud energy in the formatting iso-rotations in a galaxy, in condensed matter to sidereal objects with autonomous energy.

Then the principal equivalence requires that the fields on our manifold locally transform be as in special relativity, that is to say, if , is an element of the Lorentz group ,[16] the fields are transformed like Lorentz-vectors. Of fact this property is extended to all electro-physical modules $\mathfrak{E}$, and $\mathfrak{H}$, like $\pounds$ – modules[17].

However, the generalization to a general diffeomorphism is not unique. We could have chosen the field Ψ, as a vector field whose applications $\Gamma \in Diff(\underline{TM}, \underline{TM}^*)$ are

$$\underline{\Psi}' = (\Gamma^T)^{-1}\underline{\Psi}, \qquad \bar{\underline{\Psi}}' = \bar{\underline{\Psi}}\Gamma^T, \tag{68}$$

But as Γ, is an element of $\pounds$, that is to say $\Gamma^{-1} = \Gamma^T$, both representations (67) and (68) agree. For general diffeomorphism that will not be the case, although introducing a new field that have a modified scaling behavior, this can be possible to affected to the space-time by $\mathfrak{E} \otimes \mathfrak{H}$- fields. Then is considered $\tau \in \mathrm{Isom}(TM, \underline{TM})$, such that to fields $\underline{\Psi} = \tau\Psi$, $\Psi' = \tau'\overline{\Psi}'$, one finds the behavior

$$\tau' = (\Gamma^T)^{-1}\tau\Gamma^{-1}, \tag{69}$$

It will be useful to clarify the emerging picture of space-time properties by having a close look at a contravariant vector field Ψ^κ, as depicted in the wrapping energy around O, (see

[16] $\pounds = \{\xi \in GL(\mathbb{R}^4) \,|\, g(\xi p, \xi q) = g(p, q), \forall p, q \in \mathbb{R}^4\}$,

[17]**Proposition 2.1 (F. Bulnes) [3].** $\mathfrak{E}$, and $\mathfrak{H}$, like $\mathbb{R}$ – modules are invariant under Euclidean movements of the group $O(1,3)$, and thus are $\pounds$ – modules.

the figure 7). This field in blue is a cut in the tangent bundle, that is the set of tangent spaces $T_pM, \forall p \in M$, which describes our space-time. The field is mapped to their covariant field Ψ_ν, which is a cut in the cotangent bundle T_qM^*, , by the metric tensor [19]

$$\Psi_\nu = g_{\kappa\nu}\Psi^\nu, \tag{70}$$

Newly introducing the fields $\underline{\Psi}^\kappa$ (from here anti-graviting) this is transformed under the local Lorentz transformations like a Lorentz-vector in special relativity[18]. Then we can have (after of involve the relations of $\mathrm{Isom}(TM^*,\underline{TM}^*)$):

$$\tau = (\Lambda^T)^{-1}\hat{\tau}\Lambda^{-1} = (\Lambda\Lambda^T)^{-1}, \tag{71}$$

where $|\tau_{\kappa\underline{\kappa}}| = 1$, in the space $\mathrm{Isom}(TM,\underline{TM})$. Then $1_{SO(1,3)} = gg$, $|\tau_\kappa^{\underline{\kappa}}| = |g|$, and $|\tau_{\underline{\kappa}}^\kappa| = |\underline{g}|$, thus the properties of the vector fields are transformed directly to those of fermionic fields by using the fermionic representatives and the transformation of, in this case is the mapping $(\ )^\dagger\gamma^0$, [19] instead of the metric, is used to relate a particle to the particle transforming under the contravariant or contragradient representation.

Then using the notation ∇, to covariant derivative we have:

$$\nabla_{\underline{\kappa}} = \tau_{\underline{\kappa}}^\kappa \nabla_\kappa, \tag{72}$$

which is a new connection. Then the *Maxwell-anti-gravity Lagrangian* (that is to say, for anti-gravitational pendants A^a, of gauge fields) is introduced via the field tensors:

$$\underline{F}_{\underline{\kappa\nu}}^a = \nabla_{\underline{\kappa}} A_{\underline{\nu}}^\kappa - \nabla_{\underline{\nu}} A_{\underline{\kappa}}^a + ef^{abc}A_{\underline{\kappa}}^b A_{\underline{\nu}}^c, \tag{73}$$

---

[18] The underlined indices on these quantities do not refer to the coordinates of the manifols, but to the local basis in the tangential. All of these fields still are functions of the space-time coordinates x_ν. As diffeomorphism τ, maps the basis of one space into the other. We can expand it as $\tau = \tau_\nu^{\underline{\kappa}} dx^\nu \partial_{\underline{\kappa}}$, or $\tau = \tau_{\underline{\kappa}}^\nu dx^{\underline{\kappa}} \partial_{\underline{\nu}}$, respectively, such that (have inverses):

$\tau_\nu^{\underline{\kappa}}\tau_{\underline{\varepsilon}}^\nu = \delta_{\underline{\varepsilon}}^{\underline{\kappa}}, \quad \tau_{\underline{\kappa}}^\varepsilon\tau_\nu^{\underline{\kappa}} = \delta_\nu^\varepsilon,$

Then for completeness, let us also define the combined mappings through the relations:

$\tau_{\nu\underline{\nu}} = \tau_\nu^{\underline{\kappa}} g_{\underline{\kappa\nu}}, \quad \tau^{\nu\underline{\nu}} = g^{\nu\kappa}\tau_\kappa^{\underline{\nu}}.$

[19] γ^0, is the canonical Dirac matrix $\begin{pmatrix} 0 & I \\ I & 0 \end{pmatrix}$.

Staying an Lagrangian of the type $[\eta^{ab}\nabla_a\varphi^*\nabla_b\varphi - V(|\varphi|^2)]$ (see the section 2. 2. 1). Here f^{abc}, are the structure constants of the group and e, is the charge electron coupling with the Planck scale. Then the corresponding electro-anti-gravitational-Lie-QED-algebra is that with supercurrents

$$J^{ab}_{\underline{KV}} = \underline{E}^a_{KV} \otimes \underline{E}^b_{KV} - \underline{E}^b_{KV} \otimes \underline{E}^a_{KV} - \left\{\underline{E}^a_{KV}, \underline{E}^b_{KV}\right\}, \tag{74}$$

Then the Lagrangian of fermionic fields can now be composed from the new ingredients as [19]:

$$\mathcal{L}_{ELECTRO\text{-}ANTI\text{-}GRAVITATIONAL} = \mathcal{L}_F + \underline{\mathcal{L}}_{\underline{F}}, \tag{75}$$

where using the fields Ψ, $\overline{\Psi}$, $\underline{\Psi}$, and $\overline{\underline{\Psi}}$, the Lagrangian $\mathcal{L}_F, \mathcal{L}_{\underline{F}}$, take the form (using of Feynman symbols):

$$\mathcal{L}_F = (D\!\!\!/\,\overline{\Psi})\Psi + \overline{\Psi}(D\!\!\!/\,\Psi), \qquad \underline{\mathcal{L}}_{\underline{F}} = (D\!\!\!/\,\overline{\underline{\Psi}})\underline{\Psi} + \overline{\underline{\Psi}}(D\!\!\!/\,\underline{\Psi}), \tag{76}$$

where $\underline{\mathcal{L}} = \mathcal{L}(g_{KV} \mapsto \underline{g_{KV}}, \Psi \mapsto \underline{\Psi})$, and $\mathcal{L} = \underline{\mathcal{L}}(g_{\underline{KV}} \mapsto g_{KV}, \underline{\Psi} \mapsto \Psi)$. This prove the *theorem 5. 1.*

Testing the Lagrangian we can see that there not is direct interaction between gravitational and anti-gravitational particles. However, both of the particles-species will interact with the gravitational field, which mediates an interaction between them. But this coupling is suppressed with the Planck scale. Thus the production of anti-gravitational matter (which is not observable today) can be is explained as ones condensation matter obtained in the scattering process when the anti-gravitational wrapping is created. This usually could see as a cloud or other haze type.

What happen with the energy states Fock space?

States of the Fermionic particles entering go interact through of the corresponding C*-CAR-algebra [20, 21]. Likewise, for example if we consider the anti-symmetric Fock space $\mathcal{F}_a(\mathcal{H})$, and let p_a, the othogonal projection on to anti-symmetric vectors then C*-CAR-algebra is represented on $\mathcal{F}_a(\mathcal{H})$, by settings

$$b^*(\varphi)p_a(\psi_1 \otimes \psi_2 \otimes \ldots \otimes \psi_n) = p_a(\varphi \otimes \psi_1 \otimes \psi_2 \otimes \ldots \otimes \psi_n), \tag{77}$$

This means that the action of orthogonal group $O(2)$, stay restricted to the Hilbert space corresponding to the C*-CAR-algebra becoming the immediated finite region of the space-time in a fermionic Fock space that is mixture of particles and anti-particles (*at least until that* *is converted all space*). We could call to this restrinction of orthogonal group as $O(2,\mathcal{H})$, where a new operator is obtained by the composition $T^{\wedge} = b^*(\varphi) \circ p_a$, acting on a module Fock space that we can write as $\mathcal{A}^{a^{\otimes 2}}(\mathcal{H})$, [22] [20] which represent the new energy space whose elements are the second side of (77). Using the CAR-algebra of creation and annhilitation operators and $D_\varphi = b^\dagger(\varphi) - b(\varphi), \forall \varphi \in \mathcal{H}_{a\otimes^2}$. The canonical anti-commutation relations are equivalent to the commutator relation:

$$[D_\varphi, D_\psi] = D_\varphi D_\psi - D_\psi D_\varphi = -2i\omega(\varphi,\psi), \tag{78}$$

with the anti-symmetrical form of the Weyl relations given by $\omega(\varphi,\psi)$. If we extend the operators before to linear R-operators on Hilbert space $\mathcal{H}_{a\otimes^2}$, we obtain the relation $\omega(A\varphi, A\psi) = \omega(\varphi,\psi)$, which defines a fermionic orthogonal group

$$\pounds^{\otimes_2} = \{A \in O(2,\mathcal{H}_{\mathbb{R}}) \,|\, \omega(A\varphi, A\psi) = \omega(\varphi,\psi), \forall \varphi,\psi \in \mathcal{H}_{a\otimes^2}\}, \tag{79}$$

where appear the Bogoliubov transformation[21].

Finally, the orbital spaces created by the superconductivity in the quantum regime satisfy the corresponding orbital integrals due F. Bulnes [17] to cuspidal surfaces in the generating chirality inversion through a Dirac node(with Hamiltonian $H = \psi^\dagger(i\upsilon\nabla - \mu)\psi + \Delta\psi^\dagger{}_\uparrow\psi^\dagger{}_\downarrow + \Delta^*\psi_\uparrow\psi_\downarrow$ [23]):

$$J_t(E) = \int_{\underline{N}_F} a_F(a_t \underline{n} a_{-t})^{\rho+\mu} < \sigma(m_F(a_t \underline{n} a_{-t}))^{-1} m_F(\underline{n}) g(k(a_t \underline{n} a_{-t})) > d\underline{n}, \tag{80}$$

where E, is the total Fermi energy in all Fermi surface including the proper kinetic energies, the term $k(a_t \underline{n} a_{-t})$,is the momentum created in the chirality inversion through the node of automorphism $\underline{n} \in \underline{N}_F$, where the space $\underline{N}_F$ is the normal subgroup defined to the action created by fermions in the transit electron-phonon-electron, which is normed by the product of logarithms of the actions of their automorphisms [24].

---

[20] $\mathcal{A}^{a^{\otimes 2}}(\mathcal{H})$, is a algebra of operators from $\mathcal{H}$, in the super-algebra $a^{\otimes_2}$.

[21]The Bogoliubov transformation is a canonical transformation of these operators. To find the conditions on the constants s, and t, such that the transformation remains canonical, the commutator is expanded:

$[\hat{b},\hat{b}^\dagger] = [s\hat{a} + t\hat{a}^\dagger, s^*\hat{a}^\dagger + t^*\hat{a}] = \ldots\ldots = (|s|^2 + |t|^2)[\hat{a},\hat{a}^\dagger]$.

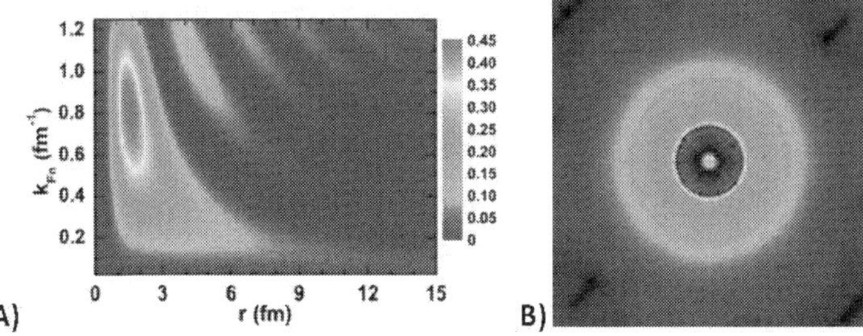

Figure 8. A). Fermionic distribution of probability density $r2|\Psi pair(r)|2$ of the neutron Cooper pairs as a function of the neutron Fermi momentum kFn and the relative distance r between the pair partners in symmetric nuclear matter [22]. This can produces Eddy currents with the property anti-gravitational current given in (83). B). The part colored in blue determines the absence of magnetic flow, such like is wanted that it happens for the super-currents existence on the surface of a levitating vehicle, making that this everything behaves as a diamagnetic, except in the central ring colored in red and yellow where exists an intense magnetic field (*this simulation was published in the Proceedings of Fluid Flow, Heat Transfer and Thermal Systems of* ASME *in the paper* IMECE2010-37107, *British Columbia, Canada with all rights reserved* ® [9]).

One example of this automorphisms $a_t \underline{n} a_{-t}$, in action are the quantum operators given by the product $a^{\dagger}_{k+q\uparrow}a^{\dagger}_{-k-q\downarrow}a_{-k\downarrow}a_{k\uparrow}$, which acts on pairs and not change the electrons in k, and $-k$, and transits to $k+q$, and $-k-q$, letting equals spins. The energy E, is given by $E=\sum_{k,k'}2\varepsilon_k e^{\dagger}_{k}e_k$, where $e^{\dagger}_k(f)=a^{\dagger}_{k\uparrow}a^{\dagger}_{-k\downarrow}$, $\quad e_k(g)=a_{-k\downarrow}a_{k\uparrow}, \forall f,g\in\mathcal{H}$.

Relating the meaning of these operators with the Debye energy to photons given by $\hbar\omega_D$,, we can to obtain a complete criteria to the energies given by (40) considering the Coulumbian repulsion, obtaining a precise wide measure of trenches where to some real superconductor we consider the term $N(0)V\approx 0.3$, is the magnetic momentum developed by the free electrons in the formatting of the Fermi liquid [25, 26, 27]. The integral $J_t(E)$, is bounded [1, 25].

5. Applications

Proposition (F. Bulnes) 6. 1. Using organized transformations as given in $\sigma_1\mathcal{T}(\mathbb{M})\otimes\ldots\otimes\sigma_n\mathcal{T}(\mathbb{M})\otimes\ldots$, we can to establish that the state of all particles in set, is their corresponding *Fock image* [15, 28].

Inside of the Fock space begins a realization of the potential of the superconductivity, since the Fock pure state involves all the states of particles of the space, object of the transformation [15], that in this case is *superconducting state*. We want organize the particles in two the phases that define our Fock space then the proposition is the shape to do it!

Theorem V. 1 (F. Bulnes, R. Goborov). The organized transformation given by [15]

$$\sigma_1 \mathcal{T}(\mathbb{M}) \otimes \ldots \otimes \sigma_n \mathcal{T}(\mathbb{M}) \otimes \ldots, \quad (81)$$

to *electro-anti-gravitational effect* produced from superconductivity must have a fermionic Fock space [10][22]

$$\mathcal{H}_1 \otimes \mathcal{H}_2 = \{ \left| \hat{e}_k^{\dagger} \hat{e}_k \right\rangle \left| \{\hat{e}_k, \hat{e}_l\} = 0, \{\hat{e}_k^{\dagger}, \hat{e}_l^{\dagger}\} = 0, \{\hat{e}_k^{\dagger}, \hat{e}_l\} = \delta_{k,l} \right. \}, \quad (82)$$

with rule of transformation in an inherent context of the space-time with Hamiltonian (*transforming each particle around of the source that produces this transformation*):

$$\hat{H} = E_{\text{Fermi Surface}} + \sum_{\mathbf{p},s} \left(\frac{\hbar^2 k_F}{m^*} - \mu > 0 \right) \times \hat{e}_{\mathbf{p},s}^{\dagger} \hat{e}_{\mathbf{p},s} + \sum_{\mathbf{p},s} \left(\mu - \frac{\hbar^2 k_F}{m^*} > 0 \right) \times \hat{e}_{\mathbf{p},s}^{\dagger} \hat{e}_{\mathbf{p},s}, \quad (83)$$

where k_F, is the Fermi sphere radios (their super-electron momentum) given by

$$k_F = \left(\frac{2mE_{\text{Fermi Surface}}}{\hbar^2} \right)^{1/2}, \quad (84)$$

Their demonstration of the theorem needs more studies and experimental results. This theorem is by way of *conjecture*. But we think that fermionic Fock space of electromagnetic nature can be who can express the phase change in all particles beginning from the structure of metal and transmitting to the immediate ambient space of the metal object (see figure 9B)).

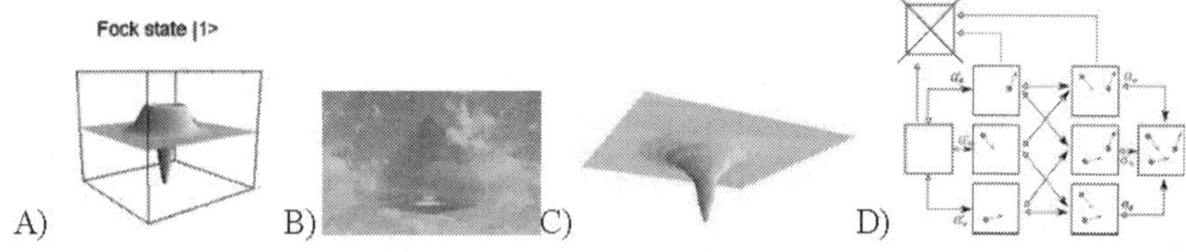

Figure 9. A) The quasi-particle region: holes. The fermionic Fock superconducting space for one particle: observe the two phases of fermion spaces, upper surface corresponds to the holes zone. Of fact this zone is like volcano, since in their interior are holes. The below surface is the free fermions whose behavior is seemed to the Bose-Einstein distribution. B). Structure of the ship transmitting the change phase of the particles that come from of the ship reactor [29, 30]. C) Electro-twistor generated by the magnetic field-superconducting interaction [9]. D) Appearing of the creation operators that shape the wrapping space over structure of the ship. This is defined by a fermionic Fock space, for example under the ship as the impeller twistor [23].

[22]A electromagnetic case is given bythe algebra: $(\mathfrak{C} \otimes \mathfrak{H})^* = \left\{ (\Psi, \Psi^*) \middle| \Psi \otimes \Psi^* - \Psi^* \otimes \Psi - [\Psi, \Psi] \right\}$.

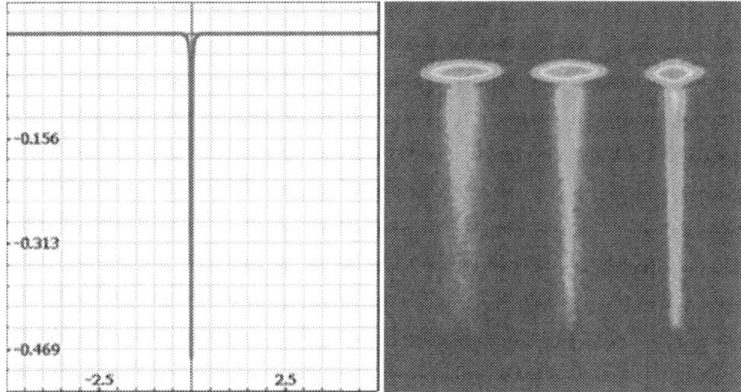

Figure 10. Electro-anti-gravitational Fields $\mathfrak{E}_t$, to levitation (see also figure 11 B)) [18]. Disk (thermal cloud forming the saucer) experiments to magnetic levitation showing the fermionic Fock space in nano-seconds [31] from the orbit-spin interaction.

6. Conclusions

The *sections 4 and 5*, establishes the general conditions to construct the fermionic Fock superconducting space which born from organized transformation of fermions and bosons from the actions of the Lie-QED-algebra $\mathfrak{E} \otimes \mathfrak{H}$, [3, 18, 32] of this way in the next section we establish this and also their transformation to obtain the two phases signed in the Hamiltonian (83) (see figure 9 A)).

Important is do note, that the energy of a quasi-particle depends of the distribution of all the other quasi-particles that haves in the system. Simplified, is can to say, that a free electron, or "naked electron", that is to say, (*outside of interactions*); have as difference with a quasi-particle, or electron with interactions the different masses. The principal effect of the interaction between electrons in normal state consists in change the effective mass of the electron; for example, the specific heat of a Fermi liquid have formally the same expression that the of a ideal Fermi gas changing so only the effective mass, m^*, for the mass of the free electron m.

The fermionic Fock space is a useful topological space picture to describe the interaction obtained for electrons and their link-wave as fonon (boson to describe the quasi-particles) inside the Fermi fluid. In the next work we need to demonstrate this interaction and related with the proposed Hamiltonian in (83). The following region (figure 9 A)), must be the free fermions that realize the transformation in the particles of immediate space moving their spins. Of fact, this change of phase happens inside the superconductor material where the superconductor phenomena happen.

One careful analysis establish certain relations between orbit-spin, saying orbit to the two surfaces that begin in certain step of superconducting process as Rasbha Effect (see figure 4 a)), from the Majorana field produced for this coupling in intermediate state of semi-conductor-superconductor. This can help to design an inter-phase with the reactor of a vehicle of magnetic levitation [23].

Appendices

A. Variation principles to EM in superconductors

Lemma. A. 1. (Bulnes, F) [24, 33]. The energy of action given by E_M, of the O, all like a diamagnetic given by $\Im_M$, satisfies to all vector magnetic potential A, ($1-$form of the corresponding Maxwell equations to the levitation: $rot\boldsymbol{B} = 4\pi \boldsymbol{j}(1/c)$, and $\boldsymbol{B} = \boldsymbol{0}$, [33]), the following Hamiltonian

$$H(A,\Im_M) = \int_O \{L_M - H^2/8\pi\} dV, \tag{85}$$

Proof. [33, 34].

B. Fermionic Fock space

Now suppose there is an infinite but discrete set of fermionic modes α, corresponding to some $1-$particle quantum states $|\alpha\rangle$, with wave functions $\psi_\alpha(\vartheta)$. In the vector ϑ, we include the spin and other non-spatial quantum numbers into $\vartheta = (x, y, z, spin, etc\ldots)$. In this case, the fermionic Hilbert space is

$$\mathcal{F} = \bigotimes_a \mathcal{H}_{\text{mode}\,\alpha}(\text{spanning}\,|n_\alpha = 0\rangle\,\text{and}\,|n_\alpha = 1\rangle), \tag{86}$$

which has infinite dimension and we may interpret this as a Fock space or arbitrary number of identical fermions. This is our space of study to organized transformations that we require [15, 35, 36]].

Author details

Francisco Bulnes*

*Address all correspondence to: francisco.bulnes@tesch.edu.mx

Research Depatment in Mathematics and Engineering, TESCHA, Federal Highway Mexico-Cuautla Tlapala "La Candelaria" Chalco, State of Mexico, Mexico

References

[1] Bulnes, F., Orbital Integrals on Reductive Lie Groups and their Algebras, Intech, Rijeka, Croatia, 2013. ISBN: 978-953-51-1007-1, http://www.intechopen.com/books/orbital-integrals-on-reductive-lie-groups-and-their-algebras/orbital-integrals-on-reductive-lie-groups-and-their-algebrasB

[2] Wilczek, F (2009). "Majorana returns," *Nature Physics*, 5 (9): 614.

[3] Bulnes, F., *Doctoral course of mathematical electrodynamics*, SEPI-IPN, Mexico, 2006, 9 (2007), p398-447.

[4] Bulnes, F., "Electromagnetic Gauges and Maxwell Lagrangians Applied to the Determination of Curvature in the Space-Time and their Applications," *Journal of Electromagnetic Analysis and Applications*, Vol. 4 No. 6, USA, 2012, pp. 252-266. doi: 10.4236/jemaa.2012.46035.

[5] Dummit, D. S. and Foote, R. M., *Abstract Algebra*, Hoboken, N. J: Wiley, USA, 2004.

[6] Strutinsky, V. M., Shell effects in nuclear physics and deformation energies, Nuclear Physics A, Vol. 95, p. 420-442 (1967).

[7] Nielsen, H. B. and Olesen, P (1973). "Vortex-line models for dual strings". *Nuclear Physics B* 61: 45–61.

[8] Alario, M. A. and Vicent, J. L., Superconductivity, Eudema Fortuny, Madrid, Spain, 1991.

[9] Bulnes, F., *Design and Development of an impeller synergic system of electromagnetic type for levitation, suspension and movement of symmetrical bodies*, IMECE/ASME, British Columbia, Canada, 2010.

[10] F. Bulnes, F., Maya, J. and Martínez, I., "Design and Development of Impeller Synergic Systems of Electromagnetic Type to Levitation/Suspension Flight of Symmetrical Bodies," *Journal of Electromagnetic Analysis and Applications*, Vol. 4 No. 1, 2012, pp. 42-52. doi: 10.4236/jemaa.2012.41006.

[11] Ginzburg, V. L. and Landau, L. D., *Zh. Eksp. Teor. Fiz.* 20, 1064 (1950).

[12] Bardeen, J.; Cooper, L. N.; Schrieffer, J. R. (April 1957). "Microscopic Theory of Superconductivity," *Physical Review* 106 (1): 162–164.

[13] Engbarth, M., Miloevic, M. V., Bending, S. L. and Nasirpouri, F (2009) Geometry-guided flux behavior in superconducting Pb microcrystals, *Journal of Physics: Conference Series*, 150 (5), 052048.

[14] Álvarez, A, (Assessor Bulnes, F), "Hilbert Inequalities and Orbital Integrals of Flux of Eddy Currents to a Disc in Levitation," XLV SMM Congress of Mathematics (Poster), Querétaro, 2012.

[15] Bulnes, F (2013) Mathematical Nanotechnology: Quantum Field Intentionality. *Journal of Applied Mathematics and Physics*, 1, 25-44. doi: 10.4236/jamp.2013.15005.

[16] Choy, T. S., 3D Fermi surface Site, http://www.phys.ufl.edu/~tschoy/r2d2/Fermi/Fermi.html

[17] Llano, M., "Unificación de la Condensación de Bose-Einstein con la Teoría BSC de Superconductores," Rev. Ciencias Exactas y Naturais, 2003;vol. 5, no. 1, pp. 9-21.

[18] F. Bulnes and A. Álvarez, "Homological Electromagnetism and Electromagnetic Demonstrations on the Existence of Superconducting Effects Necessaries to Magnetic Levitation/Suspension," *Journal of Electromagnetic Analysis and Applications*, Vol. 5 No. 6, 2013, pp. 255-263. doi: 10.4236/jemaa.2013.56041.

[19] Hossenfelder, S., Anti-gravitation, *Elsevier Science*, 2006.

[20] Emch, G. (1972), *Algebraic Methods in Statistical Mechanics and Quantum Field Theory*, Wiley-Interscience, USA.

[21] Dixmier, J. (1969), *Les C*-algèbres et leurs représentations*, Gauthier-Villars, France.

[22] Sun, B. Y., et al. Phys Lett. B683 (2010)pp134-139, arXiv:0911.2559

[23] Verkelov, I., Goborov, R, and Bulnes, F., "Fermionic Fock Space in Superconducting Phenomena and their Applications," *Journal on Photonics and Spintronics*, Vol. 2, no. 4, USA, 2013, pp19-29.

[24] Bulnes, F., *Analysis of prospective and development of effective technologies through integral synergic operators of the mechanics*, CCIA 2008, CIMM 2008, C. CIMM, ed., Vol. 1.

[25] Mahmoud, J, "Spintronics in Devices: A Quantum Multi-Physics Simulation of the Hall Effect in Superconductors," Journal on Photonics and Spintronics, Vol. 2. No. 1, pp22-27.

[26] Cooper, L (November 1956). "Bound Electron Pairs in a Degenerate Fermi Gas".*Physical Review* 104 (4): 1189–1190.

[27] Bogoliubov, N., *On the theory of superfluidity*, J. Phys. (USSR), 11, p. 23 (1947).

[28] Simon, B. and Reed, M. Mathematical methods for physics, Vol. I (functional analysis). New York: Academic Press, 1972.

[29] F. Bulnes (2013). Quantum Intentionality and Determination of Realities in the Space-Time Through Path Integrals and Their Integral Transforms, Advances in Quantum Mechanics, Prof. Paul Bracken (Ed.), ISBN: 978-953-51-1089-7, InTech, DOI: 10.5772/53439. Available from: http://www.intechopen.com/books/advances-in-quantum-mechanics/quantum-intentionality-and-determination-of-realities-in-the-space-time-through-path-integrals-and-t

[30] Bychkov, Y. A. and Rashba, E. I., Jour. Phys. C. 17, 6039 (1984).

[31] Max-Planck-Institute fur Quantenoptik, Atomlaser 2-D, http://www.quantum-munich.de/media/atomlaser/

[32] Bulnes, F., *Conference of Lie Groups,* SEPI-IPN and IM/UNAM, First Editor by Paul Cladwell, Mexico, 2005.

[33] Marsden, J. E. and Abraham, R., *Manifolds, tensor analysis and applications,* Addison Wesley, Massachusetts, USA, 1982.

[34] Bulnes, F. and Shapiro, M., *On general theory of integral operators to analysis and geometry,* IM/UNAM, SEPI/IPN, Monograph in Mathematics, 1st ed., J. P. Cladwell, Ed. Mexico: 2007.

[35] Landau, L. D. and Lifshitz, E. M., *Electrodynamics of continuous media,* Volume 8. Of Course of Theorical Physics, Pergamon Press, London, UK, 1960.

[36] F. Bulnes, *The super canonical algebra* $\mathrm{E}\otimes\mathrm{H}$, International Conferences of Electrodynamics in Veracruz, IM/UNAM, Mexico, 1998.

Section 2

Selected Topics in Applications of Quantum Mechanics

Chapter 8

The Nuclear Mean Field Theory and Its Application to Nuclear Physics

M.R. Pahlavani

Additional information is available at the end of the chapter

http://dx.doi.org/10.5772/60517

1. Introduction

Our universe consists of substance. Atoms and molecules are basic components of material. Each atom contains a nucleus which is spread in a small area of atom, and electrons. Also, a nucleus contains Neutrons and protons. It is well known today that electrons in atom and Neutrons and protons in the nucleus are interacting together through different forces. It is clear today that the source of different interactions are composed of four basic forces of the universe, namely gravitational, coulomb, strong and weak nuclear interactions.

In quantum mechanics, to study a particle, it is necessary to have knowledge about its interaction with the surrounding media. The Schrödinger equation is a second-order differential equation that is solved to obtain energy spectrum and wave functions of a particle in quantum mechanics. For a many-body system such as atom or nucleus, it is not possible to solve a set of Schrödinger equations to obtain energy spectrum and wave functions analytically. Therefore in such situations, it is necessary to use an average potential which is a mean potential of all interacting forces acting upon a single particle. Then the Schrödinger equation is should be solve for a single particle. This procedure is called the mean field method [1, 2, 3].

To review this method consider a system consisting of N identical interacting particles. The Hamiltonian of system composed of kinetic energy, T, and potential energy, V, is defined as

$$H = T + V = \sum_{i=1}^{N} t(r_i) + \sum_{\substack{i,j<1 \\ i<j}}^{N} v(r_i, r_j) = \sum_{i=1}^{N} -\frac{\hbar^2}{2m_N} \nabla_i^2 + \sum_{\substack{i,j<1 \\ i<j}}^{N} v(r_i, r_j) \tag{1}$$

where m_N is the mass of each particle, and r_i denotes the coordinates of particle i. A summed single particle potential energy, so far undefined, can be added and subtracted of the Hamiltonian to obtain the following relation,

$$\boldsymbol{H} = \left[\boldsymbol{T} + \sum_{i=1}^{N} \boldsymbol{v}(\boldsymbol{r}_i) \right] + \left[\boldsymbol{V} - \sum_{i=1}^{N} \boldsymbol{v}(\boldsymbol{r}_i) \right] = H_{MF} + \boldsymbol{V}_{\boldsymbol{RES}} \tag{2}$$

where

$$\begin{gathered} H_{MF} = T + \sum_{i=1}^{N} \boldsymbol{v}(\boldsymbol{r}_i) \equiv T + V_{MF} = \\ \sum_{i=1}^{N} \left[t(r_i) + \boldsymbol{v}(\boldsymbol{r}_i) \right] \equiv \sum_{i=1}^{N} h(r_i) \end{gathered} \tag{3}$$

is the mean field Hamiltonian of the system and

$$\boldsymbol{V}_{\boldsymbol{RES}} = V - \sum_{i=1}^{N} \boldsymbol{v}(\boldsymbol{r}_i) = \sum_{\substack{i,j<1 \\ i<j}}^{N} v(r_i, r_j) - \sum_{i=1}^{N} \boldsymbol{v}(\boldsymbol{r}_i), \tag{4}$$

is the mean residual interaction. It should be noted that the residual interaction is related to the strength of the actual interaction and can be reduced if the mean field potential is close to the actual potential of the system.

Actually, the mean field method is an approximation in which each particle of system moves under an external field generated by the remaining $N-1$ particles. This mean potential, V_{MF}, can be considered as an average of all possible interactions of nucleons during the short time interval ΔT, between the selected nucleon and its surrenders,

$$\boldsymbol{V}_{\boldsymbol{MF}} = \sum_{i=1}^{N} \boldsymbol{v}(\boldsymbol{r}_i),\ \boldsymbol{v}(\boldsymbol{r}_i) = \frac{1}{\Delta T} \int_{T}^{T+\Delta T} dt \sum_{\substack{j=i \\ j\neq i}}^{N} v(r_i(t),\ r_j(t)). \tag{5}$$

It is important to know that the time average idea was considered only for clearance of the subject and not applicable in practice unless one studies the thermo-dynamical behavior of nucleus.

Therefore the idea of using mean field theory capable of reducing many particles interacting system in to a system of non-interacting (quasi-particles) considered in an external field, VMF which is the mean potential of possible forces of interaction. The mean field potential is considered such that the stationary Schrödinger equation is solved simply to obtain single particle states and their related energy spectrum. These single-particle states are used to construct the N particle wave function as follows.

The corresponding N -particle Schrödinger equation is used to obtain solutions of the mean-field Hamiltonian H_{MF}

$$H_{MF}\Psi_0\left(r_1,r_2,....r_N\right)=E\Psi_0\left(r_1,r_2,....r_N\right). \tag{6}$$

The wave function $\Psi_0(r_1, r_2, ...r_N)$ can be separated by using the ansatz single particle wave functions

$$\Psi_0\left(r_1,r_2,....r_N\right)=\phi_{\alpha_1}\left(r_1\right)\phi_{\alpha_2}\left(r_2\right)....\phi_{\alpha_N}\left(r_N\right). \tag{7}$$

Substituting this ansatz in to the Schrödinger equation (6) yields N identical one-particle Schrödinger equations

$$\begin{gathered} h\left(r\right)\phi_\alpha\left(r\right)=\varepsilon_\alpha\phi_\alpha\left(r\right), \\ h\left(r\right)=t\left(r\right)+v\left(r\right)=-\frac{\hbar^2}{2m_N}\nabla^2+v\left(r\right). \end{gathered} \tag{8}$$

With the quasi-particle energy, ε_α, that is satisfies the following condition

$$E=\sum_{i=1}^{N}\epsilon_\alpha. \tag{9}$$

The wave function of the many- body system is thus an anti symmetric product of single-particle wave functions which are one-particle wave functions of an external potential well. In summary the mean field theory reduces the complicated many-body problem in to a simple one-particle system.

The main idea in this approach is to determine the mean field potential or in particular, an appropriate mean field potential in which the residual interactions between the quasi-particles should be optimal. To do so, one may seek an optimal set $\{\phi_\alpha(r)\}$ of one-quasi-particle states. This is a Rayleigh-Ritz variational approximation in which the variation $\phi_\alpha(r)\to\phi_\alpha(r)+\delta\phi_\alpha(r)$ of the single-particle orbital is minimized

$$\begin{gathered} E_{gs}=\left\langle\Psi_0\middle|H\middle|\Psi_0\right\rangle \\ H=T+V_{MF}+V_{RES} \end{gathered} \tag{10}$$

As a starting point, one may construct an ansatz wave function. It is customary to use a product of single particle wave functions as Eigen function of the system,

$$\Psi_0\left(r_1, r_2, \ldots. r_N\right) = \prod_{i=1}^{N} \phi_{\alpha_i}\left(r_i\right) \tag{11}$$

It is an anti-symmetrized product ansatz wave function following the Hartree-Fock method and is called the Slater determinant of the given single particle states

$$\Psi_0\left(r_1, r_2, \ldots. r_N\right) = C\left[\prod_{i=1}^{N} \phi_{\alpha_i}\left(r_i\right)\right]. \tag{12}$$

Here $\Psi_0(r_1, r_2, \ldots. r_N)$ is an anti-symmetric wave function. Also C is the normalization constant. For instance, consider a three-particles system with its single-particle Eigen states labeled 1, 2, and 3. Then the normalized anti-symmetric state, or the Slater determinant, is

$$\Psi_0\left(r_1, r_2, \ldots. r_N\right) = \frac{1}{\sqrt{6}}\begin{vmatrix} \phi_{\alpha_1}\left(r_1\right) & \phi_{\alpha_1}\left(r_2\right) & \phi_{\alpha_1}\left(r_3\right) \\ \phi_{\alpha_2}\left(r_1\right) & \phi_{\alpha_2}\left(r_2\right) & \phi_{\alpha_2}\left(r_3\right) \\ \phi_{\alpha_3}\left(r_1\right) & \phi_{\alpha_3}\left(r_2\right) & \phi_{\alpha_3}\left(r_3\right) \end{vmatrix}. \tag{13}$$

The energy E of the system has to be varied under the constraint that the normalization of Ψ_0 is preserved, i.e. $\langle \Psi_0 | \Psi_0 \rangle = 1$. This leads to the constrained variational problem,

$$\delta\left(\frac{\left\langle \Psi_0 \middle| H \middle| \Psi_0 \right\rangle}{\Psi_0 | \Psi_0}\right) = 0,$$

which can be transformed in to an unconstrained one by minimizing the energy for normalized wave function, $\Psi_0(r_1, r_2, \ldots. r_N)$. After performing the variation, the single-particle energy, ε_α, is can also be obtained.

One powerful method to address such uncertainties is the following Hartree consistent equation [4,5],

$$-\frac{\hbar^2}{2m_N}\nabla^2\phi_\alpha\left(r\right) + V_{H(F)}\left(\left\{\phi_i\right\}\right)\phi_\alpha\left(r\right) = \epsilon_\alpha\phi_\alpha\left(r\right), \\ i = 1, 2, \ldots\ldots N,\ \alpha = 1, 2, \ldots..\infty \tag{14}$$

This equation is like the Schrödinger equation except that the simple potential term, V (r), is replaced with a function of unknown wave function,

$$V(r) = V_{H(F)}\left(\{\phi_i(r)\}\right)$$

Here, the Hartree mean field potential, $V_{H(F)}$, is different from, V_{HF}, Hartree-Fock mean field.

The differential equation (14) is nonlinear and therefore, much more difficult to solve than the regular Schrödinger equation. The solution can only be carried using consistent iteration method. In this procedure, one can start using a complete set of guessed single-particle states $\{\phi_i^0\ (r)\}$, $i=1, \ldots N$ to calculate the initial potential term, $V_{H(F)}^{(0)}$. In the next step, the equation for a complete set of new wave functions $\{\phi_\alpha^{(1)}\ (r)\}$ $\alpha=1, \ldots\infty$ is solved to obtain Eigen energies $\varepsilon_\alpha^{(1)}$. The procedure is then repeated with new Eigen function $\phi_\alpha^{(1)}\ (r)$ to obtain the new potential , $V_{H(F)}^{(1)}$. This approach can be depicted through the following schematic diagram,

$$\begin{gathered}\phi_i^0(r) \rightarrow V_{H(F)}^{(0)} \rightarrow \epsilon_\alpha^{(1)}, \\ \epsilon_\alpha^{(1)} \xrightarrow{\phi_i^{(1)}} V_{H(F)}^{(1)} \rightarrow \ldots \rightarrow \phi_\alpha^{(n)},\ \epsilon_\alpha^{(n)}.\end{gathered} \tag{15}$$

This procedure is repeated to achieve self-consistency for wave functions (or Eigen energies). This means that after each loop the resultant wave function or Eigen energies compared with the starting wave function or Eigen energies and when their difference becomes less than a given preset limit, i.e.

$$\left\|\phi_\alpha^{(n-1)} - \phi_\alpha^{(n)}\right\| < preset\ limit,$$

the procedure is repeated, otherwise, it will be automatically terminated. Where the $\|\ldots\|$ denotes the norm.

The results of each run, contain a self-consistent mean field, $V_{H(F)}(r)$, the Eigen state, $\phi_\alpha\ (r)$, and its associated Eigen energies , ε_α, are all simultaneously generated. We may also note that for a finite potential-well, there will be, in addition to the bound states, an infinite number of unbound states.

In our discussions, the generated mean-field potential is a central one, that is only a function of r. Central mean field potentials describe only systems with spherical symmetry such as spherical nuclei or atoms. This is because of natural real forces that are conservative and satisfy the conservation of energy.

In some convenient way to avoid self-consistency loops, a phenomenological potential like a simple square well with finite depth, simple harmonic oscillator well and complicated Woods-Saxon with considerable parameters that can be determined using the fit of potential with experimental data, is introduced.

2. Applications of mean-field theory in nuclear physics

Over the years after Rutherford's valuable experiments that suggest nuclei for atom, many theoretical and experimental attempts have been done to obtain knowledge about the stability of nuclei. It is clear today that a nucleus of mass number A, Neutron number N and proton number (atomic number) Z, consists of A strongly interacting nucleons (protons and neutrons in the nucleus without considering their different properties called nucleon.). In addition to the strong nuclear force that is responsible for nuclear stability, the protons also sense the attractive coulomb potential because of their charge. In regular nuclear physics, the protons and neutrons are considered the point particles without any internal structure. This is an excellent approximation when the aim is to study nuclear structure at low energies. In such approach, the nuclear forces are considered a central attractive force with proper specifications like independence of charge and low range. Note that in advance models of nuclear physics such as the Yukawa Meson exchange model, it is believed that nucleons constructed quarks and interact together through the meson exchange mechanism in the base of the particle physics lows. The lightest nucleus is Deuterium with one neutron and one proton. The interaction of nucleons in the nucleus can be studied both theoretically and practically using simple Deuteron nucleus. This two-nucleon system is described by two-body interaction matrix elements, without a detailed account of the methods used to obtain them. On the other hand, the A - nucleons nucleus in quantum mechanics using the Schrödinger equation is not a solvable problem analytically at least for $A>10$. Therefore, one has to look for a reasonable approximate method to solve this many-body problem consisting of strongly interacting nucleons. A powerful approximation is to convert such many-body system in to a non-interacting system of quasi-particles using a suitable external mean field potential. The remaining interactions, called residual interaction, can be treated as a perturbation potential in the base of perturbation approximation. As discussed earlier, the transformation of system of particles in to quasi-particles is not simple, and its success depends on the nuclear system under consideration.

As mentioned above, a conventional approach is to select a particular type of mean field potential to avoid the steps leading to self-consistency. The selected mean field potential and considered remaining residual interactions as approximations produce the preciseness of the obtained results. The simplest custom potential is the three-dimensional harmonic oscillator potential well

$$\mathrm{V}_{HO}(\mathrm{r}) = -\mathrm{V}_1 + \mathrm{kr}^2 = -\mathrm{V}_1 + \frac{1}{2} m\omega^2 r^2 \tag{16}$$

where V_1 and k are the parameters to be fitted to the practical data for best result. A common, more realistic choice is the Woods–Saxon potential [6]

$$v_{WS}(r) = \frac{-V_0}{1+e^{(r-R)/a}}.$$

where V_0, R and a are the nuclear potential depth, the nuclear radius, and the surface diffuseness, respectively. They are parameterized as follows,

$$V_0 = \left(51 \pm 33\frac{N-Z}{A}\right) MeV,\ R = r_0 A^{\frac{1}{3}} = 1.27 A^{\frac{1}{3}}\ fm,\ a = 0.67\ fm.$$

The + and – signs are considered for protons and neutrons, respectively. In the case when there is no distinction between protons and neutrons a suitable average value of $V_0 = 57\,MeV$ can be used for nucleons.

The Woods–Saxon potential, v_{WS}, is a suitable choice for the mean field potential however it is a complicated function of, r, and it is not an analytically solvable one. To overcome this problem, it is possible to select the proper three- dimensional oscillator potential with energy quantum, $\hbar\omega$ and depth, V_1. The energy difference of levels, $\hbar\omega$, and depth, V_1, can be obtained with a best fit to the Woods–Saxon potential, v_{WS}, as a function of, V_0, R and a as the nuclear potential depth, nuclear radius, and surface diffuseness of the Woods–Saxon potential, respectively. The wave functions and energy spectrum of equivalent harmonic oscillator potential agreed well with the Woods–Saxon potential ones especially near the bottom of the wells in low energies. The difference of these potentials increases when the potential approaches zero. Actually the major difference of these potentials is that the harmonic oscillator potential varies more sharply than the Woods–Saxon one near the surface of the nucleus.

2.1. The spin–orbit interaction

Sometimes in 1949, Meyer and independently, Haxel, Jensen, and Swees showed that if in addition to mean field central potential, V_{MF}, a non-central potential is included in the Schrödinger equation, all closed shell nucleon numbers can be obtained successfully. These numbers 2, 8, 20, 28, 50, 82, and 126 are called magic numbers because the origin of these numbers was not known at that time. The Woods-Saxon or its equivalent harmonic oscillator central potential is not able to reproduce experimentally observed precise data of the single-particle structure energies of the nucleus using the mean field approach.

The non-central potential due to the interaction between the spin of nucleons with the angular momentum of orbital that nucleons located on it, is called spin-orbit interaction. As a result of spin-orbit interaction [7, 8], the nuclear energy level for a given l (except for $l=0$) is split in to two sublevels. The sublevels are characterized by total angular momentum numbers equal to $\left(l+\frac{1}{2}\right)$ and $\left(l-\frac{1}{2}\right)$ corresponding to whether the spin is parallel or anti-parallel to the orbital angular momentum. Each sublevel with spin j accommodates $(2j+1)$ neutrons or protons. The same interaction with a different structure is observed in atoms with a different sign as in the nucleus.

Consider that the harmonic oscillator central potential is produced only for the first three observed magic numbers 2, 8, and 20. To obtain the remaining numbers 28, 50, 82 and 126, it is necessary to add a spin-orbit interaction potential to the Schrödinger equation.

The origin of the spin–orbit interaction is not the same in atoms and nucleus. The atomic spin–orbit force is due to a well-known electromagnetic interaction, and the scale of energy separation is in the order of milli-electronvolts, while the energy difference of sublevels separated because of the nuclear spin-orbit interaction is in the order of million electronvolts and its origin is not well understood yet. In most cases, this force is considered phenomenologically. For the spin–orbit term, we use [9]

$$v_{LS}(r)=v_{LS}^{0}\left(\frac{r_0}{\hbar}\right)^2\frac{1}{r}\frac{d}{dr}\left[\frac{1}{1+e^{(r-R)/a}}\right] \tag{17}$$

The second pair of parentheses guarantees that the derivative does not operate on the wave function when substituted in the radial Schrödinger equation. The r dependence of this interaction arises from its central nature.

The derivative part of this potential is often neglected for simplicity and $v_{LS}(r)$ is replaced by a constant; however, to obtain precise results, the radial part should be considered. We have

$$v_{LS}^{0} = 0.44V0.$$

To obtain the strength of the spin-orbit part, we use

$$J^2=\left(L+S\right)^2$$
$$J^2=L^2+\ S^2+2L.S,$$

and its expectation value for the nuclear wave equation made

$$<L.S>=\frac{\hbar^2}{2}\left[j(j+1)-l(l+1)-\frac{1}{2}(\frac{1}{2}+1)\right]$$
$$<L.S>=\frac{\hbar^2}{2}l\quad for\quad j=l+\frac{1}{2},$$
$$<L.S>=\frac{\hbar^2}{2}(-l-1)\quad for\quad j=l-\frac{1}{2}.$$

In addition to the mean field plus spin–orbit interaction, protons in nuclei interact together via the coulomb force, which is defined by the following relation, considering nuclei as a sphere with a constant charge density [10]

$$V_C(r)=\frac{Ze^2}{4\pi\varepsilon_0}\begin{cases}\dfrac{3-\left(\dfrac{r}{R}\right)^2}{2R} & r\le R\ ,\quad \dfrac{1}{r}\quad r\ge R\end{cases} \tag{18}$$

To obtain the energy spectrum and wave functions for neutrons, one needs to solve the radial Schrödinger equation for the Woods-Saxon and spin-orbit potentials. Such second-order differential equation cannot be solved analytically. To solve this complicated differential equation, it is necessary to introduce some new variables and use reasonable approximations. By introducing new variable [11] $y=\frac{1}{1+\exp\frac{(r-R)}{a}}$, the Woods–Saxon potential reduces to its simple form $V_{WS}=V_{0y}$ while the spin-orbit term changes to $V_{LS}=\frac{(y-y^2)}{R_0+a\ln\left(\frac{1}{y}-1\right)}$. For orbits with small l, the Taylor expansion of the $\frac{1}{r}$ near $r=r_m$, is reasonable. According to the definition of variable y we have, $f(y)\equiv\frac{1}{1+\exp\frac{(r-R)}{a}}$ hence by expanding $f(y)$ around $y_m\equiv\frac{1}{1+\exp\frac{(r_m-R_0)}{a}}$ with $0<y_m<1$, since $0<y<1$, y^3 and the higher terms are negligible, the radial part of the spin-orbit term can be approximated using [12],

$$-\frac{a}{r}\frac{d}{dr}\left(\frac{1}{1+\exp\frac{(r-R)}{a}}\right)=\frac{1}{r_m}\left(C_0+C_1y+C_2y^2\right), \tag{19}$$

where C_0, C_1, and C_2 are dimensionless coefficients and evaluated as

$$C_0=\frac{-ar_my_m+2a^2y_m}{2r_m^2(1-y_m)},$$
$$C_1=1+\frac{ar_my_m+2a^2}{r_m^2(1-y_m)},$$
$$C_2=-1+\frac{ar_m(1+2y_m)+2a^2}{r_m^2(1-y_m)2y_m}.$$

Likewise, the Taylor expansion is applicable for $\frac{1}{r^2}$,

$$\frac{1}{r^2}\approx\frac{1}{r_m^2}\left(D_0+D_1y+D_2y^2\right),$$

where D_i 's, similar to the C_i ', are obtained through

$$D_0=1-\frac{4a}{r_m\left(1-y_m\right)}+\frac{3a^2+ar_m}{r_m^2\left(1-y_m\right)^2},$$
$$D_1=\frac{6a}{r_my_m\left(1-y_m\right)}-\frac{6a^2+2ar_m}{r_m^2y_m\left(1-y_m\right)^2},$$
$$D_2=\frac{3a^2-ar_m\left(1-2y_m\right)}{r_m^2\left(1-y_m\right)2y_m^2}.$$

This type of expansion has been widely used for differential equations resulting from the Schrödinger equation with different potentials [12].

By means of these expansions, the spin-orbit term transforms into

$$VLS = (C0 + C1y + C2y2),$$

and the centrifugal term is obtained

$$Vc.f=(D0+D1y+D2y2).$$

By using these expansions, the spin-orbit term transforms into $V_{LS} \propto (C_o + C_1 y + C_2 y^2)$ and the centrifugal term is changed to the favorable type $V_{CF} \propto (D_o + D_1 y + D_2 y^2)$. The substitution of V_{LS} and V_{CF} as a function of variable y into the Schrödinger equation transforms this equation in to the following analytically solvable differential equation

$$y(1-y)\left[y(1-y)\frac{d^2R(y)}{d\,y^2} + (1-2y)\frac{dR(y)}{d\,y^2}\right] +$$
$$+\frac{2ma^2}{\hbar^2}[V_{0y} + V_{LS}^{'(0)}\left(C_o + C_1 y + C_2 y^2\right)$$
$$-\frac{\hbar^2}{2m}\frac{l(l+1)}{r_m^2}\left(D_o + D_1 y + D_2 y^2\right) R(y) = 0$$
$$V_{LS}^{'(0)} = \frac{r_0^2 V_{LS}^{(0)}}{2ar_m}(j(j+1) - l(l+1) - 3/4)$$

This equation can be transformed into the following simple form,

$$y(1-y)\frac{d^2R(y)}{dy^2} + (1-2y)\frac{dR(y)}{dy} + \frac{-\varepsilon^2 + \beta^2 y - \gamma^2 y^2}{y(1-y)} R(y) = 0 \tag{20}$$

Equation (20) can be transformed into the well-known form of hypergeometric differential equation or, alternatively Nikiforo-Avorono (NU) type [13]. The obtained results using the NU method are

$$-\varepsilon^2 = \frac{2ma^2}{\hbar^2}\left(E + V_{LS}^{'(0)}\right) - \frac{l(l+1)}{r_m^2} D_0$$

$$\beta^2 = \frac{2ma^2}{\hbar^2}\left(V_0 + V_{LS}^{'(0)} C_1\right) - \frac{l(l+1)}{r_m^2} D_1$$

$$\gamma^2 = -\frac{2ma^2}{\hbar^2}\left(E + V_{LS}^{'(0)} C_2\right) - \frac{l(l+1)}{r_m^2} D_2$$

$$R = C\frac{\Gamma(2\varepsilon-1)\Gamma(-2i\lambda)}{\Gamma(\varepsilon+\eta'+1-i\lambda)\Gamma(\varepsilon-\eta'-i\lambda)}\left[(1-y)^{i\lambda} + \frac{\Gamma(2i\lambda)\Gamma(\varepsilon-\eta'+1-i\lambda)\Gamma(\varepsilon-\eta'-i\lambda)}{\Gamma(-2i\lambda)\Gamma(\varepsilon-\eta'+1+i\lambda)\Gamma(\varepsilon-\eta'+i\lambda)}(1-y)^{-i\lambda}\right]$$

where Γ is the well-known gamma function, and C is the normalization constant. λ, μ, and η' are defined as follows

$$\lambda = \sqrt{\beta^2 + \varepsilon^2 + \gamma^2}$$

$$\mu = i\lambda$$

$$\eta' = \sqrt{\gamma^2 - 1/4} - 1/2$$

Note that λ is valid only for the $\beta^2 > \varepsilon^2 + \gamma^2$ condition. In a special case where *l=0,* the solution reduces to its simple form. Also, the energy eigenvalues are obtained as a function of z satisfying the following relation

$$-\frac{z}{\sqrt{1-z^2}} = \tan\varnothing(z),\ z = \sqrt{1 - \frac{|E|}{V_0}}$$

which Φ (z) can be evaluated using a graphical method.

$$\varnothing(z) = k_0R_0z + \tan^{-1}\left(\frac{z}{\sqrt{1-z^2+\left(\frac{1}{k_0a}\right)}}\right) - \sum_{n=0}^{\infty}\left[\tan^{-1}\left(\frac{2k_0a}{n}z\right) - \tan^{-1}\left(\frac{z}{\sqrt{1-z^2}+\left(\frac{n}{k_0a}\right)}\right) - \tan^{-1}\left(\frac{z}{\sqrt{1-z^2}+\left(\frac{n+1}{k_0a}\right)}\right)\right].$$

Note that

$$k = \frac{\lambda}{a} = \left[\frac{2m}{\hbar^2}(V_0' - |E|\right]^{1/2},$$

and

$$k_0 = \frac{2m}{\hbar^2}(V_0^{'})^{1/2}, \; z = \frac{k}{k_0}.$$

Finally,

$$|E| = V_0\left(1 - Z^2\right)$$

The results obtained in this special case are in agreement with the results obtained using other methods [14].

3. Conclusions

In this chapter we briefly discussed the idea of mean field theory as an improvable approximation method for many-body problems of identical particles like atoms and nucleus that cannot be solved analytically. We have shown that for a system of A - nucleons nucleus by considering a suitable potential using this model, one is able to obtain energy spectrum and wave equations. However, the obtained results cannot reproduce the measured nuclear spectroscopy, but one may hope to become successful by considering an accurate potential in the Schrödinger equation.

Author details

M.R. Pahlavani

Address all correspondence to: m.pahlavani@umz.ac.ir

Department of Nuclear Physics, Faculty of Basic Science, University of Mazandaran, Babolsar, Iran

References

[1] Suhonen Jouni. From nucleon to nucleus. Springer-Verlag Berlin Heidelberg (2007).

[2] Ring P, Schuck P. The nuclear many-body problem. Springer-Verlag New York Heidelberg Berlin, pages 126, 314 and 438 (1980).

[3] Kelban M, et al., Global properties of spherical nuclei obtained from Hartree-Fock-Bogoliubov calculations with the Gogny force, Phys. Rev. C 65 (2002) 024309.

[4] Hartree D R, Hartree-Fock theory, Proc. Cambridge Phil. Soc. 24 (1928) 89.

[5] Slater J C, A simplification of Hartree-Fock method, Phys. Rev. 81 (1951) 385.

[6] Judek J, et al., Parameters of the deformed Wioods-Saxon potential outside A=110-210 nuclei, Journal of physics G: Nuclear and Particle physics, 5 (1979) 1359.

[7] Asmart Mahmoud and Ulloa Sergio, Spin-orbit interaction and isotropic electronic transport Graphen, Phys. Rev. Let. 112 (2014) 136602.

[8] Glazov M M, Sherman E ya and Dugaev V K, Two-dimensional electron gas with spin orbit disorder, Physica E 42 (2010) 2157.

[9] Hull Jr M H, Dyatt Jr K D, Fisher C R and Brelt G, Nucleon-Nucleon spin-orbit interaction potential, Phys. Rev. Let. 2 (1959) 264.

[10] Eder G and Berhummer O, The coulomb potential of spherical nuclei, Letters Al. Nuovo Cimento 15 (1976) 25.

[11] Pahlavani M R and Alavi S A, Solutions of Woods-Saxon potential with Spin-Orbit and centrifugal terms through Nikiforo-Uvarov method, Commun. Theor. Phys. 588 (2012) 793.

[12] Pahlavani M R and Alavi S A, Study of nuclear bound states using mean-field Woods-Saxon and spin orbit potentials, Modern Physics Letter A 27 (2012) 1250167.

[13] Cüneyt Berkdemir, Application of the Nikiforov-Uvarov Method in Quantum Mechanics, Theoretical Concepts of Quantum Mechanics, Prof. Mohammad Reza Pahlavani (Ed.), ISBN: 978-953-51-0088-1, InTech, (2012). DOI: 10.5772/33510.

[14] Lu G, analytic quantum mechanics of diatomic molecules with empirical potentials, Physics Scripta, 72 (2005) 349.

Chapter 9

Non-Extensive Entropies on Atoms, Molecules and Chemical Processes

N. Flores-Gallegos, I. Guillén-Escamilla and
J.C. Mixteco-Sánchez

Additional information is available at the end of the chapter

1. Introduction

During the last decade, information theory [1] as applied to the basic sciences has taken two routes in the study of physical and chemical systems, considering both extensivity and non-extensivity - these concepts are fundamental to the development of new physical theories that try to describe the behaviour of natural systems. In this sense, non-extensivity is an important concept that it is necessary to incorporate into the description of atoms, molecules and chemical processes.

At present, one of the ways to incorporate the concept of non-extensivity is by using deformed entropies, or Tsallis entropy [2]. This entropy is a generalization of Shannon entropy and has a dependency of a parameter, usually denoted by "q" and generally called a 'non-extensivity parameter', that permits us to perform a modulation between extensive and non-extensive behaviour. These new kinds of entropies are built using a new area of mathematics called "q-algebra" , or "deformed algebra" [3–5]. One important aspect to the use and application of deformed entropy is that the original definition of the entropy used for building deformed entropy needs to be strictly positive over all space and dimensionless. As such, in this work we propose a definition that fulfils this. This entropy uses the electron density obtained by the methods of quantum mechanics - this is an important point because the electron density is an observable, and so this permits us to establish a gate between the non-extensivity of classic entropies and the non-extensivity of quantum entropies. Consequently, this entropy permits us to incorporate the important concept of non-extensivity in quantum theory. In the same way, it is known that the chemistry interpretation of the same behaviour of these systems can be enriched by quantum information theory.

As we will show in this work, it is trivial to obtain some important functionals using deformed entropies. In this sense, we show how, with simple mathematical manipulations, it is possible to obtain two of the most important functionals of physics - the kinetic energy

functional of the Thomas-Fermi [6, 7] model, and the exchange energy functional of Dirac from deformed entropy. In our opinion, this opens the door to exploring the possibility of the generation of density functionals based in entropic criteria. Moreover, we will present a simple chemical process where we show the effect of the non-extensive parameter, and in the same way we present the general trends of the "q" parameter for the first 54 atoms of the periodic table. Finally, we might raise a general question that motivates this work, that is, when does a natural system become extensive or non-extensive? As we mentioned above, this parameter "q" has a strong relation with deformed algebra - in this algebra, all the operations have a dependence upon a parameter, "q", and in general when "q" is different to that of the unit, this implies that their basic properties are not completely separable and do not necessary commute. This causes us to raise a general question, namely, can nature be represented by deformed algebra? In this context, it is necessary to incorporate the concept of non-extensivity to rewrite many expressions in terms of deformed algebra and investigate their new properties.

2. Theoretical background

Since the 1980s, when the first applications of Shannon entropy to chemical systems were made, we might observe two basic definitions of it, namely

$$S = -\int \rho(\mathbf{r}) \ln \rho(\mathbf{r}) d\mathbf{r}, \tag{1}$$

and

$$S = -\int \frac{\rho(\mathbf{r})}{N} \ln \frac{\rho(\mathbf{r})}{N} d\mathbf{r}, \tag{2}$$

where $\rho(\mathbf{r})$ is the electron density subject to $\int \rho(\mathbf{r}) d\mathbf{r} = N$, and N it is the electron number of the system. However, if we perform a dimensional analysis, we immediately note that neither definition of Shannon entropy is dimensionless; in addition, and in our opinion, the more serious deficiency is that neither definition is strictly positive over all space[1]. Given this situation, it is evident that we cannot apply these two initial definitions to chemical systems. Accordingly, we propose a redefinition of Shannon entropy [8], such that

$$S = -\int \frac{\rho(\mathbf{r})}{N} \ln \frac{\rho(\mathbf{r})}{\rho_{\max}} d\mathbf{r}, \tag{3}$$

where $\rho_{\max}$ is the electron density in the nuclei position. In the case of a molecular system $\rho_{\max}$, it is necessary to take the higher value of the electron density of all the atoms that constitute the molecule. This definition fulfils the following: it is dimensionless and strictly positive over all space. In this sense, we suggest the use of the definition (3) for entropy calculations of chemical systems.

[1] A more detailed study of this aspect will be presented elsewhere.

On the other hand, in general the entropy of a composed system is very often equal to the sum of all its parts. This is fulfilled only when the energy is the sum of the parts and if the work performed by all the parts is the sum of the work performed by the system. That is,

$$S(A,B) = S(A) + S(B), \tag{4}$$

or in general,

$$S(A,B,C,\cdots) = S(A) + S(B) + S(C) + \cdots, \tag{5}$$

However, this not quite obvious, and in some cases this may not be fulfilled. For example, consider a system composed of two different homogeneous substances - in this case, it is only possible to express the energy as the sum of the individual energies if, and only if, we neglect the interaction energy of the substances or subsystems. However, this energy plays an important role in the description of natural systems; unfortunately, the mathematical development of it is, frequently, complicated. One interesting aspect of entropies involves entropic balances [9], in which is possible to write the joint entropy in terms of the subsystems' entropy and conditional entropy,

$$S_q(A+B) = S_q(A) + S_q(B|A) + (1-q)S_q(A)S_q(B|A) \tag{6}$$

where

$$S_q(A) = -\iint \frac{\rho(\mathbf{a},\mathbf{b})}{N_{AB}(q-1)}\left[1-\left\{\int \frac{\rho(\mathbf{a},\mathbf{b})}{\rho^A_{\max}N_B}d\mathbf{b}\right\}^{q-1}\right]d\mathbf{a}d\mathbf{b}, \tag{7}$$

$$S_q(B|A) = -\iint \frac{\rho(\mathbf{a},\mathbf{b})}{N_{AB}(q-1)}\left[1-\left\{\iint \frac{\frac{\rho(\mathbf{a},\mathbf{b})}{N_A+N_B}}{\int \frac{\rho(\mathbf{a},\mathbf{b})}{\rho^A_{\max}N_B}d\mathbf{b}}d\mathbf{a}d\mathbf{b}\right\}^{q-1}\right]d\mathbf{a}d\mathbf{b}, \tag{8}$$

where $\rho^A_{\max}$ and $\rho^B_{\max}$ are the maximum density values of the fragments[2] A and B, respectively. N_{AB} is the total electron number and N_A and N_B are the electron numbers of the fragments A and B respectively. The marginal densities of the probabilities are defined as

$$\int \rho(\mathbf{a},\mathbf{b})d\mathbf{a} = \rho(\mathbf{b}), \tag{9}$$

$$\int \rho(\mathbf{a},\mathbf{b})d\mathbf{b} = \rho(\mathbf{a}), \tag{10}$$

and these densities fulfil

[2] This implies that it is necessary to select an electron density partition scheme subject to the rules of information theory. In chemistry, the scheme that fulfils this is the Stock-Holder partition scheme [10, 11].

$$\iint \rho(\mathbf{a},\mathbf{b})d\mathbf{a}d\mathbf{b} = 1, \tag{11}$$

$$\int \rho(\mathbf{a})d\mathbf{a} = 1, \tag{12}$$

$$\int \rho(\mathbf{b})d\mathbf{b} = 1, \tag{13}$$

In all cases, S_q satisfies the following properties,

i) $S_q \geq 0$;

ii) S_q is a continuous function of $\rho(\mathbf{a},\mathbf{b})$, $\rho(\mathbf{a})$ or $\rho(\mathbf{b})$;

iii) S_q increases monotonically with the particle number;

iv) $S_q(A,B) = S_q(A) + S_q(B) + (1-q)S_q(A)S_q(B)$;

From the last paragraph, it is possible to think in terms of the use of linear description; in this sense, Tsallis proposes a generalization of Boltzmann-Gibbs entropy, using the so-called 'deformed functions', and substituting the original definitions by the deformed definitions. In general, two definitions are used, namely the deformed logarithm (or q-logarithm)

$$\ln_q x := \frac{1-x^{q-1}}{q-1}, \tag{14}$$

and the deformed exponential (or q-exponential)

$$\exp_q^x := [1+(1-q)x]^{\frac{1}{1-q}}, \tag{15}$$

These definitions can be obtained by solving the differential equation $\frac{dy}{dx} = y^q, y(0) = 1; q \in \Re$, see [12].

Using the deformed logarithm, we can obtain the deformed entropy, which has the following explicit form

$$\begin{aligned} S_q &= -\int \frac{\rho(\mathbf{r})}{N} \ln_q \frac{\rho(\mathbf{r})}{\rho_{\max}} d\mathbf{r}, \\ &= -\int \frac{\rho(\mathbf{r})}{N} \left[\frac{1-\left(\frac{\rho(\mathbf{r})}{\rho_{\max}}\right)^{q-1}}{1-q} \right] d\mathbf{r}, \\ &= \frac{1}{1-q} + \frac{1}{N(q-1)\rho_{\max}^{q-1}} \int \rho(\mathbf{r})^q d\mathbf{r}, \end{aligned} \tag{16}$$

where $q \in \Re$, and for a composed system, this entropy is

$$S_q(A+B) = S_q(A) + S_q(B) + (1-q)S_q(A)S_q(B), \tag{17}$$

This implies that the subsystems are correlated, and immediately we can note that for $q \neq 1$ the entropy of a composed system is non-extensive, though if we select $q \to 1$, then the definition (16) becomes the definition (3), that is,

$$\lim_{q\to 1}\left\{\frac{1}{1-q} + \frac{1}{N(q-1)\rho_{\max}^{q-1}}\int \rho(\mathbf{r})^q d\mathbf{r}\right\} = -\int \frac{\rho(\mathbf{r})}{N}\ln\frac{\rho(\mathbf{r})}{\rho_{\max}}d\mathbf{r}, \tag{18}$$

and recover the extensive behaviour.

Now, using Eq. (16), if we select that $q = \frac{4}{3}$, we obtain

$$S_{q=4/3} = -3 + \frac{3}{N\rho_{\max}^{1/3}}\int \rho(\mathbf{r})^{4/3}d\mathbf{r}, \tag{19}$$

In this expression, we can note immediately that the integral has the same form as that of the exchange functional of Dirac, and if we select that $q = \frac{5}{3}$,

$$S_{q=5/3} = -\frac{3}{2} + \frac{3}{2N\rho_{\max}^{2/3}}\int \rho(\mathbf{r})^{5/3}d\mathbf{r}, \tag{20}$$

this integral corresponds to the Thomas-Fermi kinetic energy functional. Consequently, it is trivial to obtain any density functional that involves some of the powers of electron density. Naturally, it is a simple matter to rewrite both functionals in terms of the deformed entropy - this allows us to hypothesize that the electronic energy of a system can be rewritten as a linear combination of deformed entropies. This, of course, implies that the local density functionals are a particular case of the deformed entropy. This allows us to raise the following question: for any electron density, to what does the value of q correspond?

The definition (16) can be simplified if we perform a series expansion of the term $\left(\frac{\rho(\mathbf{r})}{\rho_{\max}}\right)^{q-1}$,

$$\begin{aligned}\left(\frac{\rho(\mathbf{r})}{\rho_{\max}}\right)^{q-1} &= \frac{\rho_{\max}}{\rho(\mathbf{r})} + \frac{\rho_{\max}}{\rho(\mathbf{r})}\log\left(\frac{\rho(\mathbf{r})}{\rho_{\max}}\right) + \\ &\frac{\rho_{\max}}{2\rho(\mathbf{r})}\log\left(\frac{\rho(\mathbf{r})}{\rho_{\max}}\right)^2 + \frac{\rho_{\max}}{6\rho(\mathbf{r})}\log\left(\frac{\rho(\mathbf{r})}{\rho_{\max}}\right)^3 + \cdots + \\ &\frac{\rho_{\max}}{n!\rho(\mathbf{r})}\log\left(\frac{\rho(\mathbf{r})}{\rho_{\max}}\right)^n,\end{aligned} \tag{21}$$

and if we suppose that only the first term contributes to the general behaviour and that it is the most important term,

$$\left(\frac{\rho(\mathbf{r})}{\rho_{\max}}\right)^{q-1} \sim \left(\frac{\rho_{\max}}{\rho(\mathbf{r})}\right), \tag{22}$$

Replacing this in Eq. (16), namely

$$\begin{aligned} S_q^{approx} &= -\int \frac{\rho(\mathbf{r})}{N}\left[\frac{1-\left(\frac{\rho_{\max}}{\rho(\mathbf{r})}\right)}{1-q}\right] d\mathbf{r}, \\ &= \left[\frac{1}{q-1}+\frac{\rho_{\max}}{N(1-q)}\right]\int d\mathbf{r}, \end{aligned} \tag{23}$$

this is a good result, because it is possible to perform a simple computational implementation of S_q^{approx} and explore the behaviour of those very large systems for which the *ab initio* calculations of the electron density are very expensive (for example, for systems constituted by more than 10^4 atoms). This definition satisfies the condition of the dimensionless of the entropy.

3. Characterization of atoms in the basal state

One of the principal questions that emerges in this study is concerned with the parameter "q", namely, for an atomic system in a basal state, what is its q value? For this, we calculated the electron density in the position space using the following functionals, B3LYP, BHandH, M062x, MP2, MP3 and TPSS, and with the following *ab initio* methods, CCS, CCSD, CISD, using a standard quantum chemistry program, Gaussian 09 [13], with the basis set DGDZVP [14, 15], to obtain the energy value and the corresponding wave function. The electron density in the position space was calculated with the DGrid program [16] using the wave function obtained through several methodologies, and for the entropy calculations we used the integration algorithm designed by Pérez-Jordá *et al.* [17] with a precision of 1×10^{-5}.

In the Figures (1(a) - 1(f)), we show the results of S_q using $q = 0.9, 1.1, 1.3, 1.5, 1.7, 1.9$, in which we note that the entropy has no dependence upon the methods, and it is possible to recognize the periodicity of the elements in the periodic table. Naturally, when the q parameter changes, the difference is magnified between the different periods of the periodic table; however, with the atoms that involve d-orbitals, we note a small disruption in this tendency. In general, we can observe that the Shannon entropy increases with respect to the atomic number in a natural way, if we appeal to the interpretation of this information measure, we can specify that the content of the information tends to increase. This assumption is based on the follow interpretation: if we consider an ideal gas, this system has a uniform particle distribution; therefore, the entropy is maximum. As such, we expect that the entropy increases in proportion to the electron number -in principle- and, considering the physics of the system, when the number of particles increases the Shannon entropy tends to the Thomas-Fermi limit as a consequence of a decrease in the Wigner-Seitz radii [18]. Thus, the definition proposed in this work permits us to recover the original idea of the content of the information of a system in relation to the physical interactions between the electrons of systems, such that when the number or particles, N, tends to infinity, q will tend to the unit.

On the other hand, if we consider a system in a basal state and in equilibrium, what is the value of the parameter q? To address this question, we propose a computational form to find this value,

$$q_{i+1} = q_i + \frac{S_{q_i}}{S_{q_{i+1}}}, \tag{24}$$

This approximation requires that $S_{q_i} \approx S_{q_{i+1}}$ - when this condition occurs, it also satisfies that $\left(\frac{\partial S_{q_i}}{\partial q_i}\right) \approx \left(\frac{S_{q_{i+1}}}{\partial q_{i+1}}\right)$. With this assumption, the slope of S_q tends to zero, and this corresponds to the zone where q does not change. This will be the q value for the system; moreover, we can fix the precision with 1×10^3 and the error was calculated as

$$\%Error = \left|\frac{S_{q_{i+1}} - S_{q_i}}{S_{q_{i+1}}}\right|, \tag{25}$$

Given this consideration, we obtain the trends shown in Figure (2), where the general trend for q was calculated using the functionals B3LYP, TPSS and M062x, and with the wave function methods CCS, CCSD, MP2, MP3 and CISD with the basis set DGDZVP. In this figure, we note that only for the block d of the periodic table the tendency of q parameter, has a breaking of the tendency, this would be attributed to the basis set, but in respect to the methodologies the general trend of q permanence without considerable changes, this permit us establish that the q values has not a dependency of the methodologies. It is important to note that, according to the physics of the system, we expect that when the system sees a considerable increase in the number of electrons, the general behaviour will be like that of a Fermi gas, and consequently the system becomes an extensive system. This implies that the entropy becomes extensive, that is $q \to 1$, and that Eq. (16) becomes as in Eq. (3). From these results, we also obtain that

$$\lim_{q \to q^{op}} S_q = 1. \tag{26}$$

In the Figure (3) we present the general trend of q with CCSD/DGDZVP with dotted-crosses, and the following polynomial,

$$f(q) = C_1 + C_2 \exp\{-C_3 Z + C_4\} + C_5 \exp\{-C_6 Z^2 + C_7\}, \tag{27}$$

with a continuous line, where the coefficients have the values listed in Table (1).

Z denotes the atomic number, and in Table (2) we show the q_{opt} values for the first 54 atoms of the periodic table. Analysing the values of this table, we note that in all cases the characteristic value of each atom in the basal state is close to the unit when the electron number increases.

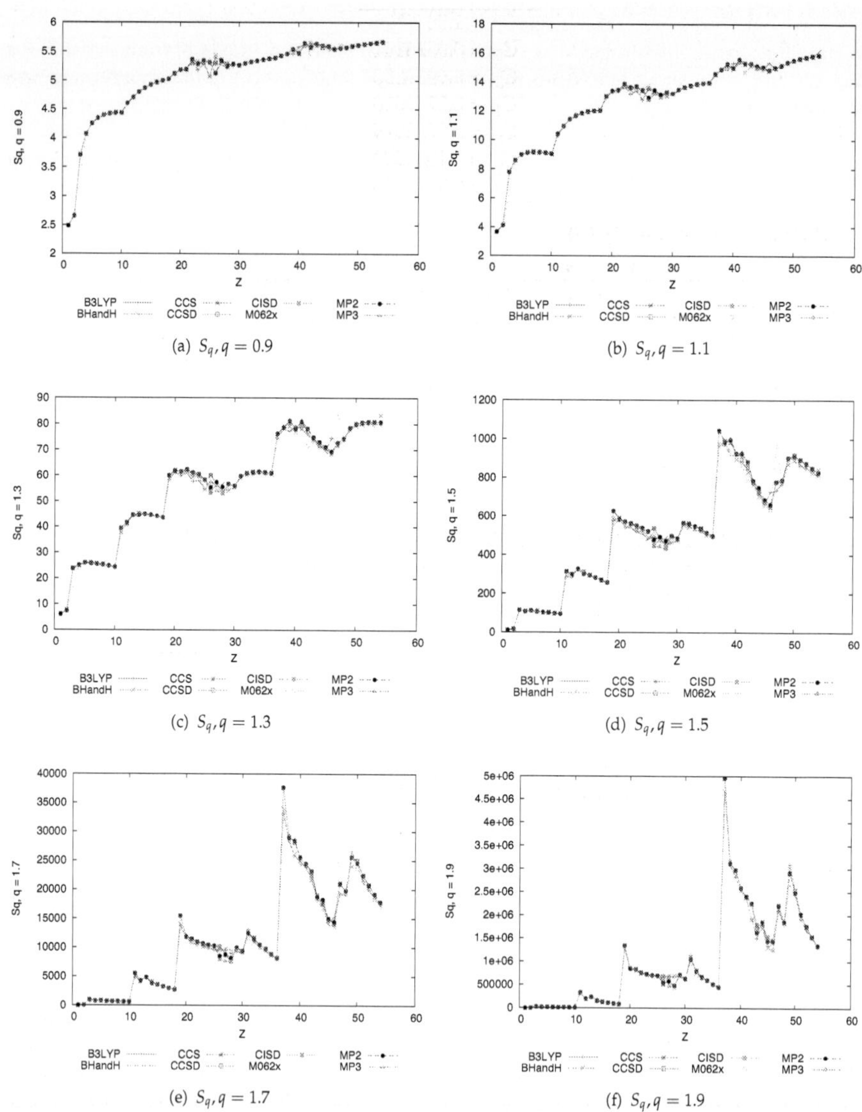

Figure 1. Effect of the variation of the parameter q for the trends of entropy, using several methodologies of quantum chemistry with the DGDZVP basis set.

$C_1 = 1.0055300$
$C_2 = 0.0526141$
$C_3 = 0.9959850$
$C_4 = 0.2739020$
$C_5 = 0.0738875$
$C_6 = 0.1080680$
$C_7 = 0.3578360$

Table 1. Values of the constants of Eq. (27).

Z	q_{opt}	Z	q_{opt}	Z	q_{opt}	Z	q_{opt}
1	1.1597116	15	1.0219020	30	1.0123013	44	0.9981021
2	1.1383104	16	1.0207019	31	1.0120013	45	1.0004019
3	1.0920058	17	1.0198019	32	1.0117012	46	1.0087011
4	1.0698041	18	1.0190017	33	1.0113013	47	1.0085011
5	1.0556034	19	1.0178018	34	1.0110012	48	1.0086010
6	1.0470029	20	1.0178016	35	1.0109012	49	1.0083010
7	1.0409027	21	1.0198014	36	1.0107012	50	1.0081010
8	1.0364026	22	1.0037026	37	1.0107012	51	1.0080010
9	1.0332024	23	1.0106020	38	1.0104011	52	1.0079010
10	1.0306023	24	1.0035025	39	1.0036017	53	1.0077009
11	1.0277023	25	1.0016026	40	1.0144007	54	1.0076009
12	1.0268022	26	1.0095018	41	0.9954024		
13	1.0246022	28	1.0132014	42	1.0024017		
14	1.0228021	29	1.0105015	43	0.9947024		

Table 2. Values of q_{opt} using CCSD(full)/DGDZVP.

4. Characterization of a Simple Chemical process

One of the interests of this work is in the study of the effect of the parameter "q" in a dissociation process. The idea is to study the effect of small interactions when a homonuclear system is dissociated.

In this case, we select the dissociation of the H_2 molecule,

$$H_2 \longrightarrow H + H.$$

The calculations were performed with Gaussian 03 [19] with CCSD(full) and the basis set cc-pVTZ [20]. For the entropy calculations, we used the wave function generated by Gaussian 03 to generate the electron density, while $\rho_{\max}$ was calculated in the position of the nuclei of each atom of the molecular system, in this case by the symmetry $\rho_{\max}(A) = \rho_{\max}(B) = \rho_{\max}$. The electron density was calculated with DGrid and the algorithm of integration that we used was designed by Peréz-Jodá *et al.* with a precision of 1×10^{-5}.

In Figure (4), we present the general trend of this simple chemical process, where we can note that the internuclear equilibrium distance is 0.754 Å, which corresponds to the minimum electronic energy; the dissociation process was carried out more than two times the van der

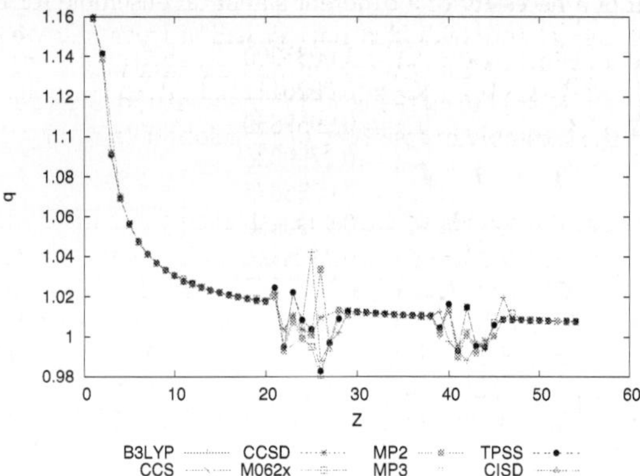

Figure 2. Trends of the parameter q for the atoms $1 < Z < 54$ using several methodologies with the basis set DGZVP; all calculations were performed in Gaussian 09.

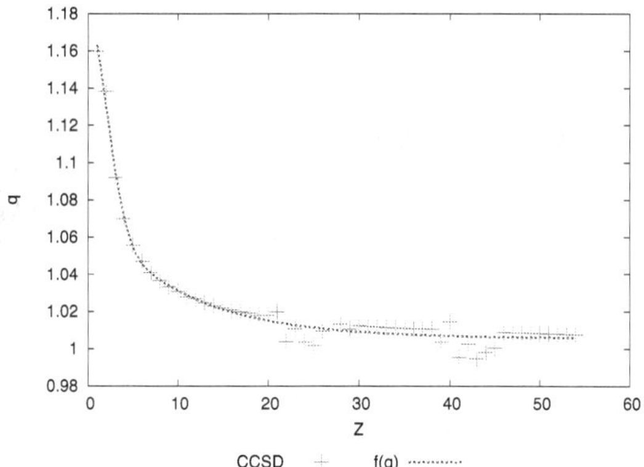

Figure 3. Trends of the parameter q for the atoms $1 < Z < 54$, with CCSD(full)/DGDZVP in Gaussian 09.

Waals radii of the hydrogen atom in the basal state (1.2 Å) to ensure that no weak chemical interactions were present. With this in mind, in this case it is natural to think that, for an internuclear distance of 3.0 Å, the electronic energy will be twice that of the electronic energy of the hydrogen atom in the basal state; however, this does not occur, and the value obtained with CCSD(full)/cc-pVTZ is -1.0007258069 a.u. Consequently, there exists a difference of 0.0007258069 a.u. (1.9056 kJ/mol) - this energy value is closer to that of hydrogen bonding. In principle, the explanation of this anomaly can be addressed in the following way: by definition, the wave function is extended over all space, and by construction the wave function used in a quantum mechanics calculation is a finite superposition of the basis set functions, $\psi = \sum_i^n \chi_i \phi_i$. However, notwithstanding that, in a limit this function will be exact, the correct description obtained with this wave function will be correct only in the equilibrium. This condition is not obvious, and how we see is not fulfil, this probable permit us talk about of a necessity of a different statistical ensemble for the more adequate description of the systems, and with this new ensembles possible we can describe of a more appropriate some phenomena present in the quantum world. The real justification for proposing (and postulating) the existence of this new set of definitions will reside in their implications, namely the incompleteness of the descriptions obtained by the actual tools and theories.

In Table (3), we present the values of the electron energy for the hydrogen molecule at an internuclear distance of 3.0 Åusing several methodologies of quantum chemistry, and with the basis sets cc-pvDZ, cc-pvTZ, cc-pvQZ and cc-pv5Z. The basis sets are designed to converge systematically on the complete basis set, such that this basis set permits us to analyse the improvement of the electron energy, and we note that the best result that we can obtain corresponds to CCSD(full)/cc-pvTZ. However, even using this sophisticated methodology and basis set, there exists an excess energy of 0.0012563951 a.u. (3.2986 kJ/mol). Here, it is convenient to observe that it is not necessary to make use of a bigger basis set corresponding to a better description (again, this is in reference to Tables (3) and (4), in which the best energy value corresponds to CISD/cc-pvTZ and not to cc-pv5Z, which is the more complete basis set of all those used in this work). In all cases, all the methodologies and basis sets overestimate the energy over large distances, but in the case of MP2, PBE and B3LYP, the overestimation it of the order of the energy of a simple covalent bond, such as an oxygen molecule (145 kJ/mol), or a simple bond of a nitrogen molecule (170 kJ/mol).

On the other side, and continuing our discussion, in Figure (4) we observe that the system can be additive but not necessarily extensive. To explain this in the Figures (6(a)-6(f)), we present a comparison of the electron energy and the deformed entropy using several q values, where we note that for different q values the minimum of the entropy change of the position, for $q = -4.1, -2.8, 0.3, 1.5, 1.8$ and a distance greater than of 2.0 Å, the slope of tendency it is zero, that is, the entropy is constant and consequently it is additive, but not extensive, because for a this values, S_q is constant, now if we consider that this system its constituted by identical subsystems we have, $S_q(A,B) = S_q(A) + S_q(B) + (1-q)S_q(A)S_q(B)$ and by the system characteristics $S_q(A) = S_q(B)$ so we have $S_q(A,B) = 2S_q(A) + (1-q)[S_q(A)]^2$. This opens the door to an interesting question: in physical systems, it is the same additive that extensive? In our opinion, they are different concepts and it is probable that the use of these concepts as synonyms is a result of the historical background. It is interesting to note that, for $q = -2.8 = -14/5$, the minimum of the entropy corresponds to the minimum of the energy (see Figure (6(b))); consequently, if we find the appropriate q value for the system, it is possible to reproduce the electron energy behaviour. The interesting aspect of this is that the deformed entropy, that it is a local functional (because has not dependency of external potential), that we can found the same tendencies of the energy in which are present the effects of an external potential. Another notable characteristic of this tendency is presented in Figure (6(c)), where in the final state of the system the slope again tends to zero but the total entropy is greater than the initial content. This implies that the term $(1-q)S_q(A)^2$ is greater than $2S_q(A)$, and if we retake the interpretation of the term $(1-q)S_q(A)^2$ then, like the degree of non-separability, we can conclude that the representation of the system with $q = 0.3$ is non-extensive over all processes, even if the internuclear distance implies that the system has no physical interactions. This is consistent with the interpretation that the wave function is extended over all space - if we accept this, probably we can establish a link between non-extensivity and quantum entanglement [21]. Consequently, it is possible to build a bridge between quantum information theory and non-extensive statistical mechanics and reclaim the idea that we can improve our understanding of nature by not only analysing behaviour exclusively in terms of matter and energy (even at the level of elementary particles), but also that study using the techniques and methods of modern physics and chemistry integrate concepts and tools that allow us to comprehensively investigate the behaviour of natural systems in order to deepen our understanding of them to incorporate information measures that take into account concepts such as entanglement, known since the early days of Quantum Mechanics, for which, however, there are no measures in many modern theories, at a more fundamental level, it has become clear that an Information Theory based on the principles of Quantum Mechanics, expands and complements the Classical Information Theory [22]. In addition to the quantum generalizations of classical notions such as sources, channels and codes, this new theory includes two complementary types of quantifiable data: classical information and quantum entanglement.

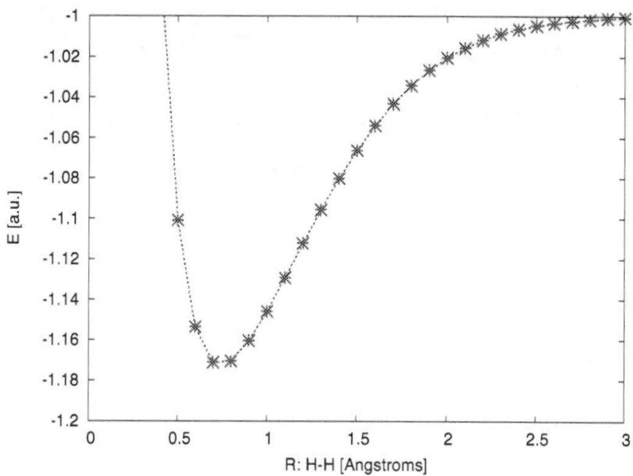

Figure 4. Trends of the electron energy of the dissociation process of H_2, with CCSD(full)/cc-pvTZ.

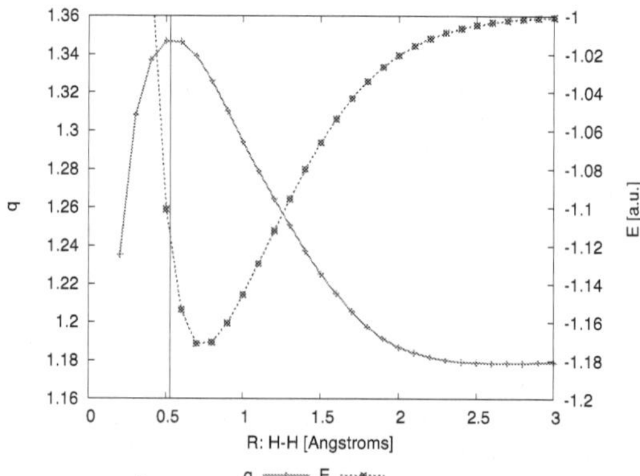

Figure 5. Comparison between the electron energy and the q_{opt} parameter for the dissociation process of H_2 with CCSD(full)/cc-pvTZ.

Basis set	CCSD	CISD	MP3
cc-pVDZ	-0.99955061881	-0.99955059355	-0.98767847037
cc-pVTZ	-1.0007258067	-1.0007257401	-0.98174230596
cc-pVQZ	-1.0010904704	-1.0010905136	-0.97990255011
cc-pV5Z	-1.0012563951	-1.0012563405	-0.97874184183
Basis set	MP2	PBE	B3LYP
cc-pVDZ	-0.92566086120519	-0.937015225619	-0.933130505886
cc-pVTZ	-0.92835582370175	-0.939927619001	-0.936467455835
cc-pVQZ	-0.92932919850342	-0.940569300858	-0.937152190049
cc-pV5Z	-0.92968788957241	-0.940901783377	-0.937514589740

Table 3. Values of the electron energy for H_2 at an internuclear distance of 3Å(in a.u.). The energy calculations were performed in Gaussian 09.

Basis set	CCSD	CISD	MP3
cc-pVDZ	1.179850	1.179916	32.350176
cc-pVTZ	1.833750	1.905430	47.935576
cc-pVQZ	2.863030	2.863143	52.765854
cc-pV5Z	3.298665	3.298521	55.813294
Basis set	MP2	PBE	B3LYP
cc-pVDZ	195.17740	165.36652	175.56585
cc-pVTZ	188.10178	157.72003	166.80469
cc-pVQZ	185.54618	156.03530	165.00692
cc-pV5Z	184.60444	155.16236	164.05544

Table 4. Absolute difference of values between the electron energy of H_2 at 3.0 Å, and $2H$ with several methodologies and with the cc-pvTZ basis set. All values are kJ/mol.

In the Figure (5), we show the general trend of q_{opt} compared with the electron energy, in which the tendency of q_{opt} has a maximum in approximately 0.529 Å, plotted as a vertical continuous black line, this value correspond at the first Bohr radii for the Hydrogen atom, this is an interesting point because it is possible talk about a non-extensive radii of the systems, where the non-extensivity it is maximum and the point of this is that the we can associate the q parameter at a physical property like the distances between the subsystems, so we suspect that the non-extensive behaviour is closely related at two characteristics; the distance and the particle number.

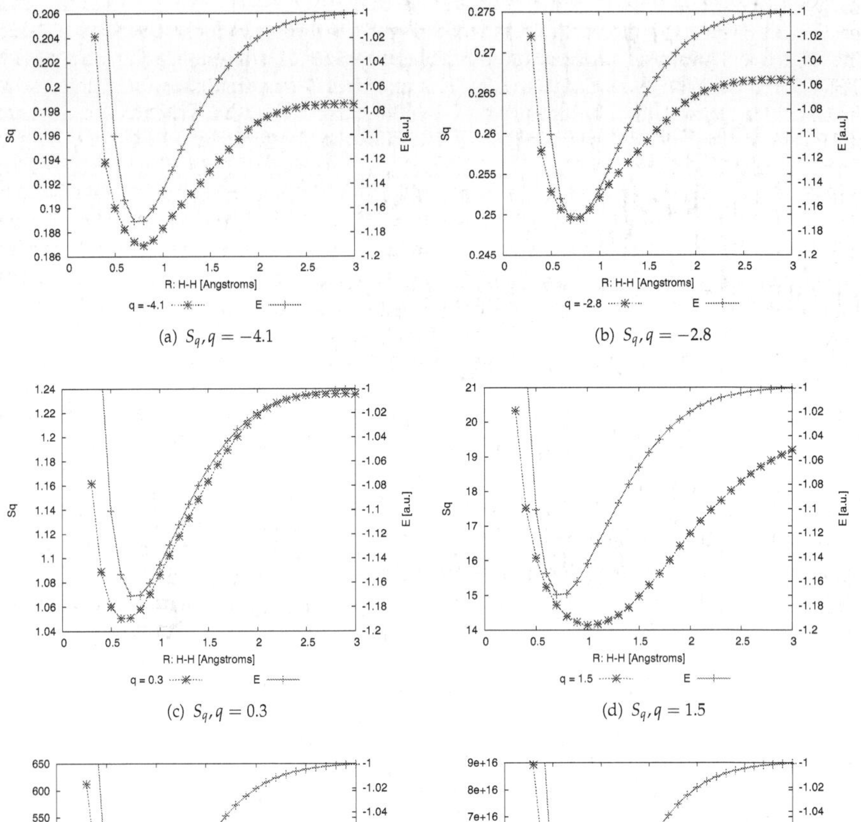

(a) $S_q, q = -4.1$

(b) $S_q, q = -2.8$

(c) $S_q, q = 0.3$

(d) $S_q, q = 1.5$

(e) $S_q, q = 1.8$

(f) $S_q, q = 2.2$

Figure 6. Comparison between the trends of the S_q entropy using several q values and the electron energy for the dissociation process of a H_2 molecule.

5. Characterization of the Chemical Reaction $H_2 + H^-$

In this section, we present the results of the reaction $H_2 + H^- \rightarrow H_2 + H^-$. This reaction is one of the more studied reactions and it is very well-characterized [23–25]. The IRC calculation was performed with MP2(full)/6-311G, and the singles points with CISD/6-311++G**, both in Gaussian 03. This reaction is symmetric, has a maximum in the transition state (which has an energy of -1.6501559031 a.u.) and an internuclear distance of 0.93236 Åbetween each hydrogen atom. Naturally, the electron energy in the reactants is the same as in their products (-1.6680093713 a.u). In the Figures (7(a))-(8(f)), we present the tendency of the deformed entropy using several values of the q parameter, with $q = -10.0$, Figure (7(a)). The entropy has a maximum value at $RX = 0.0$ and a possible local minimum in $RX = -2$ and $RX = 2$. However, is not very clear how to determine whether this q value is associated with the changes of the entropy for some other parameters related at changes physical or chemical, when we use the $q = -4.6$, Figure (7(b)), the entropy tendency has a maximum value at $RX = 0.0$ and it is similar at the tendency of the energy. The more interesting aspects of the changes in the entropy are in the interval $-1.1 \leq q \leq 1.6$, in Figure (7(c)); with $q = -1.1$, the entropy has a minimum in $RX = -1$ and $RX = 1$, and these minimums are associated at a zone where the process of the breaking and forming of the chemical bonds occurs. This zone corresponds to a zone where the normal modes of vibration have negative frequencies. In Figure (9(a)), we show this comparison, and in the same way we compare this tendency with the distances of the hydrogen's involved in the process, we labeled the atoms like like H_{in} for the Hydrogen that will be form the new bond and H_{out} for the Hydrogen that gonna be break the bond, in this case, the critic region where the physical changes occurs is $-0.5 \leq RX \leq 0.5$, see the Figure (9(b)), this only can be observed in the same zone of the entropy where has a small change in their slope, in this sense, it is possible that changing the value of q or increase the precision we can observe with more detail the changes that occurs in this zone. Figure (9(c)) presents a comparison of $S_q, q = -0.7$ with the Dipolar Moment, how occurs in the case of the frequencies this parameter has a maximums in $RX = -0.85$ and $RX = 0.85$, is it in this zone where the most important changes of the electron density occurs. In general, we can say that the changes in the deformed entropy permit us to discover some zones where the most important changes of the electron density of a system occurs; however, it is not yet known how to select the appropriate value of q, for example, when we use a value of $q = -0.7$, the tendency of the entropy has minimums in $RX \sim -0.9$ and $RX \sim 0.9$. We can say that this tendency is related to the change of the electron density, but in the case of $S_q, q = -0.1$ the entropy behaves like a specular image of the energy (see Figures (8(a)) and (8(c))). With this evidence, we believe that it is possible to derive some density functionals in which a combination of different entropic terms can be expressed, not only the deformed entropy with the form of Eq. (16), but also a contribution of a deformed Fisher entropy (for this, it will be necessary to write the gradient of the electron density in terms of deformed algebra). That is,

$$E[\rho] = \sum_i x_i S_q + x_i I_q, \tag{28}$$

where

$$I_q = \int \varrho(\mathbf{r}) |\nabla \ln_q \varrho(\mathbf{r})|^2 d\mathbf{r}, \tag{29}$$

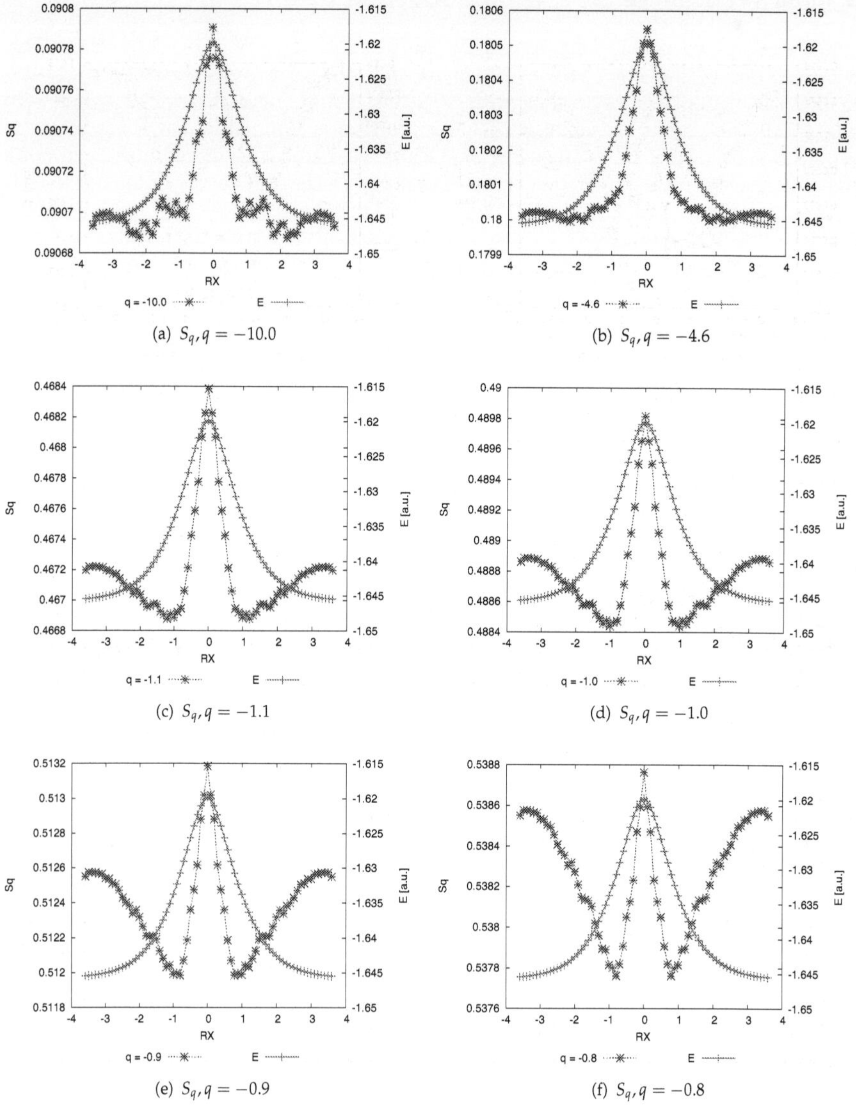

(a) $S_q, q = -10.0$

(b) $S_q, q = -4.6$

(c) $S_q, q = -1.1$

(d) $S_q, q = -1.0$

(e) $S_q, q = -0.9$

(f) $S_q, q = -0.8$

Figure 7. Comparison between the trends of the S_q entropy using several q values and the electron energy for the reaction $H_2 + H^-$.

and $\varrho(\mathbf{r})$ is the shape factor, defined as $\varrho(\mathbf{r}) = \frac{\rho(\mathbf{r})}{N}$.

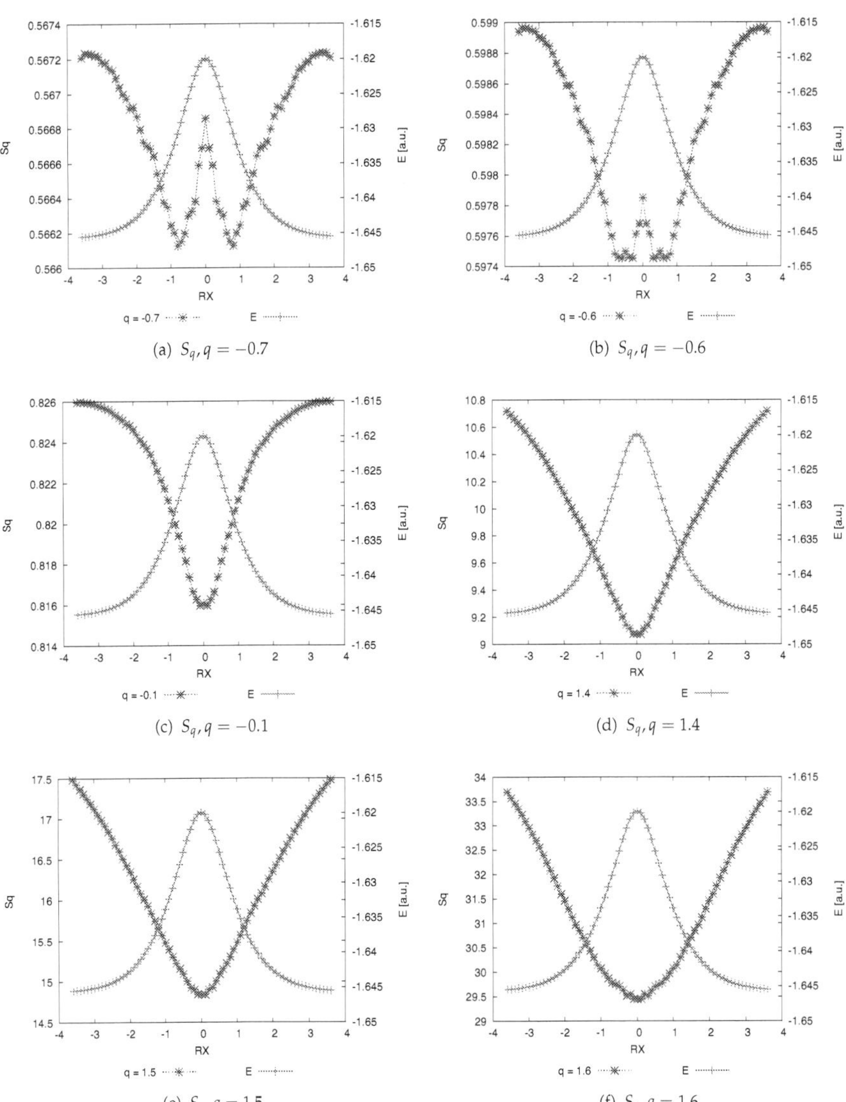

(a) $S_q, q = -0.7$

(b) $S_q, q = -0.6$

(c) $S_q, q = -0.1$

(d) $S_q, q = 1.4$

(e) $S_q, q = 1.5$

(f) $S_q, q = 1.6$

Figure 8. Comparison between the trends of the S_q entropy using several q values and the electron energy for the reaction $H_2 + H^-$.

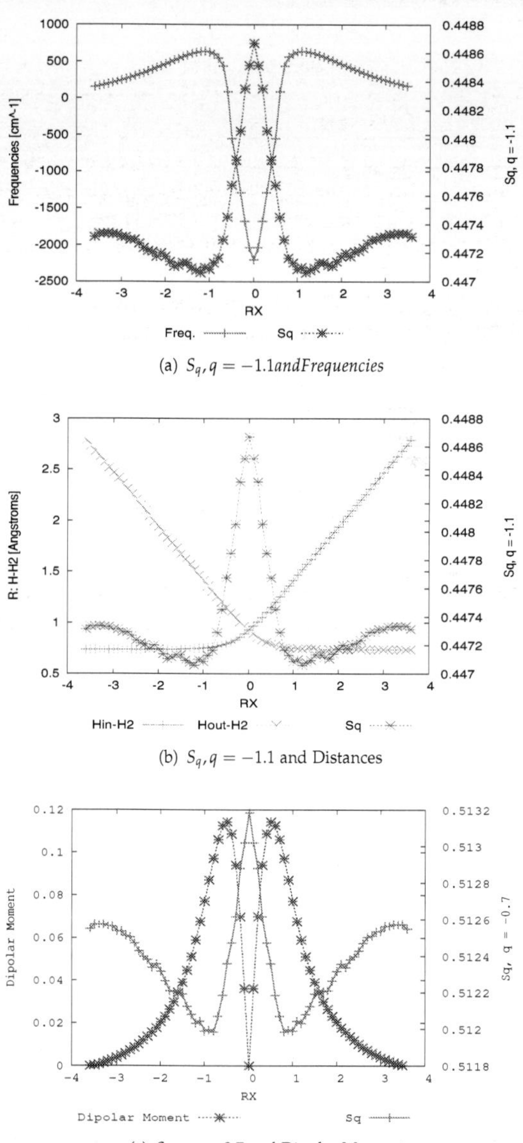

(a) $S_q, q = -1.1 and Frequencies$

(b) $S_q, q = -1.1$ and Distances

(c) $S_q, q = -0.7$ and Dipolar Moment

Figure 9. Comparison between the trends of the S_q entropy using several q values, frequencies, distances and Dipolar moment for the reaction $H_2 + H^-$.

6. Entropic profiles of atoms

In this section, we present the results of the entropic profiles from hydrogen to neon. The density was calculated with a precision of 1×10^{-5} with CCSD(full)/DGDZVP obtained in Gaussian 09. It is our particular interest to characterize the hydrogen atom, because the simplicity of this system permits us to determine and - as far as possible - try to find some periodic properties. In this sense, in the Figure (10(a)), we present the entropic profile for this system, in which we note that the maximum present is in $R \approx 1a.u.$, that is, the first Bohr radii. It is possible to speculate that it is at this distance when the system exhibits the maximum degree of non-extensivity. In the case of He, the maximum value of the entropy is displaced close to the nucleus, $R = 0.545a.u.$, Figure (10(b)). For the second period (lithium to neon, Figures (11(a)-12(b))), the maximum of the entropy coincides with the first maximum of the electron density, and the minimum of the electron density coincides with the inflection point of the entropy trend; subsequently, in the region of the maximum density the entropy there is a small change in their slope. Finally, the electron density and the entropy tends to zero. In general, we also propose verifying the changes of the entropy tendency using several values of q in the interval $0 \leq q \leq 10$ with a step size of 0.1, though we cannot observe significant changes. In this sense, it is possible that the critical points of the deformed entropy are in relation to the chemical reactivity parameters, such as the Fukui function [26–28], hardness, softness [29], [30], chemical potential, *inter alia*, [31]. However, it will be necessary to perform studies of the relation between the q parameter and these chemical descriptors. We consider that it will be important to carry this concepts into the field of deformed algebra.

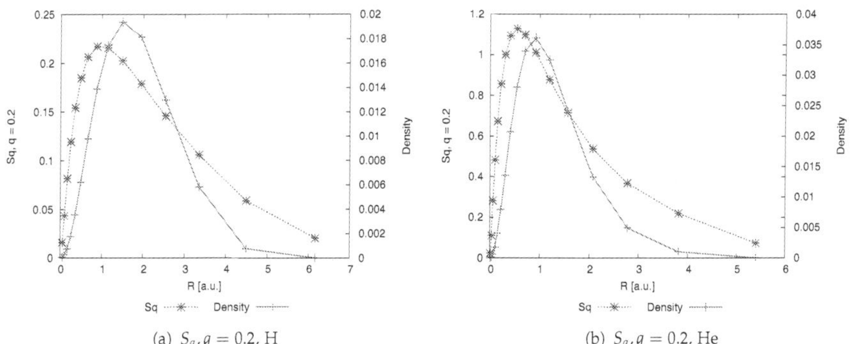

(a) $S_q, q = 0.2$, H

(b) $S_q, q = 0.2$, He

Figure 10. Comparison between the trends of the S_q entropy with $q = 0.2$ and the electron density for hydrogen and helium.

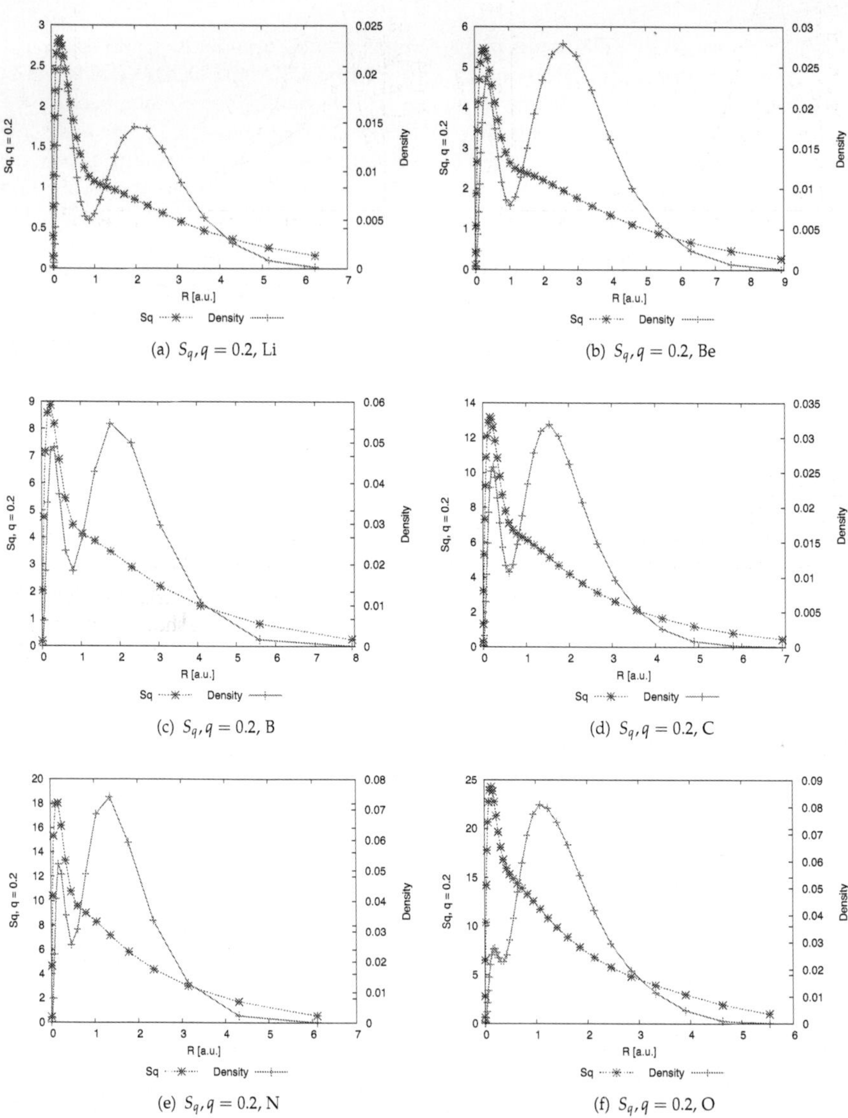

(a) $S_q, q = 0.2$, Li

(b) $S_q, q = 0.2$, Be

(c) $S_q, q = 0.2$, B

(d) $S_q, q = 0.2$, C

(e) $S_q, q = 0.2$, N

(f) $S_q, q = 0.2$, O

Figure 11. Comparison between the trends of the S_q entropy with $q = 0.2$ and the electron density for lithium to oxygen.

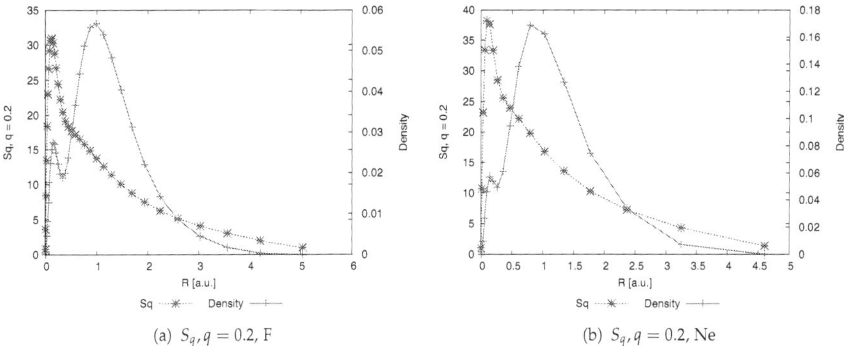

(a) $S_q, q = 0.2$, F

(b) $S_q, q = 0.2$, Ne

Figure 12. Comparison between the trends of the S_q entropy with $q = 0.2$ and the electron density for fluorine and neon.

7. Conclusion

In this work, we applied the fundamental idea of Tsallis of entropy generalization and we propose a definition of deformed entropy that is applied to the description of the first 54 atoms in the basal state of the periodic table, showing that each chemical system has a characteristic q value. Furthermore, we show that the characteristic q value can be related to the first Bohr radii, in which we suppose that there corresponds a distance where the non-extensive behaviour of the system is dominant. In the same way, we show the numerical tendencies of the deformed entropy compared to the variation of the electron density for the first 10 atoms in the basal state and observe that the changes of the deformed entropy are in relation to the significant changes in the electron density.

This has allowed us to start a new line of investigation, and with some of these results we continue with the study of the formalisms for the construction of a functional based in a principles of physics and information theory. In addition, we intend to continue with the development of models that permit us to find a direct relation between electron energy and chemical reactivity concepts with deformed entropy.

On the other hand, the application of the concepts of information theory permit us form a description that is more accurate than that based on energetic criteria alone; we speculate that it is possible to define or find a form that derives the Density Functional Theory from some fundamental expression.

Finally, with these examples we have tried to link information from a system that is subjected to a process with physical and chemical changes. Thus, we have linked the concept of *information*, which is an epistemological concept completely with ontological concepts and the interpretation of the results allows us feedback these concepts in ontological terms, according to the authors, is probable that today do not exist a orthodoxical definition of what actually is the *information*, beyond that presented by Shannon and its guidelines, criteria, characterization of it, among other things, the interpretation and the relationship with other concepts, such as energy, electron density, chemical reactivity parameters and many others need be discussed to try of establisha formal relation between concepts.

Is it clear that information concept and the model itself is interdisciplinary or transdisciplinary. The concept of information and -moreover- the model itself promote a systematic relation with causal analogies and parallelism with scientific knowledge, which transcends the framework of the source domain and extends in various directions, thus making the knowledge acquire an unusual resonance. Accordingly, we believe it is feasible to complement the explanations of natural processes and natural systems.

8. Acknowledgements

N. Flores-Gallegos wishes to thank the CONACyT for financial support through the programme Apoyos Complementarios para la Consolidación Institucional de Grupos de Investigación (Repatriación, Retención y Estancias de Consolidación) and Centro Universitario de los Valles, of Benemérita Universidad de Guadalajara.

Author details

N. Flores-Gallegos, I. Guillén-Escamilla and J.C. Mixteco-Sánchez

Centro Universitario de los Valles Benemérita Universidad de Guadalajara, Ameca, Jalisco, México

References

[1] C. E. Shannon, A mathematical theory of communication. The Bell System Technical Journal. 1948; 27: 379-423.

[2] C. Tsallis, Possible generalization of Boltzmann-Gibbs statistics. Journal of Statistical Physics. 1988; 52(1/2): 479-487.

[3] E. P. Borges, A possible deformed algebra and calculus inspired in nonextensive thermostatistics. Physica A. 2004; 340: 95-101.

[4] V. Kac, and P. Cheing, Quantum Calculus. Springer; 2000.

[5] G. Kaniadakis, M. L., A.M. Scarfone Deformed logarithms and entropies. Physica A. 2004; 340: 41-49.

[6] L. H. Thomas, The calculation of atomic fields. Mathematical Proceedings of the Cambridge Philosophical Society. 1927; 23(5): 542-548.

[7] E. Fermi, Un Metodo Statistico per la Determinazione di alcune Prioprieta dell'Atomo. Rend. Accad. Naz. 1927; 6: 602-607.

[8] N. Flores-Gallegos and A. Vela. A new analysis of Shannon entropy in atoms. *To be published*. 2014.

[9] S. Guiaşu, Information theory with applications. MacGraw-Hill International Book Company; 1977.

[10] F. L. Hirshfeld, Bonded-atom fragments for describing molecular charge densities. Theoret. Claim. Acta (Berl.) 1977; 44: 129-138.

[11] R. F Nalewajski, and E. Broniatowska, E. Entropy/information indices of the "stockholder" atoms-in-molecules. International Journal of Quantum Chemistry. 2005; 101: 349-362.

[12] C. Tsallis, Introduction to nonextensive statistical mechanics. Springer; 2009.

[13] Gaussian 09, Revision C.01, M. J. Frisch, G. W. Trucks, H. B. Schlegel, G. E. Scuseria, M. A. Robb, J. R. Cheeseman, G. Scalmani, V. Barone, B. Mennucci, G. A. Petersson, H. Nakatsuji, M. Caricato, X. Li, H. P. Hratchian, A. F. Izmaylov, J. Bloino, G. Zheng, J. L. Sonnenberg, M. Hada, M. Ehara, K. Toyota, R. Fukuda, J. Hasegawa, M. Ishida, T. Nakajima, Y. Honda, O. Kitao, H. Nakai, T. Vreven, J. A. Montgomery, Jr., J. E. Peralta, F. Ogliaro, M. Bearpark, J. J. Heyd, E. Brothers, K. N. Kudin, V. N. Staroverov, T. Keith, R. Kobayashi, J. Normand, K. Raghavachari, A. Rendell, J. C. Burant, S. S. Iyengar, J. Tomasi, M. Cossi, N. Rega, J. M. Millam, M. Klene, J. E. Knox, J. B. Cross, V. Bakken, C. Adamo, J. Jaramillo, R. Gomperts, R. E. Stratmann, O. Yazyev, A. J. Austin, R. Cammi, C. Pomelli, J. W. Ochterski, R. L. Martin, K. Morokuma, V. G. Zakrzewski, G. A. Voth, P. Salvador, J. J. Dannenberg, S. Dapprich, A. D. Daniels, O. Farkas, J. B. Foresman, J. V. Ortiz, J. Cioslowski, and D. J. Fox, Gaussian, Inc., Wallingford CT, 2010.

[14] N. Godbout, D. R. Salahub, J. Andzelm, and E. Wimmer, Optimization of Gaussian-type basis sets for local spin density functional calculations. Part I. Boron through neon, optimization technique and validation. Can. J. Chem. 1992; 70: 560-71.

[15] C. Sosa, J. Andzelm, B. C. Elkin, E. Wimmer, K. D. Dobbs, D. A. and Dixon, A local density functional study of the structure and vibrational frequencies of molecular transition-metal compounds. J. Phys. Chem. 1992; 96: 6630-36.

[16] M. Kohout, DGrid, version 4.6, Radebeul, 2011.

[17] Pérez-Jorda, José M., Becke, Axel D. and San-Fabian, Emilio. Automatic numerical integration techniques for polyatomic molecules. J. Chem. Phys. 1994; 100: 6520-6534.

[18] F. Wigner and F. Seitz, On the constitution of metallic sodium. Phys. Rev. 1933; 58: 802-810.

[19] Gaussian 03, Revision E.01, M. J. Frisch, G. W. Trucks, H. B. Schlegel, G. E. Scuseria, M. A. Robb, J. R. Cheeseman, J. A. Montgomery, Jr., T. Vreven, K. N. Kudin, J. C. Burant, J. M. Millam, S. S. Iyengar, J. Tomasi, V. Barone, B. Mennucci, M. Cossi, G. Scalmani, N. Rega, G. A. Petersson, H. Nakatsuji, M. Hada, M. Ehara, K. Toyota, R. Fukuda, J. Hasegawa, M. Ishida, T. Nakajima, Y. Honda, O. Kitao, H. Nakai, M. Klene, X. Li, J. E. Knox, H. P. Hratchian, J. B. Cross, V. Bakken, C. Adamo, J. Jaramillo, R. Gomperts, R. E. Stratmann, O. Yazyev, A. J. Austin, R. Cammi, C. Pomelli, J. W. Ochterski, P. Y. Ayala, K. Morokuma, G. A. Voth, P. Salvador, J. J. Dannenberg, V. G. Zakrzewski, S. Dapprich,

A. D. Daniels, M. C. Strain, O. Farkas, D. K. Malick, A. D. Rabuck, K. Raghavachari, J. B. Foresman, J. V. Ortiz, Q. Cui, A. G. Baboul, S. Clifford, J. Cioslowski, B. B. Stefanov, G. Liu, A. Liashenko, P. Piskorz, I. Komaromi, R. L. Martin, D. J. Fox, T. Keith, M. A. Al-Laham, C. Y. Peng, A. Nanayakkara, M. Challacombe, P. M. W. Gill, B. Johnson, W. Chen, M. W. Wong, C. Gonzalez, and J. A. Pople, Gaussian, Inc., Wallingford CT, 2004.

[20] T. H. Dunning Jr., Gaussian basis sets for use in correlated molecular calculations. I. The atoms boron through neon and hydrogen. J. Chem. Phys. 1989; 90: 1007-23.

[21] R. Horodecki, P. Horodecki, M. Horodecki, and K. Horodecki, Quantum entanglement. Rev. Mod. Phys. 2009; 8: 865-942.

[22] V. Vedral, The Role of relative entropy in quantum information theory. Rev. Mod. Phys. 2002; 74: 197-234.

[23] N. Flores-Gallegos, and R. O. Esquivel, Von Neumann entropies analysis in Hilbert space for the dissociation processes of homonuclear and heteronuclear diatomic molecules. J. Mex. Chem. Soc. 2008; 52: 19-30.

[24] Esquivel, R. O., Flores-Gallegos, N., Iuga, C., Carrera, E. M., Angulo, J. C., and Antolín, J. Phenomenological description of the transition state and the bond-breaking and bond-forming processes of selected elementary chemical reactions: an information-theoretic study. Theor. Chem. Acc. 2009; 240: 445-460.

[25] R. O. Esquivel, J. C. Angulo, J. Antolń, and S. D. Jesús, Sheila López-Rosa and Nelson Flores-Gallegos. (2010). Complexity analysis of selected molecules in position and momentum spaces. Physical Chemistry Chemical Physics. 2010; 12: 7108-7116.

[26] K. Fukui, T. Yonezawa, and H. Shingu, A molecular orbital theory of reactivity in aromatic hydrocarbons. J. Chem. Phys. 1952; 20: 722-725.

[27] K. Fukui, T. Yonezawa, and C. Nagata, Molecular orbital theory of orientation in aromatic heteroaromatic and other conjugated molecules. J. Chem. Phys. 1954; 22: 1433-1442.

[28] K. Fukui, The role of frontier orbitals in chemical reactions. Science. 1987; 218: 747-754.

[29] R. G. Pearson, Hard and soft acids and basis. J. Am. Chem. Soc. 1964; 85: 3533-3539.

[30] R. G. Parr, and W. Yang, Density functional approach to the frontier electron theory of the chemical reactivity. J. Am. Chem. Soc. 1984; 106: 4049-4050.

[31] N. Flores-Gallegos, Shannon informational entropies and chemical reactivity. Advances in Quantum Mechanics. InTech; 2012. 683-706.

Chapter 10

Computation of Materials Properties at the Atomic Scale

Karlheinz Schwarz

Additional information is available at the end of the chapter

http://dx.doi.org/10.5772/59108

1. Introduction

1.1. Inorganic solids

Many inorganic solid materials are of great interest from a fundamental point of view or for technologic applications. The challenge to theory and computation is that they are governed by different length scales, where from meters (m) down to micrometers (μm) classical mechanics and continuum models provide proper descriptions (for example using finite element methods). However, if the length scale is down to nanometers (nm) or atomic dimensions measured in Å, such as for modern devices in the electronic industry (for example in magnetic recording) or surface science and catalysis, the properties are determined (or critically influenced) by the electronic structure and thus quantum mechanics. Understanding the properties at the atomic scale is often essential for improving or designing modern materials in a systematic way. Computation has become a key element in this process, since it allows one to analyze and interpret sophisticated measurements or to plan future experiments in a rational way, replacing the old trial and error scheme. Instead of trying all kinds of elements to improve a material by preparation, characterization and functional analysis, a simulation with computers is often much more efficient and allows one to "narrow the design space". Why should one prepare or measure a sample that is not promising based on modern computation? Some facilities (for example those providing synchrotron radiation) have already adapted this concept, since the beam time for measurements is limited and thus they should be used for promising investigations only.

There is a classical treatment at the atomic scale that is often based on atomic force fields. In this case the interactions between the atoms are specified with forces which are parameterized usually in a way to reproduce a set of experimental data such as equilibrium geometries, bulk muduli or vibrational (phonon) frequencies. For a class of materials, in which good parameters are known, this can be a useful approach to answer certain questions, since force-field

calculations require little computational effort. The main drawback is that no information can be provided about the electronic structure, since force fields do not explicitly contain electrons. Therefore these approaches will not be covered here, although they have reached a high level of sophistication.

Here we focus on the electronic structure of solids (metals, insulators, minerals, etc.) and surfaces (or interfaces) which require a quantum mechanical treatment. In recent review articles [1-5] one possible approach has been described, which is based on the WIEN2k program package [6, 7] that has been developed in my group during the last 35 years (see www.wien2k.at). Several other computer codes are available which will not be covered here. Each of them may have a different focus in terms of efficiency, accuracy, sophistication, capabilities (properties), user friendliness, parallelization, documentation etc. It should be stressed that the large variety of computer codes is very beneficial for this growing field of computational material sciences, since these codes all have advantages and disadvantages compared to others. For obvious reasons we will focus on our own code to illustrate important aspects of this field.

Some basic concepts will be described below as summarized in Figure 1. As a very first step one needs to represent the atomic structure of a solid material as is outlined in Section 2. Idealized assumptions must be made which one should keep in mind when theoretical results are compared with experiments. In order to describe the electronic structure of a system of interest by means of quantum mechanics (Section 3) we first briefly mention some fundamental concepts of solid state physics (like symmetry). Next we sketch how chemists handle quantum mechanics and then focus on the most important theory for solids, namely density functional theory. Before we can describe how to solve the corresponding Kohn-Sham equations (Section 4) it is necessary to select which electrons should be included in the calculations: all of them or only the valence electrons. The form of the potential must also be chosen, where we discuss such methods as pseudo-potential, muffin-tin or full-potential approximations. This choice is important for deciding which basis set can best describe the wave functions of the electrons. A relativistic treatment will be important if the system contains heavier elements. For magnetic systems spin-polarization is essential and thus must be included. The main ideas of the all-electron full-potential linearized-augmented-plane-wave (LAPW) with local orbitals method as implemented in WIEN2k are briefly described. This should be sufficient, since there is a good documentation of the underling details.

Section 5 lists selected results which can be obtained with WIEN2k. Since about 2500 groups-both from academic and industrial institutions – are using this code there are many details available in literature (see the link papers at www.wien2k.at). Based on the experience of developing WIEN2k, it is appropriate to make some general comments about the computer code development (Section 6). As more and more experimental scientists in this field rely on computer simulations, it is useful to raise some critical questions and summarize several points, which can cause deviations between theoretical results and experiment, a topic covered in Section 7. In Section 8 concluding remarks are made and then the very appropriate acknowledgments.

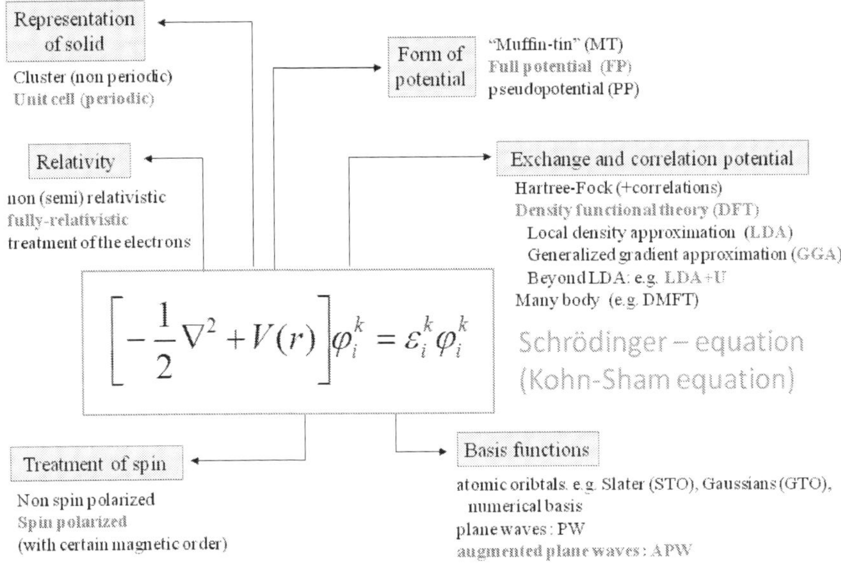

Figure 1. Possible choices for calculating the electronic structure of a material at the atomic scale (the focus is highlighted in red).

2. Atomic structure

The properties of materials at the nanometer (nm) scale or of atomic dimensions (measured in Å) are essentially determined by the electronic structure. In such a case one tries to represent the material of interest (such as a solid, a surface or a molecule) as a collection of atoms, which play the role of building blocks. It is important to realize that in practice one is forced to assume an idealized atomic structure in theoretical work, which deviates from the real structure that may be studied experimentally.

A few examples will illustrate this important point: Let us first consider a molecule which in theory is studied in vacuum, whereas in experiment it is often measured on a support (surface) or in a solution. The latter may be simulated by surrounding the molecule by a few solvent molecules or by using an embedding scheme with a dielectric constant (simulating the solvent). Recently the combination of a quantum mechanical (QM) treatment of the molecule with a cruder mechanical mechanics (MM) representation of the environment is used in QM/MM schemes. Such treatments will necessarily be approximate.

As a second example we focus on a solid. In early days one modeled a solid as a cluster of atoms, but due to size limitations this cannot represent bulk properties. With the increase in computer power we assume – especially in theory – that a solid is a perfect crystal and can be characterized by a unit cell that is repeated to infinity in all three dimensions. This means that one assumes periodic boundary conditions. A real crystal, however, is certainly finite. For experimental studies a crystalline sample is often available in the form of a powder consisting of small crystal domains. Even if experiments are carried out on a single crystal, it still has a surface and imperfections (such as defects or impurities, etc.). In compounds there are additional uncertainties, such as the stoichiometry, which may not be perfect, or the atomic arrangement, which may deviate from the underlying idealized order.

The situation of a perfect single crystal is illustrated for TiO_2 crystallizing in the rutile structure (Figure 2). The symmetry belongs to one of the 230 space groups that are tabulated in the International Tables for Crystallography [8]. Nowadays this information is also available from the Bilbao Crystallographic Server (*www.cryst.ehu.es/cryst/*). For a given crystal the unit cell must be specified with the three lattice parameters (a, b, c) and the corresponding angles (α, β, γ). For the atomic positions the Wyckoff positions must be defined, where for each type of atoms only one of the equivalent atoms needs to be specified, while all others are generated by symmetry. For known structures this type of information is available for example from the inorganic crystal structure data base [9] or can be taken from standardized CIF files, which are often available directly from literature.

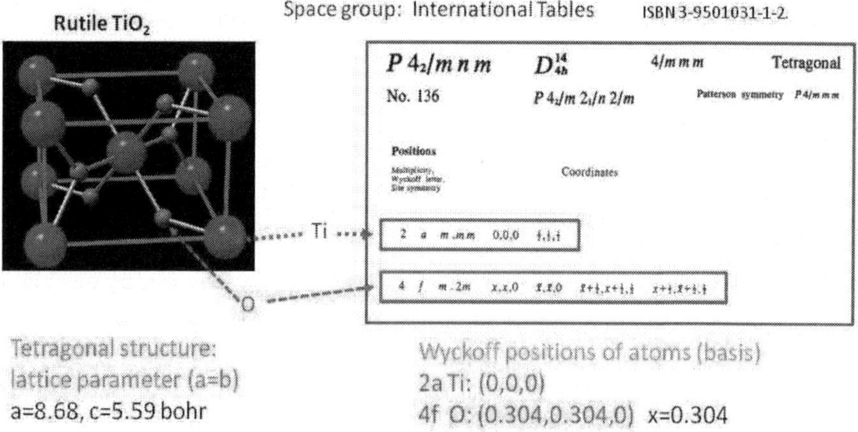

Figure 2. The crystal structure of TiO_2 in the rutile structure: It is characterized by the space group 136 which is specified in the International Table of Crystallography [8]. The two types of atoms occupy the Wyckoff positions 2a (Ti) and 4f (O) where for the latter the parameter x is not given by symmetry and thus needs to be specified. The equivalent positions are defined by symmetry. In this tetragonal structure the angles of the unit cell are all 90 degrees and for the lattice constants only a and c need to be specified, since a=b.

In Figure 3 the unit cell of a borocarbide is shown, which contains main group elements (B and C), transition metals (Ni), and rare earth elements (Nd). Often interesting materials have such complex compositions. In this case the crystal structure is well known but for an understanding

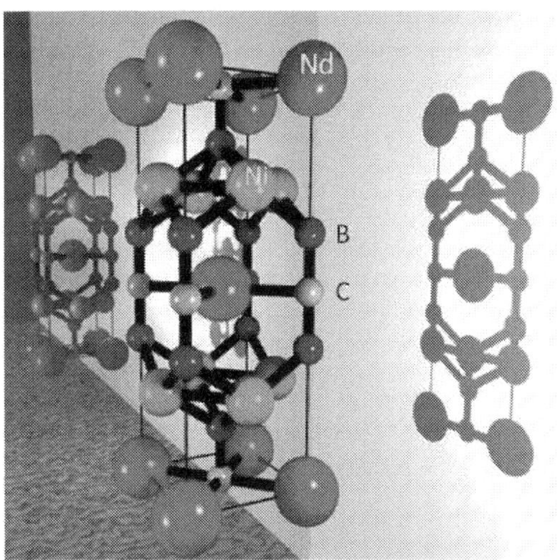

Figure 3. The unit cell of $NdNi_2B_2C$

of the properties of this compound it is essential that all these atom types (from light to heavy) can be properly treated. Details of this compound can be found in [10].

In modern material science it is often mentioned that "nano materials" have significantly different properties than their bulk analogs. A simple explanation can be provided by estimating the ratio of atoms on the surface with respect to those in the bulk of the material. The atoms on the surface have a different coordination number than the atoms in the bulk and thus have a different bonding environment. Consequently these atoms may move to a relaxed atomic position with respect to the ideal bulk crystal structure. In a nano particle a significant fraction of atoms are on the surface, whereas in a large single crystal this fraction is rather small and thus can (to a good approximation) be neglected, so that periodic boundary conditions may be used for calculating many properties.

In all the cases mentioned above the assumed atomic structure is idealized and differs from the real structure of a material that is investigated experimentally. These aspects should be kept in mind when theoretical results are compared with experimental data (as will be discussed in Section 7). This idealization is an advantage for theory with respect to experiments, since in computations the atomic structure is clearly defined as input. This is in contrast with experimental studies in which the material is often not so well characterized, in terms of stoichiometry, defects, impurities, surfaces, disorder etc. When theory simulates a structure, which is not a good representation of the real system, deviations between theory and experiment must be expected irrespective of the accuracy of the theoretical method.

With the concept of a supercell one can approximately simulate some aspects of a real system. For example one can artificially enlarge a unit cell by forming a 2 x 2 x 2 supercell, containing eight times as many atoms as the original. Figure 4 shows this case for a simple cubic case. In such a supercell one can, for example, remove an atom (representing a defect) or substitute one atom by another one (simulating a substitution) or add vacuum (about 10-15 Å) on one side of the cell (to represent a surface).

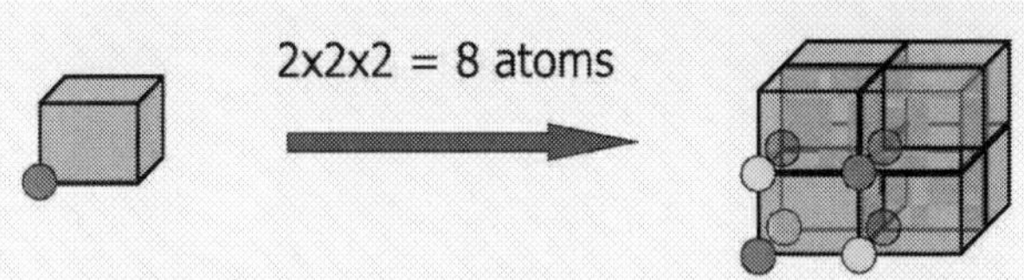

Figure 4. Schematic diagram to generate a 2 x 2 x 2 supercell from a simple cubic structure.

The possibilities of such supercells are schematically shown in Figure 5. A two-dimensional array of atoms is called slab. In a multi-layer slab the central layer approximately represents the bulk of a system, whereas the top (and bottom) layer can be used to represent a surface provided the distance to the next layer is sufficiently far away (due to the periodic boundary). On such an artificial surface one can place molecules and study for example catalytic reactions. In all such supercells one still has introduced an artificial order, since for example a defect will have a periodic image in the neighboring cells. The larger one can make the supercell the less critical the interaction between each periodic images become but this requires higher computational effort. Therefore one must make compromises.

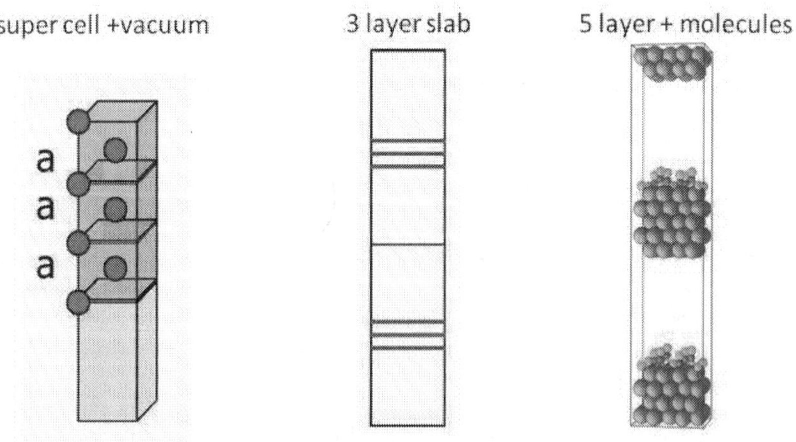

Figure 5. The construction of a supercells is schematically shown, first with three cubic cells (body-centered-cubic) plus vacuum, second a schematic supercell forming a 3-layer slab; and third a 5-layer slab (formed of metal atoms) with molecules on top.

The use of supercells is steadily increasing, since a more realistic modeling of real structures becomes attractive. In large supercells (with a few hundred atoms) one can even approximately model disorder as was recently illustrated for so called misfit layer compounds, in which the layers of PbS and TaS_2 can be stabilized by occasionally substituting Pb by Ta [11]. Such aspects will be mentioned in Section 5.

3. Quantum mechanics

3.1. DFT Fundamentals

3.1.1. Symmetry

As mentioned in the previous section we focus on materials at the atomic scale. Here we use periodic boundary conditions and start with an ideal crystal structure that is studied at zero temperature. Our unit cell (or supercell) may contain several atoms. With present computer technology unit cells with around one thousand atoms can still be simulated within a reasonable time.

Quantum mechanics (QM) governs the electronic structure which is responsible for the properties of the system, such as the relative stability, chemical bonding, relaxation of the atoms, phase transitions, electrical, mechanical or magnetic behavior, etc. In addition many quantities related to experimental data (such as spectra) are determined by QM principles.

Several basic concepts from solid state physics and group theory are needed to characterize the electronic structure of solids as summarized for example in [12]. Here we just briefly mention some of these concepts such as the Born-Oppenheimer approximation, according to which the electrons can move independent form the nuclei (which can be assumed to be at fixed positions), or the direct lattice and the Wigner-Seitz cell. Owing to the translational symmetry of a crystal, it is convenient to define a reciprocal lattice with the Brillouin zone as the unit cell. The symmetry is defined by operators for translation, rotation, reflection or inversion and leads to group theory with the space group and point group. The electronic structure of an infinite solid looks so complicated that it would seem impossible to calculate it. Two important steps make the problem feasible. The single particle approach, in which each electron moves in an average potential V(**r**) that is translational invariant V(**r**+**T**)=V(**r**) under the translation **T**. The second important concept is the Bloch theorem, which defines how the wave function (which is not translational invariant) changes under **T**, namely by a phase factor, called the Bloch factor e^{ikT}

$$\Psi_k(r+T) = e^{ikT}\psi_k(r) \tag{1}$$

where **k** is a vector in reciprocal space that plays the role of a quantum number in solids. The **k** vector can be chosen in the first Brillouin zone, because any **k**′ that differs from **k** by just a lattice vector **K** of the reciprocal lattice has the same Bloch factor and the corresponding wave function satisfies the Bloch condition again.

3.1.2. Quantum chemistry and ab initio methods

The quantum mechanical treatment of a system on the atomic scale has been discussed in many papers (for example in [5, 12]) and thus it is sufficient to summarize a few basic concepts here. According to the Pauli principle, because electrons are indistinguishable Fermions, their wave functions must be antisymmetric when two electrons are interchanged leading to the phenomenon of exchange. In a variational wave-function description this can be enforced by forming one Slater determinant (set up from one-electron wave functions), representing the well-known Hartree-Fock (HF) approximation. The HF equations have the computational disadvantage that each electron moves in a different potential (becoming orbital dependent). In HF the exchange is treated exactly but correlation effects, caused by the specific Coulomb interaction between the electrons are omitted by definition. Correlation can be included by more sophisticated approaches such as configuration interaction (CI) in which additional Slater determinants (including single, double or triple excitations into unoccupied states) are added in order to increase the variational flexibility of the basis set [13]. Another treatment of correlation effects is the coupled cluster (CC) scheme that is often used in quantum chemistry [14]. Such schemes are labeled *ab initio* (or first principles) methods and are highly accurate refinements that can reach an almost exact solution. Unfortunately the corresponding computational effort dramatically increases with N^7, where the system size is proportional to N, the number of electrons. Such nearly exact solutions would be desirable but in practice they can only be obtained for relatively small systems (atoms or small molecules). When the system size is significantly larger (as in condensed matter applications) approximations are unavoidable.

In quantum mechanics the term *ab inito* means that for a simulation of a material it is sufficient to know its constituent atoms (or isotopes) but the rest is governed by quantum mechanics. One does not need to know whether a material is insulating, metallic, magnetic, or has any other specific property. In principle an *ab initio* calculation should determine these properties from the atomic structure alone. A different situation occurs for calculations that are based on parameters that had been fitted to known properties of other systems that are similar to the material of interest. The latter type of calculations is often less demanding (in terms of computer resources) but is necessarily biased towards the related class of materials for which the parameters had been determined. Consequently one cannot find an unconventional behavior. In practice, however, it helps to know something about the system in order to choose proper approximations in the complicated quantum mechanical calculations. For example, why should one perform a spin-polarized calculation knowing that the system is not magnetic.

3.1.3. Density Functional Theory

The well-established scheme to calculate electronic properties of solids is based on density functional theory (DFT), for which Walter Kohn has received the Nobel Prize in chemistry in 1998. Fifty years ago, in 1964, Hohenberg and Kohn [15] have shown that the total energy E of an interacting inhomogeneous electron gas (as it appears in atoms, molecules or solids), in the presence of an external potential (coming from the nuclei) is a functional of the electron density ϱ which uniquely defines the total energy E of the system, i.e E[ϱ].

$$E = T_o[\rho] + \int V_{ext}\rho(\vec{r})d\vec{r} + \frac{1}{2}\int \frac{\rho(\vec{r})\rho(\vec{r}')}{|\vec{r}' - \vec{r}|}d\vec{r}d\vec{r}' + E_{xc}[\rho] \tag{2}$$

The four terms correspond to the kinetic energy (of non-interacting electrons), the nuclear-electronic interaction energy E_{ne}, the Coulomb energy (including the self-interaction) and the exchange correlation energy E_{xc}, which contains all the quantum mechanical contributions. This theorem is still exact. From a numerical point of view one can stress that the first three terms are large numbers while the last is essential but small and thus can be approximated. Thus one does not need to know the many-body electronic wave function. This is an enormous simplification. To clarify this point, consider the very simple case of a system (atom or molecule) with 100 electrons but which is still small. Each electron needs to be described by a wave function which depends on three space coordinates and the spin. Therefore the many-body wave function would depend on 400 coordinates. According to DFT all that is needed is the density ϱ(**r**) which only depends on the position **r**, i.e. on three coordinates. Unfortunately the exact form of the functional is not known but the conditions it should satisfy have been formulated, as will be discussed in Section 3.3.

3.2. The Kohn-Sham equations

From a practical point of view it was essential to formulate DFT in such a way that it could be applied. According to the variational principle a set of effective one-particle Schrödinger equations, the so-called Kohn-Sham (KS) equations [16], must be solved (Equation 3) as highlighted in Figure 1 and Figure 6. In this way DFT is a universal approach to the quantum mechanical many-body problem, where the system of interacting electrons is mapped in a unique manner onto an effective non-interacting system that has the same total density. The non-interacting particles of this auxiliary system move in an effective local one-particle potential, which consists of a classical mean-field (Hartree) part and an exchange-correlation part V_{xc} (due to quantum mechanics) that, in principle, incorporates all correlation effects exactly. Eqn.3 shows its form (written in Rydberg atomic units) for an atom with the obvious generalization to molecules and solids.

$$[-\frac{1}{2}\nabla^2 + V_{ext}(\vec{r}) + V_C[\rho(\vec{r})] + V_{xc}[\rho(\vec{r})]]\ \Phi_i(\vec{r}) = \varepsilon_i\ \Phi_i(\vec{r}) \tag{3}$$

The four terms represent the kinetic energy operator, the external potential from the nucleus, the Coulomb-, and exchange-correlation potential V_C and V_{xc}. The KS equations must be solved iteratively till self-consistency is reached (as illustrated in Figure 6). The iteration cycles are needed due to the interdependence between orbitals and potential. In the KS scheme the electron density is obtained by summing over all occupied states, i.e. by filling the KS orbitals (with increasing energy) according to the aufbau principle.

$$\rho(\vec{r}) = \sum_i^{occ} [\varphi_i(\vec{r})]^2 \tag{4}$$

A typical computation is illustrated in Figure 6. For a system of interest the unit cell must be specified by the lattice constants a, b, c and the corresponding angles (α, β, γ). In addition each atomic position is defined by the Wyckoff positions (as mentioned in Section 2). For this fixed atomic structure the self consistent field (SCF) cycle starts. As a first guess for the crystalline density one can superimpose atomic densities of neutral atoms placed at their proper positions in the unit cell. With this density one can generate a potential (within DFT). In each iteration i the DFT Kohn-Sham equations must be solved as illustrated on the right hand side.

Instead of using a uniform mesh of **k**-points s in the Brillouin zone (BZ) it is sufficient to restrict the **k**-points s to the irreducible wedge of the BZ by applying symmetry relations present in the system. From each star of equivalent **k**-points s only one must be calculated and its corresponding density is weighted according to the **k**-points symmetry (reducing the computational effort). For each **k**-points the Kohn-Sham equations must be solved.

The KS wave functions are expanded in basis sets as will be described in the next section. The expansion coefficients C_{kn} are determined by the variational method by minimizing the expectation value of the total energy with respect to these coefficients. This procedure leads to the generalized eigenvalue problem, HC=ESC, where H is the Hamiltonian, S the overlap matrix, C contains the coefficients and E the energies. After diagonalization we obtain for each energy E_{nk} the KS orbital ψ_{nk} and thus can calculate the corresponding electron density, where n is the band index.

By summing over all occupied states (with E_k smaller than the Fermi energy) the output density is obtained. This output density can be mixed with the input density of the previous iterations to obtain a new density for the next iteration. In order to reduce the number of iterations to reach self consistency, several schemes have been suggested (see section 5 in [4]). A recent mixing scheme is the multisecant version [17] which includes information from several previous iterations from the SCF cycle as samples of a higher dimensional space to generate the new density, from which the V_C (solving Poisson's equation) and V_{xc} (within DFT) potentials can be generated for the next SCF iteration. The exact functional form of the potential V_{xc} is not known and thus one needs to make approximations. With these potentials the new KS orbitals can be obtained. This closes the SCF cycle.

The SCF cycles are continued till convergence is reached, for example when the total energy of two successive iterations deviates from each other by less than a convergence criterion ε (e.g. 0.001 Ry). At this stage one can look at forces acting on the atoms in the unit cell. If symmetry allows there can be forces on the atoms which are defined as the negative gradient of the total energy with respect to the position parameters. Take for example the rutile TiO_2 structure (Figure 2), in which oxygen sits on Wyckoff position 4f which has the coordinates (x,

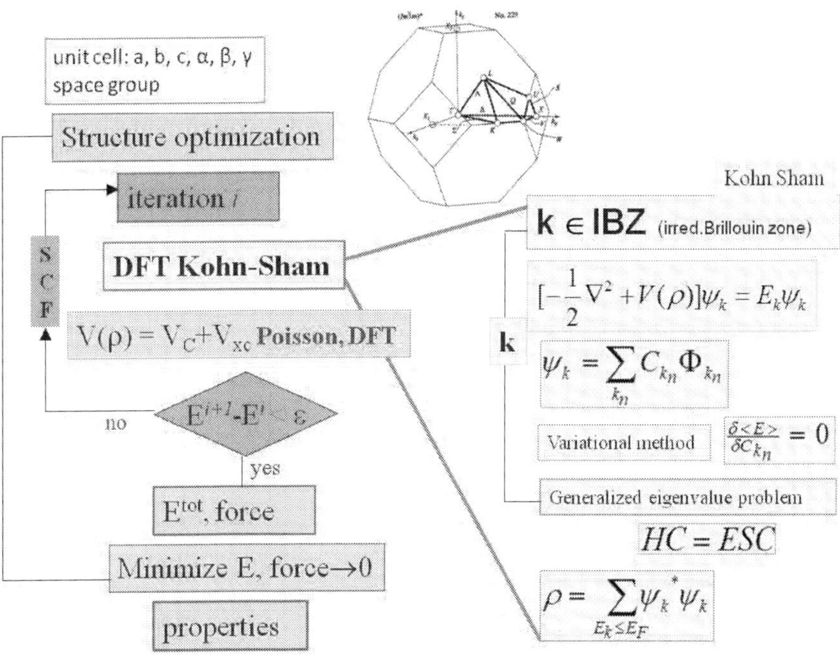

Figure 6. Major steps in DFT electronic structure calculations: self-consistent field (SCF) cycle; Kohn-Sham equations solved within a **k**-points loop; for example, a face-centered-cubic structure (space group 225) has a body-centered cubic reciprocal lattice (space group 229) as Brillouin zone with its irreducible wedge (1/48th of the BZ).

x, 0) where x is not specified by symmetry. In this case x can be varied to minimize the energy and thus a force can occur on the oxygen which vanishes at the equilibrium geometry. However, Ti is located at the Wyckoff position 2a with the fixed coordinates (0, 0, 0) and (½, ½, ½) and thus these positions are fixed and no force will act on Ti. When all atoms are essentially at their equilibrium positions (with forces around 0) then one can change the volume of the unit cell and minimize the total energy E. This would correspond to the equilibrium geometry of the system in the given structure. After this minimization is completed one can, as the last step, calculate various properties for this optimized structure.

3.3. DFT-functionals

The treatment of exchange and correlation effects has a long history and is still an active field of research. Some aspects were summarized in the review articles [1-5] but also in many other papers in this field. The reader is encouraged to look at recent developments. An excellent book [18] by Cottenier covers DFT and many aspects around the WIEN2k program package and thus is highly recommend to the reader for finding further details.

For the present presentation it is worth mentioning a few historical aspects: in 1951 Slater [19] proposed the replacement of the non-local Hartree-Fock exchange by the statistical exchange, called Slater's exchange. In the 1970ths this was modified by scaling it with the exchange parameter α (for each atom) called the Xα method [20], which was widely used for solid state calculations. It was designed to approximate Hartree-Fock, which (by construction) treats exchange exactly but neglects correlation effects completely. By making a local approximation for the potential the Xα method indirectly included correlation effects making it better than Hartree-Fock but also less accurate, since exchange is treated only approximately. This type of error cancellation is typical for many DFT functionals.

Early applications of DFT were done by using results from quantum Monte Carlo calculations [21] for the homogeneous electron gas, for which the problem of exchange and correlation can be solved exactly. Although no real system has a constant electron density, one can at each point in space use the homogenous electron gas result to treat exchange and correlation, leading to the original local density approximation (LDA). Surprisingly LDA works reasonably well but has some shortcomings mostly due to the tendency to overbind atoms, which cause e.g. too small lattice constants. The next crucial step in DFT was the implementation of the generalized gradient approximation (GGA), for example the version by Perdew, Burke, Ernzerhof (PBE) [22] which improved LDA by adding gradient terms of the electron density. For several cases this GGA gave better results and thus for a long time PBE has been a standard for many solid state calculations. During recent years, however, several improvements of GGA were proposed, which fall in two categories, both with good justifications:

- Semi-empirical GGA, which contain parameters that are fitted to accurate (e.g. experimental or ab initio) data
- ab initio GGA, in which all parameters are determined by satisfying fundamental constraints (mathematical condition) which the exact functional must obey.

One criterion for the quality of a calculation is the equilibrium lattice constant of a solid, which can be calculated by minimizing the total energy with respect to volume. By studying a large series of solids (as shown in Figure 7) some general trends can be found ([23]): LDA has the tendency of overbinding, leading to smaller lattice constants than the experiment. GGA in the version of PBE [22] always yield larger lattice constants, which sometimes are above the experimental value. The more recently suggested modifications, as discussed in [23], lead to a clear improvement at least for the lattice parameters. In addition, there are other observables (such as cohesive energy or magnetism, to mention just two), which depend on the functional. The best agreement with experiment may require different functionals for various properties. So far no functional works equally well for all cases and all systems. Therefore one must acknowledge that an optimal DFT functional has not yet been found, which is the reason why this remains an active field of research.

A systematic improvement of the exchange and correlation treatment as in quantum chemistry (section 2.3) starting from Hartree-Fock to (full) configuration interaction (CI) or coupled cluster (CC) approaches did not exist for solids and DFT. In 2005 such a scheme was proposed in [24] and was called Jacob's ladder for DFT which becomes progressively more demanding

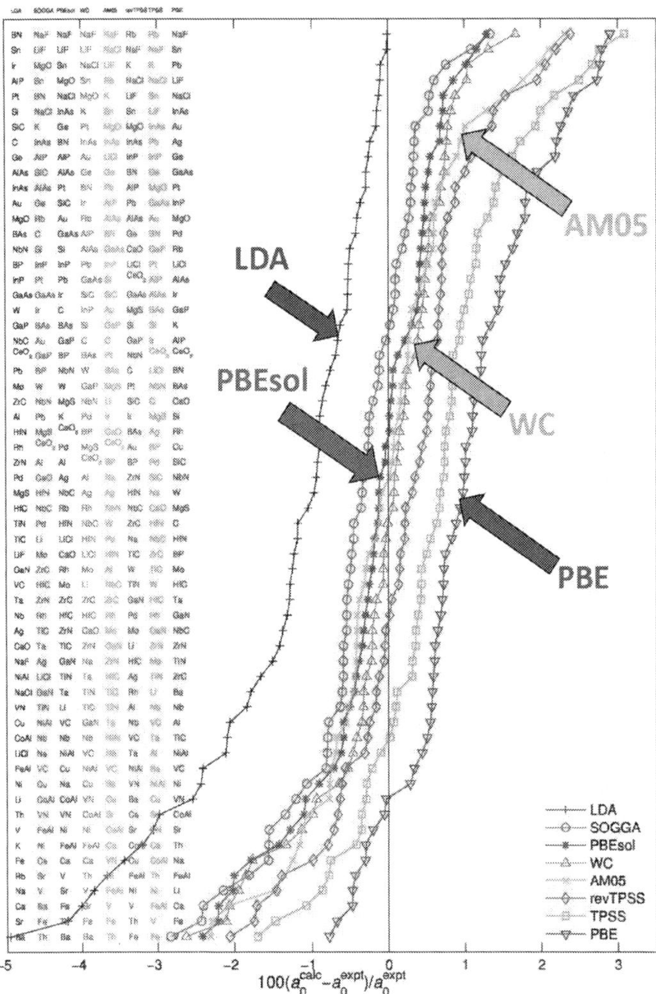

Figure 7. Comparison of several GGA functionals, showing the relative error in the equilibrium lattice constant of many solids between DFT calculations and experiment (for further details see [23]). The calculations were done with WIEN2k.

in terms of computational requirements. In Figure 8 the five rungs of this ladder ("to heaven") are briefly mentioned which indicate what is needed at each level of theory. In LDA the exchange correlation energy ε_{xc} is just a function of the density ϱ; in the next rung it depends also on the gradient of the density, in meta-GGA ε_{xc} is in addition a function of the Laplacian of the density and the kinetic energy density t (see e.g. [25]). In rung 4 one goes from the simple

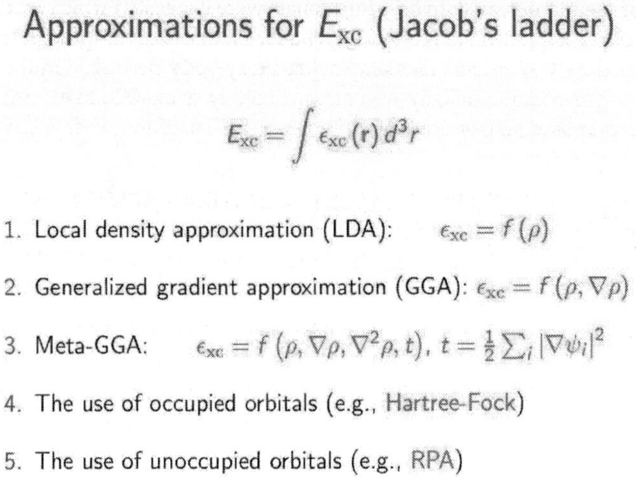

Approximations for E_{xc} (Jacob's ladder)

$$E_{xc} = \int \epsilon_{xc}(\mathbf{r})\, d^3r$$

1. Local density approximation (LDA): $\epsilon_{xc} = f(\rho)$
2. Generalized gradient approximation (GGA): $\epsilon_{xc} = f(\rho, \nabla\rho)$
3. Meta-GGA: $\epsilon_{xc} = f(\rho, \nabla\rho, \nabla^2\rho, t)$, $t = \frac{1}{2}\sum_i |\nabla\psi_i|^2$
4. The use of occupied orbitals (e.g., Hartree-Fock)
5. The use of unoccupied orbitals (e.g., RPA)

Figure 8. Jacob's ladder according to [24] with 5 rungs, demonstrating how to improve the exchange correlation treatment.

dependence on the density alone, to an orbital description, which (for occupied orbitals) allows a correct description of exchange, like in Hartree-Fock. At this level one limits the computation space to the occupied orbitals but can extend it to the hybrid functions (mixing a fraction of Hartree-Fock with a part in DFT). In the highest rung also unoccupied orbitals are included, as for example in the scheme called random phase approximation (RPA).

There are well documented cases for which conventional DFT calculations (LDA or GGA) disagree even qualitatively with experimental data and lead, for instance, to predict a metal instead of an insulator. One of the reasons can be the presence of localized states (often f-electrons or late transition metal d-orbitals) for which correlation is very strong. For these highly correlated systems one must go beyond simple DFT calculations. One simple form of improvement is to treat theses local correlations by means of a Hubbard U (see [26]) but use LDA or GGA for the rest of the electrons. With this parameter U the on-site Coulomb repulsion between the localized orbitals is included, but by introducing a parameter. This approach is generally called LDA+U. In a simple picture, U stands for the energy penalty of moving a localized electron to the neighboring site that is already occupied.

The Kohn-Sham energy eigenvalues ε_i (in equation 3) should – formally speaking – not be interpreted as excitation energies (except for the highest one). Nevertheless optical excitations are commonly described in the independent particle approximations, using these quasi particle states from DFT in the single-particle picture. One well known case is the energy gap of insulators, which in this crude single-particle picture is typically underestimated by about 50 per cent. This has been well known for some time (see e.g. section 6.7 of [4]), since even the exact Kohn-Sham gap misses the integer discontinuity Δ_{xc} between occupied and unoccupied states. It is worth considering that in Hartree Fock the gap found would typically be too large.

This is one of the reasons, why hybrid functionals were suggested which mix Hartree Fock with DFT in order to produce the correct gap. Better estimates of the quasi-particle spectrum can be obtained by GW calculations employing many-body perturbation theory, which is significantly more computationally expensive. Recently a modified Becke Johnson (mBJ) potential was proposed [27], which is still a local potential (and thus cheap) but yields energy gaps close to experiment..

When the Coulomb potential is written in terms of the density (third term in equation 2) it contains the unphysical self interaction of an electron with itself. In Hartree-Fock this term is exactly canceled by the exchange term. Due to the approximation in DFT, this cancellation is not complete and thus in some functionals a self-interaction-correction (SIC) is added [28].

The van der Waals (vdW) interaction is not described in the simple DFT approximations like LDA or GGA but can be treated with higher order treatments (rung 4 or 5 in Jacob's ladder) which become computationally rather expensive. A pragmatic solution is to add a vdW correction based on adjustable parameters (see for example Grimme's scheme [29]).

In connection with the term *"ab inito"* – as the quantum chemists define it – it is appropriate to consider the situation for large systems: the strategy differs for schemes based on HF (wave function based) or DFT. In HF based methods (including CI and CC) the Hamiltonian is well defined and can be solved almost exactly for small systems but for large cases only approximately (i.e. due to limited basis sets). In DFT, however, one must first choose the functional that is used to represent the exchange and correlation effects (or approximations to them) but then one can solve this effective Hamiltonian almost exactly. Thus in both cases an approximation enters either in the first or second step. This perspective illustrates the importance of improving the functionals in DFT calculations, since they define the quality of the calculation. The advantage for DFT is that it can treat relatively large systems.

4. Solving the Kohn-Sham equations with WIEN2k

4.1. The all electron case

A schematic summary of the main choices one has to make for computations is shown in Figure 1, where our selections are marked in red. We want to represent a solid with a unit cell (or a Supercell) as discussed in section 2 and thus invoke periodic boundary conditions. The system may contain all elements of the periodic table, from light to heavy, main group, transition metals, or rare earth atoms as for example shown in Figure 3. Let us look at Ti (with the atomic number Z=22) as an example. Its electronic configuration is $1s^2 2s^2 2p^6 3s^2 3p^6 3d^2 4s^2$ (Figure 9).

We surround the Ti nucleus by an atomic sphere with a radius of 2 Bohr (about 1Å). With respect to this sphere, the Ti electronic states can be classified in three categories:

- core states, which are low in energy, have their wave functions (or electron densities) reside completely inside this sphere.
- valence states, which are high in energy and are delocalized.

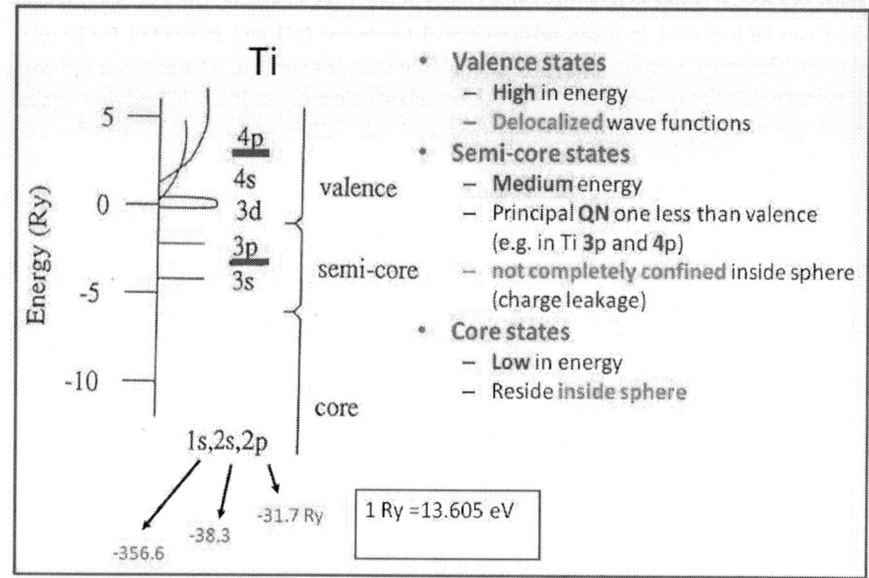

Figure 9. The electronic states of titanium: core-semi-core, and valence states.

- semi-core states, which are in between (medium energy), have a charge leakage (a few per cent of the charge density lies outside the sphere) and have a principal quantum number that is one less than the valence states (i.e. 3s vs. 4s, or 3p vs. 4p).

Traditionally the electronic properties of a material due to the chemical bonding are associated with only the valence electrons and thus the core electrons are often ignored. A typical scheme is the so called frozen core approximation, in which the electron density from the core electrons does not change during the SCF cycle (see Figure 6). This is often justified but there are cases like hyperfine interactions, where the change of the core electrons can contribute significantly and even more so the semi-core states. An all-electron treatment has therefore the advantage of being able to explore the contribution from all electrons to certain experimental data (e.g. the electric field gradient).

As long as a solid contains only light elements, non-relativistic calculations are well justified, but as soon as a system of interest contains heavier elements, relativistic effects must be included. In the medium range of atomic numbers (up to about 54) so called scalar relativistic schemes [30] are often used, which properly describe the main contraction or expansion of various orbitals (due to the Darwin s-shift or the mass velocity term) but omit spin-orbit coupling. Such schemes are computationally relatively simple and thus recommended for a standard case. The inner electrons can reach a high velocity leading to a mass enhancement. This causes a stronger screening of the nuclear charge by the relativistic core electrons with respect to a non-relativistic treatment and affects the valence electrons. The spin-orbit contribution can be included in a second–variational treatment [31] and is needed for heavier elements. The core electrons are treated by solving Dirac's equation, whereas the semi-core and valence states are described with the scalar relativistic scheme. In the latter spin remains a good quantum number and thus spin-polarized calculations are valid to treat magnetic systems.

4.2. The choice of the potential

Figure 1 schematically shows the topics where one needs to make a choice. We want to represent a solid with a unit cell (or a supercell). Relativistic and spin-polarization effects can be included as mentioned above. The next crucial point is the choice of the potential that is closely related to the basis sets. This aspect is extensively explained in [18] discussing the advantages and problems connected with pseudo potentials. The main idea is to eliminate the core electrons and replace the real wave functions of the valence states by pseudo wave functions which are sufficiently smooth so that they can be expanded in a plane wave basis set. In the outer region of an atom, where the chemical bonding occurs they should agree with the real wave function. In principle – in mathematical terms – plane waves form a complete basis set and thus should be able to describe any wave function. However, the nodal structure (for example of a 4s wave function) close to the nucleus would need to be described by extremely many plane waves.

For the all-electron case within DFT the potential looks like the one shown in Figure 10. Near each nucleus it has the form Z/r but between the atoms it is nearly flat. In the muffin-tin approximation the potential is assumed to be spherically symmetric around the atom but constant in between. In the full-potential case the potential (without any approximation of its shape) can be represented as a Fourier series in the interstitial region, but in each atomic sphere (with a radius R_{MT}) it can be expressed as a radial function $V_{LM}(r)$ multiplied by crystal harmonics, which are linear combinations of spherical harmonics having the point group symmetry of the atom α for the proper LM value. In this notation the muffin-tin case is the first term in both cases, namely the 0 0 component for LM (i.e. only the spherical part inside each atomic sphere) and a constant for the Fourier series (for the interstitial region). In the 1970ths the muffin-tin approximation was widely used because it made calculations feasible. For closely packed systems it was acceptable but for more covalently bonded systems like silicon (or even surfaces) it is a very poor approximation. Another drawback of the muffin-tin approximation was that the results depended on the choice of sphere radii, whereas in the full-potential case this dependence is drastically reduced. Due to the muffin-tin approximation different computer codes obtained results that did not agree with each other. This has changed with the use of full-potential calculations. Nowadays different codes based on the full potential yield nearly identical results provided they are carried out to full convergence and use the same structure and DFT version. This has given theory a much higher credibility and predictability (see Section 7).

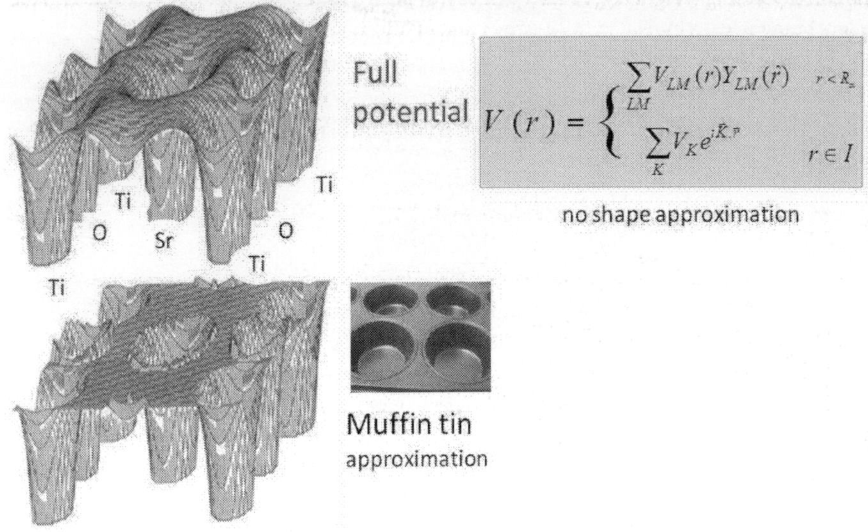

Figure 10. The full potential vs. muffin tin approximation is shown for a (110) plane of $SrTiO_3$.

4.3. The choice of the basis sets

For solving the Kohn Sham equations (see Figure 1) basis sets are needed. A linear combination of such basis functions shall describe the Kohn-Sham orbitals. One can use analytic functions-such as Slater type orbitals (STO) or Gaussian type orbitals-or just plane waves (for example in connection with pseudo potentials). Already in 1937 Slater [32] proposed the augmented plane wave (APW) method. The development of APW and its linearized version, which led to the WIEN code [6] and later to its present version WIEN2k [7] was described in detail in recent reviews [1-5]. An extensive description including many conceptual and mathematical details is given in [18]. Therefore only the main concepts will be summarized below.

4.4. The APW based method and the WIEN2k code

In the APW method one partition the unit cell into (non-overlapping) atomic spheres (type I) centered at the atomic sites and the remaining interstitial region (II) (Figure 11). Inside each atomic sphere (region I) the wave functions have nearly an atomic character and thus (assuming a muffin-tin potential) can be written as a radial function times spherical harmonics. It should be stressed that the *muffin tin* approximation (MTA) is used only for the construction of the APW basis functions and only for that. The radial Schrödinger equation is solved numerically (and thus highly accurately), but as input the energy must be provided, which makes the basis set energy dependent. In region II the potential varies only slowly and thus the wave functions can be well expressed in a series of plane waves (PW). Each plane wave is augmented by the atomic partial waves inside each atomic sphere (i.e. the PW is replaced inside the spheres).

The corresponding weight $A_{\ell m}$ of each partial wave can be fixed by a matching condition at the sphere boundary (as indicated in Figure 11 and Figure 13).

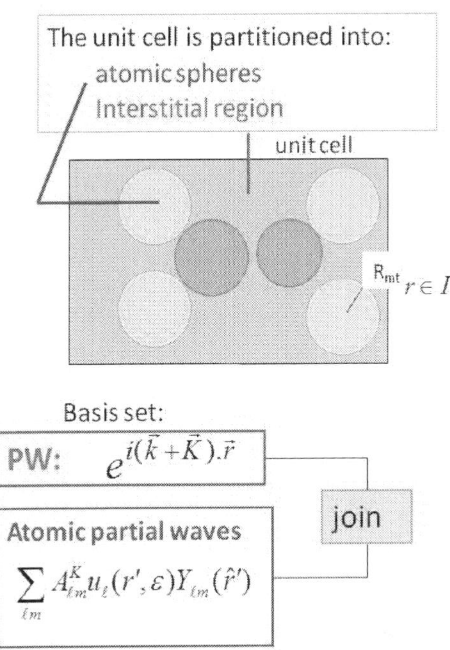

Figure 11. The Augmented Plane Wave (APW) method.

Three schemes of augmentation (APW, LAPW, APW+lo) have been suggested over the years and illustrate the progress in this development of APW-type calculations that was discussed in [18, 4, 5]. Here only a brief summary will be given. The energy dependence of the atomic radial functions $u_{\ell}(r,E)$ can be treated in different ways. In Slater's APW [32] this was done by choosing a fixed energy E, which leads to a non-linear eigenvalue problem, since the basis functions become energy dependent.

In the linearized APW, called LAPW, O. K. Andersen [33], suggested to linearize (that is treat to linear order) this energy dependence as illustrated in Figure 12. The radial Schrödinger equation is solved for a fixed linearization energy E_{ℓ} (taken at the center of the corresponding energy bands) leading to $u_{\ell}(r, E_{\ell})$ but adding an energy derivative of this function (taken at the same energy) in order to retain the variational flexibility. This linearization is a good approximation in a sufficiently small energy range around E_{ℓ}. In LAPW the atomic function inside the sphere α is given by a sum of partial waves, namely radial functions times spherical harmonics labeled with the quantum numbers (ℓ, m).

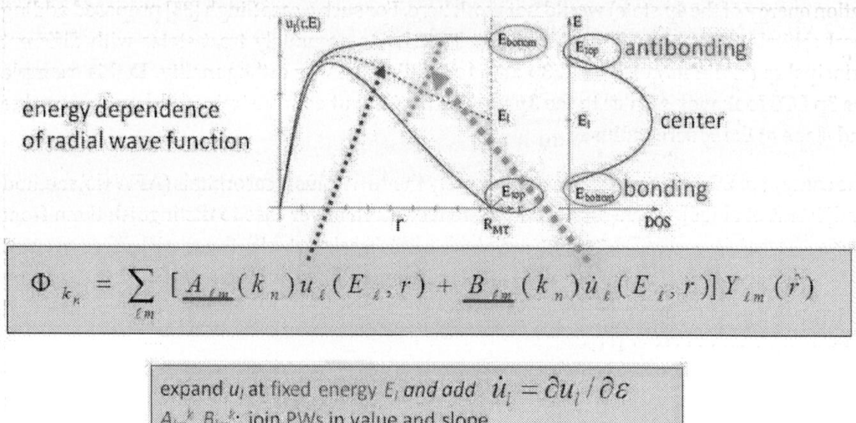

$$\Phi_{k_n} = \sum_{\ell m} [A_{\ell m}(k_n) u_\ell(E_\ell, r) + B_{\ell m}(k_n) \dot{u}_\ell(E_\ell, r)] Y_{\ell m}(\hat{r})$$

Figure 12. The energy variation of the radial wave function $u_\ell(E_\ell,r)$ according to LAPW [33] is schematically shown: i) for the center of the band (taken from a sketched density of states shown on the right), ii) for the energy E_{bottom} at the bottom of the band (bonding case where the radial wave function has zero slop at the sphere boundary R_{MT} as shown in Figure 13), and iii) for the energy E_{top} at the top of this band (antibonding case, where the wave function has a node at R_{MT}). In LAPW this energy dependence is linearized and expressed as the radial function and its energy derivative both taken at E_ℓ, where the relative weights are determined by matching (in value and slope) to plane waves at R_{MT} (as shown in Figure 13).

The two coefficients $A_{\ell m}$ and $B_{\ell m}$ (weight for function and derivative) – as given in Figure 12- can be chosen so as to match each plane wave (characterized by K) continuously (in value and slope) to the one-center solution inside the atomic sphere at the sphere boundary (for details see e.g. [18]). The main advantage of the LAPW basis set is that it allows finding all needed eigenvalues with a single diagonalization, in contrast to APW, which has the non-linear eigenvalue problem. Historically, the more strict constrain (a matching of both value and slope) had the disadvantage that in LAPW more PWs were needed to reach convergence than in APW. The LAPW basis functions u and it derivative are recalculated in each iteration cycle (see Figure 6) and thus can adjust to the chemical changes (for example due to charge transfer) requiring an expansion or contraction of the radial function. The LAPW method made it computationally attractive to go beyond the muffin-tin approximation and to treat both the crystal potential and the charge density without any shape approximation (called full-potential) as pioneered by the Freeman group [34].

In section 4.1 the partition of electronic states in core, semi-core and valence states was described and illustrated for Ti in Figure 9. Let us focus on the p-type orbitals. The 2p core state is treated fully relativistic as an atomic core state while the valence 4p state is computed within LAPW using a linearization energy at the corresponding high energy. The 3p semi-core states reside mostly inside the Ti sphere but have a "core-leakage" of a few per cent. The 3p states are separated in energy from the 4p states and thus the linearization (with the linearization energy of the 4p state) would not work here. For such a case Singh [35] proposed adding local orbitals (LO) to the LAPW basis set in order to accurately treat states with different principal quantum numbers (e.g. 3p and 4p) while retaining orthogonality. In this example the 3p LOs look very similar to the 3p radial function but are constrained to have zero value and slope at the sphere radius R_{MT}.

The concept of LO fostered another idea, namely the APW plus local orbitals (APW+lo) method by Sjöstedt et al [36]. These local orbitals are labeled in lower case to distinguish them from the semi-core LO. In APW+lo, one returns to the APW basis but with the crucial difference that each radial function is expanded at a fixed energy. The matching is again (as in APW) only made between values (Figure 13). This new scheme is significantly faster while maintaining the convergence of LAPW [37].

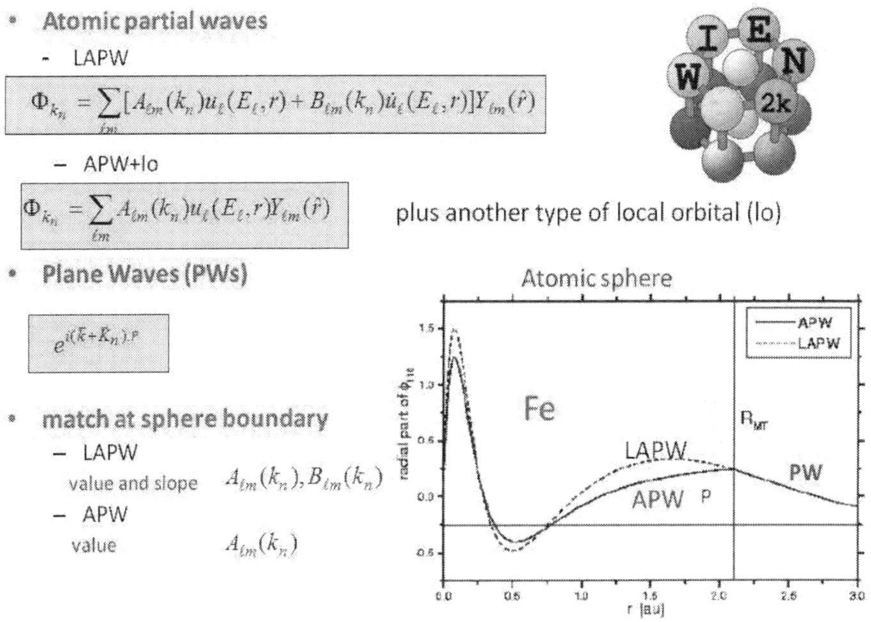

Figure 13. The (linearized) augmented plane wave method as implemented in WIEN2k [7] defining i) the different atomic partial waves in LAPW and APW+lo used inside the atomic sphere, ii) the plane waves used in the interstitial region, iii) the matching at the sphere boundary, and iv) illustrating for an Fe-4p orbital how the different matching looks at the sphere boundary for APW and LAPW.

The APW+lo scheme therefore combines the best features of all APW-based methods. It was known that LAPW converges somewhat slower than APW due to the constraint of having differential basis functions and thus it is an improvement to return to APW but only for the orbitals involved in chemical bonding. The energy-independent basis introduced in LAPW is crucial for avoiding the general eigenvalue problem of APW and thus is also used for all higher ℓ components. The local orbitals provide the necessary variational flexibility to make this new scheme efficient but they are added only where needed (to avoid any further increase in basis set). The crystalline wave functions (of Bloch type) are expanded in these APWs leading to a general eigenvalue problem. The size of the matrix is mainly given by the number of plane waves (PWs) but is increased slightly by the additional local orbitals that are used. As a rule one can say that about 50-100 PWs are needed for every atom in the unit cell in order to achieve good convergence.

5. Results with WIEN2k

The WIEN2k code is widely used and thus there is an enormous literature with many interesting results which cannot all be covered here. Many of the publications with WIEN2k can be found on the web page www.wien2k.at under the heading papers. A selected list of results, that can be obtained with WIEN2k, is provided below, where references are specified either to the original literature or in some cases to review articles [4, 5].

- After the SCF cycle has been completed one can look at various standard results: the Kohn-Sham eigenvalues E_{nk} can be shown along symmetry lines in the Brillouin zone giving the energy band structure. A symmetry analysis can determine the corresponding irreducible representation (see Fig.1 of [4]). For each of these states with E_{nk} the wave function (a complex function in three dimensions) contains information about how much the various regions of the unit cell contribute. In the APW framework this can be done by using the partial charges $q_{t\ell m}$ which define the fraction of the total charge density of this state (normalized in the unit cell) that resides in the atomic sphere t and comes from the orbital characterized by the quantum numbers ℓm. The fraction of the charge that resides in the interstitial region is contained in q_{out}. These numbers, which depend on the choice of sphere radii, help to interpret each state in terms of chemical bonding. This is an advantages of this type of basis set. There is a useful option to show the character of bands. As one example, three options of presenting the band structure are illustrated for the refractory metal titanium carbide TiC shown in Fig.1 of reference [2] showing the Ti-d (e_g symmetry) and C-p character bands, which dominate the bonding in this case. The crystal field of TiC splits the fivefold degenerate Ti-d orbitals into t_{2g} and e_g states (with a degeneracy of 3 and 2 respectively). Another example is the band structure of Cu shown in Fig. 2.2.16.1 of [12].
- The Fermi surface in a metal is often crucial for an understanding of properties (for example superconductivity). It can be calculated on a fine **k**-mesh and plotted (for example with XCrysDen [38]).
- With a calculation for a (sufficiently fine) uniform mesh of **k**-points s in the irreducible Brillouin zone as discussed in connection with Figure 6 one can determine the density of states (DOS), which gives a good description of the electronic structure. The total DOS can

be decomposed into its components by means of the $q_{t\ell m}$ values mention above. This decomposition becomes even more important in complicated cases, for example if one wants to find which state originates from an impurity atom in a supercell.

- The electron density is the key quantity in DFT and thus contains the crucial information for chemical bonding but the latter causes only small changes. Therefore it is often useful to look at difference densities, computed as difference between the SCF density of the crystal minus the superposed atomic densities (of neutral atoms), because in this presentation the changes due to bonding become more apparent. Sometimes it is useful to look at the densities corresponding to states in a selected energy window using various graphical tools (2-or 3-dimensional plots). Another possibility is a topological analysis according to Bader's theory of atoms in molecules [40]. It allows among other details, one to uniquely define atomic charges within atomic basins, a relevant quantity for charge transfer. See also chapter 6.3 of [4].
- The typical chemical bonds, like covalent, ionic or metallic bonds, can well be described within DFT. For their analysis the APW type basis is very useful because it can provide chemical interpretations in term of orbitals. Van der Waals (vdW) interactions, however, are not properly represented in conventional DFT: They can approximately be included by adding a Grimme correction, for example [29].
- The total energy of a system is the main quantity within DFT. Especially for large systems this can be a rather large number, but nowadays it can be calculated with high precision. The interest is often in total energy differences for example to find out which of two structures is more stable. In such cases the two calculations need to be done in a very similar fashion (same functional, comparable **k**-mesh and basis set, same sphere sizes, etc.). It is also possible to compare cohesive or atomization energies, where the atoms must be modeled in the same fashion as the crystal (that is in a large supercell containing just the isolated atom).
- The derivative of the total energy with respect to the nuclear coordinates yield the force acting on the atom, which is needed for structure optimization. See also chapter 6.5 of [4], in which such an optimization is discussed in connection with the bonding of hexagonal boron nitride on a Rh (111) metal substrate, where the two systems have a lattice mismatch of about 8 per cent (Figure 14). This mismatch in lattice spacing requires that 13 x 13 unit cells of h-BN are needed to match 12 x 12 unit cells of the underlying Rh (111) lattice to make it commensurate (with periodic boundary conditions). This special surface arrangement was called "nanomesh" with spacing of about 3.2 nm. In order to simulate this system with a supercell, the face-centered-cubic (fcc) metal layer is represented with three layers (with a 12 x 12 Rh lattice) which are covered (on both sides on the metal slabs) with a single layer of BN (with a 13 x 13 BN lattice) and then an empty region is added to simulate the surface. Although this is still a crude model of the real situation, it illustrates which kind of large systems can be studied nowadays. This model system contained 1108 atoms per unit cell but could be computed with WIEN2k. The corresponding calculations have shown that BN is no longer flat but becomes corrugated due to the different binding situations between BN and the metal substrate, which depends on the local geometry that is favorable ("pores" in

some regions but unfavorable "wires" in others. This corrugated BN surface was found to agree with experimental data (for further details see [41-43]). Another example is the investigation of so called misfit layer compounds [11], in which the bonding between the layers of TaS_2 and PbS required that some Pb atoms are replaced by Ta in an disordered fashion. Relatively large supercells were needed in order to represent this cross substitution. After relaxing the atomic positions the more likely arrangements have been determined on the basis of total energy differences.

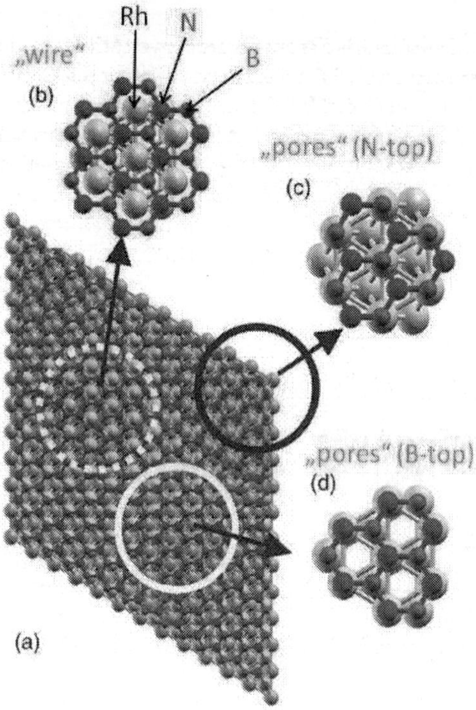

Figure 14. A hexagonal boron nitride (13 x13) is bonded to a Rh(111) surface (12 x 12) forming (a) a nanomesh; (b)N sits on an unfavorable hollow position (between three Rh) and thus BN is far away from Rh called "wire" ; (c) N is on the favorable position on top of Rh called "pores"; (d): B is on top of Rh, called "pores" (see [43]).

- In the case of magnetic systems spin-polarized calculations can provide the magnetic moments. In addition to collinear magnetic systems also non collinear magnetism can be handled, which was for example used in a study of UO_2 (see [44]. Another example is the Verwey transition that was investigated for double perovskite $BaFe_2O_5$. At low temperature this system has a charged-ordered state (with $Fe^{2+}$and Fe^{3+}at different sites) but above the Verwey transition temperature at about 309 K a valence mixed state with the formal oxidation state $Fe^{2.5+}$appears. DFT calculations made it possible to interpret this complicated situation, see [45] and section 7.4.1 of [5]. In the latter it was mentioned that it is now possible to use such calculations to look for fine details such as the magneto-crystalline anisotropy. This is defined as the total energy difference between a case, where the magnetic moment is in the y direction (with the lowest energy) or the x direction. In this case the difference in energy is found to be about 0.4 mRy but the total energy is-115,578.24065 Ry. Therefore the quantity of interest is in the tenth decimal illustrating the numerical precision that is needed for such a quantity. Needless to say that extremely well converged calculations were required, in which both cases are treated practically the same. This is necessary to have a cancellation of errors.

- The electric field gradient (EFG) is a ground state property that is sensitive to the asymmetric charge distribution around a given nucleus. By measuring the nuclear quadrupole interaction (e.g. by NMR) the EFG can be determined experimentally provided the nuclear quadrupole moment is known. This is a local probe which often helps to clarify the local atomic arrangement. See also chapter 6.4 of [4]. The EFG is a case where the semi-core states can significantly contribute as was shown for TiO_2 in the rutile structure [46]. Another important result was the nuclear quadrupole moment of ^{57}Fe, the most important Mössbauer isotope. On the basis of DFT calculations for the EFG of several iron compounds this quantity had to be adjusted by about a factor of two [47].

- Recently also the NMR shielding (chemical shifts) can be obtained [48], where the all-electron treatment opens the possibility of analyzing the dominant contributions that determine the chemical shifts as has been illustrated for fluorides [49].

- The calculation of various spectra (X-ray emission or absorption), optical spectra or energy loss near edge structure (ELNES) spectra can be performed within the independent particle model. Some structures in the excitation spectra of interacting electrons, called quasi-particle peaks, can be directly related to the excitation of independent electrons as they are treated within DFT. However, others (for example satellite structures) cannot be understood in such a simple way and require more sophisticated approaches. For example, including the electron core-hole interactions require the solution of the Bethe-Salpeter equation (BSE) as was illustrated for x-ray spectra [50]. Often such schemes are based on many-body perturbation theory. One of such approaches is the GW approximation [51]. This scheme allows calculating accurate band gaps or ionization potentials, which are not well determined by DFT eigenvalues. The GW approach is also available in connection with WIEN2k [52].

- The interpretation of scanning tunneling spectroscopy (STM) data often require a simulation by theory, which can distinguish between proposed surface structures. It is based on the Tersoff-Hamann [53] approximation, in which the images can be obtained from the charge density originating form a set of eigenstates within a certain energy window around the Fermi energy consistent with the applied voltage used in the STM measurements (see e.g. Fig. 6 in [42]).

- Phonons can be calculated based on the dynamical matrix, which is obtained by displacing one atom in a large unit cell (or supercell) in a certain direction and determining the forces on all the other atoms. The necessary independent displacements are determined by the symmetry of the cell. By diagonalizing the dynamical matrix the phonon frequencies can be determined. Such information is also useful for example in connection with ferroelectrics, structural stability, thermodynamics or phase transitions.

- For the analysis of phase transitions a fundamental understanding requires a combination of concepts, namely group theory, DFT calculations, frozen phonons, soft modes or bilinear couplings, and Landau theory. This was illustrated, for example for an Aurivillius compound [54], which shows multiple instabilities and has a phase transition to a ferroelectric state. For high pressure phase transitions a modified Landau theory was proposed and applied [55].
- Maximally localized Wannier functions can be calculated with wien2wannier [56] and provide a good starting point for more sophisticated many body theory. Dynamical mean field theory (DMFT) is one such example as is illustrated in [57]. Another extension of WIEN2k is the calculation of Berry phases with wien2kPI as modern theory of polarization in a solid (for details see ref [58]).
- Computer graphics and visualization (see [38]] can help to analyze the many intermediate results (atomic structure, character of energy bands, Fermi surfaces, electron densities, partial density of states, etc.). The more complex a case is the more support from computer graphics is needed. For an element one can plot all the energy bands, but for systems with over 1000 atoms one would be lost interpreting the band structure without the help of visualization.

6. Computer code development

From the experience of developing the WIEN2k code some general conclusions can be drawn. Some of the historical perspectives have been summarized in section 7 of [4]. During the last three to four decades it was often necessary to port the code to new architectures starting from main-frame computers, vector processors, PCs, PC-clusters, shared-memory machines, to multi-core parallel supercomputers. The power of computers has increased in several areas by many orders of magnitude such as the available memory (from kB to TB), the speed of communication (e.g. infiniband) all the way to the processors (CPU). An efficient implementation of a code made it necessary to closely collaborate with mathematicians and computer scientists in order to find the optimal algorithms, which perform well on the available hardware. One example is the idea of using the scheme of iterative diagonalization [59]. A significant portion of the computational effort in the WIEN2k calculations is the solution of the general eigenvalue problem (see Figure 6) which must be solved repeatedly within the SCF cycle. Changes from iteration to iteration are often small and thus one can use the information from the previous iteration to define a preconditioner for the next iteration and thus simplify the diagonalization and speed up the calculation.

Another aspect is the implementation of linear algebra libraries (e.g. SCALAPACK, MKL), which were highly optimized by other groups, who spend a lot of effort on these tasks, and

http://dx.doi.org/10.5772/59108

helped us significantly to speed up our code. Simultaneously, increased computer power made it possible to treat much larger systems, especially using massive parallelization. The matrix size that we could handle on the available hardware has increased by about a factor 1000 over the last several decades. Since solving the general eigenvalue problem scales as N^3, the computer power needed to solve a 1000 times bigger system must be about a factor 10^9 higher, which is available now.

Often our computational strategy had to be changed or extended. For example, to compute a metallic crystal with a small unit cell many **k**-points s in the Brillouin zone were needed to reach a good convergence. In this case **k**-points parallelization was optimal. Nowadays we can treat large unit cells (containing about 1000 atoms). In such a case, the reciprocal space is small and thus only few **k**-points s are needed for a good calculation. This requires new parallelization strategies, in which the large matrices must be distributed to many processors, where data locality and reduced communication is essential for achieving good parallel performance. Another aspect is the complexity of the code with the many tasks that need to be solved (see Figure 6). If only a small fraction is not parallelized, it may keep many processors waiting for the result that is calculated on only a single processor. This has often led to new bottle necks, which did not occur for smaller systems and thus were ignored but load balancing is important. Better computer power requires a continuous improvement of the code.

There are completely different ways of distributing a code (giving representative examples):

- open source with a free download
- use for registered user (with or without license fee), source code made available
- limited access for registered users (with a yearly license)
- software companies that distribute only executables

From a commercial point to view it is understandable that a company wants to have strict rules and do not make the source code available. From a scientific perspective, the WIEN2k group favors, the source code is made available to the registered users, who pay a small license fee once. This policy has helped to generate a "WIEN2k community", from which many researchers around the world have contributed to the development of the code and can contribute to do so in the future. It has helped in many aspects, such as to find and fix bugs, but also to add new features which are made available to all the WIEN2k users. In addition, several valuable suggestions were made, which allowed improving the documentation as well as implementing requested new features. We have organized more than 20 WIEN2k workshops worldwide, in which users are introduced to important concepts and learn how to run calculations and use kinds of associated tools. It has become a standard to help each other and thus contribute to the development of computations of solids and surfaces. In total this policy has had very positive impacts for WIEN2k and the field.

The user friendliness of WIEN2k has been improved over the years. A graphical user interface w2web was mainly developed by Luitz (see [7]) and is especially useful for novice users or in cases which are not done routinely. Later many default options were implemented, which were based on the experience of many previous calculations. This has made it much simpler to set up a calculation. For novice users or experimentalists this helps one to get started without being an expert. However, there is also a drawback, namely the danger that the code will be more used like a black box: "push a button and receive the result". In the old version the users were forced to think about how to run the calculation and thus had to look at details. This is a common problem, which all codes face.

With all the possibilities mentioned in the previous section it is often useful to combine different theories according to their advantages but keeping in mind their disadvantages. About 20 years ago the fields of quantum chemistry, DFT and many-body theory were completely separated and there was hardly any cooperation between them: this has fortunately changed. The strength and weaknesses of the different approaches are recognized and mutually appreciated. The solution of complex problems can only be found in close collaboration of the corresponding experts.

7. Theory and simulations

7.1. Theory compared to experiment

Independent of which computer code is used for computations some general questions should be asked, when theory and experiment do not agree. Some possible reasons for a disagreement are listed below (Figure 15):

- Is the atomic structure model that was chosen for the computation adequate for the experimental situation, as already discussed in Section 2. An advantage of theory is that the structure is well defined, because it is taken as input. Experiments may have uncertainties (stoichiometry, defects, impurities, disorder). It can also be that the theory is based on an idealized structure such as infinite crystal, whereas in the experiment surface effects cannot be neglected. If the latter are included in a supercell calculation, one has periodic boundary conditions and thus still assumes an ordered structure, while in the experiment the sample is disordered or contains some defects or impurities. A delicate question for the experimentalist is whether the sample that has been measured is (at least) close to the system that was assumed for the simulation.

- Is the chosen quantum mechanical treatment appropriate for the given system? Is a mean field DFT approach adequate? Are more sophisticated treatments (especially for correlation) needed or can the self-interaction within DFT cause the problem? Is there a significant dependence on the functional chosen within DFT?

- Is the performed calculation fully converged to the required accuracy, for example in terms of the basis set (for example in the number of plane waves or in other cases the choice of pseudo potential) or the underlying **k**-points mesh? In this context an evaluation of the chosen computer code can be important. Recently error estimates of solid state DFT calculations have been derived [60] in which the WIEN2k code plays the role of providing the standard (i.e. the most accurate calculation). The idea is that different implementations of the same first principles formalism (the same DFT functional) should lead to the same results or predictions. Would a different code yield other results (within a small error bar)? These tests showed that significant improvements of standard pseudo potentials were necessary, in order to reach the required accuracy. It was shown in [60] that the typical deviation (e.g. in total energy) between codes of different accuracy is an order of magnitude smaller than the typical difference with experiment.

- Is the property of interest a ground state property or are excited states involved? For example the electron core-hole interaction requires at least a treatment based on the Bethe-Salpeter equation (BSE) (see [50]).
- How about temperature and pressure? Often the calculated results corresponds to T=0 K but the experiment is carried out at room temperature. Can this difference be ignored? If phonons are included in the calculations, then at least thermodynamic estimates for a system at a higher temperature can be included. Varying the temperature is easy for experiments but difficult for theory. For pressure it is the other way around. Pressure is easy for theory but extremely difficult for experiments.

Structure model:	Quantum mechanics:	Convergence:	Other effects:
unit cell supercell surface	mean field (DFT) many body theory ground vs. excited states	basis sets k-points	temperature T>0 K pressure

Figure 15. Key aspects of modeling materials and the idealizations or approximations that must be made.

When a computation agrees with the experimental observations one should be careful because this is not a proof that the issues mentioned above are non-existent. It can be the case that there is a cancellation of errors: something like a crude model that is poorly converged. For a theorist there is always the temptation to stop improving a calculation when it already agrees with the experiments. Sometimes a deviation helps to find a better treatment in terms of atomic models, quantum mechanics (DFT functional), basis sets, temperature or some other factor that may be important in the specific case.

7.2. Results that can be provided by computations

It is appropriate to list aspects, where theory has advantages over experimental work:

- On can carry out computer experiments irrespective of the abundance, environmental effects or cost of materials. Even unstable or artificial systems (which cannot be measured) can be computed. One might wish to understand why they cannot be prepared. Sometimes several proposals (based on experiments or intuition) are under discussion: as long as they are not too many in number, theory can explore all of them and hopefully find the one which agrees with what is known about the system.

- In surface science and catalysis the calculation of observable properties are often needed, such as STM images, X-ray spectra, vibrational frequencies, electric field gradients, etc.. In addition, comparing total energies of proposed atomic structure (after a structure optimization) are often essential determining which atomic structure is likely to be correct. For an understanding of such material problems (for example at a surface) a combination of theory and experiment is essential.
- With a good calculation and by using all available tools one can gain insight and a fundamental understanding, especially in terms of trends, for which perfect agreement with experiment is not necessary. On such a basis systematic predictions can be made which can replace the trial and error scheme that is often used in materials optimization.
- One often wants to know the driving force for a certain change in properties. Is it coming from a substitution, the difference in chemical bonding, or from the related relaxation of the lattice around the impurity? Sometimes it is possible to set up artificially intermediate models which vary only one of these parameters at a time. Then an analysis can provide the answer.
- For materials with a clear structure and moderate correlation effects theory can predict experimental results. However, there are many interesting cases in material science, where the details matter. Often the interest comes, because the system is close to a transition (e.g. becoming magnetic, ferroelectric, or close to a metal-insulator transition). In such cases the two phases of the system can be rather close (e.g. in energy) and thus need special attention. Take the perovskite $SrTiO_3$ as an example. In this well known structure Ti is surrounded by six oxygen atoms forming an octahedron. Under pressure (or with temperature) these octahedra tilt in a certain fashion leading to a structural phase transition. If DFT theory yields a lattice constant that deviates by about 1 per cent from experiment, one could call this good agreement. In this case, however, this small deviation causes a difference of about 3 percent in the unit cell volume, which is sufficiently large to determine whether or not the tilt occurs (and thus such a detail matters). In such a case, one can choose another functional (for better agreement in volume) or one can carry out the calculation for the experimental volume, which can be obtained experimentally with high precision. With this choice a calculation can describe the phase transition properly.

8. Conclusion

In this chapter a selection of aspects, which play a role in modern computational theory of solids, has been given. From the atomic structure to the properties of inorganic materials a wide field of disciplines had to be included which are listed below:

- chemistry: intuition, interpretation, chemical bonding, stability
- physics: fundamentals and concepts, quantum mechanics, relativity
- crystallography: space groups, symmetry relations, group subgroup relations

- material science: understanding of trends, application, availability, environment
- mathematics: formalism, algorithms, numerics, accuracy
- computer science: data management, data bases, memory, communication, parallelization, load balancing, efficiency

It is clear that many details had to be skipped and only a few references could be given to guide the reader to the corresponding literature. Details of using the computer codes often change and thus it is highly recommended to look at the updated versions on the web (www.wien2k.at) or at the newest literature in this field. Selected results that can be obtained with WIEN2k had been summarized in this chapter. For a given atomic structure the electronic structure (band structure, density of states, and electron density) provides the basis for understanding chemical bonding. The corresponding total energy allows to judge relative stabilities of various phases or modifications. The effects of surfaces or even disordered structures can be simulated with sufficiently large supercells. Properties of insulators, metals, superconductors, or magnets etc. can be explained. Several quantities (such as spectra) can be computed which allow a direct comparison with experimental data. In some cases it is necessary to go beyond conventional DFT in order to reach agreement with experiment but DFT results are often an important and useful starting point.

It shall be stressed again that it is very useful to have a large variety of computer codes in this field. Different codes each have their emphasis on various aspects, such as accuracy, efficiency, user friendliness, robustness, portability with respect to hard-ware, features, properties and more. Some codes are more specialized for certain cases (e.g. treating only insulators) but do not work for other systems. This variety has helped to increase the importance of simulations in this field. For the comparison between theories (simulations) and experimental data several general considerations are summarized which are important for all kinds of computer codes.

Acknowledgements

I am very grateful and want to thank the many researchers who have helped developing the WIEN2k code and made many important contributions or suggestions. First of all I want to thank Peter Blaha, who has been part of the WIEN2k team for almost 35 years and who is the main person managing the code development. He has also been involved in many research projects that were carried out with this code. In addition I want to thank S.B. Trickey from the University of Florida, who has initialized the development and publication of the code (see [6], and the review [4]), as well as the other co-authors of the code [6] and the many researchers, who have contributed as acknowledged in a link at www.wien2k.at. I want to thank in particular Eamon McDermott, who has helped improving the English.

Author details

Karlheinz Schwarz*

Address all correspondence to: kschwarz@theochem.tuwien.ac.at

Institute of Materials Chemistry, Vienna University of Technology, Vienna, Austria

References

[1] Schwarz K, Blaha P, Madsen G K H, Electronic structure calculations of solids using the WIEN2k package for material science, Computer Physics Communication, 2002; 147, 71.

[2] Schwarz K, Blaha P, Solid state calculations using WIEN2k, Computational Material Science, 2003; 28, 259.

[3] Schwarz K, DFT calculations of solids with WIEN2k, J. Solid State Chemistry, 2003, 176, 319-328.

[4] Schwarz K, Blaha P, Trickey SB: Electronic structure of solids with WIEN2k. Molecular Physics 2010; 108, 3147-3166.

[5] Schwarz K, Blaha P. Electronic structure of solids and surfaces with WIEN2k. In: Leszczyncski J, Shukla M K(eds) *Practical Aspects of Computational Chemistry I: An Overview of the Last Two Decades and Current Trends.* Springer Science+Business Media B.V. 2012; Chapter 7, p191-207, ISBN 978-94-007-0918-8.

[6] Blaha P, Schwarz K, Sorantin P, Trickey S B, Full-potential linearized augmented plane wave programs for crystalline solids, Computer Physics Communication. 1990; 59, 399.

[7] Blaha P, Schwarz K, Madsen G K H, Kvasnicka D, Luitz J, *An Augmented Plane Wave Plus Local Orbitals Program for Calculating Crystal Properties,* Vienna University of Technology, 2001; ISBN 3-9501031-1-2.

[8] Hahn T (ed.), *International Tables for Crystallography, Volume A,Space-Group Symmetry* Kluwer Academic Publ.;1995; ISBN 0-7923-2950-3.

[9] Inorganic Crystal Structure Database (ICSD). URL: http://www.fiz-karlruhe.de.

[10] Diviš M, Schwarz K, Blaha P, Hilscher G, Michor M, Khmelevskyi S: Rare earth borocarbides: Electronic structure calculations and the electric field gradients. Physical Review B 2000; 62, 6774.

[11] Kabliman E, P, Blaha P, Schwarz K, Ab initio study of the misfit layer compound $(PbS)_{1.14}TaS_2$, Physical Review, 2010; 82, 024403.

[12] Schwarz K, Electrons. In: Authier A(ed.) *International Tables for Crystallography, Volume D, Physical Properties of Crystals.* Kluwer Academic Publ.; 2003, 294-313.

[13] Bartlett RJ, Musial M, Coupled cluster theory in quantum chemistry, Review Modern Physics 2007; 79, 291.

[14] Sode O, Keçli M, Hirata S, Yagi K, Coupled-cluster and many-body perturbation study of energies and phonon dispersions of solids hydrogen fluoride, International Journal Quantum Chemistry 2009; 109, 1928.

[15] Hohenberg P, Kohn W, Inhomogeneous electron gas, Physical Review, 1964; 136B, 864.

[16] Kohn W, Sham L S, Self-consistent equations including exchange and correlation effects, Physical Review, 1965; 140, A1133.

[17] Marks D L, Luke D R, Robust mixing fro quantum mechanical calculations, Physical Review B, 2008; 78, 07511.

[18] Cottenier S, *Density Functional Theory and the family of (L)APW-methods:A step-by-step introduction,* 2002-2013 (2nd edition); ISBN 978-90-807215-1-7 (freely available at http://www.wien2k.at/reg user/textbooks).

[19] Slater J C, A simplification of the Hartree-Fock method, Physical Review, 1951; 81, 385-390.

[20] Schwarz K, Optimization of the statistical exchange parameter α for the free atoms H through Nb, Physical Review B, 1972; 5, 2466-8.

[21] Ceperley C M, Alder D J, Grounds state of the electron gas by a stochastic method, Physical Review Letters. 1980; 45, 566.

[22] Perdew J P, Burke K, Ernzerhof M, Generalized gradient approximation made simple, Physical Review Letters, 1996; 77, 3865.

[23] Haas P, Tran F, Blaha P, Calculation of the lattice constant of solids with semi-local functions, Physical Review B, 2009; 79, 085104.

[24] Perdew J P, Ruzsinszky A, Tao J, Staroverov V N, Scuseria G E, Csonka G I, Prescription for the design and selection of density functional approximations: More constraint satisfaction with fewer fits, Journal of Chemical Physics, 2005; 123, 06220.

[25] Perdew J P, Kurth S, Zupan A, Blaha P, Accurate density functional with correct formal properties: a step beyond the generalized gradient approximation, Physical Review Letters,. 1999; 82, 2544.

[26] Novák P, Boucher F, Gressier P, Blaha P, Schwarz K, Electronic structure of mixed valence $(YM)_2BaNiO_3$, Physical Review, 2001; 63, 235114.

[27] Tran F, Blaha P, Accurate band gaps of semi conductors and insulators with a semilocal exchange-correlation potential, Physical Review Letters, 2009; 102, 226401.

[28] Polo V, Kraka, E, Cremer D, Electron correlation and the self-interaction error of density functional theory, Molecular Physic, 2003; 100(11) 1771-1790.

[29] Grimme S, Antony J, Ehrlich S, Krieg H, A consistent and accurate ab initio parameterization of density functional dispersion correction (DFT-D) for 94 elements H-Pu, Journal of Chemical Physics, 2010; 132, 154104.

[30] Koelling D D, Harmon B N, A technique for relativistic spin polarize calculations, Solid State Physics 1977; 10, 3107.

[31] MacDonnald A H, Picket W E, Koelling D D, A linearized relativistic augmented-plane-wave method utilizing approximate pure spin basis functions, Journal of Physics C: Solid State Physics, 1980; 13, 2675.

[32] Slater J C, Wave functions in a periodic potential Physical Review, 1937; 51, 846.

[33] Andersen OK, Linear methods in band theory, Physical Review B, 1975; 12, 3060.

[34] Weinert M, Wimmer E,Freeman A J, Total-energy all-electron density functional method for bulk solids and surfaces, Physical Review B, 1982; 24, 4571.

[35] Singh D J, Ground-state properties of lanthanum: Treatment of extended core-states, Physical Review B, 1975; 43, 6388.

[36] Sjöstedt E, Nordström L, Singh D J, An alternative way of linearizing the augmented plane wave method, Solid State Communication.2000; 114, 15.

[37] Madsen G H K, Blaha P, Schwarz K, Sjöstedt E, Nordström L, Efficient linearization of the augmented plane-wave method, Physical Review B, 2001; 64, 195134.

[38] Kokaj A, Computer graphics and graphical user interfaces as tools in simulations of matter at the atomic scale, Computational Material Science, 2003; 28, 155.

[39] Bader R W F, *Atoms in Molecules: a Quantum Theory,* Oxford university press, New York 1994.

[40] Madsen G K H, Iversen B B, Blaha P, Schwarz K, The electronic structure of sodium and potassium electro sodalites $(Na/K)_8(AlSiO_4)_6$, Physical Review B, 2001; 64, 195102.

[41] Laskowski R, Blaha P, Gallauner T, Schwarz K, Single layer model for the h-BN nanomesh on the Rh(111) surface, Physical Review Letters, 2007; 98, 106802.

[42] Laskowski R, Blaha P, Unraveling the structure of the h-BN/Rh(111) nanomesh with ab initio calculations, Journal of Physics: Condensed Matter, 2008; 064207.

[43] Koch H P, Laskowski R, Blaha P, Schwarz K, Adsorption of small gold clusters on the h-BN(Rh(111) nanomesh, Physical Review B, 2013; 86, 155404.

[44] Laskowski R, Madsen G K H, Blaha P, Schwarz K, Magnetic structure of electric-field gradients of uranium dioxide: An ab initio study, Physical Review B, 2004; 69, 140408.

[45] Spiel C, Blaha P, Schwarz K, Density functional calculations on the charge-ordered and valence-mixed modification of $YBaFe_2O_5$, Physical Review B, 2009; 79, 115123.

[46] Blaha P, Singh D J, Sorantin P I, Schwarz K, Electric field gradient calculations for systems with large extended core state contributions, Physical Review B, 1992; 46, 1321-1325.

[47] Dufek P, Blaha P, Schwarz K, Determination of the nuclear quadrupole moment of ^{57}Fe, Physical Review Letters, 1995; 75, 3545.

[48] Laskowski R, Blaha P, Calculating NMR chemical shifts using the augmented plane-wave method, Physical Review B, 2014, 89, 014402.

[49] Laskowski R, Blaha P, Origin of NMR shielding in fluorides, 2012; Physical Review B, 85, 245117.

[50] Laskowski R, Blaha P, Understanding the $L_{2,3}$ x-ray spectra of early 3d transition elements, Physical Review B, 2010, 85, 205105.

[51] Hedin L, New method for calculating the one-particle Green's function with application to the electron gas problem, Physical Review A, 1965; 139, 796.

[52] Jiang H, Gómez-Abal, Li X, Meisenbichler C, Ambrosch-Draxl C, Scheffler M, FHI-gap: A GW code based on the all-electron augmented pane wave method, Computer Physics Communication, 2013; 184, 348.

[53] Tersoff J, Hamann D R, Theory of the scanning tunneling microscope, Physical Review B, 1985; 31, 805.

[54] Perez-Mato J M, Blaha P, Schwarz K, Arroyo M, Orobengoa D, Etxebarria I, Garcia A, Multiple instabilities in $Bi_4Ti_3O_{12}$: A ferroelectric beyond the soft-mode paradigm, Physical Review B, 2008; 77, 184104-184110.

[55] Tröster A, Schranz W, Karsai F, Blaha P, Fully consistent finite-strain Landau theory for high pressure transition, Physical Review Y, 2014, 4, 03010.

[56] Kuneš J, Arita, R. Wissgott P, Toschi A, Ikeda H, Held K, Wien2Wannier: From linearized augmented plane waves to maximally localized Wannier functions, Computational Physics Communication, 2010; 181, 1888-1895.

[57] Held K, Electronic structure calculations using Dynamical Mean Field Theory, Advances in Physics 2007; 28, 155.

[58] Ahmed S J, Kibinen J, Zaporzan B, Curiel L, Pichardo S, Rubel O, BerryPI: A software for studying polarization of crystalline solids with WIEN2k density functional all-electron package, Computer Physics Communication, 2013; 184, 647-651.

[59] Blaha P, Hofstätter H, Koch O, Laskowski R, Schwarz K, Iterative diagonalization in APW-based methods in electronic structure calculations, Journal of Computational Physics, 2010; 229, 453-460.

[60] Lejaeghere K, Van Speaybroeck V, Van Oost G, Cottentier S 2013; Error estimates for solid-state density-functional theory predictions: An overview of the ground-state elemental crystals, Critical Reviews in Solid State and Materials Science 2103; 39(1) 1-24, DOI; 10.1080/1048436,2013.772503.

Chapter 11

Implications of Quantum Informational Entropy in Some Fundamental Physical and Biophysical Models

Maricel Agop, Alina Gavriluț, Călin Buzea,
Lăcrămioara Ochiuz, Dan Tesloianu,
Gabriel Crumpei and Cristina Popa

Additional information is available at the end of the chapter

http://dx.doi.org/10.5772/59203

1. Introduction

Complex systems are a large multidisciplinary research theme that has been studied using a combination of fundamental theory, derived especially from physics and computational modeling. This kind of systems is composed of a large number of elemental units that interact with each other, being called "agents" [1, 2, 62]. Examples of complex systems can be found in human societies, the brain, internet, ecosystems, biological evolution, stock markets, economies and many others.

The manner in which such a system manifests can't be predicted only by the behavior of individual elements or by adding their behavior, but is determined by the way the elements interact in order to influence global behavior. Very important properties of complex systems are those of emergence, self-organization, adaptability etc. [3, 4, 62].

An example of a complex system is represented by polymers. [Their structures present a multitude of organized networks starting from simple, linear chains of identical structural units to very complex sequences of amino acids that are chained together, thus forming the fundamental units of living fields. Probably one of the most interesting biological complex system is DNA that generates cells by employing a simple but very efficient code. It is the striking way in which individual cells organize into complex systems, such as organs and, subsequently, organisms. Research in the field of complex systems could provide new information on the realistic dynamics of polymers, solving troublesome problems such as protein folding. We note that the dynamics of such complex systems implies the quantum formalism] [1-4, 62].

Correspondingly, the theoretical models that describe the complex systems dynamics become more and more advanced [1-4]. For all that, this problem can be solved by taking into account that the complexity of the interaction process implies various temporal resolution scales, and the pattern evolution implies different degrees of freedom [5].

[In order to develop new theoretical models we must state the fact that the complex systems displaying chaotic behavior are recognized to acquire self-similarity (space-time structures can appear) in association with strong fluctuations at all possible space-time scales [1-4]. Afterwards, for temporal scales that are large with respect to the inverse of the highest Lyapunov exponent, the deterministic trajectories are replaced by a set of potential trajectories and the concept of definite positions by that of probability density] [62]. An interesting example is the collisions processes in complex systems, where the dynamics of the particles can be described by non-differentiable curves.

Since non-differentiability can be considered a universal property of complex systems, it is mandatory to develop a non-differentiable physics. In this way, by considering that the complexity of the interaction processes is replaced by non-differentiability, using the entire range of quantities from the standard physics (differentiable physics) is no longer required [19].

This topic was developed in the Scale Relativity Theory (SRT) [6, 7] and in the non-standard Scale Relativity Theory (NSSRT) [8-22]. [In the framework of SRT or NSSRT we assume that the movements of complex system entities take place on continuous but non-differentiable curves (fractal curves) so that all physical phenomena involved in the dynamics depend not only on the space-time coordinates but also on the space-time scales resolution. In this conjecture, the physical quantities that describe the dynamics of complex systems can be considered as fractal functions. In addition, the entities of the complex system may be reduced to and identified with their own trajectories. In this way, the complex system's behavior will be identical to the one of a special interaction-less "fluid" by means of its geodesics in a non-differentiable (fractal) space] [6, 7, 62].

In such context notions as informational entropy, Onicescu informational energy etc become important in the Nature description. These notions will be correlated with the fractal part of the physical quantities that describe the dynamics of complex systems.

2. Informational entropy and energy

Independently of scale resolution, the motion, either on infragalactic scale (for instance, the planetary motion), or on atomic scale (for instance, the motion of the electron around its nucleus) takes place on conics (ellipses). Such motion in invariant with respect to the SL(2R) group. In what follows, we shall consider this invariance only with respect to the motion on atomic scale.

2.1. SL(2R) invariance and canonic formalism

SL(2R) group is the group of transformations [23-26]

http://dx.doi.org/10.5772/59203

$$x' = \alpha x + \beta y, \quad y' = \gamma x + \delta y, \quad \alpha\delta - \beta\gamma = 1 \tag{1}$$

which makes invariant the areas in the phase space (x, y).

Choosing

$$\alpha = 1 + \frac{1}{2}a_2, \quad \beta = a_1, \quad \gamma = -a_3, \quad \delta = 1 - \frac{1}{2}a_2, \tag{2}$$

the infinitesimal transformations of the group have the expressions

$$x' = x + ya_1 + \frac{p}{2}a_2, \quad y' = y - \frac{1}{2}ya_2 - xa_3 \tag{3}$$

Then the Lie algebra associated to the group becomes

$$[\hat{L}_1, \hat{L}_2] = \hat{L}_1; \quad [\hat{L}_2, \hat{L}_3] = \hat{L}_3; \quad [\hat{L}_3, \hat{L}_1] = -2\hat{L}_2 \tag{4}$$

where

$$\hat{L}_1 = y\frac{\partial}{\partial x}, \quad \hat{L}_2 = \left(x\frac{\partial}{\partial x} - y\frac{\partial}{\partial y}\right), \quad \hat{L}_3 = -x\frac{\partial}{\partial y} \tag{5}$$

are the vectors of the Lie base.

The general vector of the algebra (4) is given by the linear combination

$$\hat{L} = c\hat{L}_1 + 2b\hat{L}_2 + a\hat{L}_3, \quad a, b, c = \text{const.} \tag{6}$$

The hamiltonian H results as an invariant function along the tangent trajectories to the vector (6). Precisely, it is a solution of the equation

$$\hat{L}H = 0 \tag{7}$$

According to (5), relation (7) becomes

$$(bx + cy)\frac{\partial H}{\partial x} - (ax + by)\frac{\partial H}{\partial y} = 0 \tag{8}$$

whence the characteristic differential system

$$\frac{dx}{bx+cy} = -\frac{dy}{(ax+by)} = dt \tag{9}$$

admits the integral

$$H(x,y) = \frac{1}{2}\left(ax^2 + 2bxy + cy^2\right) \tag{10}$$

We notice that the differential system (9) is Hamilton's system of equations [25]

$$\dot{x} = \frac{\partial H}{\partial y}, \quad \dot{y} = -\frac{\partial H}{\partial x} \tag{11}$$

associated to the hamiltonian (1), where the symbol " · " refers to the derivative with respect to the time.

Among the solutions of the equation (1), we have also the Gaussian

$$\rho(x,y) = A\exp[-H(x,y)], \quad A = \text{const.} \tag{12}$$

In consequence, all invariant functions on the group (7) will be functions of the hamiltonian (10) and particularly, of the Gaussian (12).

If the quadratic form (10) is positive definite, that is, the condition

$$a > 0, \quad \Omega = ac - b^2 > 0 \tag{13}$$

is fulfiled, then, by deriving the relations

$$\begin{aligned} \dot{x} &= bx + cy \\ \dot{y} &= -(ax + by) \end{aligned} \tag{14}$$

with respect to the time and eliminating $\dot{p}$ and $\dot{q}$, based on the relations (14), we obtain the symmetric equations

$$\begin{aligned} \ddot{x} + \Omega x &= 0 \\ \ddot{y} + \Omega y &= 0 \end{aligned} \tag{15}$$

These equations are formally equivalent to the equations of two linear oscillators of coordinates x, y.

Then the 2-form

$$\omega = dx \Lambda dy \tag{16}$$

has the meaning of the elementary surface in phase space (x, y) and the transformations (1) are canonic because they maintain the 2-form (16) (Liouville's theorem [25]). Simultaneously, the Gaussian (12) can be considered as a probabilistic density in phase space (x, y). In this situation, the parameters (a, b, c) can get statistical significance (see also [71]).

2.2. Shannon's informational entropy and transitivity manifolds

[In standard quantum mechanics, the impossibility of determining the variances of the position coordinate Δy_i and of the conjugate momentum component Δx_i $(x_i = -i\hbar\nabla_i)$ with arbitrary accuracy is widely accepted as being caused by the unavailable perturbation exerted on the particle by the measuring process. Because the measuring apparatus is most often not defined quantitatively and its perturbation can be very large, the uncertainty relation is formulated as a larger-than-or equal to equation

$$\Delta x_i \Delta y_i \geq \frac{1}{2}\hbar$$

Relating to this, it is unusual that the definitive nonzero variances Δx_i and Δp_i can be obtained for quantum system which are not exposed to a measuring device. This has been shown using the so called negative-result experiments. Furthermore, it can be noticed that we could theoretically obtain the nonzero variance Δx_i and Δy_i of quantum systems without including in the analysis perturbations from or in presence of a measuring device at all] [65].

In such a conjecture the uncertainty relations result in a quite natural way from the momentum perturbations associated with the fractal potential, i.e. with the Shannon's information.

Indeed, let be the probability density in the phase space, $\rho(x, y)$ with the constraints [27-32, 62]

$$\begin{aligned}
&\iint y\rho(x,y)dxdy = \langle y \rangle \\
&\iint x\rho(x,y)dxdy = \langle x \rangle \\
&\iint (y - \langle y \rangle)\rho(x,y)dxdy = (\delta y)^2 \\
&\iint (x - \langle x \rangle)\rho(x,y)dxdy = (\delta x)^2 \\
&\iint (y - \langle y \rangle)(x - \langle x \rangle)\rho(x,y)dxdy = cov(x,y)
\end{aligned} \tag{17}$$

where $\langle y\rangle$ is the mean value of the position, $\langle x\rangle$ is the mean value of the momentum, δy is the position standard deviation, δx is the momentum standard deviation and $cov(x, y)$ is the covariance of the random variables (x, y)] [62].

Now, we introduce Shannon's informational entropy [27]:

$$\bar{H} = \iint \rho(x,y)\ln[\rho(x,y)]dxdy. \tag{18}$$

Through Shannon's maximum informational entropy principle

$$\delta\bar{H} = 0 \tag{19}$$

with constraints (17), we get the normalized Gaussian distribution:

$$\rho\left(x-\langle x\rangle, y-\langle y\rangle\right) = \frac{\sqrt{ac-b^2}}{2\pi}\exp[-H\left(x-\langle x\rangle, y-\langle y\rangle\right)] \tag{20}$$

with

$$\begin{aligned} H\left(x-\langle x\rangle, y-\langle y\rangle\right) &= \frac{1}{2}[\bar{a}(x-\langle x\rangle)^2 + \\ &+ 2\bar{b}(x-\langle x\rangle)(y-\langle y\rangle) + \bar{c}(y-\langle y\rangle)^2] \\ \bar{a} = \frac{(\delta y)^2}{D},\ \bar{b} &= -\frac{cov(x,y)}{D},\ \bar{c} = \frac{(\delta x)^2}{D} \\ D &= (\delta x)^2(\delta y)^2 - cov^2(x,y) \end{aligned} \tag{21}$$

[We must note that the set of parameters $(\bar{a}, \bar{b}, \bar{c})$ has statistical significance given by relations (21).

In such context, the statistical hypothesis are specified through a particular choice of the set of parameters $(\bar{a}, \bar{b}, \bar{c})$ of the quadratic form the first Eq (21). Their class is given by the restriction [70]:

$$H(x',y') = H(x'-\langle x\rangle, y'-\langle y\rangle) = H(x-\langle x\rangle, y-\langle y\rangle) \tag{22}$$

where

$$\begin{aligned} &H(x'-\langle x\rangle, y'-\langle y\rangle) = \\ &= \frac{1}{2}\left[\bar{a}'(x'-\langle x\rangle)^2 + 2\bar{b}'(x'-\langle x\rangle)(y'-\langle y\rangle) + \bar{c}'(y'-\langle y\rangle)^2\right] \end{aligned} \tag{23}$$

If $(x'-\langle x\rangle,\ y'-\langle y\rangle)$ and $(x-\langle x\rangle,\ y-\langle y\rangle)$ are dependent through the unimodular transformations (1), we get that (22) imposes through $(\bar{a},\ \bar{b},\ \bar{c})$, the group of three parameters] (see [71] for details)

$$\begin{aligned} \bar{a}' &= \delta^2\bar{a} - 2\gamma\delta\bar{b} + \gamma^2\bar{c} \\ \bar{b}' &= -\beta\delta\bar{a} - (\beta\gamma + \alpha\delta)\bar{b} - \alpha\gamma\bar{c} \\ \bar{c}' &= \beta^2\bar{a} - 2\alpha\beta\bar{b} + \alpha^2\bar{c} \end{aligned} \tag{24}$$

If for the group (24) we choose the same parameterization as the one given by relations (2), then the corresponding infinitesimal transformations

$$\begin{aligned} \bar{a}' &= \bar{a} - \bar{a}a_2 + 2\bar{b}a_3 \\ \bar{b}' &= \bar{b} - \bar{a}a_1 + \bar{c}a_3 \\ \bar{c}' &= \bar{c} - 2\bar{b}a_1 + \bar{c}a_2 \end{aligned} \tag{25}$$

can be considered as an incompatible algebraic system in the unknowns a_1, a_2, a_3. In consequence, there cannot exist a transformation able to ensure the correspondence

$$\left(\bar{a}',\bar{b}',\bar{c}'\right) \to \left(\bar{a},\bar{b},\bar{c}\right) \tag{26}$$

Thus, the action of the group (24) in the space of variables $(\bar{a},\ \bar{b},\ \bar{c})$ is intransitive and, therefore, there exists a relation among the parameters $(\bar{a},\ \bar{b},\ \bar{c})$, which remains invariant to the action of the group (24). This relation is called transitivity manifold [26] (see also [71]).

The Lie algebra associated to the group (24) is

$$[\hat{A}_1,\hat{A}_2] = \hat{A}_1;\ \ [\hat{A}_2,\hat{A}_3] = \hat{A}_3;\ \ [\hat{A}_3,\hat{A}_1] = -2\hat{A}_1 \tag{27}$$

where

$$\begin{aligned} \hat{A}_1 &= -\bar{a}\frac{\partial}{\partial\bar{b}} - 2\bar{b}\frac{\partial}{\partial\bar{c}} \\ \hat{A}_2 &= -\bar{a}\frac{\partial}{\partial\bar{a}} + \bar{c}\frac{\partial}{\partial\bar{c}} \\ \hat{A}_3 &= 2\bar{b}\frac{\partial}{\partial\bar{a}} + \bar{c}\frac{\partial}{\partial\bar{b}} \end{aligned} \tag{28}$$

are the vectors of the base Lie. By the conditions

$$\hat{A}_1 F = 0, \quad \hat{A}_2 F = 0, \quad \hat{A}_3 F = 0 \tag{29}$$

[where F is an arbitrary function, we can obtain the transitivity manifolds of the group in the form

$$\bar{a}\,\bar{c} - \bar{b}^2 = \text{const.} \tag{30}$$

If H has energy significance, then condition (30) shows that a representative point from space (x, y) (which is in motion on a surface of constant energy (22)), can be also found on a surface of constant probabilistic density (ergodic condition) in Stoler's sense [33]:

$$\frac{\sqrt{\bar{a}'\bar{c}' - \bar{b}'^2}}{2\pi} e^{-H(x',y')} = \frac{\sqrt{\bar{a}\,\bar{c} - \bar{b}^2}}{2\pi} e^{-H(x,y)} \tag{31}$$

Therefore, the "class" of statistical hypothesis associated to the Gaussians having the same mean, is given by the ergodic condition. This highlights the strong relationship existing among the energetic issues and the probabilistic ones] (see [71]).

2.3. Informational energy in the sense of Onicescu and uncertainty relations

For the informational energy we shall use Onicescu's relation [34, 62, 71]:

$$E = \int_{-\infty}^{\infty}\int \rho^2(x,y)dxdy \tag{32}$$

Thus, the informational energy corresponding to the normed Gaussians (20), which is subject to conditions (22), becomes

$$E(\bar{a},\bar{b},\bar{c}) = \int_{-\infty}^{\infty}\int \rho^2(x,y)dxdy \tag{33}$$

where $H(x, y)>0$ is a condition imposed by the existence of the integral (33).

Thus we get

$$E(\bar{a},\bar{b},\bar{c}) = \frac{\sqrt{\bar{a}\,\bar{c} - \bar{b}^2}}{2\pi} \tag{34}$$

and therefore, if H has energetic significance, it results (see [62] and [71] for details):

http://dx.doi.org/10.5772/59203

i. The informational energy indicates the dispersion distribution (20) because the quantity

$$A=\frac{2\pi}{\sqrt{\overline{a}\ \overline{c}-\overline{b}^2}} \tag{35}$$

is a measure of the ellipses' areas of equal probability $H(p, q)=$ const., in the manner that the normed Gaussians are even more clustered the more their informational energy is higher;

ii. The class of statistical hypothesis which are specific to the Gaussians having the same mean is given by the constant value of the informational energy;

iii. The constant informational energy is equivalent to the ergodic condition;

iv. If the informational energy is constant, then the relations (21) and (34) give the egalitarian uncertainty relation

$$(\delta x)^2(\delta y)^2=\frac{1}{4\pi^2E^2\left(\overline{a},\overline{b},\overline{c}\right)}+cov^2(x,y) \tag{36}$$

or the non-egalitarian one

$$\delta x\delta y\geq\frac{1}{2\pi E\left(\overline{a},\overline{b},\overline{c}\right)}. \tag{37}$$

In such context we can show that the constant value of the Onicescu informational energy implies, in the case of a linear oscillator, the Planck's quantification condition.

2.4. Quantum mechanics and informational energy — Generalized uncertainty relations

[The original theory of de Broglie on the wave-corpuscle duality was developed using a theorem found in Lorentz's transformation [35]. This theorem interlinks the local horologes cyclic frequency (in each point of a spatial domain) with a progressive wave frequency in phase with the horologes. This wave gives determines the distribution of the oscillators' phases on the respective spatial domain. We desire to show that a distribution of this kind, in a true sense, can be determined without resorting to Lorentz's transformation [67].

The concept imagined by de Broglie, of equal pulsation horologes, can be evidenced by a periodic field, which is described by the local oscillators of equation

$$\ddot{Q}+\Omega^2Q=0 \tag{38}$$

where $Q = y + ix / m\Omega$ is the relevant complex coordinate of the field and Ω is its pulsation. The general solution of (38) can be written as [23]:

$$Q(t) = ze^{i(\Omega t+\varphi)} + \bar{z}e^{-i(\Omega t+\varphi)} \tag{39}$$

where z is a complex amplitude, $\bar{z}$ its complex conjugate and φ is a specific phase. The quantities z and $\bar{z}$ give the initial conditions, which are not the same for any point from the space. Precisely, at a time, the various oscillators corresponding to the points of the space are in different states and have different phases. A problem arises: can we apriori indicate a relationship among the parameters z, $\bar{z}$ and $e^{i(\Omega t+\varphi)}$ of the various oscillators at a given momentum? Because (39) is a solution of the equation (38) gives us an affirmative answer to this problem because (38) possesses a "hidden" symmetry that can be expressed by the homographic group: the ratio $\tau(t)$ of two solutions of the equation (38) is a solution of Schwartz's equation] [71] (see also [36, 67]).

$$\left(\frac{\tau''}{\tau'}\right)' - \frac{1}{2}\left(\frac{\tau''}{\tau'}\right)^2 = 2\Omega^2 \tag{40}$$

[This equation is invariant to the homographic transformation of $\tau(t)$: any homographic function of τ is itself a solution of (40). Since projections on the line can be characterized by the homography, we can assert that the ratio of two solutions of the equation (38) is a projective parameter for the class of the oscillators of the same pulsation from a given spatial region. Thus, one can define with ease a convenient, suitable projective parameter that should be in bi-univocal correspondence with the oscillator] [69, 71]. First, we observe a "universal" projective parameter: the ratio of the fundamental solutions of (38):

$$k = e^{2i(\Omega t+\varphi)} \tag{41}$$

Any homographic function of this ratio will be again a projective parameter [67]. Among all other, the function

$$\tau(t) = \frac{z + \bar{z}k}{1+k} \tag{42}$$

has primarily the advantage of being specific to each oscillator. But not only that: let be another function

$$\tau'(t) = \frac{z' + \bar{z}'k'}{1+k'} \tag{43}$$

which is specific to another oscillator. Since (42) and (43) are solutions of the equation (40), there exists a homographic relation between them:

$$\tau' = \frac{a'\tau + b'}{c'\tau + d'} \tag{44}$$

which, made explicit, leads to the Barbilian group equations [37]:

$$\begin{aligned} z' &= \frac{a^1 z + b^1}{c^1 z + d^1} \\ k' &= \frac{c^1 \bar{z} + d^1}{c^1 z + d^1} k \end{aligned} \tag{45}$$

[The group may be considered as a 'synchronization' group among various oscillators, a process in which the values of each take part, meaning that not only their phases, but also their amplitudes are correlated. The usual synchronization, manifested through the difference among the oscillators' phases as a whole, represents here just a particular case. Indeed, the group is involved for z, $\bar{z}$ and k, and also for (45), which indicates the fact that, indeed, the phase of k is only shifted with a value depending on the oscillator's amplitude, during passage between various members of the assembly and, moreover, the oscillator's amplitude is homographically affected.

When taking into consideration, for the group (45), the parameterization from [23], the following infinitesimal generators of the above-mentioned group will be obtained] [63]:

$$\begin{aligned} \hat{B}_1 &= \frac{\partial}{\partial z} + \frac{\partial}{\partial \bar{z}} \\ \hat{B}_2 &= z\frac{\partial}{\partial z} + \bar{z}\frac{\partial}{\partial \bar{z}} \\ \hat{B}_2 &= z^2\frac{\partial}{\partial z} + \bar{z}^2\frac{\partial}{\partial \bar{z}} + (z - \bar{z})k\frac{\partial}{\partial k} \end{aligned} \tag{46}$$

the following commutation relations

$$[\hat{B}_1, \hat{B}_2] = \hat{B}_1; \quad [\hat{B}_2, \hat{B}_3] = \hat{B}_3; \quad [\hat{B}_3, \hat{B}_1] = -2\hat{B}_2 \tag{47}$$

being involved. [Thus, a structure near-identical to group SL(2R)'s Lie algebra is shown. In consequence, the Lie algebra of the group (45) is, again, a result of group SL(2R)'s Lie algebra. Actually, as can be easily observed, the group (45) represents only another action of the group SL(2R), performed in variables z, $\bar{z}$, k] [63].

Once we fulfill the conditions of the theorem [38], the invariant functions can be found, simultaneously to the actions of the groups (5) and (46) as solutions of the equation

$$\hat{L}_i F(x,y,z,\bar{z},k) + \hat{B}_i F(x,y,z,\bar{z},k) = 0, \quad i = 1,2,3 \tag{48}$$

Explaining this equation by means of Equations (5) and (46) leads to their simple solution, by successive reduction, while simultaneously obtaining the invariant functions in the form

$$f(\mu, v) = const. \tag{49}$$

where μ and v are expressed as (see [63]):

$$\begin{aligned} \mu &= \frac{-i(z-\bar{z})}{(x-zy)(x-\bar{z}y)} \\ \nu &= k\frac{x-\bar{z}y}{x-zy} \end{aligned} \tag{50}$$

v being a unimodular complex and μ a real one. A particular class of such invariant functions is represented by linear combinations of the type

$$\bar{p}\mu = m\left(\nu + \frac{1}{\nu}\right) + 2n \tag{51}$$

where m,n and $\bar{p}$ represent three arbitrary real constants.

If considering Eq (50), then Eq (51) takes the form

$$mk^{-1}z^2 + 2nz\bar{z} + mk\bar{z}^2 \equiv \bar{p} \tag{52}$$

where the following notation has been used:

$$z = \frac{x - \bar{z}y}{\sqrt{-i(z-\bar{z})}} \tag{53}$$

We also noticed that

$$-i(z-\bar{z}) > 0 \tag{54}$$

Eq (52) represents a family of conical shapes from the phase space $(x,\ y)$. They represent ellipses if

$$m^2 - n^2 > 0 \tag{55}$$

a condition also fulfilled if

$$\begin{aligned} m &= \bar{Q}\sinh(2r) \\ n &= \bar{Q}\cosh(2r) \end{aligned} \tag{56}$$

where $\bar{Q}$ is a real constant and r is a real variable.

Quite an interesting case appears when z is completely imaginary, with no restriction concerning the generality value $z = i$. Thus, the quadratic form (52) may be identified with $H(p,\ q)$ from (10), resulting

$$\begin{aligned} \bar{a} &= \bar{Q}[\cosh(2r) + \sinh(2r)\cos\varphi] \\ \bar{b} &= -\bar{Q}\sinh(2r)\sin\varphi \\ \bar{c} &= \bar{Q}[\cosh(2r) - \sinh(2r)\cos\varphi] \end{aligned} \tag{57}$$

where φ is the value of k, assumed fixed. The square value of $\bar{Q}$ represents the value of the constant from (30), which determines the transitivity manifolds of the group (1) (see [63]).

The Gaussian distribution value obtained in such a manner represents only a particular case of the distribution that may occur, assuming in addition the obligation of satisfying the maximum principle of informational entropy under quadratic restrictions. The solutions of Eq (48) could be, however, much more general, being possibly selected from criteria involving group theory. In this context, the informational energy becomes

$$E(\bar{a}, \bar{b}, \bar{c}) = \frac{\bar{Q}}{4\pi} = const. \tag{58}$$

while the uncertainty relation (36) is

$$(\delta x)^2(\delta y)^2 = \frac{1}{\bar{Q}^2}(1 + \sinh^2(2r)\sin^2\varphi) \tag{59}$$

resulting that the concept of uncertainty is minimum only for $\varphi = 0$, i.e., all the oscillators of the assembly possess the same initial phase of zero. Based on this simplified hypothesis, at any moment of time subsequent to the initial one, the uncertainty relation (59) gives up its condition of minimum, along with the assembly's covariance which differs from zero] [63].

When the creation and annihilation operators refer to a harmonic oscillator, the uncertainty relations have the form from [33, 63] with $\bar{Q}=2/\hbar$ and $\hbar$ the reduced Planck constant. In this situation, the "synchronization" is achieved through Stoler's group [33] (the parameter r is exactly equal to the frequency ratio).

Onicescu informational energy can be correlated with the standard quantum mechanics and the second quantification (which indicates its utility, for instance in NDA-NRA dynamics) [39].

2.5. Gravity and information

The structure of the group (45) is given by the equations (46) in the manner that the only non-zero structure constants are [26]:

$$C_{12}^{1} = C_{23}^{3} = -1, C_{31}^{2} = -2 \tag{60}$$

Therefore, the invariant quadratic form is given by the "quadratic" tensor of the group,

$$C_{\alpha\beta} = C_{\alpha\nu}^{\mu} C_{\beta\mu}^{\nu} \tag{61}$$

or, more explicit, by (60),

$$C_{\alpha\beta} = \begin{pmatrix} 0 & 0 & -4 \\ 0 & 2 & 0 \\ -4 & 0 & 0 \end{pmatrix} \tag{62}$$

This yields that the invariant metric of the group is given by the relation [50]

$$\frac{ds^2}{k_0^2} = \omega_0^2 - 4\omega_1\omega_2 \tag{63}$$

where k_0 is an arbitrary factor and ω_α, three differential 1-forms, which are absolutely invariant through the group.

These 1-forms have the expressions:

$$\begin{aligned} \omega_0 &= i\left(\frac{dk}{k} - \frac{dz + d\bar{z}}{z - \bar{z}}\right) \\ \omega_1 &= \bar{\omega}_2 = \frac{dz}{k(z - \bar{z})} \end{aligned} \tag{64}$$

in which case the metric (63) becomes

$$\frac{ds^2}{k_0^2} = -\left(\frac{dk}{k} - \frac{dz + d\bar{z}}{z - \bar{z}}\right)^2 + 4\frac{dzd\bar{z}}{(z - \bar{z})^2} \tag{65}$$

It should be mentioned here a property related to integral geometry: the group (45) is measurable. Indeed, it is simply transitive and, since his structure vector $C_\alpha = C_{\nu\alpha}^{\nu}$ is identically zero, which can be seen from (60), it means that he possesses the invariant function

$$F(z, \bar{z}, k) = -\frac{1}{k(z - \bar{z})^2} \tag{66}$$

that is, the inverse of the module of the linear system's determinant obtained through the infinitesimal transformations of the group (45). Therefore, in the field variables space $(z, \bar{z}, k)$, one can build an a priori probabilities theory [40], based on the elementary probability

$$dP(z, \bar{z}, k) = -\frac{dz \Lambda d\bar{z} \Lambda dk}{(z - \bar{z})^2 k} \tag{67}$$

where Λ defines the external product of the 1-forms (64).

We now analyze the metric (65): it reduces to the metric of the Lobacevski's plan in Poincaré representation [26]:

$$\frac{ds^2}{k_0^2} = 4\frac{dzd\bar{z}}{(z - \bar{z})^2} \tag{68}$$

for $\omega_0 = 0$. Specifying ω_0 from (64) by the aid of the usual relations

$$\bar{z} = u + iv, \quad k = e^{i\varphi} \tag{69}$$

it results

$$\omega_0 = -\left(d\varphi + \frac{du}{v}\right) \tag{70}$$

and so, the condition $\omega_0 = 0$ becomes

$$d\varphi = -\frac{du}{v} \tag{71}$$

Since by this restriction, the metric (68) in the variables (69) reduces to the Lobacevski's one in Beltrami's representation:

$$\frac{ds^2}{k_0^2} = -\frac{du^2 + dv^2}{v^2} \tag{72}$$

the condition (71) defines a parallel transport of vectors in a Levi-Civita meaning: the application point of the vector moves on the geodesic, the vector always making a constant angle with the tangent to the geodesic in the current point. Truly, using the fact that the plan's metric is conformal Euclidean, the angle between the initial vector and the vector transported through parallelism can be calculated as the integral of the equation (see [36, 68, 71] for details):

$$d\varphi = \frac{1}{2}\left(\frac{\partial \ln E}{\partial v}du - \frac{\partial \ln E}{\partial u}dv\right) \tag{73}$$

along the transport curve. $E(u, v)$ denotes here the conformity factor of the respective metric, in our case $E(u, v)=1/v^2$. Substituting it in (73), we get (71).

Now the variables $(z, \bar{z}, k)$ can be considered as amplitudes of a gravitational field Thus, let us admit that we describe the gravitational field through the variables y_i for which we "discovered"' the metric

$$h_{ij}dy^i dy^j \tag{74}$$

in an ambient space of the metric

$$\gamma_{\alpha\beta}dx^\alpha dx^\beta \tag{75}$$

Then the field equations derive from the variational principle [41]

$$\delta\int L\gamma^{\frac{1}{2}}d^3x = 0 \tag{76}$$

http://dx.doi.org/10.5772/59203

relative to the Lagrange function

$$L = \gamma^{\alpha\beta} h_{ij} \frac{\partial y^i}{\partial x^\alpha} \frac{\partial y^j}{\partial x^\beta} = \gamma^{\alpha\beta} \frac{\frac{\partial z}{\partial x^\alpha} \cdot \frac{\partial \bar{z}}{\partial x^\beta}}{(z-\bar{z})^2} = \frac{\text{''}\ z \text{''}\ \bar{z}}{(z-\bar{z})^2} \tag{77}$$

In such a context, Einstein's equations with $z = i\varepsilon$ become Ernst's ones for the gravitational field of vacuum [42, 43]

$$\begin{aligned} -\frac{1}{2}(\varepsilon+\bar{\varepsilon})\text{''}\ ^2\varepsilon &= \text{''}\ \varepsilon \text{''}\ \bar{\varepsilon} \\ -(\varepsilon+\bar{\varepsilon})^2 R_{\alpha\beta}(\gamma) &= \partial_\alpha \varepsilon \partial_\beta \bar{\varepsilon} + \partial_\beta \varepsilon \partial_\alpha \bar{\varepsilon} \end{aligned} \tag{78}$$

($R_{\alpha\beta}$ is here the Ricci tensor of the three-dimensional metric $\gamma_{\alpha\beta}$).

Thus, adopting as a starting point the variational principle (76), the essential goal of the analysis in the gravitational field domain is to produce metrics of Lobacevski's plan or metrics related to them. All these can be directly related to Einstein's equations (78). Moreover, by substituting the principle of independence of the simultaneous actions, in the form of linear composition in a point of various fields intensities (through the apriori invariance of fields' action with respect to a certain group), we may conceive a gravity theory that has none of the contradictions inherently and commonly present in the current theory [44]. We observe that the SL(2R) group parameters can be interpreted as field amplitudes in a supergravitation model [23] (see also [71] for details).

2.6. Extracellular vesicles convection in haptotaxis with hydrodynamical dissipation, a novel mechanism for vesicle migration

2.6.1. On the vesicle role

In the field of cell's biology, we call vesicles those small bags wrapped in a membrane forming part of eukaryotic cell organelles. They are involved in proteins or enzymes transport and absorption, or meet other needs of the cell. Inside the membrane bag of a vesicle, there are macromolecules which require the ability to move outside the cell walls. The membrane encircling the bag merges with the outer wall of the cell to allow such macromolecules to penetrate the wall. The vesicles are important parts of the human cells, although they are also found in other multicellular organisms [66].

Cells found in humans, plants and animals use a variety of types of vesicles, depending on the type of cell and its specific intended function. For example, one type of vesicles, lysosomes, are necessary for the process of digestion. Lysosomes contain enzymes that breakdown food cells. With food absorption, a lysosome vesicle bonds to the food holding cell and releases enzymes by a process called phagocytosis. These enzymes break down food cells into smaller parts that can be better absorbed by other cells.

Secretory vesicles are frequently associated with nerve cells in humans or animals. Their membranes sacs contain neurotransmitters. Nervous system through hormonal signals activates these components. Through the process of exocytosis, the secretory vesicle's outer membrane adheres to the nerve terminal and releases neurotransmitters in the area of the nerve endings, named the synaptic cleft. Neurotransmitters transport information from one nerve terminal to the next, across the entire central nervous system, way up to the brain [66].

Vesicles, in their role as cellular mechanism are internally appointed for transport, uptake and storage of numerous imperative bodily functions. Without these tiny bags wrapped in membranes, cells could not make the exchange of materials necessary to maintain their healthy development and other crucial processes. As a conclusion, with no vesicles, humans and other pluricellular organisms could not have existed, because the essential cellular chemical processes would have no other method to pass onto another key materials [66].

Since there is increasing support that vesicle trafficking, including the release of EVs, is a highly important process in tumorigenesis, embryogenesis and tissue remodeling, in this paragraph we present an extensive discussion on the EVs convection in haptotaxis with hydrodynamical dissipation (i.e., a novel mechanism for vesicle migration).

2.6.2. Mathematical model

Vesicles are closed membranes floating in an aqueous solution (see Fig. 1). These membranes act as a barrier that efficiently controls permeability. The vesicles mimic maybe the most primitive and mechanically flexible dividing interfaces between the inside and the outside of a cell. Generally, the fluid enclosed by the membrane is incompressible in order that the vesicle evolves at a constant volume. Moreover, the membrane exchanges no phospholipid molecules with the solution, its area remaining constant as time passes [64]. Helfrich [45] described very well the vesicle's bending energy in its equilibrium state, which is compatible with the constraints above, i.e. constant volume and area. Even if the model is relatively simple it generates various equilibrium profiles, such as, discocytes (resembling red blood cells), stomatocytes, as well as forms presenting higher topologies (such as n-genus torus) that have been also observed experimentally [46]. We identify works studying alignments of vesicle in shear flows [47], fluctuations out of equilibrium [48], lift forces [49, 50], migration of vesicle in the proximity of a substrate [51, 52] or in gravity fields [53] and also vesicle tumbling [54]. One may note several recent experiments dealing with vesicle migration [55-58].

Considering the vesicle migration, we acknowledge that hydrodynamical dissipation in the neighboring fluid as well as inside the vesicle is present, and, in principle, between the two mono-layers which may glide with respect to each other. Furthermore, during motion on the substrate the dynamics of a vesicle may be restricted not only by the hydrodynamical flow but also by bonds breaking and restoring mechanisms that occur on the substrate (see [64]). It is obvious that the slowest mechanism limits the motion. Here we focus on a situation where

http://dx.doi.org/10.5772/59203

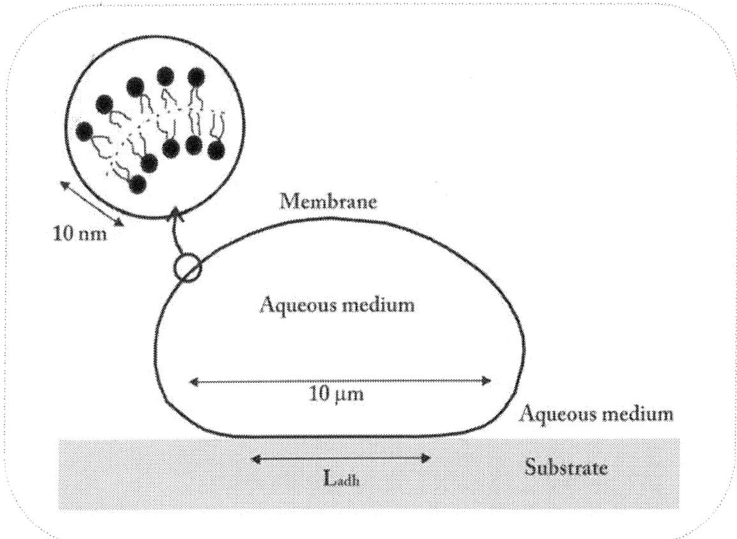

Figure 1. Schematic view of a vesicle emphasizing its microscopic structure: a bilayer of phospholipidic molecules. L_{adh} is the adherence length of the membrane

hydrodynamics are the limiting factors and we give out dissipation associated with bonds on the substrate.

Let us imagine a vesicle that initially adheres on a flat surface. We then consider an adhesion gradient along the substrate. The vesicle then moves in the direction of increasing adhesion energy (see Fig. 2) - it is named haptotaxis (a motion induced by an adhesion gradient).

A highly permeable vesicle can be pulled into a fluid without opposing any resistance (and without modifying the inner area), whereas an impermeable one would be subjected to a drag force [64]. The assumption of local impermeability is legitimate. This entails that the fluid velocity at the membrane is equal that of the membrane itself [59].

On a vesicle's scale $(R \sim 10\mu m)$ and for the expected velocities $(V \sim 1\mu m / s)$, the dissipative processes fully dominate the dynamics. The energy added instantly dissipates in various degrees of freedom. Local dissipation caused by molecular reorganization, characterized by Leslie's coefficient, is negligible with respect to the hydrodynamics modes [60].

If dissipation is dominated by bulk effects, as shown in [59], we are in the position to write down the basic governing equations for convective vesicles in a geometry depicted in Fig. 3, since we also know and it was proved the velocity field obeys Stokes equations [59].

In an original atmospheric system, the non-even distribution of ascending water droplets is determined by the interplay between solar energy-induced thermal gradients, thermal diffusivity, friction, and gravity. Ultimately, the mathematics of this model shapes the

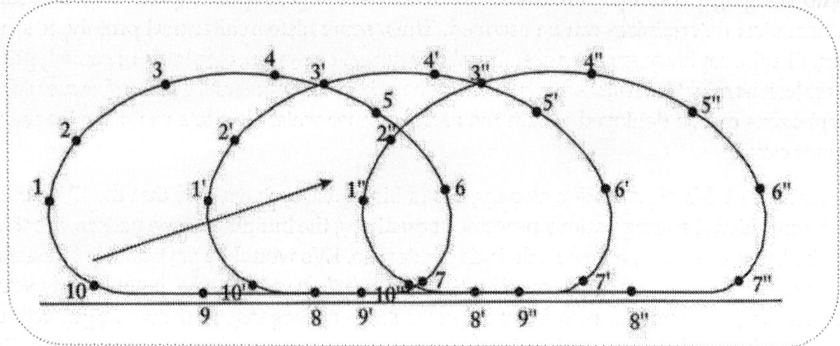

Figure 2. Stationary vesicle profiles are depicted. The vesicle is moving from the left (smaller adhesion) to the right (stronger adhesion); a few discretization points are represented and the arrow allows following one of these at three successive times. One can observe here the rolling and sliding components of the vesicle's motion.

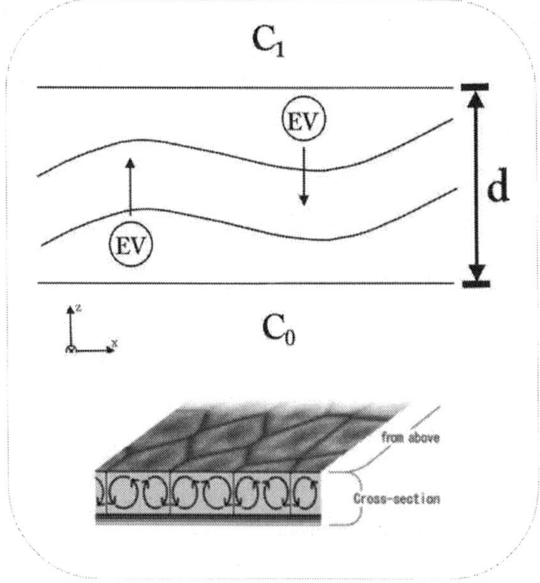

Figure 3. Convective extracellular vesicles (EVs) geometry. A fluid layer of thickness *d* of EVs, adherent on an extracellular matrix (ECM), is subjected to a gradient of concentration, where $\Delta C = C_1 - C_0 > 0$ is the difference of concentration between the front and back boundaries of the fluid layer.

umbrella-like or budding appearance of structures like cumulonimbus clouds. This model can better or uniquely describe those types of structural dynamics not explained under fractal, simple/linear and several other types of models.

Acknowledging that similar patterns occur in various biological spaces, we think that the same mathematical determinism can be ascribed. Thus, some histoarchitectural prototypic structures, like the capillary sprouting, embryologic organ, or even tumor buds of some types of cancer lesions might be in fact sculpted in that shape because gradients of molecular cues called morphogens can be deployed within the same manner water droplets can organize within nascent clouds.

Assuming that this organization also applies in biological systems, and that the EVs release can be considered among various processes organizing the budding tissue pattern, we think that the Lorenz model can govern their dynamics too. EVs would be particularly interesting as controllers of the tissue shape specification because they can include enzymatically active components (not found in conventional molecular morphogens), and thus might actively interact with the ECM fibers within their migration. Deployment of certain matrix degrading enzymes (MDEs) by EVs can modify this space while diffusing (event not produced by simple morphogens, attractive chemokines or repulsive semaphorins). This activity changes the topography of the ECM and creates spatial gradients directing the migration of subsequent EVs by haptotaxis - a mechanism better described for cell migration.

Let us consider the following thought biological experiment, equivalent to the B\'{enard experiment: a fluid layer of extracellular vesicles adherent on an ECM, in a haptotactic gradient. The fluid layer presents an unstable stratification of the potential density in a field of forces: the dense fluid is placed in front of the less dense one. We assume that in the basic state the layer of fluid of thickness d is subjected to a gradient of concentration

$$\beta = \frac{C_1 - C_0}{d} > 0 \tag{79}$$

$\Delta C = C_1 - C_0 > 0$ is the difference of concentration between the front and back boundaries of the fluid layer. The regime with the fluid at rest and a non-perturbed distribution of concentration, belongs to the thermodynamic branch, which is continuously linking the non-equilibrium stationary state ($\Delta C \neq 0$) with the equilibrium state ($\Delta C = 0$) (see Fig. 3).

We examine the evolution of a concentration fluctuation θ around the non-perturbed concentration profile $C_0(z)$.

Two dissipative processes tend to maintain the fluid at rest:

- friction (motion amortization through viscosity);
- ECM degradation subsequent to MDE's activity allowing vesicle trespassing - which lowers the concentration of the ECM, thus diminishing the forward, or advancing force.

The instability cannot be developed unless the EV is accelerated enough to overcome the effect of these dissipative processes. The gradient of concentration β which is the control parameter of this instability has to surpass a critical value β_C. Over this critical value, an organized structure of convection cells may appear.

For a one component fluid, the mass, momentum and internal energy equations are the expressions (see the fractal - nonfractal transition method [8-22]):

$$\begin{aligned} &\frac{\partial \rho}{\partial t}+\nabla\cdot(\rho\mathbf{v})=0 \\ &\frac{\partial(\rho\mathbf{v})}{\partial t}+\nabla\cdot(\overleftrightarrow{\Pi}+\rho\mathbf{v}\mathbf{v})=\rho\mathbf{g} \\ &\frac{\partial(\rho\varepsilon)}{\partial t}+\nabla\cdot(\rho\varepsilon\mathbf{v}+\mathbf{j}_d)=-\overleftrightarrow{\Pi}\otimes(\nabla\mathbf{v}) \end{aligned} \tag{80}$$

where ρ represents the mass density of the fluid, v its speed, g acceleration of a field of forces, ε the internal energy of the unit volume, and $\mathbf{j}_d$ the flux of ECM degraded by signals received from EVs. Here $\overleftrightarrow{\Pi}$ is the stress tensor and $\otimes$ denotes the product of two tensors,

$$\overleftrightarrow{A}\otimes\overleftrightarrow{B}=A_{ij}B_{ji}$$

and we use Einstein's summation convention (implicit sum over repeating indices). The stress tensor can be written

$$\overleftrightarrow{\Pi}=\overleftrightarrow{\Pi}^{e}+\overleftrightarrow{\Pi}^{v} \tag{81}$$

$\overleftrightarrow{\Pi}^{e}$ is the equilibrium part and depends on the state of the system. $\overleftrightarrow{\Pi}^{v}$ represents the non-equilibrium part and is named, viscous stress tensor. At equilibrium, this part vanishes. For an isotropic medium at rest,

$$\overleftrightarrow{\Pi}^{e}=\begin{pmatrix} p & 0 & 0 \\ 0 & p & 0 \\ 0 & 0 & p \end{pmatrix}=p\overleftrightarrow{I} \tag{82}$$

where p is the hydrostatic pressure. For viscous systems at non-equilibrium, the viscous stress tensor is not null. According to Eq. (81) and Eq. (82), the stress tensor will be, for homogeneous and isotropic viscous systems, at non-equilibrium

$$\overleftrightarrow{\Pi}=p\overleftrightarrow{I}+\overleftrightarrow{\Pi}^{v} \tag{83}$$

We start with the following assumptions:

a. the fluid is Newtonian; as a result the stress tensor is given by Eq. (83), where the viscous stress tensor is [8-22]

$$\Pi^{v}_{\alpha\beta} = -\eta\left(\frac{\partial v_\alpha}{\partial x_\beta} + \frac{\partial v_\beta}{\partial x_\alpha} - \frac{2}{3}\delta_{\alpha\beta}(\nabla\cdot\mathbf{v})\right) - \zeta\delta_{\alpha\beta}(\nabla\cdot\mathbf{v})$$

with coefficients η and ζ independent of velocity, the tangential (shear) and bulk viscosity, respectively;

b. ECM degrading by MDEs from EVs is described by the Fourier equation

$$\mathbf{j}_d = -\lambda\nabla C \tag{84}$$

where λ is the haptotactic coefficient;

c. haptotactic energy expansion is linear

$$\delta\rho = \rho - \rho_0 = -\rho_0\alpha\varepsilon = -\rho_0\alpha k_h(C - C_0) = -\rho_0\chi(C - C_0) \tag{85}$$

where we used the expression of the haptotactic energy

$$\varepsilon = k_h(C - C_0)$$

k_h being the haptotactic energy constant. In Eq. (85) α is the haptotactic energy expansion constant and $\chi = \alpha k_h$ is the haptotactic expansion constant;

d. the fluid satisfies a state equation: consequently, its internal energy is (up to a constant factor)

$$\varepsilon = k_b C \tag{86}$$

where k_b is the state constant;

e. in most liquids, thermal expansion is small. We choose everywhere a constant density, denoted by ρ_0, except the momentum equation.

With these approximations, the system of Eqs. (80) leads to the Boussinesq type system of equations

$$\begin{aligned}
&\nabla\cdot\mathbf{v} = 0\\
&\rho_0\left[\frac{\partial\mathbf{v}}{\partial t} + (\mathbf{v}\cdot\nabla)\mathbf{v}\right] + \nabla p = (\rho_0 + \delta\rho)\mathbf{g} + \eta\nabla^2\mathbf{v}\\
&\frac{\partial C}{\partial t} + (\mathbf{v}\cdot\nabla)C = \frac{\lambda}{\rho_0 k_b}\nabla^2 C
\end{aligned} \tag{87}$$

where ρ is the perturbed density

$$\rho = \rho_0 + \delta\rho \tag{88}$$

The first equation (87) represents the incompressibility condition for the fluid.

Convection occurs in the fluid layer when the forward, or advancing force, resulted from energy expansion, surpasses the viscous forces. We may define now a Rayleigh type number identical with the Eqs associated to fractal - nonfractal transition [61]

$$R = \frac{|\mathbf{F}_{asc}|}{|\mathbf{F}_{visc}|} \approx \frac{\left|\frac{\delta\rho \mathbf{g}}{\rho_0}\right|}{\left|\frac{\eta\nabla^2\mathbf{v}}{\rho_0}\right|} \tag{89}$$

The density perturbation satisfies, according to Eq. (85)

$$\frac{\delta\rho}{\rho_0} \approx \chi\Delta C \tag{90}$$

On the other side, from the internal energy equation Eq. (79), it results

$$v \approx \frac{\lambda}{\rho_0 k_b}\frac{1}{d} \tag{91}$$

Replacing Eqs. (90) and (91) in Eq. (89), and taking into account Eq. (79), we get a biological Rayleigh number

$$R = \frac{\chi\beta\rho_0 k_b g}{\nu\lambda} d^4 \tag{92}$$

where $\nu = \eta / \rho_0$ is the cinematic viscosity. For the Bénard convection, the biological Rayleigh number plays the part of a control parameter. The convection occurs for

$$R > R_{critical}$$

Most of the time, R is controlled by β, the gradient of concentration.

Within a biological context, **g** can be specified by polar/linear topography of semaphorins or/and chemokines, signals typically creating stable gradients to which EVs can respond.

We choose as reference state the rest stationary state ($v_S = 0$), for which the last two equations in the system of Eqs. (87) reduce to

$$\begin{aligned} \nabla p_S &= -\rho_S g\hat{z} = -\rho_0[1-\chi(C_S - C_0)]g\hat{z} \\ \nabla^2 C_S &= 0 \end{aligned} \tag{93}$$

where $\hat{z}$ is the versor of the vertical direction. We assume pressure and concentration varies only along the vertical direction, due to the geometry of the experiment. For concentration, the boundary conditions read

$$C(x,y,0) = C_0;\ \ C(x,y,d) = C_1$$

Integrating the second Eq. (93) with these boundary conditions, it results that, in the stationary reference state, the profile of the concentration in the vertical direction is linear

$$C_S = C_0 - \beta z \tag{94}$$

with β, the gradient of concentration. Replacing Eq. (94) in first Eq. (93) and integrating, we get

$$p_S(z) = p_0 - \rho_0 g\left(1 + \frac{\chi\beta z}{2}\right)z \tag{95}$$

The characteristics of the system in this state are independent of the kinetic coefficients η and λ which appear in Eqs. (87). We study the stability of the reference state using the small perturbations method. The perturbed state is characterized by

$$\begin{aligned} C &= C_S(z) + \theta(\mathbf{r},t) \\ \rho &= \rho_S(z) + \delta\rho(\mathbf{r},t) \\ p &= p_S(z) + \delta p(\mathbf{r},t) \\ \mathbf{v} &= \delta\mathbf{v}(\mathbf{r},t) = (u,v,w) \end{aligned} \tag{96}$$

As can be seen from Eqs. (96), the perturbations are functions of coordinate and time. Replacing Eqs. (96) in the evolution equations of the Boussinesq approximation Eqs. (87) and taking into account Eq. (94) and Eq. (95), we get, in the linear approximation, the following equations for the perturbations

$$\begin{aligned} &\nabla \cdot \delta \mathbf{v} = 0 && \text{a} \\ &\frac{\partial \delta \mathbf{v}}{\partial t} = -\frac{1}{\rho_0} \nabla \delta p + \nu \nabla^2 \delta \mathbf{v} + g\chi \theta \hat{z} && \text{b} \\ &\frac{\partial \theta}{\partial t} = \beta w + K \nabla^2 \theta && \text{c} \end{aligned} \tag{97}$$

where $K = \frac{\lambda}{\rho_0 k_b}$ is a coefficient. We pass to non-dimensional variables in Eqs. (97), using the transformations: $\mathbf{r}' = \frac{\mathbf{r}}{d}$; $t' = \frac{t}{d^2/K}$; $\theta' = \frac{\theta}{\left(\frac{\nu K}{g\chi d^3}\right)}$; $\delta \mathbf{v}' = \frac{\delta \mathbf{v}}{K/d}$ and $\delta p' = \frac{\delta p}{\left(\frac{\rho_0 \nu K}{d^2}\right)}$.

Using the standard method from [8-22], it results the biological Lorentz system.

2.6.3. Validity of theoretical model

Some results are evident:

- we build the first Lorenz model for extracellular vesicles migration;
- in [50], and similar other references ([64] etc.), the EVs behavior, under shear flow close to a substrate, was proved to be quite similar to the one encountered in two dimensional simulations, so we are confident that the 2D assumptions captures the essential features of the 3D EVs;
- different control parameter values for the Lorenz system can create shape distributions similar to the cordonal appearance of fingerprints (see Fig. 4, A), or complex skin tissue tiles like scale appendages in the amphibian covering (see Fig. 4, B);
- the biological thought experiment equivalent to Bénard's experiment, involving a fluid layer of extracellular vesicles adherent to an extracellular matrix, in a haptotactic gradient can be *checked experimentally* today to a high degree of accuracy. We think that suitable test systems would be the *embryological ones* (i.e., the development of branched vessels in membranes - avian eggshell membranes, serous membranes of the peritoneal cavity; or the budding development of lung alveoli, or of fingerprints), and, similar, *inflammatory ones* (i.e., the emergence of neoangionetic vessels driven by inflammatory proximities) - all of which apparently start as point like spots displayed in a comb-like appearance along a rectilinear or arched origin;
- we analyze the problem of EVs migration in haptotaxis, though most of the reasoning applies to chemotaxis (migration of cells biased towards a gradient of diffusible MDEs) as well as to a variety of driving forces - all of which include the possibility to specify an active parameter value within the model;
- the resulted system of equations exhibits complex behavior, hard to control, the two occurring convective rolls: either going in one direction, or in the opposite one - means patterning the EVs spreading.

http://dx.doi.org/10.5772/59203

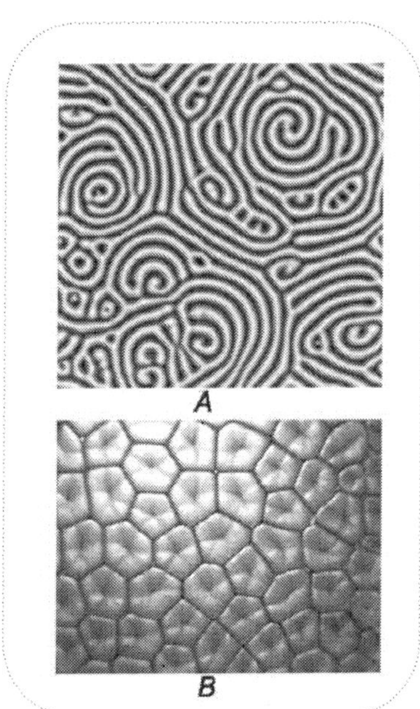

Figure 4. Bénard-Rayleigh model patterns representative for biological instances: A) for fingerprint like distribution of skin cells, B) for fish or amphibian scales.

3. Conclusions

Considering the above, we can write the following conclusions :

i. We establish a relationship between the SL(2R) group and the canonic formalism. It particularly results that all invariant functions on the SL(2R) group will be functions on the hamiltonian and on the Gausssian;

ii. We establish the statistical significations for coefficients of the hamiltonian by using Shannon's maximum informational entropy principle. Any statistical assumption is specified by particularly selecting the hamiltonian coefficients and the class of all these hypothesis by the transitivity manifolds of the group. In this manner, if the hamiltonian has energy significance, then through the transitivity manifolds, the motions of a representative point from the phase space on a surface of constant energy are in the same time on a surface of constant probability density (the ergodic hypothesis). Therefore, the class of the statistic hypothesis (which are characteristic to the Gaussians of the same average) is given by the ergodic hypothesis. In this way, we establish a fundamental relationship between energy and probability;

iii. We prove that informational energy (in the sense of Onicescu) is a measure of the dispersion of a distribution. The class of the statistic hypothesis that are characteristic to the Gaussians of the same average is characterized by the constant value of the informational energy. In addition, it is equivalent to the ergodic hypothesis. For a constant value of the informational energy, we obtain uncertainty egalitarian relationships and, particularly, for the linear harmonic oscillator, we show that the informational energy is quantified;

iv. Assuming that de Broglie's theory is materialized through a periodic field in a complex coordinate, we prove that it has a "hidden symmetry", which is expressed by the homographical transformations group in three parameters. This group (also an achievement of SL(2R)) functions as a synchronization group both in phase and in amplitude, among the oscillators of the same ensemble. The simultaneous invariance related to two different achievements of SL(2R) implies (integrally through the invariant functions on the groups) an uncertainty relation in egalitarian form and the Stoler group of synchronization among oscillators from different ensembles (i.e., the second quantification when the creation and annihilation operators refer to a harmonic oscillator);

v. The synchronization group among the oscillators of the same ensemble admits three differentiable 1-forms and one differentiable 2-form which is absolutely invariant on the group. The existence of a parallel transport in Levi-Civita's sense, in which case the 2-form in Lobacevski's metric form in Poincare representation, implies through a variation principle, equations of Ernst type for the gravitational field of vacuum;

vi. Complex measures in the study of certain physical systems dynamics need the use of a space-time endowed with a special topology, namely, the fractal space-time [6, 7] (see also [71] for details).

Author details

Maricel Agop[1,2*], Alina Gavriluț[3], Călin Buzea[4], Lăcrămioara Ochiuz[5], Dan Tesloianu[6], Gabriel Crumpei[7] and Cristina Popa[8]

*Address all correspondence to: m.agop@yahoo.com

1 Lasers, Atoms and Molecules Physics Laboratory, University of Science and Technology, Lille, France

2 Department of Physics, "Gh. Asachi"' Technical University, Iași, România

3 Department of Mathematics, "Al.I. Cuza" University from Iași, Romania

4 National Institute of Research and Development for Technical Physics, Iași, România

http://dx.doi.org/10.5772/59203

5 Faculty of Pharmacy, Department of Pharmaceutical Technology University of Medicine and Pharmacy "Gr.T. Popa", Iași, România

6 IV Internal Medicine Department, Clinical Emergency Hospital "Sf. Spiridon", University of Medicine and Pharmacy "Gr.T. Popa", Iași, România

7 Psychiatry, Psychotherapy and Counselling Center, Iași, România

8 Dentistry Department, University of Medicine and Pharmacy "Gr.T. Popa", Iași, România

References

[1] Flake G. W., *The Computational Beauty of Nature*, MIT Press, Cambridge, MA, 1998.

[2] Mitchell M., *Complexity: A Guided Tour*, Oxford University Press, Oxford, 2009.

[3] Winfree A. T., *The Geometry of Biological Time*, Springer 2nd edition, New York, 2000.

[4] Bennett C. H., *How to define complexity in physics, and why*, in W. H. Zurek (editor), Buldyrev S. V., Parshani R., Paul G., Stanley H. E., Haylin S., *Catastrophic cascade of failures in independent networks*, Nature 464, 2010

[5] Badii R., Politi A., *Complexity: Hierarchical Structure and Scaling in Physics*, Cambridge University Press, Cambridge, 1997.

[6] Nottale L., *Fractal space-time and microphysics: towards a theory of scale relativity*, World Scientific, Singapore, 1993.

[7] Nottale L., *Scale relativity and fractal space-time – a new approach to unifying relativity and quantum mechanics*, Imperial College Press, London, 2011.

[8] Agop, M., Dimitriu, D., Vrăjitoriu, L., Boicu, M., *Order to chaos transition in plasma via non-differentiability: Experimental and Theoretical Investigations*, Journal of the Physical Society of Japan 83, 054501, 2014.

[9] Agop M., Forna N., Casian-Botez I., Bejinariu C., *New theoretical approach of the physical processes in nanostructures*. Journal of Computational and Theoretical Nanoscience, 5 (4), 483-489, 2008.

[10] Agop M., Murguleț C., *El Naschie's epsilon (infinity) space-time and scale relativity theory in the topological dimension D=4*, Chaos Solitons & Fractals. 32 (3), 1231-1240, 2007.

[11] Agop M., Nica P., Gîrțu M., *On the vacuum status in Weyl-Dirac theory*, General Relativity and Gravitation, 40 (1), 35-55, 2008.

[12] Agop M., Păun V., Harabagiu A., *El Naschie's epsilon (infinity) theory and effects of nanoparticle clustering on the heat transport in nanofluids*, Chaos, Solitons and Fractals, 37 (5), 1269-1278, 2008.

[13] Casian-Botez I., Agop M., Nica P., Păun V., Munceleanu G.V., *Conductive and convective types behaviors at nano-time scales*, Journal of Computational and Theoretical Nanoscience, 7, 2271-2280, 2010.

[14] Ciubotariu C., Agop M., *Absence of a gravitational analog to the Meissner effect.* General Relativity and Gravitation, 28 (4), 405-412, 1996.

[15] Colotin M., Pompilian G. O., Nica, P., Gurlui S., Păun V., Agop M., *Fractal transport phenomena through the scale relativity model*, Acta Physica Polonica A, 116 (2), 157-164, 2009.

[16] Gottlieb I., Agop M., Jarcău M., *El Naschie's Cantorian space-time and general relativity by means of Barbilian's group. A Cantorian fractal axiomatic model of space-time*, Chaos, Solitons and Fractals, 19 (4), 705-730, 2004.

[17] Gurlui S., Agop M., Nica P., Ziskind M., Focşa C., *Experimental and theoretical investigations of transitory phenomena in high-fluence laser ablation plasma*, Phys. Rev. E 78, 026405, 2008.

[18] Gurlui S., Agop M., Strat M., Băcăiţă S., *Some experimental and theoretical results on the anodic patterns in plasma discharge*, Physics of Plasmas, 13 (6), 063503, 2006.

[19] Nedeff V., Bejenariu C., Lazăr G., Agop M., *Generalized lift force for complex fluid*, Powder Technology, 235, 685-695, 2013.

[20] Nedeff V., Moşneguţu E., Panainte M., Ristea M., Lazăr G., Scurtu D., Ciobanu B., Timofte A., Toma S., Agop M., *Dynamics in the boundary layer of a flat particle*, Powder Technology, 221, 312-317, 2012.

[21] Nica P., Agop M., Gurlui S., Bejinariu C., Focşa C., *Characterization of aluminum laser produced plasma by target current measurements*, Japanese Journal of Applied Physics, 51, DOI: 10.1143/JJAP.51.106102, 2012.

[22] Nica P., Vizureanu P., Agop M., Gurlui S., Focşa C., Forna N., Ioannou P. D., Borsos Z., *Experimental and theoretical aspects of aluminum expanding laser plasma*, Japanese Journal of Applied Physics, 48 (6), DOI: 10.1143/JJAP.48.066001, 2009.

[23] Berners-Lee T. and Kagal L., *The fractal nature of the semantic web*, AI Magazine, 29(8), 2008, 29--34.

[24] Gardiner J., et al., *The fractal nature of the brain*, NeuroQuantology, 8(2), 2010, 137-141.

[25] Agop M., Mazilu N., *Skyrmions: A Great Finishing Touch to Classical Newtonian Philosophy*, Nova Publishers, New York, 2012.

[26] Casanova G., *L'Algebre Vectorielle*, Mir, Moscou, 1979.

[27] Arnold V.I., *Metode Matematice ale Mecanicii Clasice*, Ed. Tehnică, Bucureşti, 1980.

[28] Hamermesh M., *Group Theory and its Applications to Physical Problem*, Dover Publication, Inc. New York, 1989.

[29] Alipour M., Mohajeri A., *Onicescu information energy in terms of Shannon entropy and Fisher information densities*, Molecular Physics: An International Journal at the Interface Between Chemistry and Physics, Vol. 110, Issue 7, 403-405, 2012.

[30] Berners-Lee T., Kagal L., *The Fractal Nature of the Semantic Web*, AI Magazine, 29 (8), 29-34, 2008.

[31] Jaeger G., *Fractal states in quantum information processing*, Physics Letters A, Vol. 358, Issues 56, 373-376, 2006.

[32] Agop M., Buzea C., Buzea C.Gh., Chirilă L., Oancea S., *On the information and uncertainty relation of canonical quantum systems with SL(2R) invariance*, Chaos, Solitons & Fractals, Vol. 7, Issue 5, 659-668, 1996.

[33] Agop M., Griga V., Ciobanu B., Buzea C., Stan C., Tatomir D., *The uncertainty relation for an assembly of Planck-type oscillators. A possible GR-quantum mechanics connection*, Chaos, Solitons & Fractals, Vol. 8, Issue 5, 809-821, 1997.

[34] Agop M., Melnig V., *L'énergie informationelle et les relations d'incertitude pour les systèmes canoniques SL(2R) invariants*, Entropie, no. 188/189, 119-123, 1995.

[35] Stoler D., *Equivalence Classes of Minimum Uncertainty Packets*, Phys. Rev. D, 3217-3219, 1970.

[36] Onicescu O., *Energie informationnelle*, C. R. Acad. Sci. Paris A 263 (1966), 841-842.

[37] Broglie L. de, *La Thermodynamique de la particule isolée*, Gauthier-Villars, Paris, 1964.

[38] Mihăileanu N., *Complements of Geometry: Analytical, Projective and Differential*, Ed. Didactică și Pedagogică, București, 1972.

[39] Barbilian D., *Riemannsche Raum Cubischer Binärformen*, Comptes Rendus de l'Academie Roumaine des Sciences, Vol. 12, p. 345.

[40] Stoka M.I., *Géométrie Intégrale*, Memorial des Sciences Mathématiques, Gauthier-Villars, Paris.

[41] Flores-Gallegos Nelson, *Shannon informational entropies and chemical reactivity*, in Advances in Quantum Mechanics, Edited by Paul Bracken, Intech 2013, Rijcka, Croatia, 683-706.

[42] Jeans J.H., *The Dynamical Theory of Chaos*, Dover Publication, New York, 1954.

[43] Misner C.W., *Harmonic Maps as Models for Physical Theories*, Physical Review D 18, 4510-4524.

[44] Ernst F.J., *New Formulations of the Axially Symmetric Gravitational Field Problem I*, Phys. Rev., 167, 1175-1178, *New Formulation of the Axially Symmetric Grvavitational Field Problem II*, Phys. Rev., 168, 1415-1417.

[45] Ernst F.J., *Exterior Algebraic Derivation of Einstein Field Equations Employing a Generalized Basis*, Journal of Mathematical Physics, 12, 2395-2397, 1971.

[46] Misner C.W., Thorne K.S., Wheeler J.A., *Gravitation*, W.H. Freeman and Company, San Francisco, 1973.

[47] Helfrich W., *Elastic properties of lipid bilayers-theory and possible experiments*, Z. Naturforsch. C 28, 693-703, 1973.

[48] Lipowsky, R. and Sackmann, E., *Structure and Dynamics of Membranes*, Handbook of Biological Physics. R. Lipowsky and E. Sackmann, editors. Elsevier, North-Holland, 1995.

[49] Kraus, M., Wintz, W., Seifert, U. and Lipowsky, R., *Fluid vesicles in shear flow*, Phys. Rev. Lett. 77, 3685-3688, 1996.

[50] Prost, J. and Bruinsma, R., *Shape fluctuations of active membranes*, Europhys. Lett. 33(4), 321-326, 1996.

[51] Cantat, I. and Misbah, C., *Dynamics and similarity laws for adhering vesicles in haptotaxis*, Phys. Rev. Lett. 83, 235-238, 1999.

[52] Sukumaran, S. and Seifert, U., *Influence of shear flow on vesicles near a wall: A numerical study*, Phys. Rev. E 64, 011916, 2001.

[53] Durand, I., Jönson, P., Misbah, C., Valance, A. and Kassner, K., *Adhesion-induced vesicle propulsion*, Phys. Rev. E 56, 3776, 1997.

[54] Cantat, I. and Misbah, C., *Lift Force and Dynamical Unbinding of Adhering Vesicles under Shear Flow*, Phys. Rev. Lett. 83, 880-883, 1999.

[55] Kern, N. and Fourcade, B., *Vesicles in linearly forced motion*, Europhys. Lett. 46(2), 262-267, 1999.

[56] Biben, T. and Misbah, C., *An advected-field method for deformable entities under flow*, Euro. Phys. J. B 29, 311-316, 2002.

[57] Nardi, J., Bruinsma, R. and Sackmann, E., *Vesicles as osmotic motors*, Phys. Rev. Lett. 82, 5168-5171, 1999.

[58] Abkarian, M., Lartigue, C. and Viallat, A., *Motion of phospholipidic vesicles along an inclined plane. Sliding and rolling*, Phys. Rev. E 63, 041906, 2001.

[59] Lortz, B., Simon, R., Nardi J. and Sackmann, E., *Weakly adhering vesicles in shear flow*, Tanktreading and anomalous lift force, Europhys. Lett. 51, 468, 2000.

[60] Abkarian, M., Lartigue C. and Viallat, A., *Tank treading and unbonding of deformable vesicles in shear flow. Determination of the lift force*, Phys. Rev. Lett. 88, 068103, 2002.

[61] Cantat, I., Kassner K. and Misbah, C., *Vesicles in haptotaxis with hydrodynamical dissipation*, Eur. Phys. J. E 10, 175-189, 2003.

[62] Agop, M., Gavriluţ, A., Crumpei, G., Doroftei, B., Informational Non-Differentiable Entropy and Uncertainty Relations in Complex Systems, Entropy, 2014, 16(11), 6042-6058.

[63] Agop, M., The uncertainty relation for an assembly of Planck-type oscillators. A possible GR-quantum mechanics connection, Chaos, Solitons and Fractals, 1997, Vol. 8, No. 5, 809-821.

[64] ***http://perso.univ-rennes1.fr

[65] Wilhelm, H., Hydrodynamic Model of Quantum Mechanics, Physical Review D, vol. 1, no. 8, 2278-2285.

[66] *** http://www.wisegeek.com

[67] Gottlieb, I., *El Naschie's Cantorian frames, gravitation and quantum mechanics*, Chaos, Solitons and Fractals, 2005, vol. 24, Issue 2, 391-405.

[68] Agop, M., Ciobanu, G., Zaharia, L., *Cantorian* $\varepsilon^{(\infty)}$ *space-time, frames and unitary theories*, Chaos, Solitons and Fractals, 2003, vol. 15, no. 3, 445-453.

[69] Agop, M., Abacioaie, D., El Naschie's $\varepsilon(\infty)$ space-time, interface between Weyl-Dirac bubbles and Cantorian fractal superstring, Chaos, Solitons and Fractals, 2007, vol. 34, Issue 2, 235-243.

[70] ***http://pirsquared.org

[71] Agop, M., Gavriluţ, A., Rezuş, E., Implications of Onicescu's informational energy in some fundamental physical models, International Journal of Modern Physiscs B, DOI: 10.1142/S0217979215500459, 2015.

[72] *** http://www.canaryjournal.org

Chapter 12

Physical Vacuum is a Special Superfluid Medium

V.I. Sbitnev

Additional information is available at the end of the chapter

http://dx.doi.org/10.5772/59040

1. Introduction

A dramatic situation in physical understanding of the nature emerged in the late of 19th century. Observed phenomena on micro scales came into contradiction with the general positions of classical physics. It was a time of the origination of new physical ideas explaining these phenomena. Actually, in a very short period, postulates of the new science, quantum mechanics were formulated. The Copenhagen interpretation was first who proposed an ontological basis of quantum mechanics [1]. These positions can be stated in the following points: (a) the description of nature is essentially probabilistic; (b) a quantum system is completely described by a wave function; (c) the system manifests wave–particle duality; (d) it is not possible to measure all variables of the system at the same time; (e) each measurement of the quantum system entails the collapse of the wave function.

Can one imagine a passage of a quantum particle (the heavy fullerene molecule [2], for example) through all slits, in once, at the interference experiment? Following the Copenhagen interpretation, the particle does not exists until it is registered. Instead, the wave function represents it existence within an experimental scene [3].

Another interpretation was proposed by Louis de Broglie [4], which permits to explain such an experiment. In de Broglie's wave mechanics and the double solution theory there are two waves. There is the wave function that is a mathematical construct. It does not physically exist and is used to determine the probabilistic results of experiments. There is also a physical wave guiding the particle from its creation to detection. As the particle moves from a source to a detector, the particle perturbs the wave field and gets a reverse effect from it. As a result, the physical wave guides the particle along some optimal trajectory up to its detection.

A question arises, what is the de Broglie physical wave? Recently, Couder and Fort [5] has executed the experiment with the classical oil droplets bouncing on the oil surface. A remarkable observation is that an ensemble of the droplets passing through the barrier having two

gates shows the interference fringes typical for the two slit experiment. Their explanation is that the droplet while moving on the surface induces on this surface the weak Faraday waves. The latter provide the guidance conditions for the droplets. In this perspective, we can draw conclusion that the de Broglie physical wave can be represented by perturbations of the ether when the particle moves through it. In order to describe behavior of such an unusual medium we shall use the Navier-Stokes equation with slightly modified some terms. As the final result we shall get the Schrödinger equation.

In physical science of the New time the assumption for the existence of the ether medium was originally used to explain propagation of light and the long-range interactions. As for the propagation of light, the wave ideas of Huygens and Fresnel require the existence of a continuous intermediate environment between a source and a receiver of the light - the light-bearing ether. It is instructive to compare here the two opposite doctrines about the nature of light belonging to Sir Isaac Newton and Christian Huygens. Newton maintained the theory that the light was made up of tiny particles, corpuscles. They spread through an empty space in accordance with the law of the classical mechanics. Christian Huygens (a contemporary of Newton), believed that the light was made up of waves vibrating up and down perpendicularly to the direction of its propagation, as waves on a water surface. One can imagine all space populated everywhere densely by Huygens's vibrators. All vibrators are silent until a wave reaches them. As soon as a wave front reaches them, the vibrators begin to radiate waves on the frequency of the incident wave. So, the infinitesimal volume δV is populated by infinite amount of the vibrators with frequencies of visible light. These vibrators populate the ether facilitating propagation of the light waves through the space.

In order to come to idea about existence of the intermediate medium (ether) that penetrates overall material world, we begin from the fundamental laws of classical physics. Three Newton's laws first published in Mathematical Principles of Natural Philosophy in 1687 [6] we recognize as basic laws of physics. Namely: (a) the first law postulates existence of inertial reference frames: an object that is at rest will stay at rest unless an external force acts upon it; an object that is in motion will not change its velocity unless an external force acts upon it. The inertia is a property of the bodies to resist to changing their velocity; (b) the second law states: the net force applied to a body with a mass M is equal to the rate of change of its linear momentum in an inertial reference frame

$$\vec{F} = M\vec{a} = M\frac{d\vec{v}}{dt}; \tag{1}$$

(c) the third law states: for every action there is an equal and opposite reaction.

Leonard Euler had generalized the second Newton's laws on motion of deformable bodies [7]. We rewrite this law for such media. Let the deformable body be in volume ΔV and has the mass M. We divide Eq. (1) by ΔV and determine the time-dependent mass density $\rho_M = M\Delta V$ [8]. In this case, we shall understand $\vec{F}$ as force per volume. Then the second law takes a form:

$$\vec{F} = \frac{d\rho_M \vec{v}}{dt} = \rho_M \frac{d\vec{v}}{dt} + \vec{v}\frac{d\rho_M}{dt}. \tag{2}$$

The total derivatives in the right side can be written through partial derivatives:

$$\frac{d\rho_M}{dt} = \frac{\partial \rho_M}{\partial t} + (\vec{v}\nabla)\rho_M. \tag{3}$$

$$\frac{d\vec{v}}{dt} = \frac{\partial \vec{v}}{\partial t} + (\vec{v}\nabla)\vec{v}. \tag{4}$$

Eq. (3) equated to zero is seen to be the continuity equation. As for Eq. (4) we may rewrite the rightmost term in detail

$$(\vec{v}\nabla)\vec{v} = \nabla\frac{v^2}{2} - [\vec{v}\times[\nabla\times\vec{v}]]. \tag{5}$$

As follows from this formula, the first term, multiplied by the mass, is gradient of the kinetic energy. It represents a force applied to the fluid element for its shifting on the unit of length, δS. The second term gives acceleration of the fluid element directed perpendicularly to the velocity $\vec{v}$. Let the fluid element move along some curve in 3D space. Tangent to the curve in each point points to orientation of the body motion. In turn, the vector $\vec{\omega}=[\nabla\times\vec{v}]$ is perpendicular to the plane, within which an arbitrarily small segment of the curve is situated. This vector characterizes a quantitative measure of the vortex motion. It is called *vorticity*. The vector product $[\vec{v}\times\vec{\omega}]$ is perpendicular to the both vectors $\vec{v}$ and $\vec{\omega}$. It shows the acceleration of the fluid element

The term (5) entering in the Navier-Stokes equation [9, 10] is responsible for emergence of vortex structures. The Navier-Stokes equation stems from Eq. (2) if we omit the rightmost term, representing the continuity equation, and specify forces in this equation in detail:

$$\rho_M\left(\frac{\partial \vec{v}}{\partial t} + (\vec{v}\nabla)\vec{v}\right) = \frac{\vec{F}}{\Delta V} + \mu(t)\nabla^2\vec{v} - \rho_M\nabla\left(\frac{P}{\rho_M}\right) \tag{6}$$

This equation contains two modifications represented in the two last terms from the right side: the dynamic viscosity μ depends on time and the rightmost term has a slightly modified view, namely $\rho_M\nabla(P/\rho_M)=\nabla P - P\nabla\ln(\rho_M)$. This modification will be important for us when we shall begin to derive the Schrödinger equation. However first we shall examine the Helmholtz vortices with time dependent dynamical viscosity.

2. Vortex dynamics

The second term from the right in Eq. (6) represents the viscosity of the fluid (μ is the dynamic viscosity, its units are N·s/m^2=kg/(m·s)). Let us suppose that the fluid is ideal, barotropic, and the mass forces are conservative [10]. At assuming that the external force is conservative, we apply to this equation the operator curl. We get right away the equation for the vorticity:

$$\frac{\partial \vec{\omega}}{\partial t} + (\vec{\omega} \cdot \nabla)\vec{v} = \nu(t)\nabla^2 \vec{\omega}. \tag{7}$$

Here $\nu(t) = \mu(t) / \rho_M$ is the kinematic viscosity. Its dimensionality is m^2/s what corresponds to dimensionality of the diffusion coefficient. The rightmost term describes dissipation of the energy stored in the vortex. As a result, the vortex with the lapse of time will disappear.

With omitted the term from the right (i.e., $\nu = 0$) the Helmholtz theorem reads: (i) if fluid particles form, in any moment of the time, a vortex line, then the same particles support the vortex line both in the past and in the future; (ii) ensemble of the vortex lines traced through a closed contour forms a vortex tube. Intensity of the vortex tube is constant along its length and does not change in time. The vortex tube (a) either goes to infinity by both endings; (b) or these endings lean on walls of bath containing the fluid; (c) or these endings are locked to each on other forming a vortex ring.

Assuming that the fluid is a physical vacuum, which meets the requirements specified earlier, we must say that the viscosity vanishes. In that case, the vorticity $\vec{\omega}$ is concentrated in the center of the vortex, i.e., in the point. Mathematical representation of the vorticity is δ- function. Such singularity can be a source of possible divergences of computations in further.

We shall not remove the viscosity. Instead of that, we hypothesize that even if there is an arbitrary small viscosity, because of the zero-point oscillations in the vacuum, the vortex does not disappear completely. The vortex can be a long-lived object. The foundation for that hypothesis is the observation (performed by French scientific team [5, 11, 12]) of behavior of the droplets moving on the oil surface, on which the waves Faraday exist. Here an important moment is that the Faraday waves are supported slightly below the super-critical threshold. Due to this trick the droplets can live on the oil surface arbitrary long, before they disappear in the oil. The Faraday waves that are supported near the super-critical threshold may play a role analogous to the zero-point oscillations of the vacuum.

Observe that the bouncing droplet simulates some aspects of quantum mechanics, stimulating theoretical investigations in this area [13-20]. It is interesting to note in this place that Grössing considers a quantum particle as a dissipative phase-locked steady state, where an amount of zero-point energy of the wave-like environment is absorbed by the particle, and then during a characteristic relaxation time is dissipated into the environment again [14].

Here we shall give a simple model of such a picture. Let us look on the vortex tube in its cross-section which is oriented along the axis z and its center is placed in the coordinate origin of the plane (x, y). Eq. (7), written down in the cross-section of the vortex, is as follows

$$\frac{\partial \omega}{\partial t} = \nu g(t)\left(\frac{\partial^2 \omega}{\partial r^2} + \frac{1}{r}\frac{\partial \omega}{\partial r}\right). \tag{8}$$

We do not write a sign of vector on top of ω since ω is oriented strictly along the axis z. We introduce time-dependent the kinematic viscosity. For the sake of simplicity, let it be looked as

$$g(t) = \cos(\Omega t + \phi) = \frac{e^{\,i(\Omega t+\phi)} + e^{-i(\Omega t+\phi)}}{2} \tag{9}$$

where Ω is an oscillation frequency and ϕ is the uncertain phase.

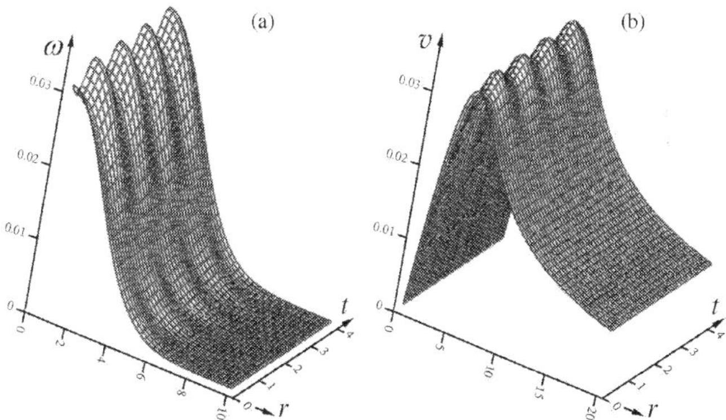

Figure 1. Vorticity $\omega(r,\ t)$ and velocity $v(r,t)$ as functions of r and t for $\Gamma = 1,\ \nu = 1,\ \Omega = \pi,$ and n=16. These parameters are conditional in order to show clearly oscillations of the vortex in time. The solution does not decay with time.

Solution of the equation (8) in this case is as follows

$$\omega(r,t) = \frac{\Gamma}{4\pi(\nu/\Omega)(\sin(\Omega t + \phi) + n)}\exp\left\{-\frac{r^2}{4(\nu/\Omega)(\sin(\Omega t + \phi) + n)}\right\}. \tag{10}$$

Here Γ is the integration constant having dimension m^2/s. An extra number $n > 1$. It prevents appearance of singularity in the cases when $\sin(\Omega t + \phi)$ tends to-1. This function at choosing the parameters $\Gamma = 1,\ \nu = 1,\ \Omega = \pi,$ and n=16 is shown in Fig. 1(a).

The velocity of the fluid matter around the vortex results from the integration of the vorticity function

$$v(r,t)=\frac{1}{r}\int_0^r \omega(r',t)r'dr'=\frac{\Gamma}{2\pi r}\left(1-\exp\left\{-\frac{r^2}{4(\nu/\Omega)(\sin(\Omega t+\phi)+n)}\right\}\right). \tag{11}$$

Fig. 1(b) shows behavior of this function at the same input parameters.

In particular, for n=0 and $\Omega t << 1$ this solution is close to the Lamb–Oseen vortex solution [21]

$$\omega(r,t)=\frac{\Gamma}{4\pi\nu t}e^{-r^2/4\nu t}, \tag{12}$$

$$v(r,t)=\frac{\Gamma}{2\pi r}\left(1-e^{-r^2/4\nu t}\right). \tag{13}$$

As seen from here, the Lamb–Oseen solution decays with time since the viscosity $\nu > 0$.

One can see from the solutions (10) and (11), depending on the distance to the center the functions $\omega(r, t)$ and $v(r,t)$ show typical behavior for the vortices. The both functions do not decay with time, however. Instead of that, they demonstrate pulsations on the frequency Ω. Amplitude of the pulsations is the smaller, the larger value of the parameter n. At n tending to infinity the amplitude of the pulsations tends to zero. At the same time the vortex disappears entirely.

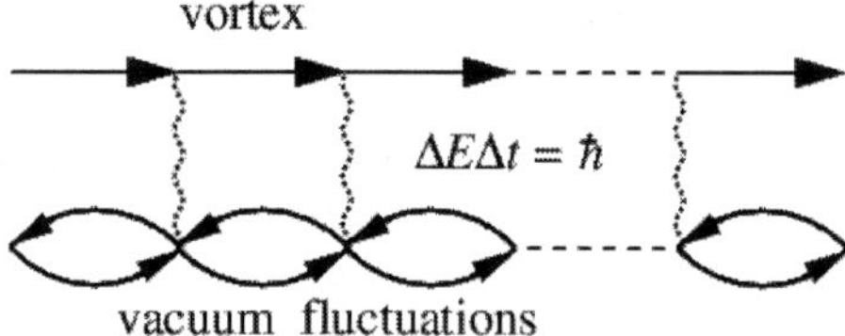

Figure 2. Periodic energy exchange between the vortex and vacuum fluctuations

The undamped solution was obtained thanks to assumption, that the kinematic viscosity is a periodic function of time, namely, $\nu g(t)=\nu\cos(\Omega t+\phi)$. The viscosity in the quantum realm is not a good concept, however. Most likely, it manifests itself through interaction of the quantum object with vacuum fluctuations. According to Eq. (9), there are half-periods when the energy of the vortex is lost at scattering on the vacuum fluctuations, and there are other half-periods when the vacuum returns this energy to the vortex, Fig. 2. On the whole, the viscosity of the fluid medium, within which the vortex tube evolves, in the average remains at zero. It can mean that this medium is superfluid. Such a scenario is not unusual. For example, at transition of helium to the superfluid phase [22] coherent Cooper pairs of electrons arise through the exchange by phonons. This attraction is due to the electron–phonon interaction. The phonons are thermal excitations of a lattice. In that case, they play a role of the background medium.

Qualitative view of the vortex tube in its cross-section is shown in Fig. 3. Values of the velocity v are shown by grey color ranging from light grey (minimal velocities) to dark grey (maximal ones). A visual image of this picture can be a hurricane (tropical cyclone [23]) shown from the top. In the center of the vortex, a so-called eye of the hurricane (the vortex core) is well viewed. Here it looks as a small light grey disk, where the velocities have small values. In the very center of the disk, in particular, the velocity vanishes. Observe that in the region of the hurricane eye a wind is really very weak, especially near the center. This is in stark contrast to conditions in the region of the eyewall, where the strongest winds exist (in Fig. 3 it looks as a dark grey annular region enclosing the light grey inner area). The eyewall of the vortex tube (a zone where the velocity reaches maximal values) has the nonzero radius.

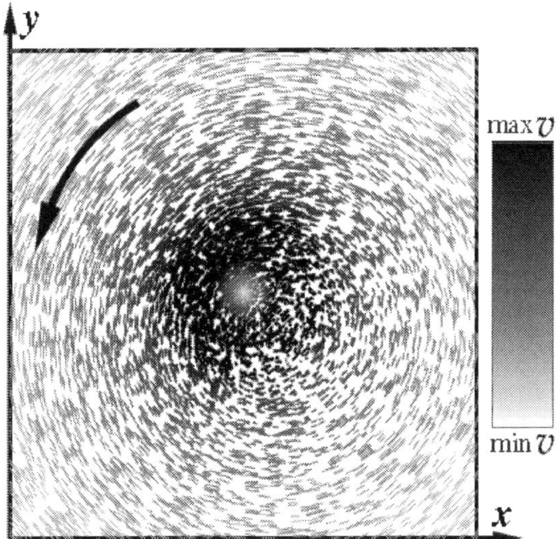

Figure 3. Cross-section of the vortex tube in the plane (x, y). Values of the velocity v are shown in grey ranging from light grey (min v) to dark grey (max v). Density of the pixels represents magnitude of the vorticity ω. Core of the vortex is well visible in the center.

Let us find the radius of the vortex core. In order to evaluate this radius we equate to zero the first derivative by r of equation (11)

$$\frac{\exp\left\{-\frac{r^2}{4(\nu/\Omega)(\sin(\Omega t+\phi)+n)}\right\}\left(2\frac{r^2}{4(\nu/\Omega)(\sin(\Omega t+\phi)+n)}+1\right)-1}{2\pi r^2}=0. \tag{14}$$

The radius is a root of this equation

$$r_v = 2\sqrt{a_0(n + \sin(\Omega t + \phi))} \cdot \sqrt{\frac{\nu}{\Omega}}. \tag{15}$$

Here $a_0 \approx 1.2564312$ is a root of the equation $\ln(2a_0 + 1) - a_0 = 0$. One can see, that r_v is an oscillating function. The larger Ω, the more quickly the vortex trembles. As Ω increases, the vortex radius decreases. However it grows with increasing the number n. Let us evaluate the radius r_v at choosing the viscosity ν equal to $\bar{\nu} = \hbar / 2m$. Here $\bar{\nu}$ is the diffusion coefficient of the Brownian sub-quantum particles wandering in the Nelson's aether [24], see Appendix A. In the case of electron $\bar{\nu} = \hbar / 2m \approx 5.79 \bullet 10^{-5}$ m²/s. As for Ω, let it be equal to $2mc^2/\hbar$, or approximately $1.6 \bullet 10^{21}$ radians per second for electron. Here c is the speed of light and m is the electron mass. Then we have $(\nu/\Omega)^{1/2} \approx 1.93 \bullet 10^{-13}$ m. This length is seen to be smaller then the Compton wavelength, $\lambda_C = 2.426 \bullet 10^{-12}$ m, in about 12 times. So, for choosing $n \approx 31$ we find from Eq. (15) that the radius of the vortex is about the Compton wavelength. From the above one can see that, on a distance about the Compton wavelength, virtual particles can be involved into a vorticity dancing around the electron core, by polarizing the electron charge. This dancing happens at trembling motion of the electron with the frequency $\Omega = 2mc^2/\hbar$. That oscillating motion has a deep relation to the so-called "Zitterbewengung" [25].

One can give a general solution of Eq. (8) which has the following presentation

$$\omega(r,t) = \frac{\Gamma}{4\pi\left(\int_0^t \nu(\tau)d\tau + \sigma^2\right)} \exp\left\{-\frac{r^2}{4\left(\int_0^t \nu(\tau)d\tau + \sigma^2\right)}\right\}, \tag{16}$$

$$v(r,t) = \frac{\Gamma}{2\pi r}\left(1 - \exp\left\{-\frac{r^2}{4\left(\int_0^t \nu(\tau)d\tau + \sigma^2\right)}\right\}\right). \tag{17}$$

The viscosity function $\nu(t)$ is a quasi-periodic function or even is represented by a color noise. The integral of the viscosity function memorizes integrally character of the viscosity of the medium. Due to this memory effect, the vortex may live a long enough. As for interpretation of these solutions with the quantum-mechanical point of view, we may say that there exists a regular exchange by quanta with the vacuum fluctuations, Fig. 2. The integral accumulates all cases of the exchange with the vacuum. The constant σ having dimension of length, prevents appearance of singularities. One can see that even at $\nu \equiv 0$, but $\sigma > 0$, abundance of long-lived vortices can exist in the vacuum. Such vortices are "ghosts" in the superfluid being invisible without interaction.

2.1. Vortex rings and vortex balls

If we roll up the vortex tube in a ring and glue together its opposite ends we obtain a vortex ring. A result of such an operation put into the (x, y) plane is shown in Fig 4. Position of points on the helicoidal vortex ring in the Cartesian coordinate system is given by

$$\begin{cases} x = (r_1 + r_0 \cos(\omega_2 t + \phi_2))\cos(\omega_1 t + \phi_1), \\ y = (r_1 + r_0 \cos(\omega_2 t + \phi_2))\sin(\omega_1 t + \phi_1), \\ z = r_0 \sin(\omega_2 t + \phi_2). \end{cases} \tag{18}$$

Here r_0 is the radius of the tube. And r_1 represents the distance from the center of the tube (pointed in the figure by arrow c) to the center of the torus located in the origin of coordinates (x,y,z). A body of the tube, for the sake of visualization is colored in cyan. Eq. (18) parametrized by t gives a helicoidal vortex ring shown in this figure. Parameters ω_1 and ω_2 are frequencies of rotation along the arrow a about the center of the torus (about the axis z) and rotation along the arrow b about the center of the tube (about the axis pointed by arrow c), respectively. Phases ϕ_1 and ϕ_2 have uncertain quantities ranging from 0 to 2π. By choosing the phases within this interval with a small increment, we may fill the torus by the helicoidal vortices everywhere densely. The vorticity is maximal along the center of the tube. Whereas the velocity of rotation about this center in the vicinity of it is minimal. However, the velocity grows as a distance from the center increases. After reaching of some maximal value the velocity further begins to decrease.

Let the radius r_1 in Eq. (18) tends to zero. The helicoidal vortex ring in this case will transform into a vortex ring enveloping a spherical ball. The vortex ring for the case r_0=4, $r_1 \approx 0$, ω_2=$3\omega_1$, and ϕ_2=ϕ_1=0 drawn by thick curve colored in deep green is shown in Fig. 5(a). Motion of an elementary vortex clot along the vortex ring (along the thick curve colored in deep green) takes place with a velocity

$$\vec{v}_R = \begin{pmatrix} v_{R,x} = -r_0\omega_2 \sin(\omega_2 t + \phi_2)\cos(\omega_1 t + \phi_1) - r_0\omega_1 \cos(\omega_2 t + \phi_2)\sin(\omega_1 t + \phi_1) \\ \quad - r_1\omega_1 \sin(\omega_1 t + \phi_1) \\ v_{R,y} = -r_0\omega_2 \sin(\omega_2 t + \phi_2)\sin(\omega_1 t + \phi_1) + r_0\omega_1 \cos(\omega_2 t + \phi_2)\cos(\omega_1 t + \phi_1) \\ \quad + r_1\omega_1 \sin(\omega_1 t + \phi_1) \\ v_{R,x} = \; r_0\omega_2 \cos(\omega_2 t + \phi_2) \end{pmatrix} \tag{19}$$

The velocity of the clot at the initial time is $v_{R,x}$=0, $v_{R,y}$=$r_0\omega_1$, $v_{R,z}$=$r_0\omega_2$=$3r_0\omega_1$ (the initial point (x, y, z)=(4, 0, 0) is on the top of the ball). We designate this velocity as $\vec{v}_+$. Through the time t=$\pi\omega_1$ the elementary vortex clot returns to the top position. The velocity in this case is equal

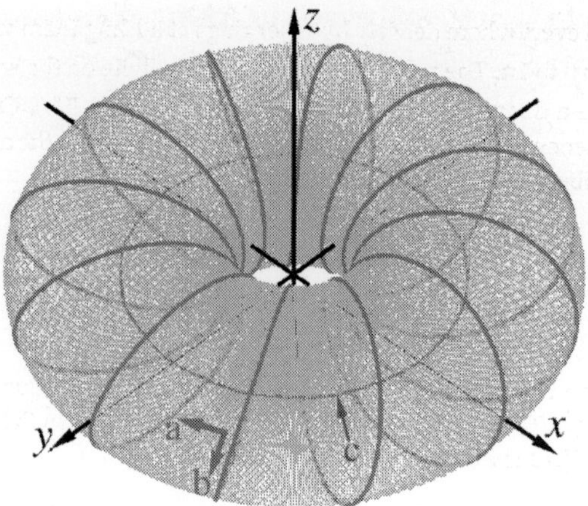

Figure 4. Helicoidal vortex ring: r_0=2, r_1=3, $\omega_2=12\omega_1$, $\phi_2=\phi_1=0$.

to $v_{R,x}=0$, $v_{R,y}=r_0\omega_1$, $v_{R,z}=-r_0\omega_2=-3r_0\omega_1$. We designate this velocity as $\vec{v}_-$. Sum of the two opposite velocities, $\vec{v}_+$ and $\vec{v}_-$, gives the velocity $\vec{v}_0=(0,\ 2r_0\omega_1,\ 0)$. During $t=(1+3k)\pi/3\omega_1$ and $t=(2+3k)\pi/3\omega_1$ (k=1, 2, ⋯) the clot travels through the positions 1 and 2 both in the forward and in backward directions, respectively. In the vicinity of these points the velocities $\vec{v}_+$ and $\vec{v}_-$ yield the resulting velocity $\vec{v}_0$ directed along the circle lying in the plane (x, y).

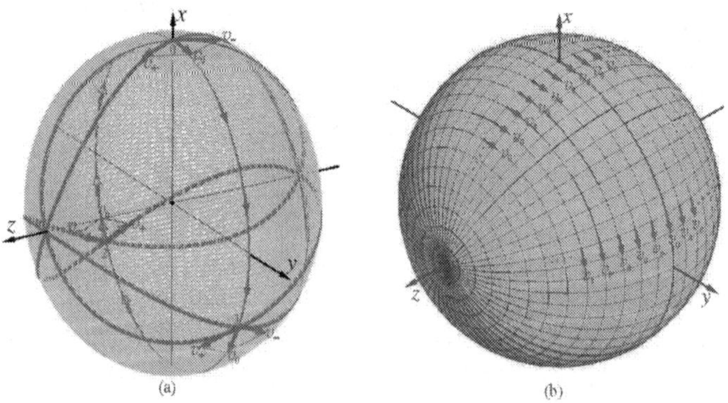

Figure 5. (a) Helicoidal vortex ring (colored in deep green) convoluted into the vortex ball. The input parameters of the ball are as follows r_0=4, r_1=0.01 $\ll$ 1, $\omega_2=3\omega_1$, $\phi_2=\phi_1=0$. The radius r_0 represents a mean radius of the ball, where the velocity v_0 reaches a maximal value. The ratio $\omega_2/\omega_1=3$ was chosen with the aim not to overload the picture by superfluous curves. (b) The vortex ball rotating about axis z with the maximal velocity v_0 that is reached on the surface of the ball.

The ball can be filled everywhere densely by other rings at adding them with other phases ϕ_1 and ϕ_2 ranging from 0 to 2π. The velocity $\vec{v}_0$ for any ring will lie on the same circles centered on the axis z. We see a dense ball that rolls along the axis y, Fig. 5(b). Observe that the ball pulsates on the frequency Ω as it rolls along its path, as it follows from the above computations. Perfect modes describing the rolling ball are spherical harmonics [26].

3.3. Derivation of the Schrödinger equation

The third term in the right side of Eq. (6) deals with the pressure gradient. One can see, however, it is slightly differ from the pressure gradient presented in the customary Navier-Stokes equation [9, 10]. One can rewrite this term in detail

$$\rho_M \nabla \left(\frac{P}{\rho_M} \right) = \nabla P - P \nabla \ln(\rho_M). \tag{20}$$

The first term, ∇P, is the customary pressure gradient represented in the Navier-Stokes equation. Whereas, the second term, $P \nabla \ln(\rho_M)$, is an extra term describing changing the logarithm of the density along increment of length (the entropy increment) multiplied by P. It may mean that change of the pressure is induced by change of the entropy per length, or else by change of the information flow [27, 28] per length. This term has signs typical of the osmotic pressure, mentioned by Nelson [24].

Let us consider in this respect the pressure P in more detail. We shall represent the pressure consisting of two parts, P_1 and P_2. We begin from the Fick's law [14]. The law says that the diffusion flux, J, is proportional to the negative value of the density gradient $J=-D\nabla \rho_M$. Here D is the diffusion coefficient $\bar{v}=\hbar/2m$ [24], see Nelson's definition in Appendix A. Since the term $\bar{v}\nabla J$ has dimension of the pressure, we define P_1 as the pressure having diffusion nature

$$P_1 = \bar{v}\, \nabla \vec{J} = -\frac{\hbar^2}{4m^2} \nabla^2 \rho_M. \tag{21}$$

Observe that the kinetic energy of the diffusion flux is $(m/2)(J/\rho_M)^2$. It means that there exists one more pressure as the average momentum transfer per unit area per unit time:

$$P_2 = \frac{\rho_M}{2} \left(\frac{\vec{J}}{\rho_M} \right)^2 = \frac{\hbar^2}{8m^2} \frac{(\nabla \rho_M)^2}{\rho_M}. \tag{22}$$

Now we can see that sum of the two pressures, P_1+P_2, divided by ρ_M gives a term

$$Q_M = \frac{\hbar^2}{8m^2}\left(\frac{\nabla \rho_M}{\rho_M}\right)^2 - \frac{\hbar^2}{4m^2}\frac{\nabla^2 \rho_M}{\rho_M}. \tag{23}$$

One can see that accurate to the divisor m this term represents the quantum potential.

To bring the expression (23) to a form of the quantum potential, we need to introduce instead of the mass density ρ_M the probability density ρ according to the following presentation

$$\rho_M = \frac{M}{\Delta V} = \frac{mN}{\Delta V} = m\rho. \tag{24}$$

Here the mass M is a product of an elementary mass m by the number of these masses, N, filling the volume ΔV. Then the mass density ρ_M is defined as a product of the elementary mass m by the density of quasi-particles $\rho = N / \Delta V$. We can imagine the quasi-particle as a long-lived local heterogeneity, which moves with the current velocity $\vec{v}$ and probably has the vorticity. Let us divide the Navier-Stokes equation Eq. (6) by the probability density ρ. We obtain

$$m\left(\frac{\partial \vec{v}}{\partial t} + (\vec{v}\nabla)\vec{v}\right) = \frac{\vec{F}}{N} + \nu(t)\nabla^2 m\vec{v} - \nabla Q \tag{25}$$

Here $\vec{F} / N$ is the force per one the quasi-particle. The kinetic viscosity $\nu(t) = \mu(t) / \rho_M$ is represented through the diffusion coefficient $\bar{v} = \hbar / 2m$ [24], $\nu(t) = 2\bar{v}\, g(t) = \nu\, g(t)$, $\nu = \hbar / m$, where $g(t)$ is the dimensionless time dependent function. The function Q here is the real quantum potential

$$Q = \frac{\hbar^2}{8m}\left(\frac{\nabla \rho}{\rho}\right)^2 - \frac{\hbar^2}{4m}\frac{\nabla^2 \rho}{\rho}. \tag{26}$$

Grössing noticed that the term ∇Q, the gradient of the quantum potential, describes a completely thermalized fluctuating force field [13, 14]. Here the fluctuating force is expressed via the gradient of the pressure divided by the density distribution of sub-quantum particles chaotically moving in the environment. Perhaps, they are virtual particle-antiparticle pairs.

Since the pressure provides a basis of the quantum potential, as was shown above, it would be interesting to interpret an osmotic nature of the pressure [24]. The interpretation can be the following (see Appendix A): a semipermeable membrane where the osmotic pressure manifests itself is an instant, which divides the past and the future (that is, the 3D brane of our being represents the semipermeable membrane in the 4D world). In other words, the thermalized fluctuating force field described by Grössing [13, 14] is asymmetric with respect to the time arrow.

3.1. Transition to the Schrödinger equation

The current velocity v contains two component – irrotational and solenoidal [10] that relate to vortex-free and vortex motions of the medium, respectively. The basis for the latter is the Kelvin-Stokes theorem. Scalar and vector fields underlie of manifestation of the irrotational and solenoidal velocities

$$\vec{v} = \vec{v}_S + \vec{v}_R = \frac{1}{m}\nabla S + \vec{v}_R. \tag{27}$$

Here subscripts S and R hint to scalar and vector (rotational) potentials underlying emergence of these two components of the velocity. These velocities are submitted by the following equations

$$\begin{cases} (\nabla \cdot \vec{v}_S) \neq 0, & [\nabla \times \vec{v}_S] = 0, \\ (\nabla \cdot \vec{v}_R) = 0, & [\nabla \times \vec{v}_R] = \vec{\omega}. \end{cases} \tag{28}$$

The scalar field is represented by the scalar function S – action in classical mechanics. Both velocities are perpendicular to each other. We may define the momentum and the kinetic energy

$$\begin{cases} \vec{p} = m\vec{v} = \nabla S + m\vec{v}_R, \\ m\dfrac{v^2}{2} = \dfrac{1}{2m}(\nabla S)^2 + m\dfrac{v_R^2}{2}. \end{cases} \tag{29}$$

Now we may rewrite the Navier-Stokes equation (25) in the more detailed form

$$\begin{aligned} \frac{\partial}{\partial t}(\nabla S + m\vec{v}_R) + \underbrace{\frac{1}{2m}\nabla\left((\nabla S)^2 + m^2 v_R^2\right) + \left[\vec{\omega} \times (\nabla S + m\vec{v}_R)\right]}_{(a)} = \\ -\nabla U - \nabla Q + \underbrace{\nu(t)\nabla^2(\nabla S + m\vec{v}_R)}_{(b)}. \end{aligned} \tag{30}$$

Note that the term embraced by the curly bracket (a) stems from $(\vec{v}\,\nabla)\vec{v} = \nabla v^2/2 + [\vec{\omega} \times \vec{v}]$, see Eq. (5). Here we take into account that the external force is conservative, i.e., $\vec{F}/N = -\nabla U$, where U is the potential energy relating to the single quasi-particle. The term embraced by the curly bracket (b) describes the viscosity of the medium. As was said above the viscosity coefficient in the average is equal to zero.

Let us rewrite Eq. (30) by regrouping the terms

$$\nabla\left(\frac{\partial}{\partial t}S+\frac{1}{2m}(\nabla S)^2+\frac{m}{2}v_R^2+U+Q-\nu(t)\nabla^2 S\right)=-m\frac{\partial}{\partial t}\vec{v}_R-\left[\vec{\omega}\times(\nabla S+m\vec{v}_R)\right]+\nu(t)m\nabla^2\vec{v}_R. \quad (31)$$

We assume that fluctuations of the viscosity about zero occur much more frequent, than characteristic time of displacements of the quasi-particles. For that reason, we omit the term $\nu(t)\nabla^2 S$ by supposing in the first approximation, that the medium is absolutely superfluid-there are no energy sources and sinks. By multiplying this equation from the left by $\vec{v}_S$ we find that the right part of this equation vanishes since $(\vec{v}_S\cdot\vec{v}_R)=(\vec{v}_S\cdot\vec{\omega})=0$. The left part vanishes if the expression under the brackets is constant. As a result, we come to the following modified Hamilton-Jacobi equation

$$\frac{\partial}{\partial t}S+\frac{1}{2m}(\nabla S)^2+\frac{m}{2}v_R^2+U(\vec{r})-\frac{\hbar^2}{2m}\left(\frac{\nabla^2\rho}{2\rho}-\left(\frac{\nabla\rho}{2\rho}\right)^2\right)=C. \quad (32)$$

The modification is due to adding the quantum potential (26). In this equation, C is an integration constant. We see that the third term in this equation represents energy of the vortex. On the other hand, we can see that the vortex given by Eq. (7) is replenished by the kinetic energy coming from the scalar field S, namely via the term $(\vec{\omega}\cdot\nabla)\vec{v}$. Solutions of these two equations, Eq. (7) and Eq. (32), describing dynamics of the vortex and scalar fields, depend on each other.

Both the continuity equation

$$\frac{\partial\rho}{\partial t}+(\vec{v}\cdot\nabla)\rho=0, \quad (33)$$

which stems from Eq. (3), and the quantum Hamilton-Jacobi equation (32) can be extracted from the following Schrödinger equation

$$\mathrm{i}\hbar\frac{\partial\Psi}{\partial t}=\frac{1}{2m}(-\mathrm{i}\hbar\nabla+m\vec{v}_R)^2\Psi+U(\vec{r})\Psi-C\Psi. \quad (34)$$

The kinetic momentum operator $(-\mathrm{i}\hbar\nabla+m\vec{v}_R)$ contains the term $m\vec{v}_R$ describing a contribution of the vortex motion. This term is analogous to the vector potential multiplied by the ratio of the charge to the light speed, which appears in quantum electrodynamics [29]. Appearance of this term in this equation is conditioned by the Helmholtz theorem.

By substituting into Eq. (34) the wave function Ψ represented in a polar form

$$\Psi=\sqrt{\rho}\cdot\exp\{\mathrm{i}S/\hbar\} \quad (35)$$

and separating on real and imaginary parts we come to Eqs. (32) and (33). So, the Navier-Stokes equation (6) with the slightly expanded the pressure gradient term can be reduced to the Schrödinger equation if we take into consideration also the continuity equation.

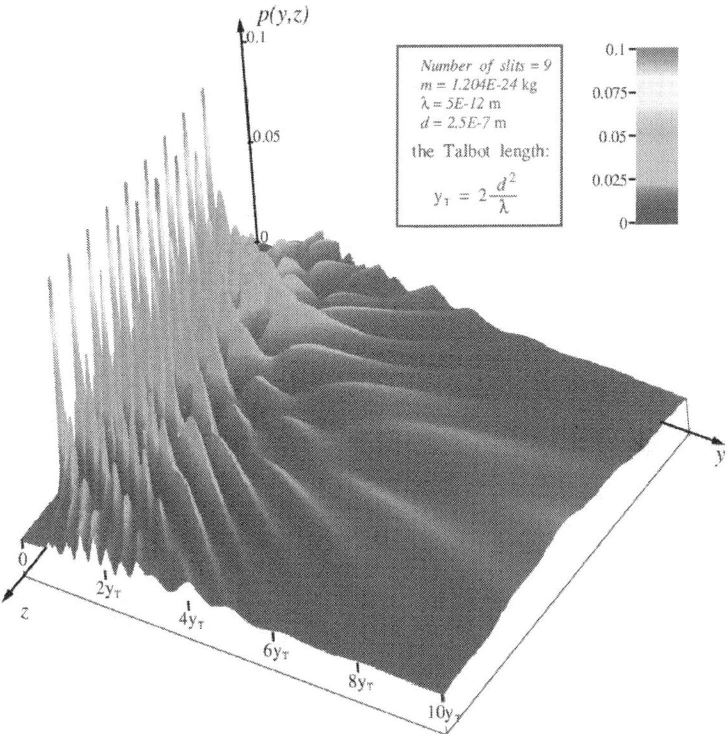

Figure 6. Probability density distribution from scattering the fullerene molecules on the grating containing 9 slits: de Broglie wavelength is 5 pm and the distance between slits is 250 nm.

There are confirmations that the Schrödinger equation is deduced from the Feynman path integral [30, 31]. Therefore, for searching solutions of the Schrödinger equation we may apply the path integral. The solution of the Schrödinger equation (34) with the potential that simulates a grating with N slits has the following view [32]

$$\left|\Psi(y,z)\right\rangle = \frac{1}{N\sqrt{1+\mathrm{i}\dfrac{\lambda y}{2\pi b^2}}}\sum_{n=0}^{N-1}\exp\left\{-\frac{\left(z-\left(n-\dfrac{N-1}{2}\right)d\right)^2}{2b^2\left(1+\mathrm{i}\dfrac{\lambda y}{2\pi b^2}\right)}\right\}. \tag{36}$$

Here λ is the de Broglie wavelength, d is the distance between slits, b is the slit width, and $\mathbf{i}=\sqrt{-1}$. In this calculation we have used $\lambda=5$ pm, $b=5\cdot10^3\lambda$, and $d=5\cdot10^4\lambda=10b=250$ nm. By choosing N=9 slits, for example, we find the interference pattern shown in Fig. 6 as the density distribution function [32]. This function is a scalar product of the wave function $|\Psi(y,z)\rangle$, namely:

$$p(y,z)=\langle\Psi(y,z)|\Psi(y,z)\rangle. \tag{37}$$

A useful unit of length at observation of the interference patterns is the Talbot length:

$$y_T=2\frac{d^2}{\lambda}. \tag{38}$$

This length bears name of Henry Fox Talbot who discovered in 1836 [33] a beautiful interference pattern, named further the Talbot carpet [34, 35].

The particles, incident on the slit grating, come from a distant coherent source. The de Broglie wavelength of the particle, $\lambda=h/p$ (h is the Planck constant, and p is the particle momentum) is a main characteristics binding the corpuscular Newtonian physics with the wave Huygens' physics. It is that we call now the wave-particle dualism. The de Broglie pilot wave being represented by the complex-valued wave function $|\Psi\rangle$ fills all ambient space, except of opaque objects, which determine the boundary conditions. The particle passes from the source to a place of detection along the optimal trajectory, Bohmian trajectory [27]. The equation describing motion of the Bohmian particle can be found, for example, in [36]. There is the unique trajectory for each the particle, the vortex ball in our case. However, an attempt to measure exact position of the ball along the trajectory together with its velocity fails. Namely, there is no way to measure simultaneously the complementary parameters, such as coordinate and velocity, what follows from the uncertainly principle [37].

Fig. 7 shows in lilac color Bohmian trajectories divergent from the slit grating. The probability density distribution is shown here in grey color ranging from white for p=0 to light grey for max p. Bundle of the Bohmian trajectories imitates a fluid flow through the obstacle, containing slits, relatively well. One can see that characteristic streamlets are formed in the flow, along which particles move. Such a vision of hydrodynamical behaviors of quantum systems is typical for many scientists since the formation of the quantum mechanics up to our days [38-42]. Principal moment is that the Schrödinger equation describes the expiration of the superfluid medium, which depend on the boundary conditions and other devices perturbing it (as, for example, the slit gratings, collimators and others). The vortex balls move along optimal directions of the flows – along the Bohmian paths.

http://dx.doi.org/10.5772/59040

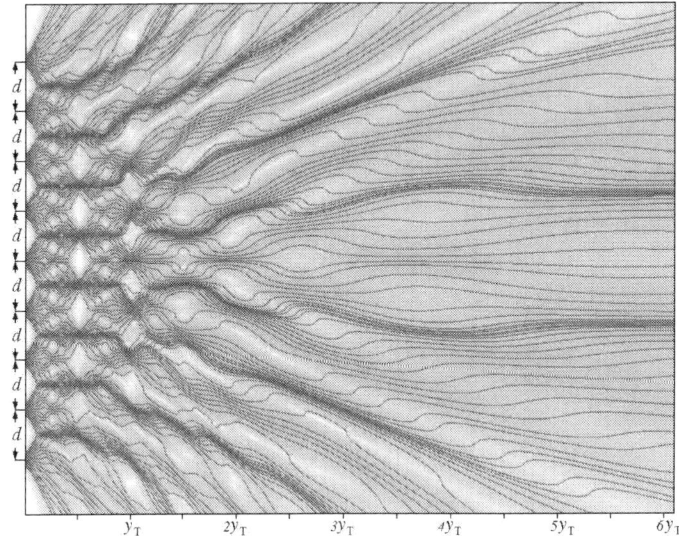

Figure 7. Interference pattern of the coherent flow of the fullerene molecules with the de Broglie wavelength $\lambda = 5$ pm within a zone $y \leq 6y_T$ from the grating containing 9 slits. Lilac curves against the grey background represent the Bohmian trajectories.

4. Physical vacuum as a superfluid medium

The Schrödinger equation (34) describes a flow of the peculiar fluid that is the physical vacuum. The vacuum contains pairs of particle-antiparticles. The pair, in itself, is the Bose particle that stays at a temperature close to zero. In aggregate, the pairs make up Bose-Einstein condensate. It means that the vacuum represents a superfluid medium [43]. A 'fluidic' nature of the space itself is exhibited through this medium. Another name of such an 'ideal fluid' is the ether [29].

The physical vacuum is a strongly correlated system with dominating collective effects [44] and the viscosity equal to zero. Nearest analogue of such a medium is the superfluid helium [22], which will serve us as an example for further consideration of this medium. The vacuum is defined as a state with the lowest possible energy. We shall consider a simple vacuum consisting of electron-positron pairs. The pairs fluctuate within the first Bohr orbit having energy about $13.6 \cdot 2\,\mathrm{eV} \approx 27\,\mathrm{eV}$. Bohr radius of this orbit is $r_1 \approx 5.29 \cdot 10^{-11}$ m. These fluctuations occur about the center of their masses. The total mass of the pair, m_p, is equal to doubled mass of the electron, m. The charge of the pair is zero. The total spin of the pair is equal to 0. The angular momentum, L, is nonzero, however. For the first Bohr orbit $L = \hbar$. The velocity of rotating about this orbit is $L\ /(r_1 m) \approx 2.192 \cdot 10^6$ m/s. It means that there exist an elementary vortex. Ensemble of such vortices forms a vortex line.

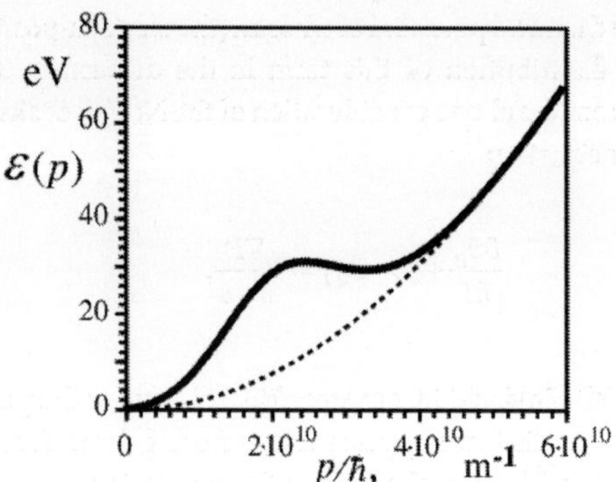

Figure 8. The dispersion relation ε vs. p. The dotted curve shows the non-relativistic square dispersion relation $\varepsilon \sim p^2$. The hump on the curve is a contribution of the roton component $p_R f(p-p_R)$, $p_R/\hbar \approx 1.89 \cdot 10^{10}$ m^{-1}, and $\sigma = 0.5 p_R$.

We may evaluate the dispersion relation between the energy, $\varepsilon(p)=\hbar\omega$, and wave number, $p=\hbar k$, as it done in [45]. As follows from the Schrödinger equation (34) we have:

$$\varepsilon(p) = \frac{1}{2m_p}\left(p + p_R f(p-p_R)\right)^2. \tag{39}$$

Here $p_R = L / r_1 = m_p v_R$ is the momentum of the rotation. The function $f(p\text{-}p_R)$ is a form-factor relating to the electron-positron pairs rotating about the center of their mass m_p. The form-factor describes dispersion of the momentum p around p_R conditioned by fluctuations about the ground state with the lowest energy. The form-factor is similar to the Gaussian curve

$$f(p-p_R) = \exp\left\{-\frac{(p-p_R)^2}{2\sigma^2}\right\}. \tag{40}$$

Here σ is the variance of this form-factor. It is smaller or close to p_R. The dispersion relation (39) is shown in Fig. 8. The hump on the curve is due to the contribution of the rotating electron-positron pair about the center of their masses. These rotating objects are named rotons [45].

Rotons are ubiquitous in vacuum because of a huge availability of pairs of particle-antiparticle. The movement of the roton in the free space is described by the Schrödinger equation

$$\mathrm{i}\hbar\frac{\partial\Psi}{\partial t} = \frac{1}{2m_p}(-\mathrm{i}\hbar\nabla + m_p\vec{v}_R)^2\Psi - C\Psi. \tag{41}$$

http://dx.doi.org/10.5772/59040

The constant C determines an uncertain phase shift of the wave function, and most possible this phase relates to the chemical potential of a boson (the electron-positron pair) [45]. We shall not take into account contribution of this term in the dispersion diagram because of its smallness. As follows from the above consideration of the Navier-Stokes equation, Eq. (41) can be reduced to the Euler equation

$$\frac{\partial \vec{v}_R}{\partial t} + [\vec{\omega} \times \vec{v}_R] = -\frac{\nabla P}{m_p \rho}, \qquad (42)$$

that describes a flow of the inviscid incompressible fluid under the pressure field P. One can see from here that the Coriolis force appears as a restoring force, forcing the displaced fluid particles to move in circles. The Coriolis force is the generating force of waves called inertial waves [45]. The Euler equation admits a stationary solution for uniform swirling flow under the pressure gradient along z.

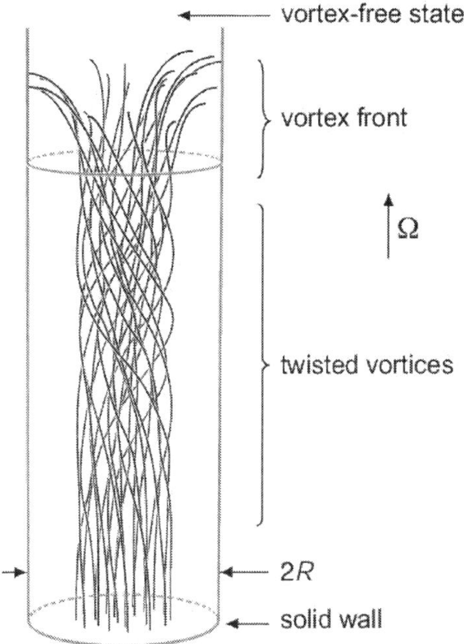

Figure 9. The formation of twisted vortex state [46]. The vortices have their propagating ends bent to the side wall of the rotating cylinder. As they expand upwards into the vortex-free state, the ends of the vortex lines rotate around the cylinder axis. The twist is nonuniform because boundary conditions allow it to unwind at the bottom solid wall. The figure gives a snapshot (at time $t = 25\Omega^{-1}$, where Ω is the angular velocity.) of a numerical simulation of 23 vortices initially generated near the bottom end (t=0). Courtesy kindly by Erkki Thuneberg.

Formation of the swirling flow, the twisted vortex state, has been studied in the superfluid ^{3}He-B [46]. These observations give us a possibility to suppose the existence of such phenomena in the physical vacuum. The twisted vortex states observed in the superfluid ^{3}He-B are closely related to the inertial waves in rotating classical fluids. The superfluid initially is at rest [46]. The vortices are nucleated at a bottom disk platform rotating with the angular velocity Ω about axis z. As the platform rotates they propagate upward by creating the twisted vortex state spontaneously, Fig. 9. The Coriolis forces take part in this twisting. The twisted vortices grow upward along the cylinder axis [47].

Analogous experiment with nucleating vortices can be realized when the lower disk A rotates in the vacuum, Fig. 10. In this case, the vortices are viewed as the dancing electron-positron pairs on the first Bohr orbit. As the vortices grow upward the spontaneous twisted vortex states arise. The latter by reaching upper fixed disk B can capture it into rotation.

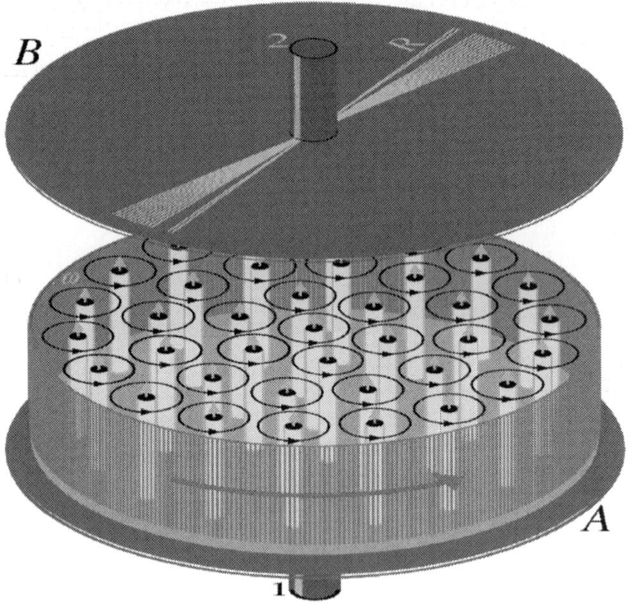

Figure 10. Rotation of the superfluid fluid is not uniform but takes place via a lattice of quantized vortices, whose cores (colored in yellow) are parallel to the axis of rotation [46, 47]. Green arrows are the vorticity ω. Small black arrows indicate the circulation of the velocity v_R around the cores. The vacuum is supported between two non-ferromagnetic disks, A and B, fixed on center shafts, **1** and **2**, of electric motors [48], see Fig. 11. Radiuses of the both disks are R=82.5 mm and distance between them can vary from 1 to 3 mm and more. The vortex bundle rotates rigidly with the disk A. As soon as the vortex bundle reaches the top disk B it begins rotation as well.

Pr. V. Samohvalov has shown through the experiment [48], that the vortex bundle induced by rotating the bottom non-ferromagnetic disk A leads to rotation of the upper fixed initially non-ferromagnetic disk B, Fig. 11. Both disks at room temperature have been placed in the container with technical vacuum at 0.02 Torr, The utmost number of the vortices that may be placed on the square of the disk A is $N_{\max}=(2\pi R^2)/(2\pi r_1^2)$=2•10[18], where R=82.5 mm is the radius of the disk and $r_1 \approx 5.29 \cdot 10^{-11}$ m is the radius of the first Bohr orbit. Really, the number of the vortices situated on the square, N, is considerably smaller. It can be evaluated by multiplying $N_{\max}$ by a factor δ. This factor is equal to the ratio of the geometric mean of the velocities $v_R = \hbar/(r_1 m) \approx 2.192 \cdot 10^6$ m/s and $V_D = R\Omega$ to their arithmetic mean. Here Ω is an angular rate of the disk A. So, we have

$$N = N_{\max} \frac{\sqrt{v_R \cdot V_D}}{v_R + V_D} = N_{\max} \sqrt{\frac{V_D}{v_R}} \approx 6 \cdot 10^{15} \tag{43}$$

at the angular rate Ω=160 1/s [48] the disk velocity V_D=13.2 m/s. Now we can evaluate the kinetic energy of the vortex bundle induced by the rotating disk A. This kinetic energy is $E = N \cdot m_p v_R^2/2 \approx 0.026$ J. This energy is sufficient for transfer of the moment of force to the disk B. Measured in the experiment [48] the torque is about 0.01 N•m. So, the disk B can be captured by the twisted vortex.

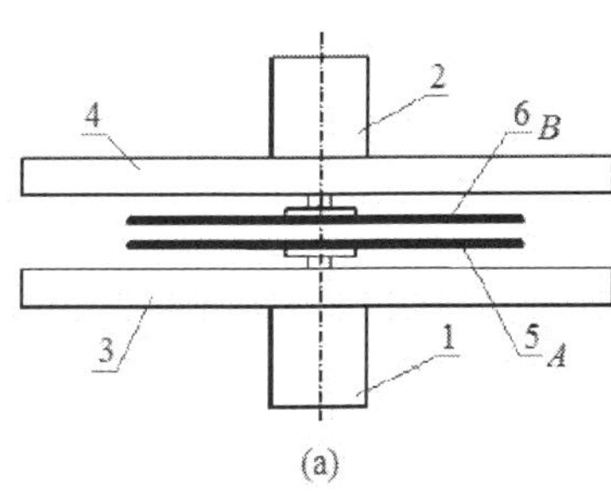

(b)

Figure 11. Basic diagram (a) and general view of the device (b) for researching mass dynamics effects [48]: 1 and 2 are shafts with mounted on them electric motors; 3 and 4 are steel plates with mounted on them electromagnetic brakes; 5 and 6 (the disks A and B, see Fig. 10) are disks rigidly fixed on flanges of the rotors of the electric motors. Courtesy kindly by Vladimir Samokhvalov.

The formation of the growing twisted vortices can be confirmed with attraction of modern methods of interference of light rays passing through the gap between the disks. Light traveling along two paths through the space between the disks undergoes a phase shift manifested in the interference pattern [29] as it was shown in the famous experiment of Aharonov and Bohm [49].

5. Conclusion

The Schrödinger equation is deduced from two equations, the continuity equation and the Navier-Stokes equation. At that, the latter contains slightly modified the gradient pressure term, namely, $\nabla P \to \rho_M \nabla (P / \rho_M) = \nabla P + P \nabla \ln(\rho_M)$. The extra term $P \nabla \ln(\rho_M)$ describes change of the pressure induced by change of the entropy $\ln(\rho_M)$ per length. In this case, the modified gradient pressure term can be reduced to the quantum potential through using the Fick's law. In the law we replace also the diffusion coefficient by the factor $\hbar / 2m$, where $\hbar$ is the reduced Planck constant and m is mass of the particle.

We have shown that a vortex arising in a fluid can exist infinitely long if the viscosity undergoes periodic oscillations between positive and negative values. At that, the viscosity, in average on time, stays equal to zero. It can mean that the fluid is superfluid. In our case, the superfluid consists of pairs of particle-antiparticle representing the Bose-Einstein condensate.

As for the quantum reality, such a periodic regime can be interpreted as exchange of the energy quanta of the vortex with the vacuum through the zero-point vacuum fluctuations. In reality, these fluctuations are random, covering a wide range of frequencies from zero to infinity. Based on this observation we have assumed that the fluctuations of the vacuum ground state can support long-lived existence of vortex quantum objects. The core of such a vortex has nonzero radius inside of which the velocity tends to zero. In the center of the vortex, the velocity vanishes. The velocity reaches maximal values on boundary of the core, and then it decreases to zero as the distance to the vortex goes to infinity.

The experimental observations of the Couder's team [5, 11, 12] can have far-reaching ontological perspectives in regard of studying our universe. Really, we can imagine that our world is represented by myriad of baryonic and lepton "droplets" bouncing on a super-surface of some unknown dark matter. A layer that divides these "droplets", i.e., particles, and the dark matter is the superfluid vacuum medium. This medium, called also the ether [24], is populated by the particles of matter ("droplets"), which exist in it and move through it [29, 50, 51]. The particle traveling through this medium perturbs virtual particle-antiparticle pairs, which, in turn, create both constructive and destructive interference at the forefront of the particle [30]. Thus, the virtual pairs interfering each other provide an optimal, Bohmian, path for the particle.

Assume next, that the baryonic matter is similar, say, on "hydrophobic" fluid, whereas the dark matter, say, is similar to "hydrophilic" fluid. Then the baryonic matter will diverge each from other on cosmological scale owing to repulsive properties of the dark matter, like soap spots diverge on the water surface. Observe that this phenomenon exhibits itself through existence of the short-range repulsive gravitational force that maintains the incompatibility between the dark matter and the baryonic matter [52, 53]. At that, the dark matter stays invisible. One can imagine that the zero-point vacuum fluctuations are nothing as weak ripples on a surface of the dark matter.

Appendix A: Nelson's derivation of the Schrödinger equation

Nelson proclaim that the medium through which a particle moves contains myriad sub-particles that accomplish Brownian motions by colliding with each other chaotically. The Brownian motions is described by the Wiener process with the diffusion coefficient

$$\bar{v} = \frac{\hbar}{2m}. \tag{44}$$

Here m is mass of the particle and $\hbar = h/2\pi$ is the reduced Planck constant. Here we use v with the upper bar in order to avoid confusion with the kinematic viscosity adopted in hydrodynamics. As seen this motion has a quantum nature [24] in contrast to the macroscopic Brownian motions where the diffusion coefficient has a view $\bar{v} = kT/m\beta^{-1}$; here k is Boltzmann constant, T is a temperature, and β^{-1} is the relaxation time.

Two equations are main in the article [24]. The position $\mathbf{x}$ (t) of the Brownian particle, being subjected either by external forces or by currents in the medium, can be written by two equivalent equations:

$$d\mathbf{x}(t) = \mathbf{b}(\mathbf{x}(t),t)dt + d\mathbf{w}(t), \tag{45}$$

$$d\mathbf{x}(t) = \mathbf{b}_*(\mathbf{x}(t),t)dt + d\mathbf{w}_*(t) \tag{46}$$

Here $w(t)$ and $w_*(t)$ are the Wiener processes, both have equivalent properties. Variables $\boldsymbol{b}$ and $\boldsymbol{b}_*$ are vector-valued forward and backward functions on space-time, respectively. In fact, they are the mean forward and mean backward measured quantities

$$\mathbf{b}(\mathbf{x}(t),t) = \lim_{\Delta t \to 0_+} E_t \frac{\mathbf{x}(t+\Delta t) - \mathbf{x}(t)}{\Delta t}, \tag{47}$$

$$\mathbf{b}_*(\mathbf{x}(t),t) = \lim_{\Delta t \to 0_+} E_t \frac{\mathbf{x}(t) - \mathbf{x}(t-\Delta t)}{\Delta t}. \tag{48}$$

Here E_t denotes the conditional expectation (average) given the state of the system at time t, and 0_+means that Δt tends to 0 through positive values. Thus $\boldsymbol{b}(x(t), t)$ and $\boldsymbol{b}_*(x(t), t)$ are again stochastic variables [54, 55]. It is instructive to compare calculus (46) and (47) with classical calculations of infinitesimal small increments

$$v(t) = \lim_{\Delta t \to 0_+} \frac{x(t+\Delta t) - x(t)}{\Delta t} = \lim_{\Delta t \to 0_+} \frac{x(t) - x(t-\Delta t)}{\Delta t}. \tag{49}$$

One can see that these calculations are symmetrical with respect to the time arrow, whereas the calculations (46) and (47) are not, in general (see below).

It should be noted that $b(x(t), t)$ and $b_*(x(t), t)$ are not real velocities. The real current velocity of the particle is calculated as

$$\vec{v}(t) = \frac{1}{2}\left(\mathbf{b}(\mathbf{x}(t),t) + \mathbf{b}_*(\mathbf{x}(t),t)\right). \tag{50}$$

There is a one more velocity, which is represented via difference of $b(x(t), t)$ and $b_*(x(t), t)$:

$$\vec{u}(t) = \frac{1}{2}\left(\mathbf{b}(\mathbf{x}(t),t) - \mathbf{b}_*(\mathbf{x}(t),t)\right). \tag{51}$$

According to Einstein's theory of Brownian motion, $\vec{u}(t)$ is the velocity acquired by a Brownian particle, in equilibrium with respect to an external force, to balance the osmotic force [19]. For this reason, this velocity is named the osmotic velocity. It can be expressed in the following form

$$\vec{u}(t) = \bar{\nu}\nabla\left(\ln(\rho(t))\right) = \frac{\hbar}{m}\frac{\nabla R(t)}{R(t)}, \tag{52}$$

where $\rho(t)$ is the probability density of $\mathbf{x}(t)$ and $R(t)=\rho(t)^{1/2}$ is the probability density amplitude. The current velocity, in turn, is expressed through gradient of a scalar field S called the action

$$\vec{v}(t) = -\frac{\hbar}{m}\nabla S(t). \tag{53}$$

The both equations, (44) and (45), introduced above are important for derivation of the Schrödinger equation. The derivation of the equation is provided by the use of the wave function presented in the polar form

$$\Psi = R\exp\{\mathrm{i}S/\hbar\}, \tag{54}$$

by replacing the velocities $\vec{v}(t)$ and $\vec{u}(t)$ in the initial equations. It should be noted that Nelson departs from two equations describing directed the forward and backward Brownian motions which are written down for real-valued functions. In order to come to the Schrödinger equation he has used a complex-valued wave function $\exp\{R+\mathbf{i}S\}$ instead of the generally accepted $R\exp\{\mathrm{i}S/\hbar\}$. Obviously, this discrepancy are eliminated by replacing $\exp\{R\}\to R$.

Observe that the wave function represented in the polar form (53) is used for getting equations underlying the Bohmian mechanics [27]. These two equations are the continuity equation and the Hamilton-Jacobi equation containing an extra term known as the Bohmian quantum potential. The quantum potential has the following view:

$$Q=-\frac{\hbar^2}{2m}\frac{\nabla^2 R}{R}=-\frac{\hbar}{2}\frac{(\nabla R\vec{u})}{R}=-\frac{m}{2}u^2-\frac{\hbar}{2}(\nabla\vec{u}). \tag{55}$$

One can see that the quantum potential depends only on the osmotic velocity, which is expressed through difference of the forward and backward averaged quantities (46) and (47). These forward and backward quantities can be interpreted as uncompensated flows through a "semipermeable membrane" which represents an instant dividing the past and the future. Following to Licata and Fiscaletti, who have shown that the quantum potential has relation to the Bell length indicating a non-local correlation [28], one can add that the non-local correlation exists also between the past and the future. E. Nelson as one can see has considered a particle motion through the ether populated by sub-particles experiencing accidental collisions with each other. The Brownian motions of the sub-particles submits to the Wiener process with the diffusion coefficient ν proportional to the Plank constant as shown in Eq. (43). The ether behaves itself as a free-friction fluid.

Acknowledgements

The author thanks Mike Cavedon for useful and valuable remarks and offers. The author thanks also Miss Pipa (quantum portal administrator) for preparing a program drawing Fig. 7.

Author details

V.I. Sbitnev[1,2]

1 St. Petersburg B.P. Konstantinov Nuclear Physics Institute, NRC Kurchatov Institute, Gatchina, Leningrad district, Russia

2 Department of Electrical Engineering and Computer Sciences, University of California, Berkeley, USA

References

[1] Hartle JB. What Connects Different Interpretations of Quantum Mechanics? In: Elitzur A., Dolev S., Kolenda N. (eds.) Quo Vadis Quantum Mechanics? Springer: Berlin Heidelberg New York; 2005. p72-82.

[2] Juffmann T, Truppe S, Geyer P, Major AG, Deachapunya S, Ulbricht H, Arndt M. Wave and particle in molecular interference lithography. Phys. Rev. Lett. 2009; 103: 263601.

[3] Brukner Č, Zeilinger A. Quantum Physics as a Science of Information. In: Elitzur A., Dolev S., Kolenda N. (eds.) Quo Vadis Quantum Mechanics? Springer: Berlin Heidelberg New York; 2005. p46-61.

[4] De Broglie L. Interpretation of quantum mechanics by the double solution theory. Annales de la Fondation Louis de Broglie. 1987; 12(4): 1-22.

[5] Couder Y, Fort E. Single-Particle Diffraction and Interference at a Macroscopic Scale. Phys. Rev. Lett. 2006; 97: 154101.

[6] Motte A. Translation of Newton's Principia (1687). Axioms or Laws of Motion. Published by Daniel Adee, 45 Liberty str., N. Y.; 1846.

[7] Frisch U. Translation of Leonhard Euler's: general principles of the motion of fluids. e-print: http://arxiv.org/abs/0802.2383 (accessed 17 Feb 2008).

[8] Sbitnev VI. From the Newton's laws to motions of the fluid and superfluid vacuum: vortex tubes, rings, and others. e-print: http://arxiv.org/abs/1403.3900 (accessed 20 Jun 2014).

[9] Landau LD., Lifshitz EM. Fluid mechanics. Pergamon Press; 1987.

[10] Kundu P, Cohen I. Fluid Mechanics. Academic Press; 2002.

[11] Couder Y, Fort E, Gautier C.-H, Boudaoud A. From bouncing to floating drops: non-coalescence of drops on a fluid bath. Phys. Rev. Lett. 2005; 94: 177801.

[12] Eddi A, Sultan E, Moukhtar J, Fort E, Rossi M, Couder Y. Information stored in Faraday waves: the origin of a path memory. J. Fluid Mech. 2011; 674: 433-463.

[13] Grössing G. On the thermodynamic origin of the quantum potential. Physica A, 2009; 388: 811-823.

[14] Grössing G. Sub-Quantum Thermodynamics as a Basis of Emergent Quantum Mechanics. Entropy, 2010; 12: 1975-2044. doi:10.3390/e12091975

[15] Grössing G., Mesa Pascasio J., Schwabl H. A classical explanation of quantization, Found. Phys. 2011: 41, 1437-1453.

[16] Grössing G. Emergence of Quantum Mechanics from a Sub-Quantum Statistical Mechanics. e-print: http://arxiv.org/abs/1304.3719 (accessed 12 Apr 2013).

[17] Harris DM, Bush WM. Droplets walking in a rotating frame: from quantized orbits to multimodal statistics. J. Fluid Mech. 2014; 739: 444-464.

[18] Oza AU, Harris DM, Rosales RR, Bush WM. Pilot-wave dynamics in a rotating frame: on the emergence of orbital quantization. J. Fluid Mech. 2014; 744: 404-429.

[19] Brady R. Anderson R. Why bouncing droplets are a pretty good model of quantum mechanics? e-print: http://arxiv.org/abs/1401.4356 (accessed 16 Jan 2014).

[20] Vervoort L. No-Go Theorems face Fluid-Dynamical Theories for Quantum Mechanics. e-print: http://arxiv.org/abs/1406.0901 (accessed 16 Jun 2014).

[21] Wu Jie-Zhi, Ma Hui-Yang, Zhou Ming-De. Vorticity and Vortex Dynamics. Springer-Verlag: Berlin Heidelberg; 2006.

[22] Volovik GE. The Universe in a Helium Droplet. Clarendon Press: Oxford; 2003.

[23] Tropical cyclone. http://en.wikipedia.org/wiki/Tropical_cyclone

[24] Nelson E. Derivation of the Schrödinger Equation from Newtonian Mechanics. Phys. Rev. 1966; 150(4): 1079-1085.

[25] Hestenes D. The Zitterbewengung interpretation of quantum mechanics. Found. Physics. 1990; 20(10): 1213-1232.

[26] Dorbolo S, Terwagne D, Vandewalle N, Gilet T. Resonant and rolling droplet. New Journal of Physics. 2008; 10: 113021. http://iopscience.iop.org/1367-2630/10/11/113021

[27] Sbitnev VI. Bohmian Trajectories and the Path Integral Paradigm – Complexified Lagrangian Mechanics. In: Pahlavani MR (ed.) Theoretical Concepts of Quantum Mechanics. Rijeka: InTech; 2012. p313-34; doi:10.5772/33064

[28] Licata I, Fiscaletti D. Bell Length as Mutual Information in Quantum Interference. Axioms 2014; (3): 153-165. doi:10.3390/axioms3020153

[29] Martins AA. Fluidic Electrodynamics: On parallels between electromagnetic and fluidic inertia. e-print: http://arxiv.org/abs/1202.4611 (accessed 21 Feb 2012).

[30] Feynman RP, Hibbs A. Quantum Mechanics and Path Integrals. McGraw Hill: N. Y.; 1965.

[31] Derbes D. Feynman's derivation of the Schrödinger equation. Am. J. Phys. 1996; 64(7) 881-884.

[32] Sbitnev VI. Matter waves in ihe Talbot-Lau interferometry. e-print: http://arxiv.org/abs/1005.0890 (accessed 17 Sep 2010).

[33] Talbot HF. Facts Relating to Optical Science. Philos. Mag. 1836; 9: 401-407.

[34] Berry MV, Klein S. Integer, fractional and fractal Talbot effects, Journal of Modern Optics. 1996; 43(10): 2139-2164.

[35] Berry M, Marzoli I, Schleich W. Quantum carpets, carpets of light. Phys. World. 2001: 6; 39-44.

[36] Sanz AS, Miret-Artés S. A causal look into the quantum Talbot effect, J. Chem. Phys. 2007; 126: 234106. http://arxiv.org/abs/quant-ph/0702224.

[37] Sbitnev VI. Generalized Path Integral Technique: Nanoparticles Incident on a Slit Grating, Matter Wave Interference. In: Bracken P (ed.) Advances in Quantum Mechanics. Rijeka: InTech; 2013. p183-211. doi:10.5772/53471

[38] Madelung E. Quantumtheorie in hydrodynamische form. Zts. f. Phys. 1926; 40: 322-326.

[39] Bohm D, Vigier JP. Model of the causal interpretation of quantum theory in terms of a fluid with irregular fluctuations. Phys. Rev. 1954; 96: 208-216.

[40] Wyatt RE. Quantum dynamics with trajectories: Introduction to quantum hydrodynamics. Springer: 2005.

[41] Jüngel A, Milišić J-P. Quantum Navier–Stokes equations. Progress in Industrial Mathematics at ECMI 2010. Mathematics in Industry. 2012; 17: p427-439.

[42] Guo Y, Bühler O. Wave–vortex interactions in the nonlinear Schrödinger equation. Physics of Fluids. 2014; 26: 027105.

[43] Sinha KP, Sivaram C, Sudarshan ECG. Aether as a Superfluid State of Particle-Antiparticle Pairs. Found. Phys. 1976; 6(1): 65-70.

[44] Roberts PH, Berloff NG. The nonlinear Schrödinger equation as a model of superfluidity. In: Barenghi CF, Donnelly RJ, Vinen WF. (eds.) Quantized vortex dynamics and superfluid turbulence. LNP 571. 2001: 235-257.

[45] Le Gal P. Waves and Instabilities in Rotating and Stratified Flows. In: Klapp J, Medina A, Cros A, Vargas CA. (eds.) Fluid Dynamics in Physics, Engineering and Environmental Applications. Springer-Verlag; 2013. p25-40.

[46] Eltsov VB, Finne AP, Hänninen R, Kopu J, Krusius M, Tsubota M, Thuneberg EV. Twisted vortex states. Phys. Rev. Lett. 2006; 96: 215302; http://xxx.lanl.gov/abs/cond-mat/0602667.

[47] Lounasmaa OV, Thuneberg E. Vortices in rotating superuid ^{3}He. PNAS, 1999; 96(14): 7760-7767.

[48] Samokhvalov VN. Nonelectromagnetic force interaction by rotating masses in vacuum. Int. J. Unconventional Science. 2013; 1(1): 6-19. http://www.unconv-science.org/n1/samokhvalov/

[49] Aharonov Y, Bohm D. Significance of electromagnetic potentials in the quantum theory. Phys. Rev. 1959; 115(3): 485-491.

[50] Pinheiro MJ, Büker M. An extended dynamical equation of motion, phase dependency and inertial back reaction. e-print: http://arxiv.org/abs/1208.3458 (accessed 16 Aug 2012).

[51] Saravani M, Afshordi N, Mann RB. Empty Black Holes, Firewalls, and the Origin of Bekenstein-Hawking Entropy. e-print: http://arxiv.org/abs/1212.4176 (accessed 2 Dec 2013).

[52] Famaey B, McGaugh S. Modified Newtonian Dynamics (MOND): Observational Phenomenology and Relativistic Extensions. Living Review Relativity. 2012; 15: 10.

[53] Chung Ding-Yu. Galaxy Evolution by the Incompatibility between Dark Matter and Baryonic Matter. Int. J. Astronomy and Astrophysics. 2014; 4: 374-383.

[54] Nelson E. Dynamical theories of Brownian motion. Princeton. New Jersey: Princeton Univ. Press; 1967.

[55] Nelson E. Quantum fluctuations. Princeton Series in Physics. Princeton, New Jersey: Princeton Univ. Press; 1985.

Chapter 13

Husimi Distribution and the Fisher Information

Sergio Curilef and Flavia Pennini

Additional information is available at the end of the chapter

http://dx.doi.org/10.5772/59126

1. Introduction

In this chapter we review some aspects of the concept of the Fisher information measure in phase space for two specific systems: the Landau Diamagnetism and the Rigid rotator. The indispensable tool in this proposal is a quasi probability called Husimi distribution [1], which is frequently employed to characterize the quantum and classical behavior [2] of systems. Also, it possesses interesting applications in several areas of physics such as Quantum Mechanics, Quantum Optics, Information Theory and Nanotechnology [3–10]. Its main properties are: 1) it is definite positive in all phase space, 2) it possesses no correct marginal properties, 3) it permits to calculate the expectation values of observables in quantum mechanics similarly to the classical case [11], and 4) it is a special type of probability that simultaneously approximate location of position and momentum in phase space. It is important to note that this quasi probability is constructed by definition as the expectation value of the density operator in a basis of coherent states [12]. Details about the formulation of coherent states and the obtaining of Husimi distribution from these can be found on our chapter that it can be read in Ref. [13]. The main propose of this chapter is to present to the reader interesting problems in physics, such as, the harmonics oscillator [5], the Landau diamagnetism model [8, 14] and, the rigid rotator [7, 15], analyzed from a point of view of the information measures. In particular, we will put emphasis in the Fisher Information measure and its construction starting from a well-defined set of coherent states.

In our previous contribution published in Ref. [13] we research about a special semi-classical measure, the Wehrl entropy, as an important application of the Husimi distribution. In the present study we analyze some consequences of obtaining the Husimi distribution; for instance, the Fisher information for fundamental problems in physics for which the coherent states formulation is well defined.

In physics, great attention has been paid to the Landau diamagnetism which consists in a particle charged in a uniform magnetic field. For our purpose we will use a complete description of the Husimi Distributionin three dimensions in order to study such system, so as it was shown in our previous contributions (see Ref. [13], where we have discussed some limiting cases as high and low temperatures. From the present analysis, when three dimensions are considered, naturally arises a lower temperature

bound, whereby it is not possible to work in all finite temperatures. Such discussion is explained with details in Ref. [13].

The other system, that we take into account here, is the linear rigid rotator and its corresponding $3D$ anisotropic version. We analyze phase space delocalization and obtain the concomitant semiclassical Fisher information measure constructed by using Husimi Distributionconstructed from suitable basis of coherent states.

In order to facilitate the understanding of this chapter to the reader, we give the following organization: in section 2 we begin introducing the concepts and methodology that will employ in the rest of the chapters. In section 3 we focus our attention on the Husimi distribution and the Fisher measure for the Landau diamagnetism. In section 4 we study the delocalization into phase space, within a semiclassical context by recourse to the Husimi distribution, for both cases of rigid rotators: linear and $3D-$anisotropic instances. Finally, some conclusions and open problems are commented in section 5.

2. Previous concepts

This section provides reference material that we consider relevant to conveniently understand the development of this chapter. These are *i)* the Husimi distribution, *ii)* Wehrl entropy, and *iii)* Fisher information measure. In all cases, we refer to the model of the harmonic oscillator in a thermal state.

2.1. Husimi distribution and Wehrl entropy

From the standard statistical mechanics, the thermal density matrix can be represented by

$$\hat{\rho} = Z^{-1} e^{-\beta \hat{H}}, \tag{1}$$

where $\beta = 1/k_B T$ the inverse temperature T, and k_B the Boltzmann constant [16], $\hat{H}$ is the Hamiltonian of the system and $Z = \mathrm{Tr}(e^{-\beta \hat{H}})$ is the partition function.

In the current strategy, the expectation value of the density operator in a basis of coherent states is related to the Husimi distribution as [1]

$$\mu(z) = \langle z|\hat{\rho}|z\rangle, \tag{2}$$

where the set $\{|z\rangle\}$ denotes the eigenstates of the annihilation operator $\hat{a}$, i.e., $\hat{a}|z\rangle = z|z\rangle$ defined for all $z \in \mathbb{C}$ [12] and they are the coherent states for the system. Therefore, the normalization of the distribution is given by

$$\int \frac{\mathrm{d}^2 z}{\pi} \mu(z) = 1, \tag{3}$$

where the integration is over the complex plane z and the element of integration is an area proportional to phase space element given by $\mathrm{d}^2 z = \mathrm{d}x\mathrm{d}p/2\hbar$.

The set $\{E_n\}$ stands for the spectrum of an arbitrary Hamiltonian $\hat{H}$, where n is a positive integer. With these elements the Husimi distribution takes the form

$$\mu(z) = \frac{1}{Z}\sum_{n} e^{-\beta E_n} |\langle z|n\rangle|^2, \tag{4}$$

where the set $\{|n\rangle|\}$ represents energy eigenstates with eigenvalues E_n [4, 5].

A direct application that is additionally a useful measure of localization in phase-space [17, 18] is the Wehrl entropy, which is suitably defined as

$$W = -\int \frac{\mathrm{d}^2 z}{\pi} \mu(z) \ln \mu(z). \tag{5}$$

As a consequence of the uncertainty principle, Lieb [4] proved the inequality $W \geq 1$ which was previously conjectured by Wehrl [17].

For the Hamiltonian $\hat{H} = \hbar\omega[\hat{a}^\dagger \hat{a} + 1/2]$ of the harmonic oscillator, it is obtained a basis $\{|n\rangle\}$ and the spectrum $E_n = \hbar\omega(n + 1/2)$, with $n = 0, 1, \ldots$ from the complete orthonormal set of eigenstates and eigenvectors, respectively. The algebra allows us to define the following elementary properties:

1. A set of Glauber coherent states is given by [19]

$$|z\rangle = e^{-|z|^2/2} \sum_{n=0}^{\infty} \frac{z^n}{\sqrt{n!}} |n\rangle. \tag{6}$$

2. The normalization,

$$\langle n|n'\rangle = \delta_{n,n'} \tag{7}$$

   where $\delta_{n,n'}$ is the Kronecker delta function.

3. The completeness property is contained in the relation

$$\sum_{n=0}^{\infty} |n'\rangle\langle n| = \hat{1}, \tag{8}$$

   where $\hat{1}$ represents the identity operator in the defined space of eigenvectors.

Now, a suitable application of the present theoretical characterization [4] comes from certain calculations of the harmonic oscillator as the Husimi distribution

$$\mu_{HO}(z) = \left(1 - e^{-\beta\hbar\omega}\right) e^{-\left(1 - e^{-\beta\hbar\omega}\right)|z|^2}, \tag{9}$$

and the Wehrl entropy

$$W_{HO} = 1 - \ln(1 - e^{-\beta\hbar\omega}), \tag{10}$$

which are respectively known and useful analytical expressions [4].

2.2. Fisher information measure

A pertinent quantifier of information, which possess innumerable applications in several fields of Physics, is the Fisher information measure [20]. The last years have witnessed a great deal of activity revolving around physical applications of Fisher information measure [20, 21] providing tools to yield most of the canonical Lagrangians of theoretical physics [20, 21] related properly to the Boltzmann entropy [22, 23]. The Fisher information connected with translations of an observable x with the consistent probability density $\rho(x)$ is given by [24]

$$\mathcal{F} = \int \mathrm{d}x \rho(x) \left[\frac{\partial \ln \rho(x)}{\partial x}\right]^2, \tag{11}$$

and the Cramer–Rao inequality is given by [24]

$$\Delta x \geq \mathcal{F}^{-1} \tag{12}$$

where Δx is the variance for the stochastic variable x which is of the form [24]

$$\Delta x^2 = \langle x^2 \rangle - \langle x \rangle^2 = \int \mathrm{d}x \rho(x) x^2 - \left(\int \mathrm{d}x \rho(x) x\right)^2. \tag{13}$$

In particular, it is interesting to study its representation appealing to the semiclassical approach (see, for example, Ref. [25] and references therein), whose main tool is a distribution function in phase space in the basis of coherent states. Specially, in this proposal, we pay attention to a particular distribution, the well-known $Q-$function or Husimi distribution.

An original, compact expression in phase space is advanced for the "semiclassical" Fisher information measure, that can be easily derived from the Wehrl-methodology described in Refs. [5] and [6]. The appearance for this measures reads

$$\mathcal{F} = \frac{1}{4} \int \frac{\mathrm{d}^2 z}{\pi} \mu(z) \left\{\frac{\partial \ln \mu(z)}{\partial |z|}\right\}^2, \tag{14}$$

which will be used in the following sections.

Inserting the $\mu-$expression for the harmonic oscillator into Eq. (14) we find its anlytical form

$$\mathcal{F}_{HO} = 1 - e^{-\beta \hbar \omega}, \tag{15}$$

leading to the following limits:

$$\begin{aligned} &\text{for } T \to 0 \text{ one has } \mathcal{F}_{HO} = 1 \\ &\text{for } T \to \infty \text{ one has } \mathcal{F}_{HO} = 0, \end{aligned} \tag{16}$$

as it should be expected.

3. Landau Diamagnetism: Charged particle in a uniform magnetic field

Diamagnetism is a problem firstly appointed by Landau who showed the discreteness of energy levels for a charged particle in a magnetic field [26]. By the observation of the diverse scenarios in the framework provided by the Landau diamagnetism we can study some relevant physical properties [27–29] as the role of the size of systems or the influence of boundaries, also the thermodynamic limit or quasi-stationary states. The primary motivation even today for several specialists is to find a useful measure to characterize theoretically every practical consequence of the system and its behavior.

In the past, Feldman and Kahn calculated the proper partition function for this system by appealing to the concept of Glauber coherent states from a set of basis states [30]. This formulation uses classical concepts as electron orbits, even though it contains all quantum effects [30]. This approach was previously used to obtain measures as the Wehrl entropy [17, 18] and Fisher information [31] with the purpose of studying the thermodynamics of the free spinless charged particle in a uniform magnetic field [32], this is the Landau diamagnetism problem. As observed, in such contribution the formulation is not completely consistent because it was necessary to normalize the Husimi distribution in order to arrive to reliable expressions for semiclassical measures [9, 32, 33].

Certainly, because the relevant effects seem to come only from the transverse motion, several efforts are made to describe this problem in two dimensions [9, 28, 29, 32–35]. Furthermore, the discovery of the quantum Hall effect has aroused much interest in understanding the behavior of electrons moving in a plane perpendicular to the magnetic field [35]. The confinement is possible at the *interface* typically between a semiconductor and an insulator, where a quantum well that traps the particles is formed, allowing their motion just in the direction of the interface plane at low energies, forbidding the motion in any other directions.

Conversely, we discuss here this problem in three dimensions, the most complete formulation. However, if the length of the cylindrical geometry of the system is large enough the results are close to those in two dimensions. Despite this latter, it is suggested that the formulation in two dimensions is not sufficient to explain the whole problem. As suggested before, electronic devices are based in interfaces. As a consequence of this line of reasoning, a natural lower temperature bound is theoretically imposed, that appears from the analysis in three dimensions.

3.1. The model of one charged particle in a magnetic field

We introduce the present application giving the essential ingredients of the well-known Landau model for diamagnetism: a spinless charged particle in a magnetic field B. Consider the kinetic momentum

$$\vec{\pi} = \vec{p} + \frac{q}{c}\vec{A}, \tag{17}$$

where m_q is the mass of a particle of charge q, the vector $\vec{p}$ is the linear momentum subject to the action of $\vec{A}$, the vector potential.

If we follow the presentation of Feldman et al. [30]), the Hamiltonian reads [30]

$$H = \frac{\vec{\pi} \cdot \vec{\pi}}{2m_q}, \tag{18}$$

and the magnetic field is $\overrightarrow{B}=\overrightarrow{\nabla}\times\overrightarrow{A}$. The vector potential is chosen in the symmetric gauge as $\overrightarrow{A}=(-By/2,Bx/2,0)$, which corresponds to a uniform magnetic field along the $z-$direction.

By using the formulation of the step-ladder operators [30], one needs to define the step operators as follows [30]

$$\hat{\pi}_{\pm}=\hat{p}_x\pm i\hat{p}_y\pm\frac{i\hbar}{2\ell_{\mathrm{B}}^2}(\hat{x}\pm i\hat{y}), \tag{19}$$

where the length

$$\ell_{\mathrm{B}}=(\hbar c/qB)^{1/2} \tag{20}$$

is the classical radius of the ground-state Landau orbit [30]. Motion along the $z-$axis is free [30]. For the transverse motion, the Hamiltonian specializes to [30]

$$\hat{H}_t=\frac{\hat{\pi}_+\hat{\pi}_-}{2m_q}+\frac{1}{2}\hbar\Omega\hat{1}, \tag{21}$$

where an important quantity characterizes the problem, namely,

$$\Omega=qB/m_qc, \tag{22}$$

the cyclotron frequency [36]. The set of eigenstates $\{|N,m\rangle\}$ is characterized by two quantum numbers: N related to the energy and m wuth the $z-$ projection of the angular momentum. They are consequentely eigenstates of both $\hat{H}_t$, the Hamilronian, and $\hat{L}_z$, the angular momentum operator [30], thus

$$\hat{H}_t|N,m\rangle=\left(N+\frac{1}{2}\right)\hbar\Omega|N,m\rangle=E_N|N,m\rangle \tag{23}$$

and

$$\hat{L}_z|N,m\rangle=m\hbar|N,m\rangle. \tag{24}$$

The eigenvalues of $\hat{L}_z$ are not bounded by below, because m takes the values $-\infty,\ldots,-1,0,1,\ldots,N$ [30]. This fact agrees with the energies $(N+1/2)\hbar\Omega$ that are infinitely degenerate [36]. As seen below, for estimation purposes, the physical relevance of phase-space localization is diminished by this fact. In addition, L_z is not an independent constant of the motion [36].

There exists a analogous formulation of an charged particle in a magnetic field by Kowalski that takes into account the geometry of a circle [33] (and for a comparison with the Feldman formulation see Ref. [9]), but at this point, we choose the Feldman formulation to work because the measure is easily defined and the normalization condition and other semiclassical measures are well described.

3.2. Husimi distribution and Wehrl entropy

We will start our present endeavor defining the Hamiltonian $\hat{H} = \hat{H}_t + \hat{H}_l$ where $\hat{H}_t = \hbar\Omega(\hat{N} + 1/2)$ to describe the transverse motion, being Ω the cyclotron frequency as defined by the Eq. (22) and $\hat{N}$ the number operator; the Hamiltonian $\hat{H}_l = \hat{p}_z^2/2m_q$ to represent the longitudinal one-dimensional free motion, for a particle of mass m_q and charge q in a magnetic field B. A possible way to define the Husimi function η is given by

$$\eta(x, p_x; y, p_y; p_z) = \langle \alpha, \xi, k_z | \hat{\rho} | \alpha, \xi, k_z \rangle, \tag{25}$$

where $\hat{\rho}$ is the thermal density operator and the set $\{|\alpha, \xi, k_z\rangle\}$ stands for the coherent states for the description in three dimensions. By the direct product $|\alpha, \xi, k_z\rangle \equiv |\alpha, \xi\rangle \otimes |k_z\rangle$, the set $\{|\alpha, \xi\rangle\}$ corresponds to the coherent states of the transverse motion and $\{|k_z\rangle\}$ to the longitudinal motion. Therefore, the thermal density operator is given by

$$\hat{\rho} = \frac{1}{Z} \mathrm{e}^{-\beta(\hat{H}_l + \hat{H}_t)}, \tag{26}$$

where $\beta = 1/k_B T$, T is the temperature and k_B the Boltzmann constant. In addition, Z is the partition function for motion in three dimensions of the particle. Now, if Z can be separated by using Z_t (the contribution for the transverse motion) and Z_l (the contribution for the one-dimensional free motion), then the partition function could be written as $Z = Z_l Z_t$. Thus, the Husimi function [1] is expressed as

$$\eta = \frac{\mathrm{e}^{-\beta p_z^2/2m_q}}{Z_l Z_t} \sum_{n,m} \mathrm{e}^{-\beta\hbar\Omega(n+1/2)} |\langle n, m | \alpha, \xi \rangle|^2. \tag{27}$$

where

$$Z_l = (\mathcal{L}/h)(2\pi m_q k_B T)^{1/2} \qquad \text{and} \tag{28}$$

$$Z_t = \mathcal{A} m_q \Omega / (4\pi\hbar \sinh(\beta\hbar\Omega/2)), \tag{29}$$

being $\mathcal{L}$ the length of the cylinder, $\mathcal{A} = \pi R^2$ the area for cylindrical geometry [30]. In addition, the matrix element $|\langle n, m | \alpha, \xi \rangle|^2$ describes the probability of finding the particle in the coherent state $|\alpha, \xi\rangle$. Its expression was defined previously [37].

The distribution η is written as:

$$\eta = \eta_l(p_z)\, \eta_t(x, p_x; y, p_y), \tag{30}$$

where η has been separated as a function of two distributions, namely, $\eta_l = \eta_l(p_z)$ and $\eta_t = \eta_t(x, p_x; y, p_y)$. The explicit form of the Hamiltonian $\hat{H}_l$ makes to miss the dependence on the variable z. Therefore, summing in Eq. (27) we solve

$$\eta_l = \frac{e^{-\beta p_z^2/2m_q}}{Z_l}, \tag{31}$$

$$\eta_t = \frac{2\pi\hbar}{\mathcal{A} m_q \Omega}\left(1 - e^{-\beta\hbar\Omega}\right) e^{-(1-e^{-\beta\hbar\Omega})|\alpha|^2/2\ell_B^2}, \tag{32}$$

where the length ℓ_B is defined by the Eq. (20). From expressions (31) and (32), we emphasize again that $\eta_l(p_z)$ describes the free motion of the particle in the magnetic field direction and $\eta_t(x, p_x; y, p_y)$ the Landau levels due to the circular motion in a transverse plane to the magnetic field, similar to the harmonic oscillator of Eq. (9) since $|z|^2 \to |\alpha|^2/2\ell_B^2$. Consequently Eqs. (30), (31) and (32) together contain the complete description of the system. We noticed both distributions are naturally normalized in a standard form, i.e.,

$$\int \frac{dz dp_z}{h} \eta_l(p_z) = 1, \tag{33}$$

and

$$\int \frac{d^2\alpha d^2\xi}{4\pi^2 \ell_B^4} \eta_t(x, p_x; y, p_y) = 1. \tag{34}$$

In consequence, both Eqs. (31) and (32), under conditions (33) and (34), allow us to get a close form for the Wehrl entropy. Furthermore, using one of the most basic property of the entropy, the additivity, we can state $W_{\text{total}} = W_l + W_t$. Hence,

$$W_l = -\int \frac{dz dp_z}{h} \eta_l(p_z) \ln \eta_l(p_z), \tag{35}$$

$$W_t = -\int \frac{d^2\alpha d^2\xi}{4\pi^2 \ell_B^4} \eta_t(x,p_x;y,p_y) \ln \eta_t(x,p_x;y,p_y), \tag{36}$$

again, the subindexes t and l represent respectively the transverse and longitudinal motions.

As a consequence of solving the integrals (35) and (36) we can identify the two entropies, they are

$$W_l = \frac{1}{2} + \ln\left(\frac{\mathcal{L}}{\lambda}\right), \tag{37}$$

$$W_t = 1 - \ln\left(1 - e^{-\beta\hbar\Omega}\right) + \ln(g), \tag{38}$$

where $\lambda = h/(2\pi m_q k_B T)^{1/2}$ represents the mean thermal length of the particle and $g = \mathcal{A}/2\pi\ell_B^2$ the degeneracy of a Landau level [38].

3.3. Semiclassical behavior

In fact, the classical entropy for a free particle in one dimension and Eq. (37) are coincident. Furthermore, the Eq. (38) is the Wehrl entropy for the transverse motion and possesses a form close to the harmonic oscillator entropy given by the Eq. (10), with the exception of a term associated with the degeneracy. Some properties of entropies that can be directly derived from Eqs. (37) and (38) are:

1. As commented before, W_l and the classical entropy for the free motion in one dimension coincide between them. Furthermore, this part of the entropy has to be nonnegative at all temperatures, this is $W_l \geq 0$. This condition imposes a minimum to the temperature, given by

$$T_0 = \frac{h^2}{2\pi m_q e k_B \mathcal{L}^2}, \tag{39}$$

where $e = 2.718281828$. Due to this basic property of W_l, the system is forced to take high values of temperature, being $T > T_0$, where the behavior of the system is classical. Equivalently, it is possible to assure that, if $T/T_0 \geq 1$, the length of a thermal wave λ lower than the average of the spacing among particles and quantum considerations are not relevant [39]. In addition, T_0 does not depend on external or internal physical parameters related to the system, as the transverse area, external magnetic field, charge of the particle, etc, practically depends only on the size of the system. If the system is large enough, the minimum temperature is low. However, modern electronic systems possess junctions where $\mathcal{L}$ can be considered almost zero. Thus, minimum temperature required to make applicable the present description is enough high [40].

2. The Wehrl entropy that is associated with transverse motion satisfies $W_t \geq 1 + \ln(g)$ for all temperatures of the system, which is very nearly the Lieb condition in one dimension [41] with an additional term given by the logarithm of g, the degeneracy. The transverse motion is approximately bi-dimensional, but the Landau approach reduces the quantum motion of the particle in a magnetic field to a degenerate spectrum in one dimension essentially recovering the physics of the missing dimension. Therefore, the discussion about the behavior of the Wehrl entropy in light of the Lieb condition does not increase any applicability of the present treatment because the latter is always satisfied. The main problem that appears from the emphasis on the transverse motion is the restricted vision that is obtained of the behavior of the system. [9, 30, 32, 33], which represents the main difference with other contributions that discuss this topic. The combination of reasoning including both motions has sense when the imposition over the temperature is satisfied. For values of the temperature lower than T_0, the behavior is essentially anomalous, thus this proposal is not applicable.

Additionally, the total entropy is expressed simply as follows

$$W_{\text{total}} = \frac{3}{2} - \ln(1 - \mathrm{e}^{-\beta\hbar\Omega}) + \ln(g) + \ln\left(\frac{\mathcal{L}}{\lambda}\right). \tag{40}$$

Now, we can discuss some approximate and limiting cases.

In first order of approximation, for $k_B T \gg \hbar\Omega$, we have $\ln(g/(1 - e^{-\beta\hbar\Omega})) \approx \ln(\mathcal{A}T/T_0\mathcal{L}^2)$. If we write the thermal wave length in terms of the temperature T_0, as $\lambda = \mathcal{L}(eT_0/T)^{1/2}$ and considering that $\mathcal{V} = \mathcal{A}\mathcal{L}$, the entropy (40) is rephrased as follows

$$W_{\text{total}}^{(1)} = \frac{3}{2} + \ln\left(\frac{\mathcal{V}}{\lambda^3}\right). \tag{41}$$

This is a particular expression for the entropy of a free particle in three dimensions related to the motion of a charged particle into a region of the magnetic field making mention of some geometrical properties of the system.

In second order of approximation, considering the special condition $\mathcal{A} \sim \mathcal{L}^2$, Wehrl entropy is expressed as follows

$$W_{\text{total}}^{(2)} \approx W_{\text{total}}^{(1)} + \frac{T_0}{T} g. \tag{42}$$

As explained before, the Wehrl entropy takes values that are permitted by the Lieb condition, namely, $W \geq 1$. According to Eq. (42) the slope decreases as temperature increases. This fact also illustrates why the disorder increases as the magnetic field increases too.

The lower bound of temperature is related to values of T greater than T_0, because this approach does not consider any temperature less than T_0. In addition to this, the behavior of the total Wehrl entropy is reduced to the logarithm of the magnetic field. In order to see what occurs in the limiting case of the lowest temperature, according to Eq. (39), we take systems with $L \to \infty$; thus the transverse entropy of Eq. (38) is rewritten as follows

$$W_t^{T \to 0^+} = 1 + \ln(g). \tag{43}$$

As aforementioned, the Wehrl entropy is similar to the entropy of the harmonic oscillator and the lowest temperature comes being greater than the bound temperature, thus $W \geq 1$ [41] as it was conjectured by Wehrl and shown by Lieb. From this condition, it must arrive to the following inequality for the magnetic field

$$g \geq 1, \tag{44}$$

where $g = q\mathcal{A}B/hc$ also accounts for the ratio between the flux of the magnetic field $\mathcal{A}B$ and the quantum of the magnetic flux given by $hc/q = 4.14 \times 10^{-7} [gauss\ cm^2]$ [14]. Then the inequality (44) adopts the form

$$B \geq \frac{1}{\mathcal{A}} \frac{hc}{q} = B_0. \tag{45}$$

Moreover, the magnetic field $B_0 = hc/\mathcal{A}q$ becomes to take a bound limiting value representing a minimum value for the external magnetic field. If $\mathcal{A} \to \infty$, we can study what occurs to the system when the magnetic field close to zero.

Now, we add two comments about the quantum description of particles in magnetic field close to limiting values of temperatures and magnetic fields, respectively:

1. The quantum Hall effect is observed in two-dimensional electron systems subjected to low temperatures and strong magnetic fields and emerges from the Landau quantization [42, 43] which corresponds to a quantum version of the Hall effect [35]. The degeneracy is given by [14]

$$\phi = \nu\phi_0, \tag{46}$$

where $\phi_0 = hc/q$ is the minimum quantity (or quantum) of the magnetic flux. The factor ν takes integer values as $\nu = 1, 2, 3, \ldots$ and it is related to the "*filling factor*" and simply with the conductivity quantization as $\sigma = \nu q^2/h$. The subsequent discovery of the fractional quantum Hall effect [34] expand the values ν to rational fractions as $\nu = 1/3, 1/5, 5/2, 12/5, \ldots$. Thus the fractional quantum Hall effect relies on other phenomena associated with interactions. In any case, the degeneracy is ν greater than 1 due to the inequality (44), as before, the transverse entropy always satisfies the Lieb bound for all temperatures and large enough systems, obtaining an infinite family of Wehrl entropies

$$W_t = 1 - \ln(1 - e^{\beta\hbar\Omega}) + \ln\nu. \tag{47}$$

The limiting value of ν provides a good descriptor for the integer quantum Hall effect. Conversely, for the fractional values of ν less than 1 are left out the present approach.

2. The Haas-van Alphen effect is other phenomenon that we can discuss. It is observed at low enough values of temperatures, describing oscillations in the magnetization, because the particles tend to occupy the lowest energy states. In the present description it is manifest for finite values of $\mathcal{A}$ and B lower than B_0. Whereas if the value of the magnetic field decreases a less number of particles can be in the lowest state due to degeneracy is directly proportional to B [38]. Then, the transverse Wehrl entropy W_t is well defined for values of the magnetic field over B_0, this is $B/B_0 \geq 1$ and/or $g \to 1^+$.

3.4. Fisher Information Measure versus degeneracy

In the present subsection we propose a compact expression for the transverse Fisher information measure, taking into account a special way formerly developed in Ref. [6], which is given by

$$\mathcal{F}_t = \int \frac{\mathrm{d}^2\alpha \mathrm{d}^2\xi}{4\pi^2\ell_{\mathrm{H}}^4}\,\eta_t(\alpha)\left(\frac{\partial \ln \eta_t(\alpha)}{\partial\alpha}\right)^2. \tag{48}$$

After introducing the known expression for η_t, we arrive to

$$\mathcal{F}_t = \frac{2}{\ell_{\mathrm{H}}^2}(1 - \mathrm{e}^{-\beta\hbar\Omega}). \tag{49}$$

Fisher measure $\mathcal{F}_t$ has space dimension $(L)^{-2}$ and quantifies the ability for estimating the parameter α [44]. This parameter corresponds to the radio of a circular orbit of coherent states. By combining Eqs. (49) and (46) with the definition of ℓ_{H} we obtain

$$\mathcal{F}_t = \frac{4\pi\nu}{A}(1 - \mathrm{e}^{-\beta\hbar\Omega}), \tag{50}$$

which represents the linear dependence of the measure $\mathcal{F}_t$ with the magnetic field through the constant ℓ_{H}^2 at low temperature.

The inverse exponential dependence on the temperature, of the Fisher information, is clear from Eq. (50). Further, the initial value directly depends on the factor ν.

Now, to complete the description of the movement, we consider the Fisher information measure for the longitudinal motion, this is

$$\mathcal{F}_l = \int \frac{dz dp_z}{h} \eta_l(p_z) \left(\frac{\partial \ln \eta_l(p_z)}{\partial p_z} \right)^2, \tag{51}$$

where p_z is the variable that we contain in the present discussion, which was previously ignored [32], making a great difference when the results are compared. The function η_l is included into the above equation to get

$$\mathcal{F}_l = \frac{\beta}{m}. \tag{52}$$

As seen before, the Fisher measure in one dimension coincides with the classical one for the free particle [45]. As expected, the total Fisher measure is constructed multiplying Eqs. (50) and (52).

3.5. Additional appointments and consequences

The Wehrl entropy, which we obtain here, depends on multiple parameters, for instance, the degeneracy g, and the ratio between the cylinder and thermal lengths, *i.e.*, $\mathcal{L}$ and λ. The combination of these parameters can effectively give some interesting results. Therefore, in a especial perspective we can see that the harmonic oscillator is behaved as a particular case of the charged particle in a magnetic field. Thus, we can consider, for example, the following relation among parameters:

$$g = \frac{\lambda}{\mathcal{L}} \exp\left(\frac{1}{2}\right), \tag{53}$$

which leads the Wehrl entropy from the Landau diamagnetism to the one-dimensional harmonic oscillator (15). This is a nontrivial approach because the nature of problems are radically different. For instance, the harmonic oscillator, that we use here, is a one-dimensional system, but the Landau diamagnetism is three-dimensional. Consequently, phase spaces are not coincident and measures are not the same.

Besides, we know $g \geq 1$. But, if we consider the minimum value $g = 1$, we can obtain the bound value of the temperature T_0, given by Eq. (39), above this value, the present approach is valid. Afterward, we obtain a relationship between both lengths involved into the problem, this is a bound value for the length of the cylinder, $\mathcal{L} \geq \lambda/e$. Thus, for values where this condition is violated, this approach is not valid.

The comparison between the Fisher information measures, for both cited problems, is also possible. Hence, in the same previous line we can propose a comparison of the Fisher information measures, considering measures dimensionally compatible. Originally, the classical Fisher information (11) accounts the localization of the corresponding probability density $\rho(x)$, which is approached by

Cramer-Rao inequality (12), where Δx is the variance for the stochastic variable x. However, the variation of the definition (11) takes into account the localization, not in the variable x or any other coordinate, but the localization into phase space. This is well defined for the transverse motion. Moreover, the longitudinal motion is classical, not quantized, and any coherent state formulation is proposed. The quantum counterpart can be defined as a problem of continuous spectrum [46], and a suitable formulation of coherent states is still unknown; for the time, this continues being an open problem. Thus, the classical formulation is used and we have decided to advance evaluating the classical distribution for the longitudinal motion.

In addition, with the purpose of describing the complete motion, we consider now the Fisher information measure for the movement, this is

$$\mathcal{F}' = \frac{\lambda^2}{\ell_{\mathrm{H}}^2}(1 - \mathrm{e}^{-\beta\hbar\Omega}). \tag{54}$$

where $\mathcal{F}'$ is defined as $\mathcal{F}' = h^2 I_t I_l / 4\pi$ in order to compare the trend of this Fisher measure with corresponding one of the harmonic oscillator. These cases are comparable with the harmonic oscillator only if $\lambda^2 = \ell_{\mathrm{H}}^2$ and are depicted with red-solid-line in Fig. 1.

4. Description of the molecular rotation: Rigid rotator

There are few physical systems whose spectrum is analytically known, aside from the previous one we have the anisotropic rigid rotator, which is a system of a single particle that can rotate in several ways. Thermodynamic properties can be analytically described [47]. It is expected that this treatment can characterize important features of molecular systems [48] to apply such concepts to several aspects related to materials [49].

4.1. Linear rigid rotator

We begin exploring the linear rigid rotator based on the excellent discussion made in Ref. [50] about the coherent states for angular momenta. The Hamiltonian of this simple system is [16]

$$\hat{H} = \frac{\hat{L}^2}{2I_{xy}}, \tag{55}$$

where the operator $\hat{L}^2$ is associated with the angular momentum and the parameter I_{xy} is the corresponding inertia momentum. The set $\{|IK\rangle\}$ is the set of eigenstates of the Hamiltonian, where we can verify the following relations

$$\begin{aligned} \hat{L}^2|IK\rangle &= I(I+1)\hbar^2|IK\rangle \\ \hat{L}_z|IK\rangle &= K\hbar|IK\rangle, \end{aligned} \tag{56}$$

with $I = 0, 1, 2\ldots$, for $-I \leq K \leq I$. Additionally, the energy spectrum is given by eigenstates of the operator H

$$\varepsilon_I = \frac{I(I+1)\hbar^2}{2I_{xy}}. \tag{57}$$

A suitable construction of coherent states is found in Ref. [51, 52] for the lineal rigid rotator, using Schwinger oscillator model of angular momentum, in the fashion

$$|IK\rangle = \frac{(\hat{a}_+^\dagger)^{I+K}(\hat{a}_-^\dagger)^{I-K}}{\sqrt{(I+K)!(I-K)!}}|0\rangle, \tag{58}$$

where $\hat{a}_+$, $\hat{a}_-$ are the corresponding creation and annihilation operators, respectively, and they show the following basic properties

1. The vacuum state

$$|0\rangle \equiv |0,0\rangle$$

.

2. Orthogonality is satisfied by

$$\langle I^{'}K^{'}|IK\rangle = \delta_{I^{'},I}\delta_{K^{'},K},$$

3. The completeness property is contained in the relation

$$\sum_{I=0}^{\infty}\sum_{K=-I}^{I}|IK\rangle\langle IK| = \hat{1}.$$

Due to we are interested in two degrees of freedom, the resulting coherent states come from the tensor product of $|z_1\rangle$ and $|z_2\rangle$ [50, 53], where

$$|z_1 z_2\rangle = |z_1\rangle \otimes |z_2\rangle, \tag{59}$$

and

$$\hat{a}_+|z_1 z_2\rangle = z_1|z_1 z_2\rangle, \tag{60}$$

$$\hat{a}_-|z_1 z_2\rangle = z_2|z_1 z_2\rangle. \tag{61}$$

Therefore, $|z_1 z_2\rangle$ is the coherent state written [50] as

$$|z_1 z_2\rangle = e^{-\frac{|z|^2}{2}} e^{z_1\hat{a}_+^\dagger} e^{z_2\hat{a}_-^\dagger}|0\rangle, \tag{62}$$

with

$$|z_1\rangle = e^{-\frac{|z_1|^2}{2}} e^{z_1\hat{a}_+^\dagger}|0\rangle, \tag{63}$$

$$|z_2\rangle = e^{-\frac{|z_2|^2}{2}} e^{z_2\hat{a}_-^\dagger}|0\rangle. \tag{64}$$

We need to introduce the suitable notation

$$|z|^2 = |z_1|^2 + |z_2|^2. \tag{65}$$

Using Eqs. (58) and (62) we easily calculate $|z_1 z_2\rangle$ and, after a bit of algebra, find

$$|z_1 z_2\rangle = e^{-\frac{|z|^2}{2}} \sum_{n_+,n_-} \frac{z_1^{n_+}}{\sqrt{n_+!}} \frac{z_2^{n_-}}{\sqrt{n_-!}} |IK\rangle, \tag{66}$$

where $n_+ = I + K$ and $n_- = I - K$. Thus, the probability of obtaining the state $|IK\rangle$ in the coherent state $|z_1 z_2\rangle$ is of the form

$$|\langle IK|z_1 z_2\rangle|^2 = e^{-|z|^2} \frac{|z_1|^{2n_+}}{n_+!} \frac{|z_2|^{2n_-}}{n_-!}. \tag{67}$$

The present coherent states satisfy resolution of unity

$$\int \frac{\mathrm{d}^2 z_1}{\pi} \frac{\mathrm{d}^2 z_2}{\pi} |z_1 z_2\rangle\langle z_1 z_2| = 1. \tag{68}$$

Furthermore, z_1 and z_2 are continuous variables.

The procedure developed by Anderson *et al.* [4] is easily followed and used to assess the Husimi distribution [1]. In our approach this is defined, from Eq. (4), as

$$\mu(z_1, z_2) = \langle z_1, z_2|\hat{\rho}|z_1, z_2\rangle, \tag{69}$$

where the density operator is

$$\hat{\rho} = Z_{2D}^{-1} \exp(-\beta \hat{H}). \tag{70}$$

The concomitant rotational partition function Z_{2D} is given in Ref. [16]

$$Z_{2D} = \sum_{I=0}^{\infty} (2I+1)\, e^{-I(I+1)\frac{\Theta}{T}}, \tag{71}$$

with $\Theta = \hbar^2/(2I_{xy}k_B)$. We emphasize that in the present context the performing of the sum $\mathrm{Tr} \equiv \sum_{I=0}^{\infty}\sum_{K=-I}^{I}$ corresponds to the "operation trace" . Using now the completeness property into Eq. (69) and theEq. (67), we fobtain the Husimi distribution in the form

$$\mu(z_1, z_2) = e^{-|z|^2} \frac{\sum_{I=0}^{\infty} \frac{|z|^{4I}}{(2I)!} e^{-I(I+1)\frac{\Theta}{T}}}{\sum_{I=0}^{\infty} (2I+1)\, e^{-I(I+1)\frac{\Theta}{T}}}. \tag{72}$$

It is easy to show that this distribution is normalized to unity

$$\int \frac{d^2z_1}{\pi}\frac{d^2z_2}{\pi}\mu(z_1,z_2) = 1, \tag{73}$$

where z_1 and z_2 are given by Eqs. (60), (61), and (65). We must employ the binomial expression $(|z_1|^2 + |z_2|^2)^{4I}$ and then integrate over the whole complex plane in two dimensions to verify the normalization condition. The differential element of area in the $z_1(z_2)$ plane is $d^2z_1 = dxdp_x/2\hbar$ $(d^2z_2 = dydp_y/2\hbar)$ [19]. Moreover, we have the phase-space relationships

$$|z_1|^2 = \frac{1}{4}\left(\frac{x^2}{\sigma_x^2} + \frac{p_x^2}{\sigma_{p_x}^2}\right), \tag{74}$$

$$|z_2|^2 = \frac{1}{4}\left(\frac{y^2}{\sigma_y^2} + \frac{p_y^2}{\sigma_{p_y}^2}\right), \tag{75}$$

where $\sigma_x \equiv \sigma_y = \sqrt{\hbar/2m\omega}$ and $\sigma_{p_x} \equiv \sigma_{p_y} = \sqrt{m\omega\hbar/2}$.

The profile of the Husimi function is similar to that of a Gaussian distribution.

As before, a semiclassical measure of localization is the Wehrl entropy [17], and the Fisher [5] as well. For the present model in two dimensions, the Wehrl entropy reads

$$\mathcal{W} = -\int \frac{d^2z_1}{\pi}\frac{d^2z_2}{\pi}\mu(z_1,z_2)\ln\mu(z_1,z_2), \tag{76}$$

where $\mu(z_1,z_2)$ is given by Eq. (72).

4.1.1. Fisher information measure

The Fisher measure [5, 20, 21] regards as a semiclassical counterpart of Wehrl entropy [5]. Now, extending the ideas developed in Ref. [5] for the case of the harmonic oscillator in one dimension to the present case in two dimensions, we can define the shift invariant Fisher measure in the fashion

$$\mathcal{F}_{2D} = \frac{1}{4}\int \frac{d^2z_1}{\pi}\frac{d^2z_2}{\pi}\mu(z_1,z_2)\left(\frac{\partial\ln\mu(z_1,z_2)}{\partial|z|}\right)^2. \tag{77}$$

From Eq. (72) it is easy to prove that

$$\eta(z_1,z_2) = \frac{1}{2}\frac{\partial\ln\mu(z_1,z_2)}{\partial|z|} = \frac{\sum_{I=0}^{\infty}\left[\frac{|z|^{4I-1}}{(2I-1)!} - \frac{|z|^{4I+1}}{(2I)!}\right]e^{-I(I+1)\Theta/T}}{\sum_{I=0}^{\infty}\frac{|z|^{4I}}{(2I)!}e^{-I(I+1)\Theta/T}}. \tag{78}$$

Therefore, the corresponding Fisher measure acquires the simpler appearance

$$\mathcal{F}_{2D} = \int \frac{d^2 z_1}{\pi} \frac{d^2 z_2}{\pi} \mu(z_1, z_2)\, \eta(z_1, z_2)^2, \tag{79}$$

i.e.,

$$\mathcal{F}_{2D} \equiv \langle \eta(z_1, z_2)^2 \rangle, \tag{80}$$

where with the notation

$$\langle \mathcal{G} \rangle = \int \frac{d^2 z_1}{\pi} \frac{d^2 z_2}{\pi} \mu(z)\, \mathcal{G}, \tag{81}$$

we refer to the *semi-classical expectation value* of $\mathcal{G}$. In Fig. 1 we plot the Fisher information and the Wehrl entropy as a function of the temperature (black-dashed-line), which we compare with the same measures for the transverse Landau diamagnetism (blue-solid-line). At low temperatures, the Fisher information measure describes the inverse-delocalization and takes its maximum value when the Wehrl entropy is minimum. This behavior is reversed for high temperatures. Every curve can be compared with the respective counterpart shown for the harmonic oscillator in one dimension (red-solid-line).

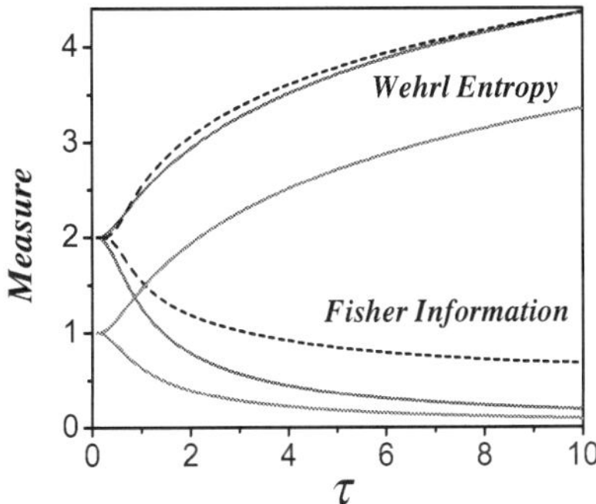

Figure 1. Trends of Fisher Information and Wehrl entropy for the rotator (black-dashed-line) in two dimensions is compared with the transverse Landau diamagnetism (blue-solid-line), the horizontal axis is the normalized temperature $\tau = kT(2Ixy)/\hbar^2$ and $\tau = kT/\hbar\Omega$, respectively. Additionally, we show a case where the Landau diamagnetism dimensionally coincides with the one-dimensional harmonic oscillator (red-solid-line). The Wehrl entropy starts in W=1. If the normalized temperature increases, the Fisher information decreases while Wehrl entropy increases.

4.2. Rigid rotator in three dimensions

In the present section we consider a more general problem, the model of the rigid rotator in three dimensions, whose Hamiltonian writes [54]

$$\hat{H} = \frac{\hat{L}_x^2}{2I_x} + \frac{\hat{L}_y^2}{2I_y} + \frac{\hat{L}_z^2}{2I_z}, \tag{82}$$

where the parameters I_x, I_y, and I_z are the inertia momenta. The set $\{|IMK\rangle\}$ corresponds to a complete set of eigenvectors of the operator $\hat{H}$. The following relations are additionally applied

$$\begin{aligned} \hat{L}^2|IMK\rangle &= I(I+1)\hbar^2|IMK\rangle \\ \hat{L}_z|IMK\rangle &= K\hbar|IMK\rangle \\ \hat{J}_z|IMK\rangle &= M\hbar|IMK\rangle, \end{aligned} \tag{83}$$

with $-I \leq K \leq I$ and $-I \leq M \leq I$, where $I = 0,\ldots,\infty,$. The elements of set $\{|IMK\rangle\}$ satisfy orthogonality and completeness property [54]

$$\langle I^{'}M^{'}K^{'}|IMK\rangle = \delta_{I^{'},I}\delta_{M^{'},M}\delta_{K^{'},K} \tag{84}$$

$$\sum_{I=0}^{\infty}\sum_{M=-I}^{I}\sum_{K=-I}^{I}|IMK\rangle\langle IMK| = \hat{1}. \tag{85}$$

If we take $\hat{L}^2 = \hat{L}_x^2 + \hat{L}_y^2 + \hat{L}_z^2$ and assume axial symmetry, i.e., $I_{xy} \equiv I_x = I_y$, we can recast the Hamiltonian as

$$\hat{H} = \frac{1}{2I_{xy}}\left[\hat{L}^2 + \left(\frac{I_{xy}}{I_z} - 1\right)\hat{L}_z^2\right], \tag{86}$$

where the operator $\hat{L}_z$ represents the projection on the rotation axis z of the $\hat{L}^2$, which is the angular momentum operator. The concomitant spectrum of energy becomes

$$\varepsilon_{I,K} = \frac{\hbar^2}{2I_{xy}}\left[I(I+1) + \left(\frac{I_{xy}}{I_z} - 1\right)K^2\right], \tag{87}$$

where the number I is integer and non-negative and it stands for the eigenvalue of the operator $\hat{L}^2$, the angular momentum. The range of the other quantum number $-I \leq m \leq I$ represents the projections on the intrinsic rotation axis of the rotator. Every state has a degeneracy $(2I+1)$. The inertia momenta are quantified by the parameters $I_x = I_y \equiv I_{xy}$ and I_z. The ratio I_{xy}/I_z characterizes different "geometrical" issues. For instance, some typical values of I_{xy}/I_z are 1, 1/2 and ∞, which correspond to the spherical, the extremely oblate and prolate cases, respectively.

4.2.1. Construction of coherent states

Again, we cite the work of Morales *et al.* where they construct a suitable set of coherent states for the rigid rotator in Ref. [54] and kindly discuss their mathematical foundations. First, they start introducing the auxiliary quantity

$$X_{I,M,K} = \sqrt{I!(I+M)!(I-M)!(I+K)!(I-K)!} \tag{88}$$

to obtain [54]

$$|z_1 z_2 z_3\rangle = e^{-\frac{|u|^2}{2}} \sum_{IMK} \frac{[(2I)!]^2 z_1^{(I+M)} z_2^I z_3^{(I+K)}}{X_{I,M,K}} |IMK\rangle, \tag{89}$$

where Morales *et al.* introduced the following supplementary variable

$$|u|^2 = |z_2|^2 (1+|z_1|^2)^2 (1+|z_3|^2)^2. \tag{90}$$

These coherent states comply at least two requirements: continuity of labeling and resolution of unity. In relation to this latter property, we add

$$\int d\Gamma |z_1 z_2 z_3\rangle\langle z_1 z_2 z_3| = 1, \tag{91}$$

where the measure of integration dΓ is given by [54]

$$\mathrm{d}\Gamma = \mathrm{d}\tau \left\{ 4[(1+|z_1|^2)(1+|z_3|^2)]^4 |z_2|^4 - 8[(1+|z_1|^2)(1+|z_3|^2)]^2 |z_2|^2 + 1 \right\} \tag{92}$$

with

$$\mathrm{d}\tau = \frac{\mathrm{d}^2 z_1}{\pi} \frac{\mathrm{d}^2 z_2}{\pi} \frac{\mathrm{d}^2 z_3}{\pi}. \tag{93}$$

In accordance with this requirement on coherent states, we can assert that the present formulation satisfy the weaker version, because the measure is non-positive definite [54].

4.2.2. *Husimi function, Wehrl entropy*

In order to get a valid expression for the Husimi distribution and the Wehrl entropy, a proper formulation of coherent states is essential. Using now Eq. (89) we find

$$|\langle IMK|z_1 z_2 z_3\rangle|^2 = \frac{e^{-|u|^2}}{X^2_{I,M,K}} [(2I)!]^2 |z_1|^{2(I+M)} |z_2|^{2I} |z_3|^{2(I+K)}. \tag{94}$$

Therefore, the rotational partition function is given by

$$Z_{3D} = \sum_{I=0}^{\infty} \sum_{K=-I}^{I} \sum_{M=-I}^{I} e^{-\beta \varepsilon_{I,K}}, \tag{95}$$

i.e.,

$$Z_{3D} = \sum_{I=0}^{\infty} (2I+1)\, e^{-I(I+1)\frac{\Theta}{T}} \sum_{K=-I}^{I} e^{-\left(\frac{I_{xy}}{I_z}-1\right)K^2\frac{\Theta}{T}}. \tag{96}$$

We see that Z_{2D} is recovered from Z_{3D} for the limiting case defined as the extremely prolate. The Husimi distribution yields

$$\mu(z_1,z_2,z_3) = \frac{e^{-|u|^2}}{Z_{3D}} \sum_{I=0}^{\infty} \frac{(2I)!}{I!} |v|^{2I} e^{-I(I+1)\frac{\Theta}{T}} \times g(I), \tag{97}$$

where

$$g(I) = \sum_{K=-I}^{I} \frac{|z_3|^{2(I+K)}}{(I+K)!(I-K)!} e^{-\left(\frac{I_{xy}}{I_z}-1\right)K^2\frac{\Theta}{T}}, \tag{98}$$

with

$$|v|^2 = (1+|z_1|^2)^2 |z_2|^2, \tag{99}$$

$$|u|^2 = |v|^2(1+|z_3|^2)^2. \tag{100}$$

Other relevant property that it is easily verified for $\mu(z_1,z_2,z_3)$ is normalization in the fashion

$$\int d\Gamma\, \mu(z_1,z_2,z_3) = 1. \tag{101}$$

Now, we obtain the Wehrl entropy in the form

$$\mathcal{W} = \int d\Gamma\, \mu(z_1,z_2,z_3) \ln \mu(z_1,z_2,z_3). \tag{102}$$

The spherical rotator, that corresponds to another special case, we explicitly obtain

$$\mu(z_1,z_2,z_3) = e^{-|u|^2} \frac{\sum_{I=0}^{\infty} \frac{|u|^{2I}}{I!} e^{-I(I+1)\frac{\Theta}{T}}}{\sum_{I=0}^{\infty} (2I+1)^2 e^{-I(I+1)\frac{\Theta}{T}}}. \tag{103}$$

Having the Husimi functions the Wehrl entropy is straightforwardly computed.

In order to emphasize some special cases associated to possible applications we consider several possibilities.

1. The spherical rotator $I_{xy} = I_x = I_y = I_z$, which corresponds to $I_{xy}/I_z = 1$ (*e.g.* CH_4).
2. The oblate rotator $I_{xy} = I_x = I_y < I_z$, being $1/2 \leq I_{xy}/I_z < 1$ (*e.g.* C_6H_6).
3. The prolate rotator $I_{xy} = I_x = I_y > I_z$, thus $I_{xy}/I_z > 1$ (*e.g.* PCl_5).
4. The extremely prolate rotator is equivalent to the linear case (all diatomic molecules, $I_z = 0$, this is $I_{xy}/I_z \to \infty$ (*e.g.* CO_2, C_2H_2).

4.2.3. Fisher information measure

In this circumstance we define the shift invariant Fisher measure in $3D-$dimensions as

$$\mathcal{F}_{3D} = \frac{1}{4}\int d\Gamma\,\mu(z_1,z_2,z_3)\left(\frac{\partial \ln\mu(z_1,z_2,z_3)}{\partial |u|}\right)^2. \tag{104}$$

Thus, from Eq. (97) we get

$$\varphi(z_1,z_2) = \frac{1}{2}\frac{\partial \ln\mu(z_1,z_2)}{\partial |u|} = \frac{\sum_{I=0}^{\infty}\left[\frac{|u|^{2I-1}}{(I-1)!} - \frac{|u|^{2I+1}}{(I)!}\right] e^{-I(I+1)\Theta/T}}{\sum_{I=0}^{\infty}\frac{|u|^{2I}}{(I)!}\, e^{-I(I+1)\Theta/T}}, \tag{105}$$

and, the corresponding Fisher measure can be expressed as

$$\mathcal{F}_{3D} = \int d\Gamma\,\mu(z_1,z_2)\,\varphi(z_1,z_2,z_3)^2 = \langle\varphi(z_1,z_2,z_3)^2\rangle. \tag{106}$$

5. Final remarks

In this chapter, we have described some elements to motivate possible and future applications in condensed matter and information theory. Our fundamental discussion is devoted to two interesting systems, those are: the Landau diamagnetism and the rigid rotator in three dimensions. We choose these systems because the quantum mechanics is analytically solved. Specifically, the spectrum and a suitable formulation of coherent states are known without approximations.

In general, quantum distributions as the Husimi distribution, have long been seen as powerful tools for studying the quantum-classical correspondence and semi-classical aspects of quantum mechanics. Then, a crucial starting point in the present strategy, to evaluate some theoretical measures, is to get the Husimi distribution. This is made evoking a convenient set of coherent states in every system. As introduced by Gazeau and Klauder in the context of the harmonic oscillator, we use the same formal perspective of general requirements for formulations of coherent states that we use in the current contribution. Additionally, we have included some mathematical and practical details of the the present formalisms in order to make it instructive in courses of quantum mechanics (for graduates) and easy to apply to specific calculations of theoretical measures.

The present derivation of Husimi distributions is based on the evaluation of the mean value of the density operator in the basis of a single-particle coherent state. Then, after defining the Husimi distribution we are ready to make a possible semiclassical description evaluating (i) the semiclassical Wehrl entropy and (ii) the phase-space location via measures as Fisher information.

Furthermore, we evaluate the probability of observing a quantum state in a coherent state, by projecting the quantum states over the coherent states, as a function of a variable related to the coherent states. We see that the localization of probability and correspondingly the Husimi distribution in the phase space decreases as temperature increases.

As known, while the coherent states are independent-particle states, the Husimi function takes into account collective and environmental effects being necessary many wave packets of independent-particle states to represent them. Furthermore, the thermodynamics of particles in systems does not depend on any coherent states formulation.

Finally, we remark, all results presented here were kindly obtained in an analytical fashion. We show some instances where the Landau diamagnetism is equivalent to the harmonic oscillator and, in the other example, where the linear rigid rotator is reobtained as a particular instance of the formulation in three dimensions. Some indications given in the present work lead to the conclusion that Fisher measure is a better indicator of the delocalization than Wehrl entropy.

We acknowledge partial financial support by FONDECYT 1110827 and CONICYT PSD065.

Author details

Sergio Curilef[1] and Flavia Pennini[1,2]

1 Departamento de Física, Universidad Católica del Norte, Antofagasta, Chile

2 Instituto de Física La Plata–CCT-CONICET, Fac. de Ciencias Exactas, Universidad Nacional de La Plata, La Plata, Argentina

References

[1] K. Husimi *Proceedings of the Physico-Mathematical Society of Japan* **22** (1940) 264.

[2] K. Takahashi and N. Saitô, *Physical Review Letters* **55** (1985) 645.

[3] M.C. Gutzwiller, *Chaos in Classical and Quantum Mechanics*, (Springer-Verlag, New York, 1990).

[4] A. Anderson and J.J. Halliwell, *Physical Review* **D 48** (1993) 2753.

[5] F. Pennini and A. Plastino, *Physical Review* **E 69** (2004) 057101.

[6] F. Olivares, F. Pennini, G.L. Ferri, A. Plastino, Brazilian Journal of Physics **39**, 2A (2009).

[7] S. Curilef, F. Pennini, A. Plastino and G.L. Ferri, *J. Phys. A: Math. Theor.* **40** (2007) 5127.

[8] S. Curilef, F. Pennini and A. Plastino, *Physical Review* **B 71** (2005) 024420.

[9] D. Herrera, A.M. Valencia, F. Pennini and S. Curilef, *European Journal of Physics* **29** (2008) 439.

[10] M. Janssen, *Fluctuations and Localization in Mesoscopic Electron Systems*, World Scientific Lecture Notes in Physics Vol. 64., 2001

[11] W.P. Scheleich, *Quantum Optics in phase space*, (Wiley VCH-Verlag, Berlin, 2001).

[12] J.P. Gazeau and J.R. Klauder, *Journal of Physics A: Math Gen.* **32** (1999) 123.

[13] S. Curilef and F. Pennini (2013). The Husimi Distribution: Development and Applications, Advances in Quantum Mechanics, Prof. Paul Bracken (Ed.), ISBN: 978-953-51-1089-7, InTech,

DOI: 10.5772/53846. Available from: http://www.intechopen.com/books/advances-in-quantum-mechanics/the-husimi-distribution-development-and-applications

[14] F. Olivares, F. Pennini, S. Curilef, *Physical Review* **E 81** (4) (2010) 041134.

[15] S. Curilef, F. Pennini, A. Plastino, G.L. Ferri, *Journal of Physics: Conference Series* **134** (2008) 012029.

[16] R.K. Pathria, *Statistical Mechanics*, (Pergamon Press, Exeter, 1993).

[17] A. Wehrl, *Reviews of Modern Physics* **50** (1978) 221.

[18] A. Wehrl, *Reports on Mathematical Physics* **16** (1979) 353.

[19] R.J. Glauber, *Physical Review* **131** (1963) 2766.

[20] B.R. Frieden. *Physics from Fisher information*, Cambridge University Press, Cambridge, England, 1998.

[21] B.R. Frieden. *Science from Fisher information*, Cambridge University Press, Cambridge, England, 2003.

[22] A.R. Plastino, A. Plastino, *Phys. Rev.* **E 54** (1996) 4423.

[23] A. Plastino, A.R. Plastino, H.G. Miller, *Phys. Lett. A* **235** (1997) 129.

[24] M.J.W. Hall, *Phys. Rev.* **A 62** (2000) 012107.

[25] F. Pennini, A. Plastino, G.L. Ferri, *Entropy* **14** (2012) 2081.

[26] L.D. Landau, *Z. Physik* **64** (1930) 629.

[27] S. Dattagupta, A.M. Jayannvar, N. Kumar, *Current Science* **80** (2001) 861.

[28] J. Kumar, P.A. Sreeram, S. Dattagupta, *Physical Review* **E 79** (2009) 021130.

[29] A. M. Jayannavar, M. Sahoo, *Physical Review* **E 75** (2007) 032102.

[30] A. Feldman, A.H. Kahn, *Physical Review* **B 1** (1970) 4584.

[31] B.R. Frieden, (Cambridge University Press, Crambridge, England, 1998).

[32] S. Curilef, F. Pennini, A. Plastino, *Physical Review* **B 71** (2005) 024420.

[33] K. Kowalski, J. Rembielínski, *Journal of Physics A: Math. Gen.* **38** (2005) 8247.

[34] D.C. Tsui, H.L. Stormer, A.C. Gossard, *Physical Review Letters* **48** (1982) 1559.

[35] K. v.Klitzing, G. Dorda, M. Pepper, *Physical Review Letters* **45** (1980) 494.

[36] M.H. Johnson and B. A. Lippmann, *Physical Review* **76** (1949) 828.

[37] See Eq. (16) in Ref. [32].

[38] K. Huang, *Statistical Mechanics*, Wiley, New York, Second edition (1963).

[39] L.E. Reichl, *A Modern course in statistical physics*, Wiley, New York, Second edition, (1998).

[40] M. Janssen, *Fluctuation and localization in messoscopic electronic systems*, World Scientific, (2001).

[41] E.H. Lieb, *Communications in Mathematical Physics* **62** (1978) 35.

[42] F. Lado, *Physics Letters* **A 312** (2003) 101.

[43] T. Chakraborty, P. Pietiläinen, *The Quantum Hall Effects*, Springer-Verlag, Berlin, (1995).

[44] A. Feldman, A.H. Kahn, Phys. Rev. **B 1** (1970) 4584.

[45] F. Pennini, A. Plastino, *Physics Letter* **A 349** (2006) 15.

[46] F. Pennini and S. Curilef Commun. Theor. Phys. **53** (2010) 535.

[47] N. Ullah, *Physical Review* **E 49** (1994) 1743.

[48] F.J. Arranz, F. Borondo, R. M. Benito, *Physical Review* **E 54** (1996) 2458.

[49] J.T. Titantah, M. N. Hounkonnou, *Journal of Physics A: Math. Gen.* **32** (1999) 897; **30** (1997) 6347; **30** (1997) 6327; **28** (1995) 6345.

[50] M.M. Nieto, *Physical Review* **D22** (1980) 391.

[51] J.J. Sakurai, Modern Quantum Mechanicsed S.F. Tuan (Reading, MA: Addison-Wesley, 1994) p 217.

[52] J. Schwinger, In Quantum Theory of Angular Momentumed, L C Biedenharn and H van Dam (New York: Academic, 1965) p 229.

[53] C. Cohen-Tannoudji, B. Diu, F. Laloe, Quantum Mechanics vol.1 John Wiley & Sons, (1977).

[54] J.A. Morales, E. Deumens and Y Öhrn, *Journal of Mathematical Physics* **40** (1999) 776.

Chapter 14

Implications of the "Subquantum Level" in Carcinogenesis and Tumor Progression via Scale Relativity Theory

Daniel Timofte, Lucian Eva, Decebal Vasincu,
Călin Gh. Buzea, Maricel Agop and Radu Florin Popa

Additional information is available at the end of the chapter

http://dx.doi.org/10.5772/59233

1. Introduction

The last 25 years witnessed tremendous achievements in cancer diagnose and treatment. Technology currently permits small size tumors (like breast cancers) diagnosis and treatment, ductal cancer in situ currently including 25...30% of all freshly diagnosed breast cancers at the majority of medical centers [1]. Thus, early detection now allows the understanding of growth patterns. Surgeons are in the front line of technological and basic scientific medical advances. Current ideas, such as the physiological characteristics of shock, organ transplantation, antisepsis, wound healing, or sequence medical care, are cast by surgical investigators.

The field of mathematics suffered an identical evolution. Revolutionary mathematical branches such as topology, fractals, chaos theory, and development of nonlinear descriptive strategies have provided mathematicians new inventive tools to create growth models and to behaviors at the small environmental level [2, 3]. Growth, angiogenesis [4], cell-to-cell adhesion [5], hydrogen ion concentration regulation and drug delivery [6] can now systematically be described using specific formulas. From a clinical viewpoint several of these formulas could seem simple, however they put together a very important foundation for descriptive insight.

What is currently lacking is a connection between these two naturally and mutual analysis endeavors. For oncology surgeons, the ability to mathematically analyze and predict patterns of growth provides precise techniques that are beneficent for both current and future therapies. For mathematicians, defining the clinical factors essential for growth development and metastasis can provide realistic insight into these biological processes, successively allowing the event of correct, clinically relevant mathematical formulas. In almost every dedicated

medical institutions, the teams that have comprehensive cancer are being led by surgeons. Mathematics that can be applied in the oncology field provide a chance to expand the leadership role of the surgeons and to raise awareness about the significance of understanding growth behavior and, also improve cancer treatments.

The present study aims at defining a new concept of carcinogenesis and tumor progression. Consequently, we use the natural 'environment' where malignant tumors grow, space(-time) with non-integer fractal dimension, questing for further applications of the newly discovered and intriguing phenomenon of tumor self-seeding by circulating cancer cells (CTC). More precisely, we assume that the metastatic tumor cells move (through the systemic circulation, yet not necessarily only there) as a coherent wave, or even more precisely, a chemically pumped travelling laser wave with oxygen. The extracellular matrix (ECM) and in particular, the tumor microenvironment (TME) are assumed as non-differential media endowed with holographic properties and may be good candidates for "recording" materials. As a result, the tumor self-seeding by CTC may be proved mathematically, the fact that the CTC returning to the initial tumor site and fueling the primary tumor growth or even grow a new tumor is a particular case of complete holography (i.e. a hologram which does not represent only the virtual object's image, but it becomes the very object - which we believe, is a characteristic of the living organisms). We believe our findings may provide new opportunities to set up new targeted therapies that may slow down or even prevent tumor progression.

According to the Pribram-Bohm's holographic theory (http://en.wikipedia.org/wiki/Holonomic_brain_theory) of the brain, the intercellular and intracellular communication implies the existence of a fractal medium equivalent to the vacuum between the elementary particles. Since vacuum dynamics is studied using quantum mechanics, it is only natural that the status of the fractal medium implies the use of a quantum type mathematical formalism (i.e. Scale Relativity theory either in its Schrödinger type representation or the fractal hydrodynamic one) applicable to different resolution scales (mesoscopic for intercellular communications or nanoscopic for intracellular ones). In Nottale's interpretation, each resolution scale is characterized by a Planck type constant $\bar{h}$

$$\bar{h} = \hbar\left(\frac{dt}{\tau}\right)^{(2/D_F)-1}$$

where $\hbar$ is the standard Planck's constant, dt is time's resolution scale, τ the time's reference scale and D_F is the fractal dimension of the motion curve.

2. The biology of cancer

2.1. Cancer, what should be noticed

Cancer or malignant neoplasm is a class of diseases that rises from the anomalous behavior of normal tissue. Cancer cells are aberrant cells which have acquired malignant traits such as uncontrolled growth (cells continuously proliferate), tissue invasion (they intrude into normal tissue and destroy it) and metastasis (they spread outside the location of the body where they were originally generated). Additionally, the term tumor or neoplasm is used to indicate an abnormal swelling of tissue caused by an excessive cell proliferation.

A tumor can be of benign or malignant nature, while benign tumors are self-limiting, do not express patterns of invasion, and they do not metastasize, malignant tumors do possess all these characteristics. The term malignant tumor is also used as synonym for cancer, although some cancers, such as leukemia, do not form tumors.

Cancer cells develop these malignant features because of genetic mutations, accumulated during the organism lifetime. Cancer is in fact a multi-step chance process that transforms a normal cell into a tumor cell, after having collected a set of 5...8 crucial genetic alterations [7-9] as schematically shown in Fig. 1.

A newborn malignant cell, expressing aberrant traits, can lead to the formation of cancer and, in most of the cases, of a tumor. Without treatment, the destructive behavior of such colony of cells is usually lethal for the patient. The probabilistic nature of this disease and the increase in life expectancy had made cancer the second cause of death in the industrialized countries (see any cancer statistics). Nevertheless, cancer it is not a modern disease and it was known since the antiquity: Egyptians of the New Kingdom [10], Greeks [11] and Romans [12] accurately described medical treatments for tumor removal. It is only within the last two centuries however, that due to the higher standards of living, cancer has become one of the main life-threatening diseases.

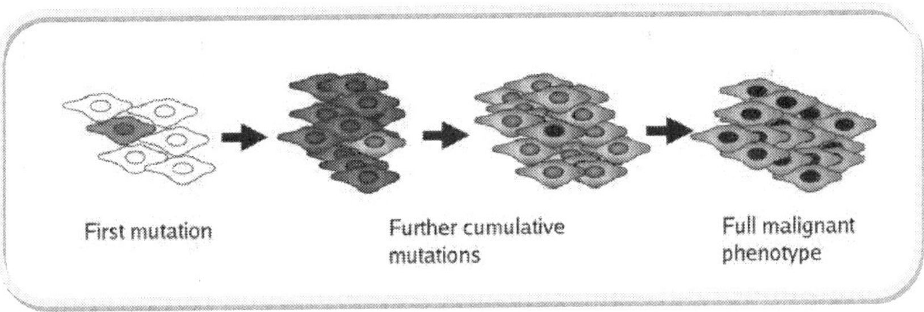

Figure 1. Acquisition of the tumorigenic phenotype by a population of normal cells through multiple genetic mutations.

2.2. Distinguishing traits of cancer

The tumorigenic properties, generically discussed in the previous section, have shown to be common to almost all cancers. They have been studied since the dawn of cancer research and they can be enumerated and defined with a relatively high accuracy. These hallmarks are a set of characteristic traits typical of cancer cells that are essential for the formation of a macroscopic malignant neoplasm [13]:

Self-Sufficiency in Growth Signals. All cells communicate through signals. A biological signal is, in most of the cases, a protein able to deliver a particular piece of information by binding uniquely to specific receptors on the cell surface. Normal cells need mitogenic growth signals to proliferate (signals that allow and stimulate cell proliferation). Those signals are regulated by the homeostasis of the tissue and they guarantee a correct balance between cell proliferation and death, according to the needs of the organism. In order to lead to cancer, tumor cells may develop the ability of self-generating such signals in one way or another. One possible way is a genetic aberration in one of the fundamental genes responsible for the building of the signaling pathway, for instance the RAS oncogene [9,14]. As consequence, the associated component of the signaling system would become constitutively active and hence, independent by the signal molecule. A second option is the self-production of growth factors that would stimulate growth by paracrine signaling, where a cells stimulates the neighbors and vice-versa or even autocrine signaling, when the cell stimulates its own receptors as shown in Fig. 2.

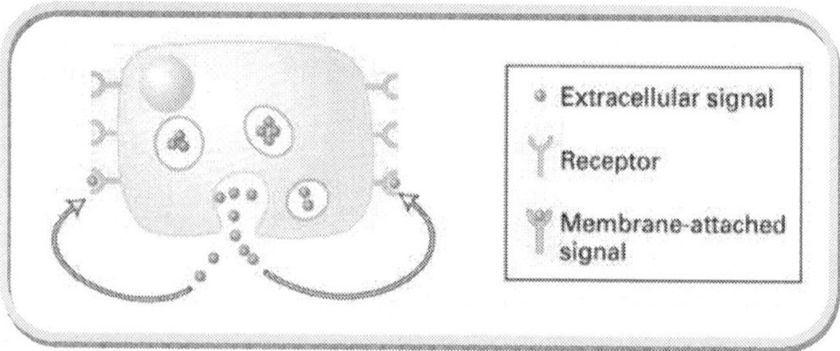

Figure 2. Example of self-signaling (autocrine): the cell produces its own growth factors which stimulate the growth receptors on the surface (Dr. W.H. Moolenaar, Netherlands Cancer Institute).

Insensitivity to Antigrowth Signals. As counterparts of growth factors, homeostasis employs growth inhibiting signals as well. These signals act similarly to their antagonists but they promote cell cycle arrest or cell quiescence, rather than proliferation. An example of a crucial gene involved in anti-growth pathways is the retinoblastoma protein (pRb). The retinoblastoma protein is capable of altering the function of the E2F transcription factors and control the expression of the bank of genes essential for the transition from GAP-1 phase to DNA Synthesis phase of the cell cycle [15]. The disruption of such pathway results in the insensitivity of the cell to anti-growth signals.

Evading Apoptosis. Apoptosis is a mechanism of controlled cell death. Through special signals, a cell has the capacity of terminating itself in a highly regulated way. A normal cell dying by apoptosis undergoes a sequence of events such as condensation, fragmentation and phagocytosis. This avoids the cell to free potentially dangerous enzymes and proteins stored inside its cytoplasm and its nucleus. During apoptosis the cell membrane is kept intact while, in 30...120 minutes the cell is fragmented in small parts or apoptotic bodies, still protected by pieces of membrane. Those cell leftovers are successively phagocytated by macrophages within the next 24 hours [16]. Apoptosis is a common mechanism of cell death and takes part in the homeostasis of a healthy organism as well as in its embryogenesis and in its morphogenesis. When any cell violates such homeostasis, an apoptotic signal is delivered to it. Therefore, in order for cancer cells to develop into a malignant lesion, it is necessary to deactivate apoptotic signal pathways. A mutation in the p53 tumor suppressor gene (TSG) is one of the most common ways to acquire resistance to apoptosis because p53 regulates the whole signaling process of programmed cell death. Indeed, more than 50% of human cancers carry a mutation in the p53 tumor suppressor gene [17].

Limitless Replicative Potential. Even with all the anti-growth and anti-apoptosis pathways triggered off, a cell could not generate a vast population able to form a tumor. That is because of the intrinsic proliferation limit of all mammalian cells. All chromosomes have an ending cap called telomere, a T-loop non-coding DNA sequence (2...50 Kb) that prevents the end of the chromosomes from attaching to other genetic material. At every mitosis the cell loses a small part of its telomeres because of the impossibility for DNA duplication enzymes, for instance DNA-polymerase, to continue working until the very end of the genome (Fig. 3). This limitation is due to the fact that enzymes like DNA-polymerase always move in the 5′...3′ direction of the DNA sequence, so when the side of the replication is opposite, a small part of the genome is lost. The shortening of the telomeres induces cell senescence, a state of cellular elderly where division no longer occurs. This avoids genetically unstable cells to replicate. Senescence starts after the so called Hayflick limit [18] of about 50 cell divisions. In cancer cells instead, the disabling of the pRb and the p53 pathways allows unlimited replication, until the point when the telomeres are completely absent. Once having entirely consumed the telomeres, the cell population is believed to undergo a phase of massive genomic instability, causing extended cell death. The high selective pressure induced by this crisis may permit specific resistant clones to emerge (Fig. 4). Those survivor cells would be immortalized (unlimited proliferative potential) by finding ways to maintain their telomeres long enough. A possible way is the over expression of the telomerase gene [19] which appears to take place in 85...90 % of cancers. Telomerase is a telomere-rebuilding enzyme normally expressed in germ line cells and stem cells, in which immortalization is an essential feature. Once immortalized, malignant cells have made a further step towards the formation of cancer.

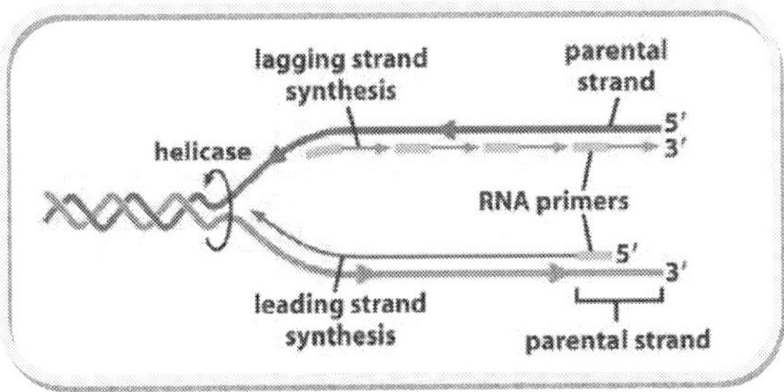

Figure 3. Illustration of the end replication problem: at both sides of the copying, the leading DNA strand has lost part of the telomeric sequence, which stops at the 5′ end of the parental strand, whereas the lagging strand results completed until the very end (Dr. R. Beijersbergen, Netherlands Cancer Institute).

Sustained Angiogenesis. In order for a cell to survive normally, it must rely within 100 μm from a capillary blood vessel [13]. For this reason, the initial exponential growth of a newborn malignant neoplasm causes a shortage of nutrients among cancer cells. Local pre-existent vascularisation is never enough to sustain growth for more than 10^8 cells. The colony must therefore develop angiogenesis-triggering capabilities [20, 21]. Angiogenesis is the process of formation of new blood vessels in response to a stimulus secreted by poor vascularised tissues. Angiogenesis is important for the organism morphogenesis and even later maintains the correct supply of nutrients for all tissues. Fast growing cells, such as cancer cells, start soon to starve, and have the need of additional blood supply in order to keep expanding. A possible solution is the production by cancer cells of vascular endothelial growth factors (VEGF) and fibroblast growth factors (FGF1/2) which bind to the transmembrane receptors of endothelial cells (cells covering the interior surface of blood vessels) stimulating their growth towards the signal concentration gradient [22]. Angiogenesis is the principal mechanism that transforms a microscopic malignancy into a macroscopic tumor and, also in the later stages, it is necessary for a lesion to grow and sustain itself. This implies that angiogenesis is an important target for anti-cancer drugs like thrombospondin-1 [23] and bevacizumab [24], also known as avastin.

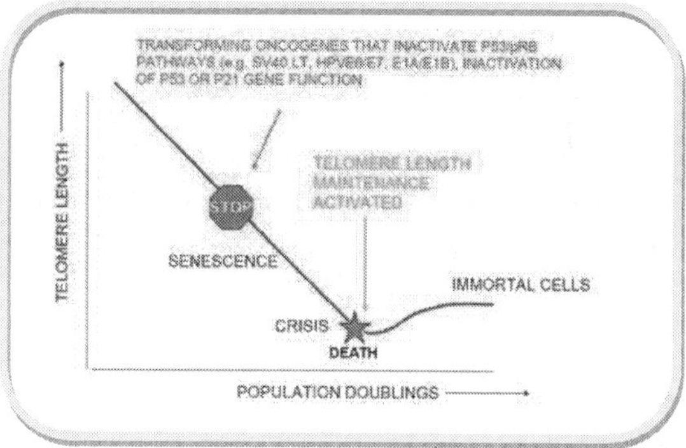

Figure 4. The progressive shortening of the telomeres leads to a massive cell death due to the induced genomic instability (death by genomic catastrophe while duplicating). From such process of intense genetic mutation and selection an immortalized clone could emerge (Dr. R. Beijersbergen, Netherlands Cancer Institute).

Tissue Invasion and Metastasis. The most dangerous and destructive features of cancer are tissue invasion and the consequent metastasis. Its ability of forming distant colonies or metastases all over the body represents the cause of 90% of all cancer related deaths [25]. Normal cells are usually unable to travel outside their own tissue due to their necessity to be anchored and reside among similar cells. An eventual detachment from the extracellular matrix or ECM (a complex structure of proteins and specific cells forming the tissue scaffold and microenvironment – see Sec. 6.1) would occur in a form of apoptosis called anoikis [26]. Contrary to their normal counterparts, cancer cells are able to survive the loss of anchorage, to travel through the vascular system and form distant tumors elsewhere (Fig. 5). The traits expressed by invasive and metastatic cancer cells are principally loss of cell-to-cell adhesion, anchorage-independence, chemotaxis (migration towards a diffusible substance gradient), haptotaxis (migration towards a non-diffusible substance gradient) and production of matrix degrading enzymes (*e.g.* Matrix metalloproteinase) which cleave the extracellular matrix [27-29] making space for invasion and freeing growth and angiogenic factors trapped inside.

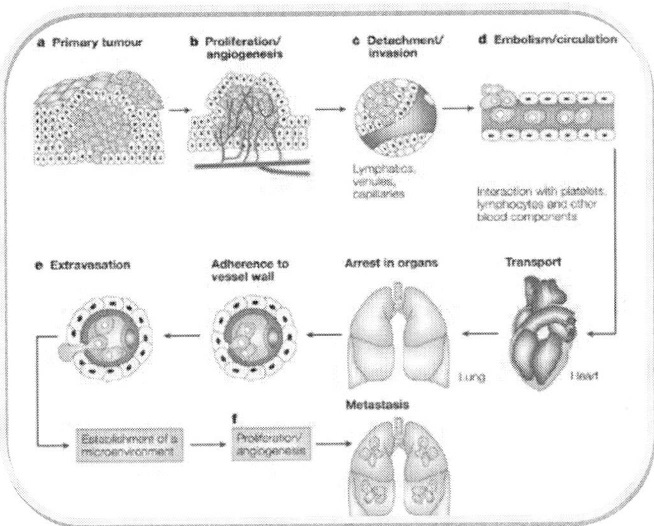

Figure 5. Tissue invasion is a multi-step process that requires the cancer cell to have developed many malignant traits, necessary for the formation of new distant colonies called metastases [27].

3. Mathematics of cancer

In comparison to biology, cell biology, and drug delivery analysis, mathematics has, to date, made comparatively very few contributions to this field of research. A statistical analysis of the PubMed platform list information (http://www.ncbi.nlm. nih.gov/PubMed/) showed that out of 1.5 million works that deal with cancer analysis, only 5% are associated with mathematical modeling. However, it is clear that mathematics could contribute significantly to areas of experimental cancer analysis since there is currently a wealth of experimental information which needs a systematic analysis.

Even in these conditions, in the last decade, mathematical modeling and machine simulation of cancer has multiplied dramatically (e.g., reviews like [30-35]. A broad range of strategies were developed, specializing in one or additional aspects of cancer. For example, genetic instability, natural selection or interactions of individual cell with each other or the environment have been modeled using methods of cellular automata and agent-based modeling.

These discrete methods have the disadvantage of being difficult to use when we deal with tumors of significant size. (see [36-38] for samples of cellular automata modeling and [39,40] for samples of agent-based modeling). In systems at larger scales, the neoplasic cell population is of the order of 10^6 or more, making these discrete methods unfitted. For these situations, the continuum methods provide the best approach. Early work, as well as [41-43], used ODE to model cancer as a uniform population and partial differential equation models restricted to spherical geometries. To assess the stability of spherical tumors to asymmetric perturbations and to characterize the degree of aggression [31, 44-48] use linear and weakly nonlinear analysis. The interactions of a growth with the microenvironment, like stress-induced limitations to growth, are studied in [30,49-54]. For the sake of simplicity, most of the modeling has considered single-phase (e.g., single cell species) tumors. To provide an elaborate account of growth non-uniformity, [50, 55, 56] have been developed a mixture of models.

The results of morphology instabilities on each avascular and vascular solid neoplasm growth have been recently studied using non-linear modeling. With the help of boundary integral methods, Cristini et al [47] performed the first absolutely nonlinear simulations of a time model of neoplasm growth within the avascular and vascular growth stages with arbitrary boundaries. The model from [47] has been extended in 3D by Li et al. [48] via adaptive boundary integral technique. The inclusion of angiogenesis and extratumoral environment has been performed by Zheng et al., [57]. By developing and coupling a level set implementation with a hybrid continuum-discrete growing model originally developed by Anderson & Chaplain [58]they found that low-nutrient (e.g., hypoxic) conditions could lead to morphological instability. Their work served as a building block for recent studies of the impact of therapy on neoplasm growth [59] and for studies of morphological instability and invasion [60-62]. Macklin & Lowengrub used a ghost cell/level set technique for evolving interfaces to check neoplasm growth in heterogeneous tissue and additional studied neoplasm growth as a function of the microenvironment [63]. Wise et al. [64] and Frieboes et al. [65] have developed a diffuse interface implementation of solid neoplasm growth for the study of the evolution of multiple neoplasm cell species, that was used in [65] to model the 3-D vascularised growth of malignant gliomas (brain tumors).

In biological systems, the fractal structure of area in which cells act and differentiate is important for their organization and emergence of the hierarchical network of multiple cross-interacting cells, sensitive to external and internal conditions. The biological phenomena occur within the area whose dimensions aren't represented solely by integers (1, 2, 3, etc.) of Euclidean space. Particularly, malignant tumors [53-56] grow in a space with non-integer dimension, i.e. fractal dimension. The analytical formulae describing the time-dependence of the temporal fractal dimension and scaling reproduce the expansion of the Flexner–Jobling rat's neoplasm in particular and growth of different rat's tumors generally. The results of some calculations indicated that the formula derived for the time-dependent temporal fractal dimension and the scaling factor describe the experimental data obtained by Schrek for the Brown-Pearce rabbit's neoplasm growth within the fractal time-space [3, 66-68].

In our assertion, fractal space(-time) consists in developing the consequences of the withdrawal of space(-time) differentiability's hypothesis and acquiring a fractal geometry, namely space(-time) becomes explicitly dependent on the observation scale [69].

On the other hand, of great use in our further reasonings will be the fact that in many biological systems it is possible to empirically demonstrate the presence of attractors that operate starting from different initial conditions (Ivancevic). Some of these attractors are points, some are closed curves, while the others have non–integer, fractal dimension and are termed "strange attractors" [70]. It has been proposed that a prerequisite for proper simulating tumor growth by computer is to establish whether typical tumor growth patterns are fractal. The fractal dimension of tumor outlines was empirically determined using the box-counting method [71]. In particular, fractal analysis of a breast carcinoma was performed using a morphometric method, which is the box-counting method applied to the mammogram as well as to the histological section of a breast carcinoma [72].

If tumor growth is chaotic, this could explain the unreliability of treatment and prediction of tumor evolution. More importantly, if chaos is established, this could be used to adjust strategies for fighting cancer. Treatment could include some form of chaos control and/or anti-control.

4. A few words about holography

"Although it generates a three-dimensional image, a hologram is most often recorded on a photographic plate or a flat piece of film. Moreover, producing a hologram does not imply, in the conventional sense, the recording of an image. To better understand this apparent paradox and, as a result, the way holography works, we have to begin with the main principles.

In conventional imaging techniques, e.g. photography, what is being recorded is merely the intensity distribution in the original scene. Thus, all information about the optical paths to different parts of the scene is lost.

The unique property of holography is the method of recording both the phase and the amplitude of the light waves from an object. Since all recording materials respond only to the intensity in the image, it is mandatory to convert the phase information into intensity variations. Holography accomplishes this by using coherent illumination and introducing, as shown in Fig. 6, a reference beam derived from the same source. The photographic film records the interference pattern produced by this beam and the light waves scattered by the object in cause.

Since the intensity at any given point in this pattern of interference also depends on the phase of the object wave, the resulting recording (the hologram) contains information on the phase as well as the amplitude of the object wave. If the hologram is illuminated once again with the original reference wave, as shown in Fig. 7, it reconstructs the original object wave.

An observer looking through the hologram sees a perfect three-dimensional image. This image exhibits all the effects of perspective, and depth of focus when photographed, that characterized the original object.

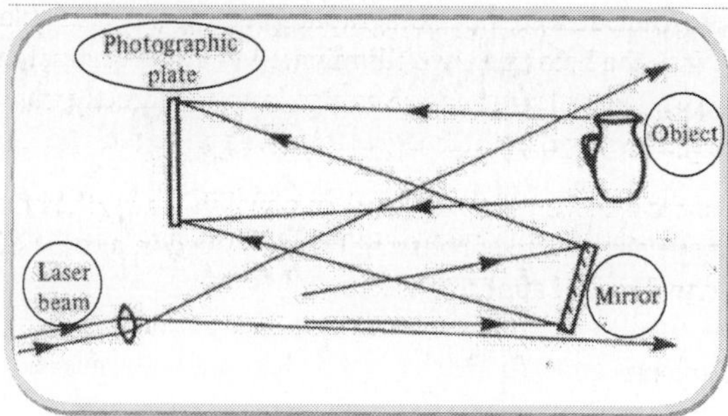

Figure 6. Hologram recording: the interference pattern produced by the reference wave and the object wave is recorded.

4.1. Early development

Gabor's historical experiment of holographic imaging [73] consisted in a transparency formed of opaque lines on a clear background which was illuminated with a collimated beam of monochromatic light, the interference pattern produced by the directly transmitted beam (the reference wave) and the light scattered by the lines on the transparency being recorded on a photographic plate. When the hologram (i.e. a positive transparency made from this photographic negative) was illuminated with the original collimated beam, it produced two diffracted waves, one which reconstructed an image of the object in its original location, and the other, with identical amplitude but an opposite phase, which formed a second, *conjugate* image.

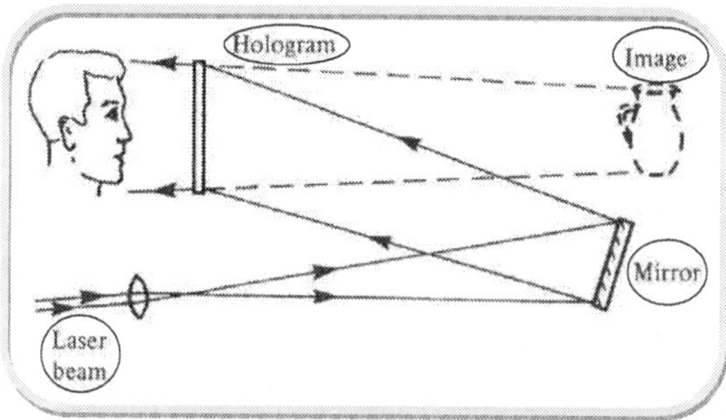

Figure 7. Image reconstruction: light diffracted by the hologram reconstructs the object wave.

An important flaw of this method of image reconstruction was the poor quality of the resulting image, due to the fact that it was degraded by the conjugate image that was superimposing on it as well as by the scattered light from the directly transmitted beam.

Leith and Upatnieks [74-76] found a solution to the above-mentioned problem developing a off-axis reference beam technique presented schematically in Figs. 6 and 7. They used a

http://dx.doi.org/10.5772/59233

separate reference wave incident on the photographic plate at an appreciable angle to the object wave. As a result, when the hologram was illuminated with the original reference beam, the two images were separated by large enough angles from the directly transmitted beam, and from each other, thus ensuring that the images not overlap.

The improvement of the off-axis technique, and, in equal measure, the invention of the laser, which provided a powerful source of coherent light, resulted in a surge of activity in holography that led to several crucial applications.

4.2. The in-line hologram

Let us now look upon the optical system presented in Fig. 8 in which the object (a transparency containing small opaque details on a clear background) is illuminated by a collimated beam of monochromatic light along an axis normal to the photographic plate.

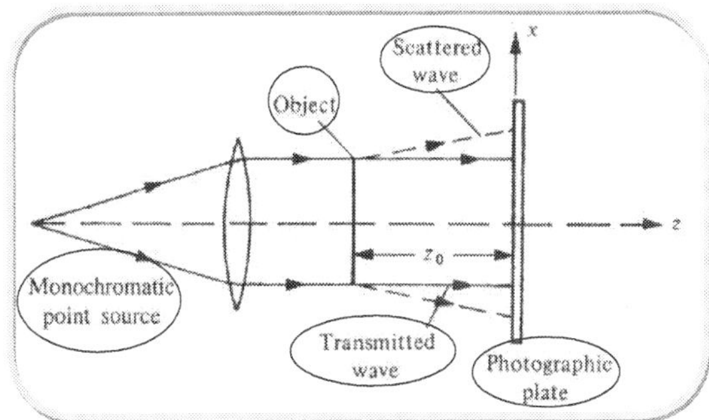

Figure 8. Optical system used to record an in-line hologram.

We can observe two components of the incident light. The first is the directly transmitted wave, which is a plane wave whose amplitude and phase do not vary across the photographic plate. Thus, its complex amplitude can be noted as a real constant r. The second one is a weak scattered wave whose complex amplitude at any point (x, y) on the photographic plate can be noted as $o(x, y)$, where $|o(x, y)| << r$.

From these it can be shown that the resulting complex amplitude is the sum of these two complex amplitudes, and because of that the intensity at this point is

$$I(x,y) = |r + o(x,y)|^2 = r^2 + |o(x,y)|^2 + ro(x,y) + ro^*(x,y) \tag{1}$$

where $o^*(x, y)$ is the complex conjugate of $o(x, y)$.

A 'positive' transparency (the hologram) is then made by contact printing from this recording. Therefore it can be assumed that this transparency is processed so that its amplitude transmittance (the ratio of the transmitted amplitude to that incident on it) can be written as

$$t = t_0 + \beta T I \tag{2}$$

where t_0 is a constant background transmittance, T is the exposure time and β is a parameter determined by the photographic material used and the processing conditions, the amplitude transmittance of the hologram is

$$t(x,y) = t_0 + \beta T[r^2 + |o(x,y)|^2 + ro(x,y) + ro^*(x,y)] \tag{3}$$

Then, the hologram is illuminated, as shown in Fig. 9, with the same collimated beam of monochromatic light employed to produce the original recording. Since the complex amplitude at any point in this beam is, aside from a constant factor, the same as that in the original reference beam, the complex amplitude transmitted by the hologram can be written as

$$u(x,y) = rt(x,y) = r(t_0 + \beta T r^2) + \beta T r|o(x,y)|^2 + \beta T r^2 o(x,y) + \beta T r^2 o^*(x,y) \tag{4}$$

The right-hand side of (4) contains four terms. The first, $r(t_0+\beta Tr^2)$, which represents a uniformly attenuated plane wave, corresponds to the directly transmitted beam.

The second, $\beta Tr\ |o(x, y)|^2$, can be neglected, because is extremely small, compared to the other terms.

The third term, $\beta Tr^2 o(x, y)$, is, except for a constant factor, identical with the complex amplitude of the scattered wave from the object and has the property of reconstructing an image of the object in its original position. Due to the fact that this image forms behind the hologram, and the reconstructed wave appears to diverge from it, it is a virtual image.

The fourth term, $\beta Tr^2 o^*(x, y)$, represents a wave similar to the object wave, but having an opposite curvature. This wave converges to form a real image (the conjugate image) at the same distance in front of the hologram.

With an in-line hologram, an observer viewing one image sees it superimposed on the out-of-focus twin image as well as a strong coherent background. Another problem is that the object must have a high average transmittance in order for the second term on the right-hand side of (4) to be negligible. Thus, it is possible to form images of fine opaque lines on a transparent background, but not *vice versa*. Finally, the hologram must be a 'positive' transparency. If the initial recording is used directly, β in (2) is negative, and the reconstructed image can be considered a photographic negative of the object.

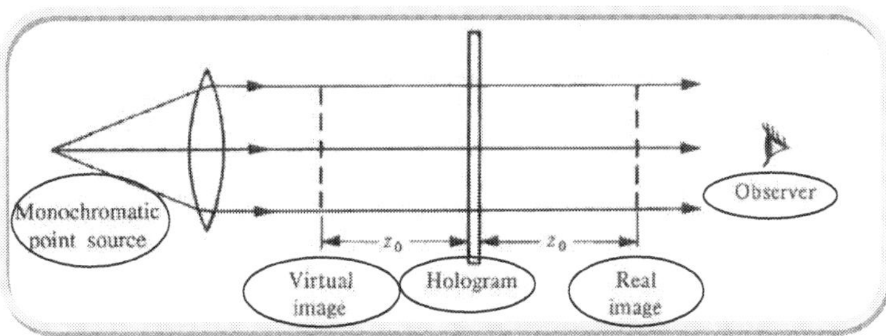

Figure 9. Optical system used to reconstruct the image with an in-line hologram, showing the formation of the twin images.

4.3. Off-axis holograms

In order to understand the formation of an image by an off-axis hologram, we must consider the recording arrangement shown in Fig. 10, in which (for simplicity) the reference beam is a collimated beam of uniform intensity, derived from the same source as the one used to illuminate the object.

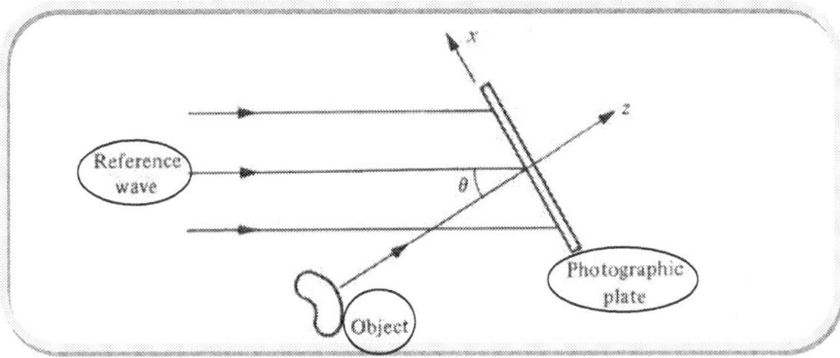

Figure 10. The off-axis hologram: `recording.

The complex amplitude at any point (x, y) on the photographic plate due to the reference beam can then be written as

$$r(x,y) = r\exp(i2\pi\xi x) \tag{5}$$

where ξ=(sinθ)/λ. Since only the phase of the reference beam varies across the photographic plat and because of the object beam, for which both the amplitude and phase vary, we can write the following:

$$o(x,y) = \left|o(x,y)\exp\left[-i\phi(x,y)\right]\right| \tag{6}$$

The resultant intensity is, therefore,

$$\begin{aligned} I(x,y) &= \left|r(x,y) + o(x,y)\right|^2 = \left|r(x,y)\right|^2 + \left|o(x,y)\right|^2 + \\ &+ r\left|o(x,y)\right|\exp\left[-i\phi(x,y)\right]\exp(-i2\pi\xi x) + r\left|o(x,y)\right|\exp\left[i\phi(x,y)\right]\exp(i2\pi\xi x) = \\ &= r^2 + \left|o(x,y)\right|^2 + 2r\left|o(x,y)\right|\cos[2\pi\xi x + \phi(x,y)] \end{aligned} \tag{7}$$

The amplitude and phase of the object wave are encoded as amplitude and phase modulation, respectively, of a set of interference fringes equivalent to a carrier with a spatial frequency of ξ.

If, as in (2), we assume that the amplitude transmittance of the processed photographic plate is a linear function of the intensity, the resultant amplitude transmittance of the hologram is

$$\begin{aligned} t(x,y) &= t_0' + \beta T\left|o(x,y)\right|^2 + \beta Tr\left|o(x,y)\right|\exp[-i\phi(x,y)]\exp[-i2\pi\xi x] + \\ &+ \beta Tr\left|o(x,y)\right|\exp[i\phi(x,y)]\exp[i2\pi\xi x] \end{aligned} \tag{8}$$

where $t_0' = t_0 + \beta Tr^2$ is a constant background transmittance.

When the hologram is illuminated for a second time with the original reference beam, as shown in Fig. 11, the complex amplitude of the transmitted wave can be written as

$$\begin{aligned} u(x,y) &= r(x,y)t(x,y) = t_0' r\exp(i2\pi\xi x) + \beta Tr\left|o(x,y)\right|^2\exp(i2\pi\xi x) + \\ &+ \beta Tr^2 o(x,y) + \beta Tr^2 o^*(x,y)\exp(i4\pi\xi x) \end{aligned} \tag{9}$$

The first term on the right-hand side of (9) corresponds to the directly transmitted beam, while the second term generates a halo surrounding it, with approximately twice the angular spread of the object. The third term is identical to the original object wave, except for a constant factor βTr^2, and produces a virtual image of the object in its original position. The fourth term corresponds to the conjugate image which, in this case, is a real image. If the offset angle of the reference beam is taken large enough, the virtual image can be separated from the directly transmitted beam and the conjugate image.

In this setup, corresponding points on the real and virtual images are located at equal distances from the hologram, but on opposite sides of it. Because the depth of the real image is inverted, it is called a pseudoscopic image, as opposed to the normal, or orthoscopic, virtual image. It should also be mentioned that the phase of the reconstructed image is influenced only by the sign of β, always resulting in a "positive" image, even in the case of the hologram recording being a photographic negative.

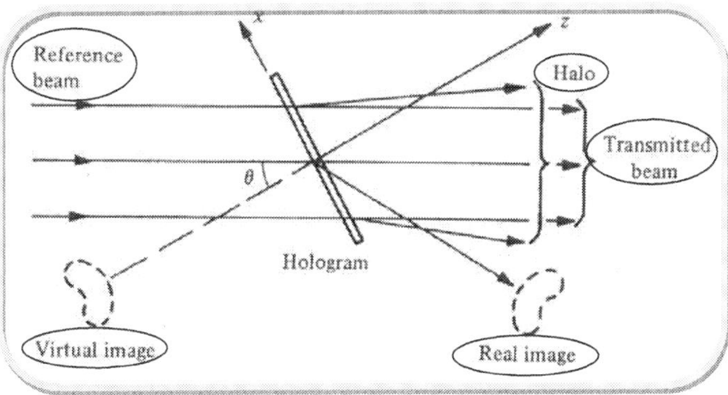

Figure 11. The off-axis hologram: image reconstruction.

4.4. Recording materials

Several recording materials have been used for holography [77]. Table 1 lists the principal characteristics of those that have been found most useful.

Material	Exposure J/m^2	Resolution mm^{-1}	Processing	Type	ηmax (diffraction efficiency)
Photographic	≈1.5	≈ 5000	Normal	Amplitude	0.06
			Bleach	Phase	0.60
DCG (dichromated gelatin)	10^2	10000	Wet	Phase	0.90
Photoresists	10^2	3000	Wet	Phase	0.30
Photopolymers	$10\text{-}10^4$	5000	Dry	Phase	0.90
PTP (photothermoplastics)	10^{-1}	500-1200	Dry	Phase	0.30
BSO ($Bi_{12}SiO_{20}$ photorefractive crystals)	10	10000	None	Phase	0.20

Table 1. Recording materials for holography

High-resolution photographic plates and films were the first materials used to record holograms. These are used widely even now, due to the fact that they exhibit relatively high sensitivity when compared to other hologram recording materials [78]. Moreover, they can be dye sensitized so that their spectral sensitivity matches the most commonly used laser wavelengths.

Combining the high sensitivity of photographic materials with the high diffraction efficiency, low scattering and high light-stability of DCG (dichromated gelatin) [79] was made possible by the silver-halide sensitized gelatin technique.

In positive photoresists, such as Shipley AZ-1350, the areas exposed to light become soluble and are washed away during development to produce a relief image [80].

Several organic materials can be activated by a photosensitizer to produce refractive index changes, because they suffer photopolymerization, when exposed to light [81]. A commercial photopolymer, coated on a polyester film base (DuPont OmniDex) that can be used to produce volume phase holograms with high diffraction efficiency is being currently produced [82].

Photothermoplastics (PTP) - a hologram can be recorded in a multilayer structure consisting of a glass or Mylar substrate coated with a thin, transparent, conducting layer of indium oxide, a photoconductor, and a thermoplastic [83,84].

When a photorefractive crystal is exposed to a spatially varying light pattern, electrons are liberated in the illuminated areas. These electrons migrate to adjacent dark regions, being trapped there. The spatially varying electric field produced by this space-charge pattern modulates the refractive index through the electro-optic effect, producing the equivalent of a phase grating. The space charge pattern can be removed by uniformly illuminating the crystal, after which another recording can take place [85,86].

It is essential to use coherent illumination for maximizing the visibility of the interference fringes formed by the object and reference beams, in the process of recording a hologram. In addition to being spatially coherent, the coherence length of the light must be much greater than the maximum value of the optical path difference between the object and the reference beams in the recording system. Lasers are, as a result, employed almost universally as light sources for recording holograms." (The text in quotation marks was reproduced from Hariharan P. [165]).

Consequently, to get a hologram, one needs a laser, which provides a powerful source of coherent light, and a 'recording material' which records the interference pattern produced by a reference beam and the light waves scattered by the object.

5. Tumor-associated ECM or tumor microenvironment, nonlinear medium with holographic properties

5.1. Extracellular matrix and tumor microenvironment

Within tissue, cells are surrounded by a meshwork of proteins and proteoglycans collectively called the extracellular matrix (ECM), which compartmentalizes tissues. The ECM is divided into two distinct layers:

i. the basement membrane, which is composed of sheet-like layers of ECM and lies under epithelial cells segregating tissues into functionally distinct regions;

ii. the interstitial matrix, which exists within intercellular space. The ECM serves multiple functions that are critical for embryonic development and wound repair. These functions include providing tissues with shape and flexibility and acting as a cushion to absorb external pressure. The ECM also serves as a base for cell anchorage, which mediates cell polarity, intercellular signaling, and assists in migration. The key to the ECM function lies in its unique composition and structure. The ECM is constructed in a specific pattern that is critical to its ability to carry out these functions and alterations in the expression level or arrangement of proteins within the ECM can be used to manipulate its function.

The most obvious function of the ECM is to provide structural support, shape, and stability for tissues. It does this by functioning as a base for cell anchorage. This base consists of three main structural components collagen, fibronectin, and elastic fibers, which bind to one another building a protein lattice upon which cells adhere.

Cell adherence to the ECM lattice provides cells support needed for cell migration. This is particularly important during embryonic development when cells are required to migrate into surrounding regions and differentiate into specific tissues [87]. A less obvious yet possibly more important function of the ECM in regards to tissue homeostasis and disease is its ability to mediate intracellular signaling. The ECM affects signaling through three main mechanisms:

i. cell – ECM interaction;

ii. regulation of the bioavailability of growth factors;

iii. the function of matricellular proteins. Cell attachment to the ECM via integrins induces signaling cascades that promote survival. Loss of cell-ECM contact can result in a form of apoptosis termed anoikis [88]. Anchorage-dependent survival is observed in most cells with the exception of red blood cells and inflammatory cells. However, tumor cells are often resistant to anoikis and can survive without a physical attachment to the ECM allowing them to successfully metastasize to distant tissues [89].

The ECM also affects cellular activity by serving as a reservoir for proteins required for proper tissue function and repair. This includes a plethora of growth factors and proteases. These pleiotropic molecules have been shown to robustly affect proliferation, survival and migration in numerous cell types. Once growth factors are secreted from cells, they often become embedded within the ECM and require ECM degradation by proteases such as elastase to release the active protein allowing it to interact with surrounding and transduce downstream signaling. The ability of the ECM to control the bioavailability of growth factors provides another means of regulating cellular activities and further explains how alterations in the makeup of the ECM as observed in diseases such as cancer affect cell response.

Matricellular proteins also reside in the ECM. They are a unique family of proteins that do not function as structural proteins but rather orchestrate the deposition of the ECM and mediate cell-cell and cell-ECM interactions. To do this, matricellular proteins interact directly with cell surface receptors, structural proteins, growth factors and proteases found within the ECM [90]. Their expression is found in every tissue begins early in development, persists throughout adulthood and is increased during tissue remodeling events. Matricellular proteins are critical regulators of many aspects of cell function including differentiation, survival, proliferation and migration making them necessary for proper tissue function. Not surprisingly, given their affect on cell-ECM mediated signaling pathways, matricellular proteins have been shown to strongly influence tumor growth.

For tumor cells to metastasize, the local ECM must be remodeled to create an environment conducive to tumor survival and progression. This includes altering the architecture and composition of the tumor-associated ECM or tumor microenvironment (TME) to facilitate tumor cell dissemination [91]. Changes in ECM architecture are primarily carried out by enzymes such as MMPs which assist in remodeling of the TME by degrading structural proteins such as collagen and fibronectin allowing tumor cells to freely navigate through the surrounding ECM. MMPs and other proteases assist in destruction of the first barrier tumor cells face to successful metastasis, the basement membrane. They degrade the underlying basement membrane allowing tumor cells to escape the primary tumor and invade into surrounding non-neoplasic tissues. MMPs continue to breakdown barriers in the surrounding ECM clearing a path to blood vessels where tumor cells will intravasate into the circulatory system and seed secondary tumors [92]. Destruction of the ECM by proteases also promotes tumor progression by facilitating the release of angiogenic and mitogenic factors bound within the ECM [93]. In a surprising unexpected twist, studies revealed that the breakdown of ECM proteins by MMPs was more complex than anticipated. In fact it is a highly organized process which results in the generation of both protumor and antitumor cleavage products [94].

Presence of the ECM is required for cellular survival therefore increased degradation of the ECM within the TME must be balanced by an increase in ECM synthesis. The development of a tumor, much like a wound, provokes a robust inflammatory response causing an influx of mast cells, macrophages and neutrophiles into the TME [95].

We may summarize that the extracellular matrix generates signaling cues that regulate cell behavior and orchestrate functions of cells in tissue formation and homeostasis. Microenvironmental signaling, a process that determines cell shape, motility, growth, survival and differentiation is highly influenced by the ECM properties: composition, three-dimensional organization and proteolytic remodeling. Recent studies have shown that misregulation of cell–ECM interactions can contribute to many diseases, including developmental, immune, haemostasis, degenerative and malignant disorders.

Consequently, the structure and the behavior of the tumor-associated ECM allows us to think of it as a non-differential medium, and as will be shown below, a medium which holds the properties of a hologram (capacity to memorize, interference abilities) and may become a source of forces. In other words, ECM and TME are very suitable candidates for a 'recording material'.

5.2. Tumor-associated ECM as a non-differential fractal medium

We can simplify the dynamics of a biological system supposing that the motions on ECM take place on continuous but non-differentiable curves, i.e. fractal curves (for example, the Peano curve, the Koch curve or the Weierstrass curve [69,96,97].

Once this hypothesis has been accepted, some consequences of non-differentiability by SRT are evident [69,96]: i) the physical quantities that are used in describing the biological system dynamics are fractal functions, i.e. functions dependent both on spatial coordinates and time as well as on the scale resolution, $\delta t/\tau$ (identified here with dt/τ by means of the substitution principle [69,96]. We mention that in the standard biophysics, the physical quantities describing the dynamics of a biological system are continuous, but differentiable functions depending only on spatial coordinates and time; ii) the dynamics of the biological systems are given by the fractal operator $\hat{d}/dt$ [98]:

$$\frac{\hat{d}}{dt}=\frac{\partial}{\partial t}+\hat{V}\cdot\nabla-i\frac{\lambda^2}{\tau}\left(\frac{dt}{\tau}\right)^{\left(\frac{2}{D_F}\right)-1}\Delta \tag{10}$$

where

$$\hat{\mathbf{V}}=\mathbf{V}_D-i\mathbf{V}_F \tag{11}$$

is the complex velocity, V_D is the differentiable and resolution scale independent velocity, V_F is the non-differentiable and resolution scale dependent velocity, $\hat{V}\ \nabla$ is the convective term,

$$\frac{\lambda^2}{\tau}\left(\frac{dt}{\tau}\right)^{\left(\frac{2}{D_F}\right)-1}\Delta=\frac{\lambda^2}{\tau}\left(\frac{dt}{\tau}\right)^{\left(\frac{2}{D_F}\right)-1}\left(\frac{\partial^2}{\partial x^2}+\frac{\partial^2}{\partial y^2}+\frac{\partial^2}{\partial z^2}\right) \tag{12}$$

is the dissipative term, D_F is the fractal dimension of the movement curve, λ is the space scale, τ is the time scale and λ^2/τ is a coefficient specific to the fractal – non - fractal transition. For D_F any definition can be used (the Hausdorff – Besikovici fractal dimension, the Kolmogorov fractal dimension, etc. [97], but once such definition is accepted for D_F, it has to remain constant over the entire analysis of the complex fluid dynamics. In a particular case, for motions on Peano curves, $D_F = 2$ [97] of the complex fluid entities, the fractal operator (1) is reduced to Nottale's operator $(\hat{d}/dt)_N$

$$\frac{\hat{d}}{dt}=\frac{\partial}{\partial t}+\hat{\mathbf{V}}\cdot\nabla-iD_N\Delta$$

where $D_N = \lambda^2/\tau$ is the Nottale's coefficient associated to the fractal-non-fractal transition.

Applying the fractal operator (10) to the complex velocity (11) and accepting the principle of scale covariance [69,96] in the form:

$$\frac{\hat{d}\hat{\mathbf{V}}}{dt} = -\nabla U \tag{13}$$

we obtain the motion equation:

$$\frac{\hat{d}\hat{\mathbf{V}}}{dt} = \frac{\partial \hat{\mathbf{V}}}{\partial t} + \left(\hat{\mathbf{V}} \cdot \nabla\right)\hat{\mathbf{V}} - i\frac{\lambda^2}{\tau}\left(\frac{dt}{\tau}\right)^{\left(\frac{2}{D_F}\right)-1} \Delta\hat{\mathbf{V}} = -\nabla U \tag{14}$$

where U is an external scalar potential. Equation (14) is a Navier – Stokes type equation. It means that at any point of a fractal path, the local acceleration term, $\partial_t \hat{V}$, the non-linearly (convective) term, $(\hat{V}\ \nabla)\ \hat{V}$, the dissipative term, $(\lambda^2/\tau)(dt/\tau)^{\left(\frac{2}{D_F}\right)-1} \Delta \hat{V}$, and the external free term ∇U make their balance. Therefore, the biological fluid is assimilated to a "rheological" fractal fluid, whose dynamics are described by the complex velocities field, $\hat{V}$, and by the imaginary viscosity type coefficient, $i(\lambda^2/\tau)(dt/\tau)^{\left(\frac{2}{D_F}\right)-1}$. The "rheology" of the fractal fluid can provide hysteretic properties to the biological fluid (the fractal fluid has a hysteresis cycle, memory, etc. [98-100].

For irrotational motions of the biological system entities

$$\begin{aligned} &\nabla \times \hat{\mathbf{V}} = 0, && \text{a} \\ &\nabla \times \hat{\mathbf{V}}_D = 0, && \text{b} \\ &\nabla \times \hat{\mathbf{V}}_F = 0 && \text{c} \end{aligned} \tag{15}$$

we can choose $\hat{V}$ of the form

$$\hat{\mathbf{V}} = -i\frac{\lambda^2}{\tau}\left(\frac{dt}{\tau}\right)^{\left(\frac{2}{D_F}\right)-1} \nabla \ln \psi \tag{16}$$

where $\phi \equiv \ln\psi$ is the velocity scalar potential. By substituting (16) in (14) and using the method described in [98-100], it results

$$\frac{\hat{d}\hat{\mathbf{V}}}{dt} = -i\frac{\lambda^2}{\tau}\left(\frac{dt}{\tau}\right)^{\left(\frac{2}{D_F}\right)-1} \nabla\left[\frac{\partial \ln\psi}{\partial t} - i\frac{\lambda^2}{\tau}\left(\frac{dt}{\tau}\right)^{\left(\frac{2}{D_F}\right)-1} \frac{\nabla\psi}{\psi} + U\right] = 0 \tag{17}$$

This equation can be integrated in a universal way and yields

$$\frac{\lambda^4}{\tau^2}\left(\frac{dt}{\tau}\right)^{\left(\frac{4}{D_F}\right)-2}\Delta\psi + i\frac{\lambda^2}{\tau}\left(\frac{dt}{\tau}\right)^{\left(\frac{2}{D_F}\right)-1}\frac{\partial\psi}{\partial t} - \frac{U}{2}\psi = 0 \tag{18}$$

up to an arbitrary phase factor which may be set to zero by an appropriate selection of the phase of ψ. Relation (18) is a Schrödinger type equation. For motions on Peano curves, $D_F = 2$ [97] at Compton scale, which implies $\lambda^2/\tau = \hbar/2m_0$ [69,96], with $\hbar$ the reduced Plank constant and m_0 the rest mass of the biological entities, the relation (18) becomes the standard Schrödinger equation:

$$\frac{\hbar^2}{2m_0}\Delta\psi + i\hbar\frac{\partial\psi}{\partial t} - \frac{U}{2m_0}\psi = 0$$

If $\psi = \sqrt{\rho}\, e^{iS}$, with $\sqrt{\rho}$ the amplitude and S the phase of ψ, the complex velocity field (16) takes the explicit form:

$$\begin{aligned}
\hat{\mathbf{V}} &= \frac{\lambda^2}{\tau}\left(\frac{dt}{\tau}\right)^{\left(\frac{2}{D_F}\right)-1}\nabla S - i\frac{\lambda^2}{2\tau}\left(\frac{dt}{\tau}\right)^{\left(\frac{2}{D_F}\right)-1}\nabla \ln\rho \quad &\text{a}\\
\mathbf{V}_D &= \frac{\lambda^2}{\tau}\left(\frac{dt}{\tau}\right)^{\left(\frac{2}{D_F}\right)-1}\nabla S &\text{b}\\
\mathbf{V}_F &= \frac{\lambda^2}{2\tau}\left(\frac{dt}{\tau}\right)^{\left(\frac{2}{D_F}\right)-1}\nabla \ln\rho &\text{c}
\end{aligned} \tag{19}$$

By substituting (19a-c) in (14) and separating the real and the imaginary parts, up to an arbitrary phase factor which may be set to zero by appropriate selection of the phase of ψ, we obtain:

$$\frac{\partial \mathbf{V}_D}{\partial t} + (\mathbf{V}_D\cdot\nabla)\mathbf{V}_D = -\nabla(Q+U) \tag{20}$$

$$\frac{\partial\rho}{\partial t} + \nabla\cdot(\rho\mathbf{V}_D) = 0 \tag{21}$$

with Q the specific fractal potential

$$Q = -2\frac{\lambda^4}{\tau^2}\left(\frac{dt}{\tau}\right)^{\left(\frac{4}{D_F}\right)-2}\frac{\Delta\sqrt{\rho}}{\sqrt{\rho}} = -\frac{\mathbf{V}_F^2}{2} - \frac{\lambda^2}{\tau}\left(\frac{dt}{\tau}\right)^{\left(\frac{2}{D_F}\right)-1}\nabla\cdot\mathbf{V}_F \tag{22}$$

Equation (20) represents the specific momentum conservation law, while equation (21) represents the states density conservation law. By means of the fractal velocity, V_F, the specific fractal potential Q is a measure of non-differentiability of the biological entities trajectories, *i.e.* of their chaoticity. The equations (20)-(22) define the fractal hydrodynamics model (FHM). In such a context, the biological system can be considered a fractal fluid.

Thus, it can be concluded that: i)Any entity of the biological system is in a permanent interaction with the fractal medium by means of the specific fractal potential; ii) The fractal medium is identified with a non-relativistic fractal fluid described by equations (20)-(22); iii) For motions on Peano curves at Compton scale [69,96,97], the FHM reduces to a quantum hydrodynamic model (QHM). Indeed, according to our previous considerations the relations (19a-c) become

$$\hat{\mathbf{V}} = \frac{\hbar}{m_0}\nabla S - i\frac{\hbar}{2m_0}\nabla \ln \rho$$

$$\mathbf{V}_D = \frac{\hbar}{m_0}\nabla S$$

$$\mathbf{V}_F = \frac{\hbar}{m_0}\nabla \ln \rho$$

in order that the momentum and density conservation laws are given by (20) and (21), respectively, with V_D and V_F previously defined, and the specific fractal potential by the expression

$$Q = -\frac{\hbar^2}{2m_0}\frac{\nabla^2\sqrt{\rho}}{\sqrt{\rho}} = -\frac{\mathbf{V}_F^2}{2} - \frac{\hbar}{2}\nabla \cdot \mathbf{V}_F$$

Moreover the fractal medium is assimilated to Bohm subquantum level [96]; iv) The fractal velocity V_F cannot be regarded as actual mechanical motion; it contributes to the transfer of the specific momentum and the concentration of energy. This fact can easily be deduced from the absence of V_F in the states density conservation law, and from its role in the variational principle. Any interpretation of Q should take into account the "self" or internal nature of the specific momentum transfer. While the energy is stored in the form of mass motion and potential energy (as it is classically), a part of it is available elsewhere and only the total is conserved. Reversibility and the existence of eigenstates is ensured by the conservation of energy and specific momentum, but this also means that a Brownian motion [97] form of interaction with an external medium is denied; v) For Peano curves motions [96,97], at spatial scales higher than the dimension of the boundary layer and at temporal scales higher than the oscillation periods of the pulsating velocities which overlaps the average velocity of the biological fluid motions (for details see [101-103], the FHM reduces to the standard hydrodynamical model [104]; vi) Since the position vector of the biological system entity is assimilated with a stochastic Wiener type process [96,97], ψ is not only the scalar potential of a complex velocity (through $\phi \equiv \ln\psi$) in the frame of FHM, but represents also the states density (through ψ^2) in the frame of a Schrödinger type model. Thus it can be seen that the formalism of the FHM and the one of Schrödinger type are equivalent. In addition, the chaoticity, either by means of turbulence in the fractal hydrodynamics approach, or by means of stochasticization in the Schrödinger type approach, is generated only by the non-differentiability of the movement trajectories in a fractal space; vii) In the standard model (Landau's scenario [104]) the Fourier spectrum is always discrete and cannot approximate a continuum spectrum that in case of a large number of frequencies will generate an unlimited number of spectral components as a result of their beats which appear due to the presence of nonlinearities in the biological fluid. Still, taking into account the standard model, the flow can never be exactly chaotic because, in case of multiple periodic functions, correlations tend to be null, although having an oscillating character. As a result, the transition towards chaotic behavior can be described by Landau's scenario only in a biological system with an infinite number of degrees of freedom. In our case, when $\delta t/\tau \to 0$ for $D_F \neq 2$ the physical quantities that describe the dynamics of the biological system are no longer defined. So, in this approximation, a simulation of a system with an infinite number of degrees of freedom is used. Moreover, the possibility of the dynamic states generation should be noted, which is characterized by windows of regular oscillations interrupted by chaotic bursts, the transition between the two states being spontaneous, unpredictable and independent of any of the control parameters variation (turbulence through intermittency); viii) The fractal medium and in particular the subquantum level has some computational properties: viii1) bistability, which implies the existence of its fractality and in particular, for motions on Peano curves at Compton scales, of the quantum bit. And from here, the entire fractal logics and in particular the quantum one; viii2) the self-replication, which implies the existence of some specific self-copying mechanisms; viii3) memory, which implies hysteresis type mechanisms; viii4) self-similarity, which implies the holographic type behavior; viii5) polarization, which implies mechanisms of changing the "computational state" from a given to a desired one; viii6) depositing and transmitting the information etc. For details see [105].

6. Tumor self-seeding by CTC and hypoxia support the idea of complete holography

6.1. The self-seeding hypothesis of tumor growth

The unfolding of cancer cells from their original sites to different ones within the body, i.e metastasis, has, for many years, been regarded as a unidirectional journey. However some researchers conjointly consider that metastatic cancer cells can increase primary tumor growth, this fact being crucial for the planning and type of the cancer treatment.

The concept of growing self-metastasis, or tumor "self-seeding," was first introduced at Memorial Sloan-Kettering Cancer Center, in a range of studies conducted by Drs. Joan Massagué, head of the Metastasis research facility, and Larry Norton, deputy physician-in-chief of the center's breast cancer programs. In the studies conducted on mice, Dr. Massagué discovered that breast tumors express genes related to metastasis were growing quicker than tumors that didn't express these genes, even if the genes had no apparent role in increased cellular division or decreased cell death (Fig. 12). These results did not fit within the standard tumor growth theories. In 2006, the researchers theorized that cells that become independent from a tumor and colonize distant tissues may also return home to the microenvironment within which they initial developed via the cardiovascular system [106]. They tested their hypothesis in a mouse model of cancer and revealed their findings in 2009 in Cell [107].

In one particular experiment, they selected a non-metastatic breast cancer cell line and an isolated set of daughter cells from that line that had gained the ability over time to metastasize to the lungs. Consequently, they implanted the parent cells in one mammary gland and the metastatic daughter cells in the opposite gland to serve as "donor tumors". They noticed that the daughter cells migrated to the lungs and to the tumors that were being formed by the parent cells in the opposite organ, accounting for 5 to 30% of the size of the parent tumors. Also, it was obvious that parent tumors seeded by daughter cells grew quicker than parent tumors that were implanted without daughter cells within the opposite gland.

This specific seeding behavior with daughter cells that spread to the bones and brain was noticed in the studies of colon cancer and skin cancer cell lines, but not when non-metastatic daughter cells were transplanted.

Furthermore, in various related follow-up laboratory experiments, the researchers demonstrated that cells from primary tumors can attract circulating metastatic tumor cells, and found several proteins that probably encourage this migration. They also found that the "come-back" metastatic cells mainly influenced primary tumor growth through the release of proteins that modify the tumor microenvironment, as well as blood vessels and immune cells.

Their hypothesis started to obtain support from different researchers at the United Nations agency. In 2009, Dr. Philip Hahnfeldt and his colleagues published the results of computer modeling studies conceived to search at the intersection of two biological phenomena found in tumors [108]. One of these phenomena consists in a small population of cancer cells acting like stem cells; thus, they could possess the ability to reproduce an infinite number of times, and also to generate secondary cancer cells that, with timė, lose the ability to divide. The second phenomena is that tumor growth is restricted by the available space for growing. Normally, healthy cells are spatially separated by a space that is not available in the tumor, due to the fact that cancer cells grow tightly in a dense mass till all the available has been occupied and the cellular division stops. At the periphery of the tumor, where the normal tissues density decreases, cancer cells continue to multiply and expand, increasing the size of the tumor.

Their models showed a crucial relation between cell migration, cell death, and tumor growth. When the offspring of a cancer stem cell within the model did not migrate or die spontaneously, tumor growth remained constant at around 110 cells. On the other hand, a high death rate among the non-stem cell progeny combined with a high cell migration rate produced the biggest tumors within the shortest amount of your time, to virtually 100,000 cells in over three years.

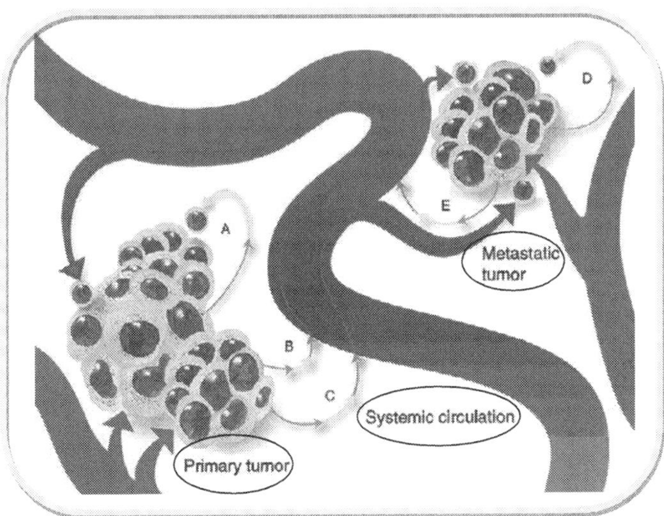

Figure 12. In the self-seeding concept of cancer growth and metastasis, a mobile tumor cell can take one of five different pathways in the body. *A* – evade and return to the primary tumor, using only the close ECM and not the systemic circulation; *B* – escape into the systemic circulation and then return to the original tumor; *C* – migrate through the systemic circulation and grow a metastatic tumor elsewhere in the organism; *D* – evade and return to the metastatic tumor, not using the systemic circulation; *E* –escape and return to the metastatic tumor through the systemic circulation.

This theoretical phenomena – accelerated growth jump-started by a high rate of growth death – has potential implications for the clinical treatment of cancer. Traditional cytotoxic therapy medicine kill massive numbers of speedily dividing cancer cells, however might not have an effect on cancer stem cells in each tumor type.

In the light of the above, we think of the CTC returning to the initial tumor site and fueling the primary tumor growth or even grow a new tumor as a particular case of complete holography (*i.e.* a hologram which does not represent only the virtual object's image, but becomes the very object - which we believe, is a characteristic of the living organisms).

6.2. Hypoxia and cancer

Vascularized tissues is the trigger factor for a large number of cellular processes, combined with an adaptive response. [109,110]. Following the drop in oxygen supplies, cells start to adapt to the less favorable environment and to initiate a vascularization process in order for them to raise the local oxygen supply. In the center of the hypoxic response is the angiogenic shift, with production of potent angiogenic factors such as vascular endothelial growth factor (VEGF). Hypoxia, which is present in many solid malignancies, means that oxygen level in tumors corresponds to around 1.5% [109]. This is caused on one hand by the result of the abnormal vascularisation in tumors, that is short in supply oxygen to the sometimes rapidly expanding malignant lesion and on the other hand, the existence of areas with acute lack of oxygen, resulting in necroses and cell death on a large scale. In breast cancer these forms are usually related to clinically aggressive behavior. Markers for hypoxia like HIF-1a have been connected to extremely malignant features and could be relevant prognostic markers for distinguishing subgroups of breast cancer with certain malignant properties [111,112]. There is a current discussion whether or not hypoxia contributes to increase the aggressiveness of tumors or if aggressive tumors have more widespread hypoxia, but, apparently, one explanation doesn't essentially exclude the opposite. Recent analysis has shown that 'the hypoxia response' in tumors may be used to conceive new treatment methods [113]. Emerging cancer therapies will most definitely put more focus on specific targeting of hypoxic processes.

Human cancers are characterized by intratumoral hypoxia that results from the proliferation of deregulated cell and the physiological responses that is triggered by it have impact on all aspects of cancer progression, together with im mortalization, transformation, differentiation, genetic instability, ontogeny, metabolic adaptation, autocrine protein communication, invasion, metastasis, and resistance to therapy.

We assume the relationship between hypoxia and aggressive tumors may be due to the presence of the coherent wave laser with oxygen of metastatic tumor cells in the area, where the produced oxygen gradients lead to oxygen consumption. It has been already shown that laser photocoagulation is effective in the treatment of diabetic retinopathy, in a series of major studies [114-117]. The oxygen-consumption may be based on a multilayer solution to Fick's law of diffusion, yet the essence is that the oxygen consumption is greatest where the oxygen gradient changes most rapidly [118-120].

All the above considerations and hypoxia's impact on all critical aspects of cancer progression support the idea that, the metastatic tumor cells moving through the systemic circulation (and not necessarily in there), may be considered a travelling wave chemically pumped type laser with oxygen.

7. Basic model

7.1. The PDE cancer-invasion model

We consider and present in what follows in extenso, the basic mathematical model of growth of a generic solid tumor, which is assumed just been vascularised, i.e. a blood supply has been established. Let us focus on four key variables involved in tumor cell invasion, in order to produce a minimal model, namely tumor cell density (denoted by n), matrix-degradative enzymes (MDE) concentration (denoted by m), the complex mixture of macromolecules from the extracellular material's (MM) concentration (denoted by f) and the oxygen concentration (denoted by c). Each of the four variables (n, m, f, c) is a function of the spatial variable $\mathbf{x}$ and time t. Firstly, we have to define a system of coupled non-linear partial differential equations to model tumor invasion of surrounding tissue.

We make the assumption that the ECM is a mixture of MM (*e.g.* collagen, fibronectin, laminin and vitronectin) only and not any other cells. Most of the MM of the ECM which are important for cell adhesion, spreading and motility are *fixed* or *bound* to the surrounding tissue. MDEs are important at many stages of tumor growth, invasion and metastasis, and they interact with inhibitors, growth factors and tumor cells in a very complex way. Yet it is widely accepted that tumor cells produce MDEs which locally degrade the ECM. As well as creating space into which tumor cells may be transported by simple diffusion (random motility), we can assume that this also results in a gradient of these bound cell-adhesion molecules, such as fibronectin. As a result, while the ECM may be a barrier to normal cell movement, it also represents a substrate to which cells may adhere and move upon. The presence of a minimum of ECM elements is a requirement for the growth and survival of most mammalian cells, and indeed these cell will migrate up a gradient of bound (*i.e.* non-diffusible) cell-adhesion molecules in the *in vitro* cultures [121-126].

We can define haptotaxis as a directed migratory response of the cells to gradients of fixed or bound chemicals (i.e. non-diffusible chemicals). While studies have not yet clearly shown haptotaxis occurs in an *in vivo* situation, given the structure of human tissue, it is not without reason to assume that haptotaxis is a major component of directed movement in tumor cell invasion. Indeed, there has been much recent effort to characterize such directed movement [125-127]. We therefore will treat this directed movement of tumor cells in this model as haptotaxis, i.e. a response to gradients of bound MM such as fibronectin. To incorporate this response in the mathematical model, we take the haptotactic flux to be $\mathbf{J}_{\text{hapto}} = \chi\, n\, \nabla f$, where $\chi > 0$ is the (constant) haptotactic coefficient.

As we stated early, the only other contribution to tumor cell motility in this model is the random motion. To describe the random motility of the tumor cells, we assume a flux of the form $\mathbf{J}_{\text{rand}} = -D_n \nabla n$, where D_n is the constant random motility coefficient.

We only model the tumor cell migration at this level, as all other tumour cell processes, such as proliferation, adhesion and death will be treated at a single cell level within the hybrid discrete-continuum model. The conservation equation for the tumour cell density n can therefore be written as

$$\frac{\partial n}{\partial t} + \nabla\left(\mathbf{J}_{rand} + \mathbf{J}_{hapto}\right) = 0$$

and hence the partial differential equation governing tumor cell motion (in the absence of cell proliferation) is

$$\frac{\partial n}{\partial t} = D_n \nabla^2 n - \chi \nabla\left(n \nabla f\right) \tag{23}$$

The ECM is known to contain many MM, including fibronectin, laminin and collagen, which can be degraded by MDEs [128,129]. We assume that the MDEs degrade ECM upon contact and hence the degradation process is modeled by the following simple equation

$$\frac{\partial f}{\partial t} = -\delta m f \tag{24}$$

where δ is a positive constant.

Active MDEs are produced (or activated) by the tumor cells, diffuse throughout the tissue and undergo some form of decay (either passive or active). The equation governing the evolution of MDE concentration is therefore given by

$$\frac{\partial m}{\partial t} = D_m \nabla^2 m + g(n,m) - h(n,m,f) \tag{25}$$

where D_m is a positive constant, the MDE diffusion coefficient, g is a function modeling the production of active MDEs by the tumor cells and h is a function modeling the MDE decay. For simplicity we assume that there is a linear relationship between the density of tumor cells and the level of active MDEs in the surrounding tissues (not taking into consideration the amount of enzyme precursors secreted and the presence of endogenous inhibitors) and so these functions will be $g = \mu\, n$ (MDE production by the tumor cells) and $h = \lambda\, m$ (natural decay), respectively.

The fact solid tumors need oxygen to grow and invade is a well-known one. Oxygen is assumed to diffuse into the MM, decay naturally and be consumed by the tumor. We assume that oxygen production is proportional to the MM density. This is a crude way of modeling an angiogenic oxygen supply for a more appropriate way of modeling the angiogenic network. The oxygen equation then has the form,

$$\frac{\partial c}{\partial t} = D_c \nabla^2 c + \beta f - \gamma n - \alpha c \tag{26}$$

where D_c, β, γ, α are positive constants representing the oxygen diffusion coefficient, production, uptake and natural decay rates, respectively.

The complete system of equations describing the interactions of the tumor cells, MM, MDEs and oxygen as detailed above, is

$$\begin{aligned}
\frac{\partial n}{\partial t} &= \overbrace{D_n \nabla^2 n}^{\text{random motility}} - \overbrace{\chi \nabla \left(n \nabla f \right)}^{\text{haptotaxis}} && \text{a} \\
\frac{\partial f}{\partial t} &= -\overbrace{\delta m f}^{\text{degradation}} && \text{b} \\
\frac{\partial m}{\partial t} &= \overbrace{D_m \nabla^2 m}^{\text{diffusion}} + \overbrace{\mu n}^{\text{production}} - \overbrace{\lambda m}^{\text{decay}} && \text{c} \\
\frac{\partial c}{\partial t} &= \overbrace{D_c \nabla^2 c}^{\text{diffusion}} + \overbrace{\beta f}^{\text{production}} - \overbrace{\gamma n}^{\text{uptake}} - \overbrace{\alpha c}^{\text{decay}} && \text{d}
\end{aligned} \tag{27}$$

http://dx.doi.org/10.5772/59233

where D_n, D_m and D_c are the tumor cell, MDE and oxygen diffusion coefficients, respectively, χ is the haptotaxis coefficient and δ, μ, λ, β, γ and α are positive constants. We should also note that cell–matrix adhesion is modeled here by the use of haptotaxis in the cell equation, i.e. directed movement up gradients of MM. Therefore, χ maybe considered as relating to the strength of the cell–matrix adhesion.

7.2. The PDE cancer-invasion model via scale relativity theory

The presence of the fractal medium implies the substitution of the standard derivative d/dt with the fractal operator (10). Then the system (27a-d) becomes

$$\begin{aligned}
&\frac{\hat{d}n}{dt}=\frac{\partial n}{\partial t}+\left(\hat{\mathbf{V}}\cdot\nabla\right)n-i\frac{\lambda^2}{\tau}\left(\frac{dt}{\tau}\right)^{(2/D_F)-1}\Delta n=D_n\nabla^2 n-\chi\nabla\left(n\nabla f\right) && \text{a}\\
&\frac{\hat{d}f}{dt}=\frac{\partial f}{\partial t}+\left(\hat{\mathbf{V}}\cdot\nabla\right)f-i\frac{\lambda^2}{\tau}\left(\frac{dt}{\tau}\right)^{(2/D_F)-1}\Delta f=-\delta mf && \text{b}\\
&\frac{\hat{d}m}{dt}=\frac{\partial m}{\partial t}+\left(\hat{\mathbf{V}}\cdot\nabla\right)m-i\frac{\lambda^2}{\tau}\left(\frac{dt}{\tau}\right)^{(2/D_F)-1}\Delta m=D_m\nabla^2 m+\mu n-\lambda m && \text{c}\\
&\frac{\hat{d}c}{dt}=\frac{\partial c}{\partial t}+\left(\hat{\mathbf{V}}\cdot\nabla\right)c-i\frac{\lambda^2}{\tau}\left(\frac{dt}{\tau}\right)^{(2/D_F)-1}\Delta c=D_c\nabla^2 c+\beta f-\gamma n-\alpha c && \text{d}
\end{aligned} \tag{28}$$

or more explicitly, by separating the scales of interaction, for the differentiable scale

$$\begin{aligned}
&\frac{\partial n}{\partial t}+\left(\hat{\mathbf{V}}_D\cdot\nabla\right)n=D_n\nabla^2 n-\chi\nabla\left(n\nabla f\right) && \text{a}\\
&\frac{\partial f}{\partial t}+\left(\hat{\mathbf{V}}_D\cdot\nabla\right)f=-\delta mf && \text{b}\\
&\frac{\partial m}{\partial t}+\left(\hat{\mathbf{V}}_D\cdot\nabla\right)m=D_m\nabla^2 m+\mu n-\lambda m && \text{c}\\
&\frac{\partial c}{\partial t}+\left(\hat{\mathbf{V}}_D\cdot\nabla\right)c=D_c\nabla^2 c+\beta f-\gamma n-\alpha c && \text{d}
\end{aligned} \tag{29}$$

and for the fractal scale

$$\begin{aligned}
&\left(\hat{\mathbf{V}}_F\cdot\nabla\right)n=\frac{\lambda^2}{\tau}\left(\frac{dt}{\tau}\right)^{(2/D_F)-1}\Delta n && \text{a}\\
&\left(\hat{\mathbf{V}}_F\cdot\nabla\right)f=\frac{\lambda^2}{\tau}\left(\frac{dt}{\tau}\right)^{(2/D_F)-1}\Delta f && \text{b}\\
&\left(\hat{\mathbf{V}}_F\cdot\nabla\right)m=\frac{\lambda^2}{\tau}\left(\frac{dt}{\tau}\right)^{(2/D_F)-1}\Delta m && \text{c}\\
&\left(\hat{\mathbf{V}}_F\cdot\nabla\right)c=\frac{\lambda^2}{\tau}\left(\frac{dt}{\tau}\right)^{(2/D_F)-1}\Delta c && \text{d}
\end{aligned} \tag{30}$$

Thus, the transport equations (27a-d) are generalized by involving the convective terms $(\hat{V}_F \cdot \nabla)n$, $(\hat{V}_F \cdot \nabla)f$, $(\hat{V}_F \cdot \nabla)m$, $(\hat{V}_F \cdot \nabla)c$ at differentiable scale. Moreover, at the fractal level one specifies new transport mechanisms where the convective effects are balanced by dissipative ones.

Now, the transport equations for the fractal to non-fractal transition are obtained by substracting the relations (29a) and (30a), (29b) and (30b), (29c) and (30c), (29d) and (30d) and using the substitution $\mathbf{V}=\mathbf{V}_D-\mathbf{V}_F$. One gets

$$\begin{aligned}
&\frac{\partial n}{\partial t}+\left(\hat{\mathbf{V}}\cdot\nabla\right)n=\left[D_n-\frac{\lambda^2}{\tau}\left(\frac{dt}{\tau}\right)^{(2/D_F)-1}\right]\Delta n-\chi\nabla\left(n\nabla f\right) && \text{a}\\
&\frac{\partial f}{\partial t}+\left(\hat{\mathbf{V}}\cdot\nabla\right)f=-\frac{\lambda^2}{\tau}\left(\frac{dt}{\tau}\right)^{(2/D_F)-1}\Delta f-\delta mf && \text{b}\\
&\frac{\partial m}{\partial t}+\left(\hat{\mathbf{V}}\cdot\nabla\right)m=\left[D_m-\frac{\lambda^2}{\tau}\left(\frac{dt}{\tau}\right)^{(2/D_F)-1}\right]\Delta m+\mu n-\lambda m && \text{c}\\
&\frac{\partial c}{\partial t}+\left(\hat{\mathbf{V}}\cdot\nabla\right)c=\left[D_c-\frac{\lambda^2}{\tau}\left(\frac{dt}{\tau}\right)^{(2/D_F)-1}\right]\Delta c+\beta f-\gamma n-\alpha c && \text{d}
\end{aligned} \tag{31}$$

Assuming now both the coherence fractal to non-fractal and harmonic type behavior for the f field the system of equations (31a-d) becomes

$$\begin{aligned}
&\frac{\partial n}{\partial t}=\left[D_n-\frac{\lambda^2}{\tau}\left(\frac{dt}{\tau}\right)^{(2/D_F)-1}\right]\Delta n-\chi\nabla\left(n\nabla f\right) && \text{a}\\
&\frac{\partial f}{\partial t}=-\delta mf && \text{b}\\
&\frac{\partial m}{\partial t}=\left[D_m-\frac{\lambda^2}{\tau}\left(\frac{dt}{\tau}\right)^{(2/D_F)-1}\right]\Delta m+\mu n-\lambda m && \text{c}\\
&\frac{\partial c}{\partial t}=\left[D_c-\frac{\lambda^2}{\tau}\left(\frac{dt}{\tau}\right)^{(2/D_F)-1}\right]\Delta c+\beta f-\gamma n-\alpha c && \text{d}
\end{aligned} \tag{32}$$

7.3. Non-dimensionalization and parameterization

For us to utilize realistic parameter values, we must first non-dimensionalize the equations in the standard formalism. We therefore rescale the distance with an appropriate length scale L (*e.g.* the maximum invasion distance of the cancer cells at the first stage of invasion, approximately 1 cm), time with τ (*e.g.* the average time taken for mitosis to occur, approximately 8...24 h [130], tumor cell density with n_0, ECM density with f_0, MDE concentration with m_0 and oxygen concentration with c_0 (where n_0, f_0, m_0 and c_0 are appropriate reference variables). Therefore, setting

$$\tilde{n}=\frac{n}{n_0},\quad \tilde{f}=\frac{f}{f_0},\quad \tilde{m}=\frac{m}{m_0},\quad \tilde{c}=\frac{c}{c_0},\quad \tilde{\mathbf{x}}=\frac{\mathbf{x}}{L},\quad \tilde{t}=\frac{t}{t_0}$$

in (27) and dropping the tildes for notational convenience, we obtain the scaled system of equations

$$\begin{aligned}
&\frac{\partial n}{\partial t}=\overbrace{d_n\nabla^2 n}^{\text{random motility}}-\overbrace{\rho\nabla\left(n\nabla f\right)}^{\text{haptotaxis}} && \text{a}\\
&\frac{\partial f}{\partial t}=-\overbrace{\eta m f}^{\text{degradation}} && \text{b}\\
&\frac{\partial m}{\partial t}=\overbrace{d_m\nabla^2 m}^{\text{diffusion}}+\overbrace{\kappa n}^{\text{production}}-\overbrace{\sigma m}^{\text{decay}} && \text{c}\\
&\frac{\partial c}{\partial t}=\overbrace{d_c\nabla^2 c}^{\text{diffusion}}+\overbrace{\nu f}^{\text{production}}-\overbrace{\omega n}^{\text{uptake}}-\overbrace{\varphi c}^{\text{decay}} && \text{d}
\end{aligned}\qquad(33)$$

where $d_n=\tau\, D_n-\frac{\lambda^2}{\tau}\left(\frac{dt}{\tau}\right)^{(2/D_F)-1}/L^2$, $\rho=\tau\chi f_0/L^2$, $\eta=\tau m_0\delta$, $d_m=\tau\, D_m-\frac{\lambda^2}{\tau}\left(\frac{dt}{\tau}\right)^{(2/D_F)-1}/L^2$, $\kappa=\tau\mu n_0/m_0$, $\sigma=\tau\lambda$, $d_c=\tau\, D_c-\frac{\lambda^2}{\tau}\left(\frac{dt}{\tau}\right)^{(2/D_F)-1}/L^2$, $\nu=\tau f_0\beta/c_0$, $\omega=\tau n_0\gamma/c_0$, $\phi=\tau\alpha$.

The cell cycle time can be highly variable (particularly the G1 phase) and in fact depends on the specific tumor taken under consideration. As an approximate reference time we take τ = 16 h, halfway between 8...24 h [130]. The cell motility parameter $D_n \sim 10^{-9}$ cm^2 s^{-1} was estimated from available experimental evidence [131]. Tumor cell diameters again will vary depending on the type of tumor being considered but are in the range 10...100 μm [132] with an approximate volume of 10^{-9} to 3×10^{-8} cm^3 [133,134]. We will assume that a tumor cell has the volume 1.5×10^{-8} cm^3 and therefore take $n_0 = 6.7 \times 10^7$ cells cm^{-3}. The haptotactic parameter $\chi \sim 2600$ cm^2 s^{-1} M^{-1} was estimated to be in line with that calculated in [135] and the parameter $f_0 \sim 10^{-8}$ to 10^{-11} M was taken from the experiments of [136]. We took D_m to be 10^{-9} cm^2 s^{-1}, which is perchance small for a diffusing chemical, but current studies imply that it is in fact a combination of the MDE and MM, and, as a result, the MM degrades and diffuses very little [137]. An *in vivo* estimate for the MDE concentration m_0 is rather difficult to obtain since no value (that we are aware of) has been currently determined and we also know that certain inhibitors (*e.g.* tissue inhibiting metalloproteases) are produced within the ECM which affects the MDE concentration. Plasma levels of specific MDEs have been measured (e.g. MMP-2 [138]) and are approximately 130 ng ml^{-1} with further increases observed in patients with cancer [139]. How is this related to the MDE concentration within the ECM is not clear and we have therefore left this parameter undefined. Estimates for the kinetic parameters μ, λ and δ were not available since these are rather hard to obtain experimentally – and thus we use the values of [135]. The diffusion rate of oxygen through water is $D_c = 10^{-5}$ cm^2 s^{-1} and also, the oxygen consume rate of the cells is 6.25×10^{-17} M cells^{-1} s^{-1} [134]. The background oxygen concentration estimation within the tissue was somehow difficult to be done as it depends on the tissue vascularization. If we set the value of the oxygen concentration in the blood supplying the tumor/tissue to be 0.15 ml O_2 per ml of blood, since we know that 1 M of oxygen occupies 22400 ml then there is 0.15/22400 M O_2 ml^{-1} = 6.7×10^{-6} M O_2 ml^{-1}, and since 1 ml = 1 cm^3 then we calculate $c_0 = 6.7 \times 10^{-6}$ M O_2 cm^{-3} [140]. Obviously this would be an overestimate, due to the fact that not all of the domain will be fully vascularised but at least we have obtained a reference value. The values of the non-dimensional parameters were given as

$$\begin{array}{ll} d_n = 0.0005, & \text{a} \\ r = 0.01, & \text{b} \\ h = 50, & \text{c} \\ d_m = 0.0005, & \text{d} \\ k = 1, & \text{i} \\ s = 0, & \text{f} \\ d_c = 0.5, & \text{g} \\ n = 0.5, & \text{h} \\ w = 0.57, & \text{i} \\ f = 0.025. & \text{j} \end{array} \qquad (34)$$

7.4. Laser beam in a multiscale diffusion cancer-invasion model

7.4.1. Laser as a lorenz system

A laser system is the result of an interaction between the electromagnetic field and the substance, under certain circumstances. The Lorenz form of laser equations may be obtained using a semi-classical reasoning where the environment is analyzed quantically using the formalism of density matrix, and the electromagnetic field is treated classically, by means of Maxwell's equations [141,142]. Here we consider only two energy levels of the involved microscopic systems (atoms, molecules, ions).

The first treatment of a two levels system was made by Bloch who analyzed the interaction of electrons with an oscillatory magnetic field superposed over a static magnetic field, in the framework of a magnetic resonance phenomenon. Due to the similarity of treatments and form of the obtained equations for the laser system, one can say that it forms the Maxwell-Bloch system.

Note that the density matrix method is applied in the treatment of laser systems, no matter how many energy levels, or number of oscillating modes are considered, as well as, in the consequent quantum treatment, where the electromagnetic field is quantized [143,144].

We start by discussing the effect of the electromagnetic field on the atoms of the environment. In the simplest situation, the electric field will induce in each atom an electric dipole whose moment is proportional to the field and is oriented along its direction. Neglecting the vectorial character, we have

$$\mu = \alpha E \tag{35}$$

where α is a constant characteristic to the type of the atom considered. If the concentration of (identical) atoms in the considered environment is N_a, then the polarization vector of the environment, equals the vectorial sum of all the dipole moments from the unit volume, and will be given by

$$P = N_a \mu = \varepsilon_0 \chi E \tag{36}$$

where ε_0 is the empty space permittivity and χ represents the electric susceptibility of the environment.

The problem of the induced dipole moment must be solved quantically using Schrödinger's equation (see paragraph 5.2)

$$i\hbar \frac{\partial \psi}{\partial t} = \hat{H}\psi \tag{37}$$

where $\hat{H}$ is the Hamiltonian operator and ψ the wave function of the atom.

Since a monochromatic field of frequency ω_0, not very intense, interacts with the atom inducing transitions between two of its energetic levels, E_1 and E_2 *i.e.* $E_2 - E_1 = \hbar\omega_0$, it is usual to neglect the other levels and to approximate the atom as a system with two energy levels. If the wave functions of the atom in the two states are ψ_1 and ψ_2, respectively, then we have

$$\psi = C_1\psi_1 + C_2\psi_2 \tag{38}$$

where C_1, C_2 are the time dependent complex amplitude probabilities for the atom to find itself on the energy levels E_1 and E_2, respectively. In other words $\rho_{11}=C_1^*C_1 \equiv |C_1|^2$ represents the probability of the atom to find itself in the state ψ_1, and $\rho_{22}=C_2^*C_2 \equiv |C_2|^2$ the probability of the atom to find itself in the state ψ_2. The combinations $\rho_{12}=C_1C_2^*$ and $\rho_{21}=C_2C_1^*$ are transition probabilities between the two states. The four numbers ρ_{ij} $(i, j = 1, 2)$ forms the so called density matrix for the 2-levels system. The asterisk attached to a parameter means the complex conjugate of the respective parameter. Obviously, $\rho_{21} = \rho^*_{12}$.

The Hamiltonian operator of the system will consist of a sum between the Hamiltonian of the nonperturbed atom $\hat{H}_0$ and a term which describes the interaction of the atom with the field, $\hat{H}'$,

$$\hat{H} = \hat{H}_0 + \hat{H}' \tag{39}$$

The functions ψ_1 and ψ_2 are eigenfunctions of the nonperturbed Hamiltonian, *i.e.*

$$\hat{H}_0\psi_1 = E_1\psi_1, \hat{H}_0\psi_2 = E_2\psi_2, \tag{40}$$

Replacing (38), (39) and (40) into the Schrödinger equation, and after some standard calculus, we get the equations

$$\begin{aligned} i\hbar\dot{C}_1 &= E_1C_1 + H'_{12}C_2, \quad &\text{a} \\ i\hbar\dot{C}_2 &= E_2C_2 + H'_{21}C_1, \quad &\text{b} \end{aligned} \tag{41}$$

where we introduced the notations

$$\begin{aligned} H'_{12} &= \int \psi_1^* \hat{H}' \psi_2 dV, \text{a} \\ H'_{21} &= \int \psi_2^* \hat{H}' \psi_1 dV, \text{b} \end{aligned} \tag{42}$$

It has been taken into account the orthonormal property of the wave functions ψ_1 and ψ_2

$$\int \psi_i^* \psi_j dV = \delta_{ij} \quad (i,j = 1,2) \tag{43}$$

where δ_{ij} is Kroeneker's symbol, and the fact that the interaction matrix H'_{ij} has no diagonal elements.

It is common to introduce the following simplification: if one chooses the zero energy value at the center of the interval between the two energies, then they become

$$\begin{aligned} E_2 &= (1/2)\hbar\omega_0, \quad &\text{a} \\ E_1 &= -(1/2)\hbar\omega_0, \quad &\text{b} \end{aligned} \tag{44}$$

and Eqs. (41) rewrite

$$i\hbar\dot{C}_1 = -\frac{\hbar\omega_0 C_1}{2} + H'_{12}C_2, \quad a$$
$$i\hbar\dot{C}_2 = \frac{\hbar\omega_0 C_2}{2} + H'_{21}C_1, \quad b \tag{45}$$

In quantum mechanics, the electric dipole momentum is calculated as the expectation value of the classical electric dipole momentum $\mu = ex$, where e is the electron charge and x is its displacement along the direction of the electric field.

For an atom in the state ψ this is given by

$$\mu = \int \psi^* ex\psi dV = \left(C_1^* C_2 + C_1 C_2^*\right)\int \psi_1^* ex\psi_2 dV \tag{46}$$

The integral $\mu_{12} = \int \psi_1^* ex\psi_2 dV$ represents the electric dipole momentum of the interaction. In the considered approximation, the interaction Hamiltonian is identical to the classical expression of the interaction energy between an electric field and the induced electric dipole: $U = -\mu E$, but with $\mu \rightarrow \mu_{12}$ (the dipole momentum of the interaction), i.e.

$$H'_{12} = -\mu_{12}E \tag{47}$$

By choosing a convenient phase relation between the wave functions, we can make μ_{12} real so H'_{12}, H'_{21} to be also real.

Eq. (46) suggests considering the expression

$$X = C_1 C_2^* + C_1^* C_2 \equiv \rho_{12} + \rho_{21} \tag{48}$$

We remark that the polarization (36) is expressed as a function of X through the equation

$$P = N_a \mu_{12} X = N_a \mu_{12}\rho_{12} + c.c. \tag{49}$$

where by c.c. we denote the complex conjugate of the previous expression.

We further consider the combinations

$$Y = i\left(C_1^* C_2 - C_1 C_2^*\right) \equiv -i\left(\rho_{12} - \rho_{21}\right) \tag{50}$$

$$Z = |C_2|^2 - |C_1|^2 \equiv \rho_{22} - \rho_{11} \tag{51}$$

For Z we also have a simple interpretation. If we multiply (51) by N_a we get the expressions $N_a\rho_{22}$ and $N_a\rho_{11}$, which represents the populations from the unit volume of the two levels, in other words, $N_a Z \equiv N$ represents the difference of population between the levels (inversion of population) from the unit volume.

All the three expressions X, Y, Z are functions depending only on time. Their time derivatives are easily calculated using Eqs. (45) and their complex conjugates. The following relations result:

$$\dot{X} = -\omega_0 Y - \frac{i}{\hbar}\left(H'_{12} - H'_{21}\right)Z = -\omega_0 Y \tag{52}$$

$$\dot{Y} = \omega_0 X + \frac{2\mu_{12}}{\hbar} EZ \tag{53}$$

$$\dot{Z} = -\frac{2\mu_{12}}{\hbar} EY \tag{54}$$

where in Eq. (52) the second form was obtained taking into account $H'_{12} = H'_{21}$.

By multiplication of Eq. (52) with $N_a\mu_{12}$, it transforms into an equation for $\dot{P}$, namely

$$\dot{P} = N_a\mu_{12}\dot{X} = N_a\mu_{12}\left(\dot{\rho}_{12} + \dot{\rho}_{21}\right) = N_a\mu_{12}\dot{\rho}_{12} + c.c. \tag{55}$$

The equation for $\dot{\rho}_{12}$ is obtained from Eqs. (52) and (53), taking into account Eqs. (48) and (50). It results

$$\dot{\rho}_{12} = i\omega_0\rho_{12} + i\frac{\mu_{12}}{\hbar} EZ \tag{56}$$

Another equation for polarization is obtained by deriving Eq. (52) once again and using (53). It results

$$\ddot{P} + \omega_0^2 P = -\frac{2\omega_0}{\hbar}\mu_{12}^2 EN \tag{57}$$

It is interesting to note that, in the absence of the electromagnetic field, *i.e.* for $E = 0$, Eq. (57) describes a harmonic oscillator. This is unacceptable, since polarization is induced by the field, so it must attenuate after the field cancels. Physically, it occurs both because of the internal dynamics of the atomic (molecular) systems and of the dephasing between the oscillations of different dipoles by means of their self-interaction or their interaction with the crystal lattice (for a solid environment). This phenomenon is taken into consideration through phenomenological reasonings by introducing an amortization term of the form $\dot{P} / T_2$ in the equation. Eq. (57) transforms into the equation of a forced dumped oscillator:

$$\ddot{P} + \frac{\dot{P}}{T_2} + \omega_0^2 P = -\frac{2\omega_0}{\hbar}\mu_{12}^2 EN \tag{58}$$

Usually, the time T_2 is named *transversal relaxation time*. It is characteristic to the non-diagonal elements of the density matrix, so Eq. (56) must be rewritten

$$\dot{\rho}_{12} + (\gamma_{12} - i\omega_0)\rho_{12} = i\frac{\mu_{12}}{\hbar} EZ \tag{59}$$

where $\gamma_{12} = 1/T_2$.

By multiplication of Eq. (54) with N_a the left side becomes $\dot{N}$. Replacing Y from (52) in (54), we get

$$\frac{\partial N}{\partial t} = \frac{2}{\hbar\omega_0} E\dot{P} \tag{60}$$

Eq. (60) shows that, at the disappearance of the electromagnetic field, the inversion of population must remain constant. However, an electromagnetic field resonant with the considered transition ($\omega \approx \omega_0$) is composed of quanta which can be absorbed by atoms, so may have the effect of a transfer of population between the two levels. It is obvious that, at the canceling of the field, the inversion of population must evolve towards an equilibrium value N^e which is obtained by a process of pumping and by spontaneous relaxation processes. They imply the presence of other energetic levels besides those already considered. We proceed again by phenomenological reasonings. We suppose that this evolution is again exponentially, thus we add a term of the form $(N - N^e)/T_1 \equiv \gamma_{11}(N - N^e)$, where T_1 is named *longitudinal relaxation time* (it is characteristic to the diagonal elements of the density matrix). The equation for the inversion of population becomes

$$\frac{\partial N}{\partial t} + \gamma_{11}(N - N^e) = \frac{2}{\hbar\omega_0} E\dot{P} \tag{61}$$

Eq. (61), together with (55) coupled with (59), or with (58) represents the substance equations. They must be associated with the electromagnetic field equation which we transcribe here

$$\nabla^2 \mathbf{E} - \frac{\eta^2}{c^2}\frac{\partial^2 \mathbf{E}}{\partial t^2} = \mu_0 \frac{\partial^2 \mathbf{P}}{\partial t^2} \tag{62}$$

where $c^2 = 1/\varepsilon_0\mu_0$, $\eta = \sqrt{\varepsilon/\varepsilon_0}$ is the refraction index of the environment (without the contribution of the transition between the two levels), P is the resonant part of the induced polarization and two non-conductive and non-magnetic laser environments were considered. In this case also, the energy losses produced by different mechanisms will be considered phenomenologically by means of an attenuation term introduced in the final form of the equation.

If we have a laser oscillator, the laser medium is placed between two mirrors which form an open cavity, and the oscillations may be triggered only by modes of oscillation characteristic to the cavity. They must satisfy the Helmholtz equation

$$\nabla^2 U + k^2 U = 0 \tag{63}$$

where $k = \eta\omega_c/c$, ω_c being the frequency of the considered mode (index c from cavity). For simplification, in what follows, we consider $\eta=1$ (gaseous environment).

We suppose the oscillation is produced on a single mode described by a spatial dependence of the form $W(x, y, z)$. This dependence will characterize both the field and polarization, so we can write

$$\mathbf{E} = \tilde{\mathbf{E}}(t) W(x,y,z) \exp(i\omega_c t) + c.c. \tag{64}$$

and

$$\mathbf{P} = \tilde{\mathbf{P}}(t) W(x,y,z) \exp(i\omega_c t) + c.c. \tag{65}$$

respectively, where $\tilde{E}(t)$ and $\tilde{P}(t)$ are slowly time-varying (complex) amplitudes.

Since the vectors E and P have the same directions, we can neglect the vectorial aspect. Introducing (64) and (66) in (62), applying the slowly varying amplitude approximation, i.e. taking $d^2\tilde{E}/dt^2 \approx 0$, $d^2\tilde{P}/dt^2 \approx -\omega_c P$ and having in view that $W(x, y, z)$ satisfies Eq. (63), we get

$$\frac{d\tilde{E}}{dt} = \frac{i\omega_c}{2\varepsilon_0}\tilde{P} \tag{66}$$

It is necessary to include the loss by radiation which are due to, in the first place, mirrors imperfections. We do that by introducing a term $\kappa\tilde{E}$ in the left hand side, so the field equation becomes

$$\frac{d\tilde{E}}{dt}+\kappa\tilde{E}=\frac{i\omega_c}{2\varepsilon_0}\tilde{P} \tag{67}$$

The equation for polarization is obtained comparing Eqs. (49) and (65). It results

$$\begin{aligned} \rho_{12} &= \frac{\tilde{P}W\exp(i\omega_c t)}{N_a\mu_{12}}, && \text{a} \\ \dot{\rho}_{12} &= \frac{1}{N_a\mu_{12}}\left(\frac{d\tilde{P}}{dt}+i\omega_c\tilde{P}\right)W\exp(i\omega_c t) && \text{b} \end{aligned} \tag{68}$$

which introduced in Eq. (59), and after multiplying both sides with $W^{*}\exp(-i\omega_c t)$ and integrating over the entire volume (mode) of the cavity, leads to

$$\frac{d\tilde{P}}{dt}+(\gamma_{12}+i(\omega_c-\omega_0))\tilde{P}=\frac{i\mu_{12}^2}{\hbar}\tilde{E}N \tag{69}$$

where, in the left hand side, we introduced the first term from (64) and where N represents the inversion of population from the volume occupied by the cavity mode, defined by the relation

$$N=\frac{N_a\int(\rho_{22}-\rho_{11})W^{*}WdV}{\int W^{*}WdV} \tag{70}$$

We then introduce the relations for the field and polarization (64) and (65) into the equation for the inversion (61). Using the approximation $\dot{P}=i\omega_c P$, neglecting the rapidly varying terms (which contain exp ($\pm 2i\omega_c t$)), multiplying by WW^{*} and integrating over the volume of the cavity, we get

$$\frac{\partial N}{\partial t}+\gamma_{11}(N-N^{e})=\frac{2iA_0}{\hbar}(\tilde{E}^{*}\tilde{P}-\tilde{E}\tilde{P}^{*}) \tag{71}$$

where N is given by Eq. (70). We consider $\omega_c/\omega_0=1$, and note

$$A_0=\frac{N_a\int(W^{*}W)^2dV}{\int W^{*}WdV} \tag{72}$$

Eqs. (67), (69) and (71) form the Bloch-Maxwell system and describe a unimodal laser oscillator. If we make the change of variables

$$
\begin{aligned}
t &= \frac{1}{\gamma_{12}} t', && \text{a} \\
\tilde{E} &= \frac{i\hbar\gamma_{12}}{2\sqrt{A_0\mu_{12}}} A, && \text{b} \\
\tilde{P} &= \frac{\hbar\gamma_{12}\varepsilon_0\kappa}{\omega_c\sqrt{A_0\mu_{12}}} R, && \text{c} \\
N - N^e &= \frac{2\hbar\gamma_{12}\varepsilon_0\kappa}{\omega_c\mu_{12}^2} n && \text{d}
\end{aligned}
\tag{73}
$$

the Bloch-Maxwell equations gets the form of the Lorenz system. They become

$$
\begin{aligned}
\frac{dA}{dt'} &= -\sigma A + \sigma R && \text{a} \\
\frac{dR}{dt'} &= rA - R(1 + i\Omega) - An && \text{b} \\
\frac{dn}{dt'} &= -bn + \frac{1}{2}(AR^* + A^*R) && \text{c}
\end{aligned}
\tag{74}
$$

where the following notations were used

$$
\begin{aligned}
\sigma &= \frac{\kappa}{\gamma_{12}}, && \text{a} \\
\Omega &= \frac{\omega_c - \omega_0}{\gamma_{12}}, && \text{b} \\
b &= \frac{\gamma_{11}}{\gamma_{12}}, && \text{c} \\
r &= \frac{\omega_c\mu_{12}^2 N^e}{2\varepsilon_0\hbar\kappa\gamma_{12}} && \text{d}
\end{aligned}
\tag{75}
$$

In the form (74) the equations make up a complex Lorenz system. This was discussed in detail in the paper [145]. The complex Lorenz system transforms into the well known real Lorenz system if the resonance is exact ($\omega_c = \omega_0 \Rightarrow \Omega = 0$) and the phases of the amplitudes $\tilde{E}$ and $\tilde{P}$ are chosen so the functions A and R to be real

$$
\begin{aligned}
\frac{dA}{dt'} &= \sigma(R - A) && \text{a} \\
\frac{dR}{dt'} &= A(r - n) - R && \text{b} \\
\frac{dn}{dt'} &= AR - bn && \text{c}
\end{aligned}
\tag{76}
$$

The fact that a unimodal laser oscillator is described by the Lorenz system was remarked for the first time by Haken (Haken, 1975). Therefore, it was demonstrated that the immense variety of dynamical behaviors, including the chaotic ones, presented by a Lorenz system, must be expected to occur in a laser. Among the first who reported Lorenz type chaotic behaviors in a laser, were Weiss and Brock [147].

Equations of the same form are obtained also when one considers a travelling wave laser, such as laser amplificators where the wave passes only one time the environment, or a circular unidirectional laser [148,149].

7.5. A chaotic multi-scale cancer-invasion model

From the non–dimensional space-time model (33), discretization was performed by neglecting all the spatial derivatives resulting in the following simple 4D temporal dynamical system

$$\begin{aligned} &\frac{dn}{dt}=0 && \text{a}\\ &\frac{df}{dt}=-\eta mf && \text{b}\\ &\frac{dm}{dt}=\kappa n-\sigma m && \text{c}\\ &\frac{dc}{dt}=\nu f-\omega n-\varphi c && \text{d} \end{aligned} \tag{77}$$

When simulated, the temporal system (77) with the set of parameters (34) exhibits a virtually linear temporal behavior with almost no coupling between the four concentrations that have very different quantitative values (all phase plots between the four concentrations, not shown here, are virtually one-dimensional). To see if a modified version of the system (77) could lead to a chaotic description of tumor growth, and following the method in [150], four new parameters, a_1, a_2, a_3, and a_4 are introduced. The resulting model is

$$\begin{aligned} &\frac{dn}{dt}=0 && \text{a}\\ &\frac{df}{dt}=a_1\eta(m-f) && \text{b}\\ &\frac{dm}{dt}=f\left(a_3-c\right)-m+a_2\kappa n && \text{c}\\ &\frac{dc}{dt}=\nu fm-a_4\varphi c-\omega n && \text{d} \end{aligned} \tag{78}$$

The introduction of the parameters (a_1, a_2, a_3, a_4) was motivated by the fact that tumor cell shape represents a visual manifestation of an underlying balance of forces and chemical reactions [151]. Specifically, the parameters represent the following quantities: a_1 = tumor cell volume (proliferation/non-proliferation fraction), a_2 = glucose level, a_3 = number of tumor cells, a_4 = diffusion from the surface (saturation level).

A tumor is composed of proliferating (P) and quiescent (or non-proliferating) (Q) cells. Tumor cells shift from class P to class Q as the tumor grows in size [152]. Model dependence on the ratio of proliferation to non-proliferation is introduced via the first parameter, a_1. The discretization of Eq. (33a) leads to cell density being modeled as a constant in Eq. (78a). Accordingly, cell density does not play a role in the dynamics. In (78) the cell density is re-introduced into the dynamics via the cell number, a_3. The importance of introducing a_3 also appears in connection with the cyclin-dependent kinase (Cdk) inhibitor p27, the level and activity of which increase in response to cell density. Levels and activity of Cdk inhibitor p27 also increase with differentiation following loss of adhesion to the ECM [153].

The ability to estimate the growth pattern of an individual tumor cell type on the basis of morphological measurements should have general applicability in cellular investigations, cell–growth kinetics, cell transformation and morphogenesis [154].

Cell spreading alone is conducive to proliferation and increases in DNA synthesis, indicating that cell morphology is a critical determinant of cell function, at least in the presence of optimal growth factors and extracellular matrix (ECM) binding [155]. The varying morphology of most cells can stimulate cell proliferation through integrin-mediated signaling indicating that cell shape may govern how individual cells will respond to chemical signals [156].

Parameters (a_1, a_2, a_3, a_4), introduced in connection with cancer cells morphology and dynamics could also influence the very important factor chromatin associated with aggressive tumor phenotype and shorter patient survival time.

For computations, the parameters were set to a_1=0.06, a_2=0.05, a_3=26.5 and a_4=40. Small variation of these chosen values would not affect the qualitative behavior of the new temporal model (78). Simulations of (78), using the same initial conditions and the same non-dimensional parameters as before, show chaotic behavior in the form of Lorenz-like strange attractor in the 3D (f–m–c) subspace of the full 4D (n–f–m–c) phase-space (Figs. 13-Figs. 16).

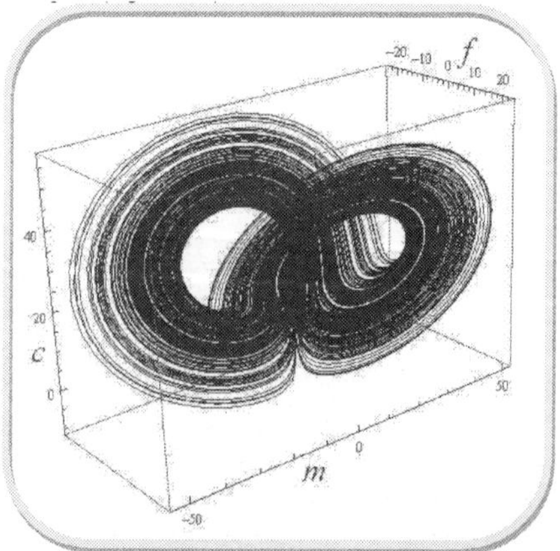

Figure 13. A 3D Lorenz-like chaotic attractor from the modified tumor growth model (78 *b-d*). The attractor effectively couples the MM–concentration *f*, the MDE–concentration *m*, and the oxygen concentration *c* in a mask–like fashion.

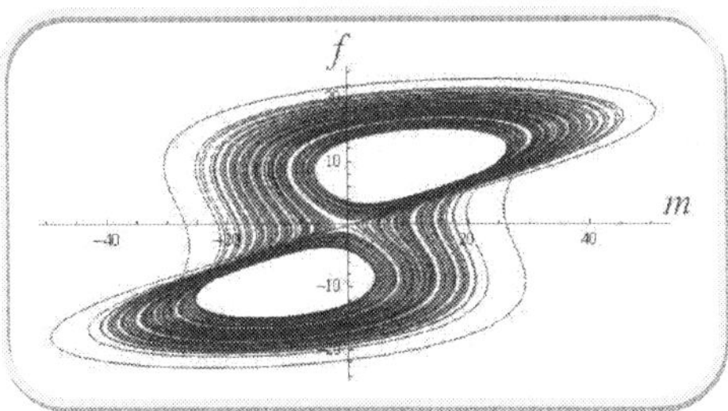

Figure 14. The $m - f$ phase plot of the 3D attractor.

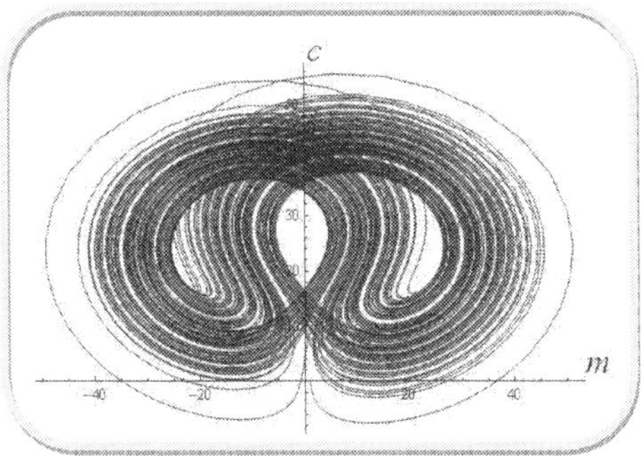

Figure 15. The $m - c$ phase plot of the 3D attractor.

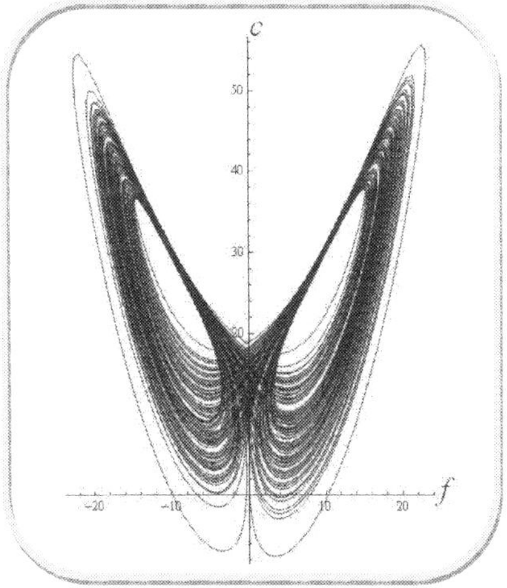

Figure 16. The $f - c$ phase plot of the 3D attractor.

The space-time system of rate PDEs corresponding to the system in (33) provides the following multi-scale cancer invasion model

$$\begin{aligned} \frac{\partial n}{\partial t} &= d_n \nabla^2 n - \rho \nabla \cdot \left(n \nabla f \right) && \text{a} \\ \frac{\partial f}{\partial t} &= a_1 \eta (m - f) && \text{b} \\ \frac{\partial m}{\partial t} &= d_m \nabla^2 m + a_2 kn + f\left(a_3 - c\right) - m && \text{c} \\ \frac{\partial c}{\partial t} &= d_c \nabla^2 c + \nu fm - \omega n - a_4 \varphi c && \text{d} \end{aligned} \qquad (79)$$

The new tumor–growth model (79) retains all the qualities of the original model (33) plus includes the temporal chaotic 'butterfly'–attractor. This chaotic behavior may be a more realistic view on the tumor growth, including stochastic–like long–term unpredictability and uncontrollability, as well as sensitive dependence of a tumor growth on its initial conditions.

Now, if we compare (78) with (76) and make a one-to-one correspondence between these two systems of equations, we see that A which is the electric field amplitude corresponds to f, the MM concentration, R which is the polarization amplitude corresponds to m, the MDE concentration and n the inversion of population corresponds to c, the oxygen concentration. Since both systems, the laser and the tumor invasion can be written in the form of a Lorenz system, we can suppose that the metastatic cancer cells moving through the systemic circulation form a coherent wave, i.e. a particular type of chemically pumped (since it may obtain its energy from chemical reactions) laser with oxygen. In the following section we show moreover, that this coherent wave can be identified with a travelling wave laser with oxygen.

7.6. Travelling waves in the multiscale diffusion cancer-invasion model

Let us write the system (28) again. We assume the model refers to the the averaged behavior of the tumor cells in the direction of invasion only and ignores variations in a plane normal to the direction of invasion.

Invasive cells. Since in their experiments Aznavoorian *et al.* [157] reported minimal chemokinetic movement, a key feature of the following model is the absence of the term for random cell motility. Also, we introduce a term of increased proliferation of malignant cells relative to normal cells, $F(n)$, which will be initially modeled as a logistic type growth of the form $k_1 n (k_2 - n)$ which has been shown [158], in order for us to describe adequately the growth of human tumors grown [159].

Extracellular matrix. The motility of extracellular matrix elements, unlike the one of malignant cells and oxygen, is negligible due to the fact that these elements are much longer than cells. The dynamics of connective tissue can therefore be modeled as a simple passive degradation by the activity of the tissue proteases; this proteolysis can now be described by $-G(f, m)$, since it depends on the amount of collagen f still present as well as on the protease m.

Proteases. Proteases generation is narrowly confined to the interface invading tumor and receding connective tissue interface. In some cases it is possible to localize the interstitial collagenase production to the stromal fibroblasts immediately adjacent to the site of tumor invasion, possibly leading to the fact that invasive cells release a stimulus for induction of interstitial collagenase by fibroblasts. Nabeshima *et al.* [160] managed to sequence a tumor cell derived collagenase stimulatory factor. Protease generation located only at the invading front can be explained however in other ways. In their work, Xie *et al.* [161] have revealed that the induction of 92-kd type IV collagenase activity in cultures of A431 human epidermoid carcinoma cells is density dependent. They showed that only dividing cells stained positive when treated with anti-MMP antibodies and as a consequence only noncontact-inhibited tumor cells produce protease. Many proteases are predominantly membrane bound (e.g. uroplasminogen activator), but even when the protease is secreted into the extracellular space, activation occurs only on the cell surface, so as a result the behavior closely resembles that for membrane bound proteases [162]. Therefore protease diffusion in the model is not included. We must then define the function $H(n,f)$ to represent the dependence of this tightly regulated protease production on the local concentrations of the melanoma cells and collagen. In addition we assume that the protease decays linearly, with half-life K.

Oxygen. As in the original model, we presuppose it diffuses into the MM, decays naturally, is consumed by the tumor and for simplicity, oxygen production is proportional to the MM density. Therefore, we introduce the function $I(n,f)$ and c decays linearly, with half-life Λ. The parameter c does not appear anywhere else in the system, so this equation will be easily separated.

Combining all of the above, we are now ready to write the model as

$$\begin{aligned}
&\frac{\partial n}{\partial t} = \overbrace{F(n)}^{\text{invasive cell proliferation}} - \overbrace{k_3 \frac{\partial}{\partial x}\left(n\frac{\partial f}{\partial x}\right)}^{\text{haptotactic cell movement}} && \text{a}\\
&\frac{\partial f}{\partial t} = -\overbrace{G(f,m)}^{\text{proteolysis}} && \text{b}\\
&\frac{\partial m}{\partial t} = \overbrace{H(n,f)}^{\text{protease production}} - \overbrace{Km}^{\text{natural decay}} && \text{c}\\
&\frac{\partial c}{\partial t} = \overbrace{d_c \frac{\partial^2 c}{\partial x^2}}^{\text{diffusion}} + \overbrace{I(f,n)}^{\text{production and uptake}} - \overbrace{\Lambda c}^{\text{decay}} && \text{d}
\end{aligned} \tag{80}$$

where F, G, H and I are functions of n, f and m. Compared to previous work on the modeling of cell movement, this model is unusual in that there is no cellular diffusion. This case has been considered previously [163] in the very different context of cellular aggregation, where they obtained conditions for blow-up in the absence of cell kinetics.

Before proceeding further, we must eliminate m from the equations as follows. The time scales associated with protease production and protease decay are much shorter than a typical timescale for the invading cells. Hence writing $H(n, f) = K\overline{H}(n, f)$, where we assume $K >> 1$, and multiplying through Eq. (80c) by the small parameter K^{-1}, we deduce that to leading order $m = \overline{H}(n, f)$. Henceforth no reference to m is needed: this expression may be used to eliminate m from Eqs. (80a) and (80*b*). In the same way, writing $I(f, n) = \Lambda\bar{I}(f, n)$, assuming $\Lambda >> 1$ and multiplying through Eq. (80d) by the small parameter Λ^{-1}, we deduce that to leading order $c = \bar{I}(f, n)$. This type of quasi-steady state assumption is a common one in enzyme kinetics [164], and numerical simulations of the four equations (80*a-d*) are in good accordance with the simplified system of two equations; a strong point of the two equation case is that it is amenable to detailed mathematical analysis.

We examine the model using the simple functional forms

$$\begin{aligned} &F(n) = k_1 n(k_2 - n), && \text{a} \\ &G(f,m) = k_4 mf, && \text{b} \\ &H(n,f) = k_5 nf, && \text{c} \\ &I(f,n) = k_6 f - k_7 n && \text{d} \end{aligned} \tag{81}$$

7.7. Nondimensionalization

After making the substitutions for F, G, H and I from (81) into equations (80a-d) and eliminating m using $m = H(n, f)$ and c using $c = I(f, n)$ we nondimensionalize the resulting equations using

$$\tilde{n} = \frac{n}{n^*}, \quad \tilde{f} = \frac{f}{f^*}, \quad \tilde{t} = \frac{t}{T}, \quad \tilde{x} = \frac{x}{L}, \quad L = \left[\frac{k_3}{k_2 k_4 k_5}\right]^{1/2}, \quad T = \frac{1}{k_1 k_2}, \quad n^* = k_2, \quad f^* = \frac{k_1}{k_4 k_5}$$

Dropping tildes for notational convenience then gives rise to the system

$$\begin{aligned} &\frac{\partial n}{\partial t} = n(1-n) - \frac{\partial}{\partial x}\left[n\frac{\partial f}{\partial x}\right] && \text{a} \\ &\frac{\partial f}{\partial t} = -nf^2 && \text{b} \end{aligned} \tag{82}$$

7.8. Spatially homogeneous system

Setting $\partial/\partial x = 0$ in (82a) we note that the spatially homogeneous system has two steady states:

i. $n = 0$, f arbitrary – this is a continuum of (unstable) steady states parameterized by the (variable) amount of connective tissue in different tissues;

ii. $n = 1$, $f = 0$ – this (stable) steady state corresponds to complete replacement of the normal tissue by invading malignant cells.

With $\partial/\partial x = 0$, (82a) and (82b) can be solved explicitly giving

$$n(t)=\left[1+\exp(-t+c_2)\right]^{-1} \quad \text{a}$$
$$f(t)=[t-c_1+\log[1+\exp(-t+c_2)]]^{-1} \quad \text{b} \qquad (83)$$

where c_1 and c_2 are arbitrary constants. The behavior as $t \to \infty$ shows that $n \to 1$ and $f \to 0$, hence justifying our classification of the steady state $(n, f) = (1, 0)$ as stable.

7.9. Travelling wave analysis

The invasion process should normally correspond to the travelling wave solutions of the model (82a) and (82b) with the normal tissue steady state $n = 0$ ahead of the wave and the fully malignant state $n = 1, f = 0$ behind the wave. This fact is verified by numerical solutions of (82a) and (82b), which are not detailed here. These travelling wave solutions can be studied analytically using the travelling wave differential equations. We look for constant shape travelling wave front solutions of (82a) and (82b) by setting

$$n(x,t)=N(z), \quad \text{a}$$
$$f(x,t)=F(z), \quad \text{b} \qquad (84)$$
$$z=x-\xi t \quad \text{c}$$

where ξ is the positive wave speed which has to be determined. When solutions of the type (84) exist, they represent travelling waves moving in the positive x - direction. Substitution of (84) into (82a) and (82b) followed by simple algebraic manipulation gives

$$\frac{dN}{dz}\left(-\xi+\frac{2NF^2}{\xi}\right)=N(1-N)-\frac{2N^3F^3}{\xi^2} \quad \text{a}$$
$$\frac{dF}{dz}=\frac{NF^2}{\xi} \quad \text{b} \qquad (85)$$

The analysis of (85a,b) involves the study of the (N, F) phase plane. Since we are looking for travelling waves connecting $(1, 0)$ and $(0, \hat{F})$ in the (N, F) phase plane we look for solutions of (85a,b) with boundary conditions

$$N(-\infty)=1, \quad \text{a}$$
$$F(-\infty)=0, \quad \text{b}$$
$$N(\infty)=0, \quad \text{c} \qquad (86)$$
$$F(\infty)=\hat{F}, \quad \text{d}$$

which requires $(1, 0)$ to have an unstable manifold while $(0, \hat{F})$ must have a stable manifold. In order to study this we look at the stability of the system (85a,b).

7.10. Stability analysis

The steady states (N_0, F_0) of (85a) and (85b) are $(0, \hat{F})$ and (1, 0), where $\hat{F}$ represents a steady states continuum. We study their stability by analyzing the eigenvalues of the stability matrix linearized about the steady states. The eigenvalues about $(0, \hat{F})$ are $-1/\xi$ and 0. The corresponding eigenvectors are $(1, -\hat{F})$ and (0, 1). The negative eigenvalue shows that there is a stable manifold along $(1, -\hat{F})$. The zero eigenvalue represents translations along the steady states continuum.

The eigenvalues about (1, 0) are $1/\xi$ and 0. The eigenvector corresponding to $1/\xi$ is (1, 0) and represents movement along the N axis. The eigenvector corresponding to the zero eigenvalue is (0, 1) which is in the direction normal to the N axis. The trajectory leaving this steady state leaves along the eigenvector corresponding to the zero eigenvalue of the linearized system, as shown by the numerical solutions of (85a,b). This zero eigenvalue is a result of (85b). In order to get a more detailed and clear image of the behavior close to (1, 0) we must look at the nonlinear terms in (85b). One way to do this is to use the techniques of the centre manifold theory which shows that as $z \to -\infty$, $N(z)$ approaches 1 exponentially while $F(z)$ tends zero as z^{-1}.

The existence of an unstable manifold about $(0, \hat{F})$ as $z \to \infty$ and a stable centre manifold about (1, 0) as $z \to -\infty$ is consistent with the existence of a travelling wave orbit connecting the two steady states.

8. Conclusions

1. Cancer cannot be reduced to simple mathematical principles. Its irregular mode of carcinogenesis, erratic tumor growth, variable response to tumoricidal agents, and less-known metastatic patterns constitute highly variable clinical behavior. Characterizing this process requires an accurate understanding of tumor cells and host tissues interactions and ultimately determines prognosis. Applying time-tested and evolving mathematical methods to oncology may provide new methods, with inherent advantages, for the description of tumor behavior, selection of therapeutic modes, prediction of metastatic patterns, and the defining of an inclusive basis for prognostication. Mathematicians describe equations that define tumor growth and behavior, whereas surgeons actively deal with biological processes. Mathematics in oncology applies these principles to clinical settings.

2. The main conclusions of this work are as follows:

   i. mathematics of cancer proves to be chaotic and highly non-linear, justifying the use of space(-time) non-differentiability as a starting base model;

   ii. a chaotic multi-scale cancer-invasion model is manufactured, which embeds a Lorenz attractor in its solutions;

iii. since laser can be expressed as a Lorenz system, we may assume some correspondences between the laser and the above mentioned chaotic multi-scale cancer-invasion model;

iv. the basic model for solid tumor growth admits a travelling wave solution;

v. we suggest that metastatic tumor cells which move through the systemic circulation are similar to a coherent wave, i.e. a travelling wave chemically pumped type laser with oxygen;

vi. we assume the extracellular matrix and in particular, the tumor microenvironment are non-differential media endowed with holographic properties (capacity to memorize, interference abilities and source of forces);

vii. the two well-known phenomena: tumor self-seeding by CTC and hypoxia, in our opinion, both support the idea of complete holography (a hologram which becomes the very object in the particular case of living organisms).

3. Experimentally testable, mathematics applied in oncology may provide a framework to determine clinical outcome on a patient-specific basis and increase the growing awareness that mathematical models help simplify seemingly complex and random tumor behavior.

Author details

Daniel Timofte[1], Lucian Eva[2], Decebal Vasincu[3], Călin Gh. Buzea[4*], Maricel Agop[5,6] and Radu Florin Popa[7]

*Address all correspondence to: calinb2003@yahoo.com

1 Surgery Dept., Hospital "Sf. Spiridon", University of Medicine and Pharmacy "Gr. T. Popa", Iaşi, Romania

2 Emergency Clinical Hospital "Prof. Dr. Nicolae Oblu", Iaşi, Romania

3 Surgery Dept., Biophysical Section, University of Medicine and Pharmacy "Gr. T. Popa", Iaşi, Romania

4 National Institute of Research and Development for Technical Physics – IFT Iasi, Iaşi, Romania

5 Dept. of Physics "Gh. Asachi" Technical University, Iaşi, Romania

6 Lasers, Atoms and Molecules Physics Lab., University of Science and Technology, Lille, France

7 Surgery Dept., University of Medicine and Pharmacy "Gr. T. Popa", Iaşi, Romania

References

[1] Armstrong K., Eisen A., Weber B., *Assessing the Risk of Breast Cancer*. N. Engl. J. Med., 342, 564-571 (2000).

[2] Friedman A., Reitich F., *Analysis of a Mathematical Model for the Growth of Tumors*. J. Math. Biol., 38, 262-284 (1999).

[3] Waliszewski P., Molski M., Konarski J., *On the Holistic Approach in Cellular and Cancer Biology: Nonlinearity, Complexity, and Quasi-determinism of the Dynamic Cellular Network*. J. Surg. Oncol., 68, 70-78 (1998).

[4] Orme M. E., Chaplain M. A. J., *Two-dimensional Models of Tumor Angiogenesis and Antiangiogenesis Strategies*. IMA J. Math. Appl. Med. Biol., 14, 189-205 (1997).

[5] Perumpanani A.J., Sherratt J. A., Norbury J. et al., *Biological Inferences from a Mathematical Model for Malignant Invasion*. Invasion Metastasis, 16, 209-221 (1996).

[6] Secomb et al 2001

[7] Fodde R., Smits R., Clevers H., *Apc, Signal Transduction and Genetic Instability in Colorectal Cancer*. Nature Reviews Cancer, 1, 55 (2001).

[8] Hahn W. C., Counter C. M., Lundberg A. S., Beijersbergen R. L., Brooks M. W., Weinberg R. A., *Creation of Human Tumour Cells with Defined Genetic Elements*. Nature, 400, 464-468 (1999).

[9] Kinzler K., Vogelstein B., *Lessons from Hereditary Colorectal Cancer*. Cell, 87(2), 159-170 (1996).

[10] Olson J., *Bathsheba's Breast: Women and Cancer and History*. John Hopkins University Press, Baltimore, 2002.

[11] Porter R., *The Greatest Benefit to Mankind: A Medical History of Humanity from Antiquity to the Present*. Harper Collins Publishers, London, 1997.

[12] Hajdu S. I., *Greco-roman Thought about Cancer*. Cancer, 100(*10*). 2048-2051 (2004).

[13] Hanahan D., Weinberg A., *The Hallmarks of Cancer*. Cell, 100, 57-70 (2000).

[14] Medema R., Bos J., *The Role of p21ras in Receptor Tyrosine Kinase Signaling*. Crit. Rev. Oncog., 4(*6*) 615-661 (1993).

[15] Weinberg R., *The Retinoblastoma Protein and Cell Cycle Control*. Cell, 81, 323-330 (1995).

[16] Wyllie A., Kerr J., Curri A., *Cell Death: The Significance of Apoptosis*. Int. Rev. Cytol., 68, 251-306 (1980).

[17] Harris C., *P53 Tumor Suppressor Gene: from the Basic Research Laboratory to the Clinic - An Abridged Historical Perspective*. Carcinogenesis, 17, 1187-1198 (1996).

[18] Hayflick L., *Mortality and Immortality at the Cellular Level. A Review*. Biochemistry, 62, 1180-1190 (1997).

[19] Shay J., Bacchetti S., *A Survey of Tolomerase Activity in Human Cancer*. Eur. J. Cancer, 33, 787-791.

[20] Bouck N., Stellmach V., Hsu S., *How Tumors Become Angiogenic*. Adv. Cancer Res. 69, 135–174 (1997).

[21] Hanahan D., Folkman J., *Patterns and Emerging Mechanisms of the Angiogenic Switch During Tumorigenesis*. Cell, 86, 353-364 (1996).

[22] Veikkola T., Alitalo K., *Vegfs and Receptors and Angiogenesis*. Semin. Cancer Biol., 9, 211-220 (1999).

[23] Bull H., Brickell P., Dowd P., *Src-related Protein Tyrosine Kinases are Physically Associated with the Surface Antigen cd36 in Human Dermal Microvascular Endothelial Cells*. FEBS Lett., 351, 41-44 (1994).

[24] Shih T., Lindley C., Bevacizumab, *An Angiogenesis Inhibitor for the Treatment of Solid Malignancies*. Clinical Therapeutics, 28(*11*), 1779-1802 (2006).

[25] Sporn M., *The War on Cancer*. Lancet, 347, 1377-1381 (1996).

[26] Frisch S., Screaton R., *Anoikis Mechanisms*. Current Opinion in Cell Biology, 13(5), 555-562 (2001).

[27] Fidler I., *The Pathogenesis of Cancer Metastasis: The "Seed and Soil" Hypothesis Revisited*. Nature Reviews Cancer, 3 1-6 (2003).

[28] Matrisian L., *The Matrix-degrading Metalloproteinases*. Bioessays, 14, 455-463 (1992).

[29] Mignatti P., Rifkin D., *Biology and Biochemistry of Proteinases in Tumor Invasion*. Physiology Rev., 73, 161-195 (1993).

[30] Araujo R. P., McElwain D. L. S., *A Linear-elastic Model of Anisotropic Tumor Growth*. Eur. J. Appl. Math., 15, 365-384 (2004).

[31] Byrne H. M., Alarcon T., Owen M. R., Webb S. D., Maini P. K., *Modelling Aspects of Cancer Dynamics: A Review*. Phil. Trans. Roy. Soc., A364, 1563-1578 (2006).

[32] Adam J., *General Aspects of Modeling Tumor Growth and the Immune Response*. In J. Adam, N. Bellomo (Eds.), *A Survey of Models on Tumor Immune Systems Dynamics*, Birkhäuser, Boston, MA, 1996.

[33] Bellomo N., de Angelis E., Preziosi L., *Multiscale Modelling and Mathematical Problems Related to Tumor Evolution and Medical Therapy*. J. Theor. Med., 5, 111-136 (2003).

[34] Quaranta V., Weaver A. M., Cummings P. T., Anderson A. R. A., *Mathematical Modeling of Cancer: The Future of Prognosis and Treatment*. Clin. Chem. Acta, 357, 173-179 (2005).

[35] Sanga S., Sinek J.P., Frieboes H. B., Fruehauf J. P., Cristini V., *Mathematical Modeling of Cancer Progression and Response to Chemotherapy*. Expert. Rev. Anticancer Ther., 6, 1361-1376 (2006).

[36] Alarcón T., Byrne H. M., Maini P. K., *A Cellular Automaton Model for Tumour Growth in Inhomogeneous Environment*. J. Theor. Biol., 225, 257-274 (2003).

[37] Anderson A. R. A., A *Hybrid Mathematical Model of Solid Tumour Invasion: The Importance of Cell Adhesion*. IMA Math. Appl. Med. Biol., 22, 163-186 (2005).

[38] Mallett D. G., de Pillis L. G., *A Cellular Automata Model of Tumor Immune System Interactions*. J. Theor. Biol., 239, 334-350 (2006).

[39] Abbott R. G., Forrest S., Pienta K. J., *Simulating the Hallmarks of Cancer*. Artif. Life, 12, 617- 634 (2006).

[40] Mansury Y., Kimura M., Lobo J., Deisboeck T.S., *Emerging Patterns in Tumor Systems: Simulating the Dynamics of Multicellular Clusters with an Agent-based Spatial Agglomeration Model*. J. Theor. Biol., 219, 343-370 (2002).

[41] Byrne H. M., Chaplain M. A. J., *Growth of Necrotic Tumors in the Presence and Absence of Inhibitors*. Math. Biosci., 135, 187-216 (1996).

[42] Byrne H. M., Chaplain M. A. J., *Modelling the Role of Cell-cell Adhesion in the Growth and Development of Carcinomas*. Math. Comput. Model., 24, 1-17 (1996).

[43] Greenspan H. P., *On the Growth and Stability of Cell Cultures and Solid Tumors*. J. Theor. Biol., 56, 229-242 (1976).

[44] Araujo R., McElwain D., *A History of the Study of Solid Tumour Growth: The Contribution of Mathematical Modelling*. Bull. Math. Biol., 66, 1039-1091 (2004).

[45] Byrne H. M., Matthews P., *Asymmetric Growth of Models of Avascular Solid Tumors: Exploiting Symmetries*. IMA J. Math. Appl. Med. Biol., 19, 1-29 (2002).

[46] Chaplain M. A. J., Ganesh M., Graham I. G., *Spatio-temporal Pattern Formation on Spherical Surfaces: Numerical Simulation and Application to Solid Tumour Growth*. J. Math. Biol., 42, 387-423 (2001).

[47] Cristini V., Lowengrub J. S., Nie Q., *Nonlinear Simulation of Tumor Growth*. J. Math. Biol., 46, 191-224 (2003).

[48] Li X., Cristini V., Nie Q., Lowengrub J. S., *Nonlinear Three Dimensional Simulation of Solid Tumor Growth. Disc.* Cont. Dyn. Sys., B7, 581-604 (2007).

[49] Ambrosi D., Mollica F., *On the Mechanics of a Growing Tumor*. Int. J. Eng. Sci., 40, 1297-1316 (2002).

[50] Ambrosi D., Preziosi L., *On the Closure of Mass Balance Models for Tumor Growth*. Math. Mod. Meth. Appl. Sci., 12, 737-754 (2002).

[51] Ambrosi D, Guana F., *Stress-modulated Growth.* Math. Mech. Solids (in press) doi: 10.1177/1081286505059739.

[52] Araujo R. P., McElwain D. L. S., *A Mixture Theory for the Genesis of Residual Stresses in Growing Tissues.* (II) *Solutions to the Biphasic Equations for a Multicell Spheroid.* SIAM J. Appl. Math., 66, 447-467 (2005).

[53] Jones A. F., Byrne H. M., Gibson J. S., Dold J. W., A *Mathematical Model of the Stress Induced During Avascular Tumor Growth.* J. Math. Biol., 40, 473-499 (2000).

[54] Roose T., Netti P. A., Munn L. L., Boucher Y., Jain R., *Solid Stress Generated by Spheroid Growth Estimated Using a Linear Poroelastic Model.* Microvasc. Res., 66, 204-212 (2003).

[55] Byrne H., Preziosi L., Modelling *Solid Tumour Growth Using the Theory of Mixtures.* Math. Med. Biol., 20, 341-366 (2003).

[56] Chaplain M. A. J., Graziano L., Preziosi L., *Mathematical Modelling of the Loss of Tissue Compression Responsiveness and its Role in Solid Tumour Development.* Math. Med. Biol., 23, 192-229 (2006).

[57] Zheng X., Wise S. M., Cristini V., *Nonlinear Simulation of Tumor Necrosis, Neo-vascularization and Tissue Invasion Via an Adaptive Finite Element/Level Set Method.* Bull. Math. Biol., 67, 211-259 (2005).

[58] Anderson A. R. A., Chaplain M. A. J., *Continuous and Discrete Mathematical Models of Tumor-induced Angiogenesis.* Bull. Math. Biol., 60, 857-900 (1998).

[59] Sinek J., Frieboes H., Zheng X., Cristini V., *Two-dimensional Chemotherapy Simulati-ons Demonstrate Fundamental Transport and Tumor Response Limitations Involving Nanoparticles.* Biomed. Microdev., 6, 197-309 (2004).

[60] Cristini V., Frieboes H. B., Gatenby R., Caserta S., Ferrari M., Sinek J., *Morphological Instability and Cancer Invasion.* Clin. Cancer Res., 11, 6772-6779 (2005).

[61] Frieboes H. B., Zheng X., Sun C-H., Tromberg B., Gatenby R., Cristini V., *An Integrated Computational/Experimental Model of Tumor Invasion.* Cancer Res., 66, 1597-1604 (2006).

[62] Hogea C. S., Murray B. T., Sethian J. A., *Simulating Complex Tumor Dynamics from Avascular to Vascular Growth Using a General Level-set Method.* J. Math. Biol., 53, 86-134 (2006).

[63] Macklin P., Lowengrub J., *Nonlinear Simulation of the Effect of Microenvironment on Tumor Growth.* J. Theor. Biol., 245, 677-704 (2007).

[64] Wise S. M., Lowengrub J. S., Frieboes H. B., Cristini V., *Three Dimensional Diffuse-interface Simulation of Multispecies Tumor Growth- I: Numerical Method.* Bull. Math. Biol. (in review).

[65] Frieboes H. B., Lowengrub J. S., Wise S., Zheng X., Macklin P., Bearer E. L., Cristini V., *Computer Simulation of Glioma Growth and Morphology*. NeuroImage, 37, S59-S70 (2007).

[66] Waliszewski P., Molski M., Konarski J., *Self-similarity, Collectivity and Evolution of Fractal Dynamics During Retinoid-induced Differentiation of Cancer Cell Population*. Fractals, 7, 139-149 (1999).

[67] Waliszewski P., Konarski J., Molski M., *On the Modification of Fractal Self-space During Cell Differentiation or Tumor Progression*. Fractals, 8, 195-203 (2000).

[68] Waliszewski P., Molski M., Konarski J., *On the Relationship Between Fractal Geometry of Space and Time in which a Cellular System Exists and Dynamics of Gene Expression*. Acta Biochimol., 48, 209-220 (2001).

[69] Nottale L., *Fractal Space-Time and Microphysics: Towards a Theory of Scale Relativity*. World Scientific, Singapore, 1993.

[70] Guarini G., Onofri E., Menghetti E., *New Horizons in Medicine. The Attractors*. Recenti Prog. Med., 84(*9*), 618-623 (1993).

[71] Sedivy R., *Fractal Tumours: Their Real and Virtual Images*. Wien Klin. Wochenschr., 108(*17*), 547-551 (1996).

[72] Sedivy R., Windischberger C., *Fractal Analysis of a Breast Carcinoma – Presentation of a Modern Morphometric Method*. Wien Klin. Wochenschr., 148(*14*), 335-337 (1998).

[73] Gabor D., *A New Microscopic Principle*. Nature, 161, 777-778 (1948).

[74] Leith E. N., Upatnieks J., *Reconstructed Wavefronts and Communication Theory*. Journal of the Optical Society of America, 52, 1123-30 (1962).

[75] Leith E. N., Upatnieks J., *Wavefront Reconstruction with Continuous-tone Objects*. Journal of the Optical Society of America, 53, 1377-81 (1963).

[76] Leith E.N., Upatnieks J., *Wavefront Reconstruction with Diffused Illumination and Three-dimensional Objects*. Journal of the Optical Society of America, 54, 1295-301 (1964).

[77] Smith H.M. (Ed.), *Holographic Recording Materials*. Springer-Verlag, Berlin, 1977.

[78] Bjelkhagen H. I., *Silver Halide Materials for Holography & Their Processing*. Springer-Verlag, Berlin, 1993.

[79] Pennington K. S., Harper J. S., Laming F. P., *New Phototechnology Suitable for Recording Phase Holograms and Similar Information in Hardened Gelatine*. Applied Physics Letters, 18, 80-84 (1971).

[80] Bartolini R. A., *Photoresists, in: Holographic Recording Materials*. In *Topics in Applied Physics*, Vol. 20, H. M. Smith (Ed.), Springer-Verlag, Berlin, 1977, pp. 209-27.

[81] Booth B. L., *Photopolymer Laser Recording Materials*. Journal of Applied Photographic Engineering, 3, 24-30 (1977).

[82] Smothers W. K., Monroe B. M., Weber A. M., Keys D. E., *Photopolymers for Holography*. In *Practical Holography IV*, Proceedings of the SPIE, Vol. 1212, S. A. Benton (Ed.), SPIE, Bellingham, 1990, pp. 20-29.

[83] Lin L. H., Beauchamp H. L., *Write-read-erase in Situ Optical Memory Using Thermoplastic Holograms*. Applied Optics, 9, 2088–92 (1970).

[84] Urbach J. C., *Thermoplastic Hologram Recording*. In: *Holographic Recording Materials, Topics in Applied Physics*, Vol. 20, H. M. Smith (Ed.), Springer-Verlag, Berlin, 1977, pp. 161–207.

[85] Huignard J. P., Micheron F., *High Sensitivity Read-write Volume Holographic Storage in Bi12SiO20 and Bi12GeO20 Crystals*. Applied Physics Letters, 29, 591–593 (1976).

[86] Huignard J. P., *Phase Conjugation, Real Time Holography and Degenerate Four-wave Mixing in Photoreactive BSO Crystals*. In *Current Trends in Optics*, Arecchi & Aussenegg (Eds.), Taylor & Francis, London, 1981, pp. 150–60.

[87] Svoboda K. K., Fischman D. A., Gordon M. K., *Embryonic Chick Corneal Epithelium: A Model System for Exploring Cell-matrix Interactions*. Developmental Dynamics, 237, 2667-2675 (2008).

[88] Giannoni E., Buricchi F., Grimaldi G. et al., *Redox Regulation of Anoikis: Reactive Oxygen Species as Essential Mediators of Cell Survival*. Cell Death and Differentiation, 15, 867-878 (2008).

[89] Chiarugi P., Giannoni E., *Anoikis: A Necessary Death Program for Anchorage-dependent Cells*. Biochemical Pharmacology, 76, 1352-1364 (2008).

[90] Framson P. E., Sage E. H., *SPARC and Tumor Growth: Where the Seed Meets the Soil*. Journal of Cellular Biochemistry, 92, 679-690 (2004).

[91] Pupa S. M., Menard S., Forti S., Tagliabue E., *New Insights into the Role of Extracellular Matrix During Tumor Onset and Progression*. Journal of Cellular Physiology, 192, 259-267 (2002).

[92] Hofmann U. B., Houben R., Brocker E. B., Becker J. C., *Role of Matrix Metallo-proteinases in Melanoma Cell Invasion*. Biochimie, 87, 307-314 (2005).

[93] von Kempen L. C., Ruiter D. J., van Muijen G. N., Coussens L. M., *The Tumor Microenvironment: a Critical Determinant of Neoplastic Evolution*. European Journal of Cell Biology, 82, 539-548 (2003).

[94] Lopez-Otin C., Palavalli L. H., Samuels Y., *Protective Roles of Matrix Metallo-proteinases: From Mouse Models to Human Cancer*. Cell Cycle, 8, 3657-3662 (2009).

[95] Wu Y., Zhou B. P., *Inflammation: A Driving Force Speeds Cancer Metastasis.* Cell Cycle, 8, 3267-3273 (2009).

[96] Nottale L., *Scale Relativity and Fractal Spece-Time,* Imperial College Press, London (2011).

[97] Mandelbrot B., *The Fractal Geometry of Nature,* W.H. Freeman, New York (1983).

[98] Casian Botez I., Agop M., Nica P., Paun V., Munceleanu G.V., *Conductive and Convective Type Behavior at Nano-Time Scales,* J. Comput. Theor. Nanosci., 7, 2271-228 (2010).

[99] Agop M., Forna N., Casian-Botez I., Bejenariu I. C., *New Theoretical Approach of the Physical Processes in Nanostructures,* J. Comput. Theor. Nanosci., 5, 483-489 (2008).

[100] Munceleanu G.V., Paun V.P., Casian Botez I., Agop M., *The Microscopic-macroscopic Scale Transitions through a Chaos Scenario in the Fractal Space-Time Theory,* International Journal of Bifurcation and Chaos, 21, 603-618 (2011).

[101] Luis G., *Complex Fluids,* Springer, Vol. 415 (1993).

[102] Mitchell M., *Complexity: A Guided Tour,* Oxford Univ. Press, Oxford (2009).

[103] Thomas Y. Hou, Multiscale Phenomena in Complex Fluids: Modeling, Analysis and Numerical Simulations, World Scientific Publishing Company, Singapore (2009).

[104] Landau L., Lifshitz E.M., *Fluid Mechanics,* 2-nd Edition, Butterworth Heinemann, Oxford (1987).

[105] Birlescu V.S., Agop M., M. Craus, *Computational Properties of a Fractal Medium,* International Journal of Quantum Information, 12, 4 (2014).

[106] Norton L., Massagué J., *Is Cancer a Sisease of Self-seeding*? Nature Medicine, 12, 875-878 (2006).

[107] Kim Mi-Y., Oskarsson T., Acharyya S., Nguyen D. X., Zhang X. H.-F., Norton L., Massagué J., *Tumor Self-seeding by Circulating Cancer Cells.* Cell, 139, 1315-1326 (2009).

[108] Enderling H., Anderson A.R.A., Chaplain M.A.J., Beheshti A., Hlatky L., Hahnfeldt P., *Paradoxical Dependencies of Tumor Dormancy and Progression on Basic Cell Kinetics.* Cancer Research, 69(22), 8814-8821 (2009).

[109] Vaupel P., Kelleher D. K., Hockel M., *Oxygen Status of Malignant Tumors: Pathogenesis of Hypoxia and Significance for Tumor Therapy.* Semin. Oncol., 28, 29-35 (2001).

[110] Vaupel P., *The Role of Hypoxia-induced Factors in Tumor Progression.* Oncologist, 9, Suppl. 5, 10-17 (2004).

[111] Harris A. L., *Hypoxia – A Key Regulatory Factor in Tumour Growth.* Nat. Rev. Cancer, 2, 38-47 (2002).

[112] Kimbro K.S., Simons J.W., *Hypoxia-inducible Factor-1 in Human Breast and Prostate Cancer.* Endocr. Relat. Cancer, 13, 739-749 (2006).

[113] O'Donnell J. L., Joyce M. R., Shannon A. M., Harmey J., Geraghty J., Bouchier-Hayes D., *Oncological Implications of Hypoxia Inducible Factor-1alpha (HIF-1alpha) Expression.* Cancer Treat. Rev., 32, 407-416 (2006).

[114] Lovestam M.–A, Agardh E., *Photocoagulation of Diabetic Macular Oedema: Complications and Visual outcome.* Acta Ophthalmolm. Scand., 78, 667-671 (2000).

[115] McCarty C. A., McKay R., Keeffe J. E., *Management of Diabetic Retinopathy by Australian Ophthalmologists.* Working Group on Evaluation of the NHMRC Retinopathy Guideline Distribution, National Health and Medical Research Council. Clin. Exp. Ophthalmol., Vol. 28, 2000, pp. 107-112.

[116] Fong D. S., Ferris F. L. III, Davis M. D., Chew E. Y., *Causes of Severe Visual Loss in the Early Treatment Diabetic Retinopathy Study.* ETDRS Report no. 24, *Early Treatment Diabetic Retinopathy Study Research Group.* Am. J. Ophthalmol., 127, 137-141 (1999).

[117] Dogru M., Nakamura M., Inoue M., Yamamoto M., *Long-term Visual Outcome in Proliferative Diabetic Retinopathy Patients after Panretinal Photo-coagulation.* Jpn. J. Ophthalmol., 43, 217-224 (1999).

[118] Yu D.-Y., Cringle S. J., Su E., Yu P. K., Humayun M. S., Dorin G., *Laser-Induced Changes in Intraretinal Oxygen Distribution in Pigmented Rabbits.* Invest. Ophthalmol. Vis. Sci., 46(3), 988-999 (2005).

[119] Cringle S..J., Yu D.-Y., Yu P. K., Su E.-N., *Intraretinal Oxygen Consumption in the Rat in Vivo.* Invest. Ophthalmol. Vis. Sci., 43, 1922-1927(2002).

[120] Cringle S..J., Yu D-Y., A *Multi-layer Model of Retinal Oxygen Supply and Consumption Helps Explain the Muted Rise in Inner Retinal PO_2 During Systemic Hyperoxia.* Comp. Biochem Physiol., 132, 61-66 (2002).

[121] Carter S.B., *Principles of Cell Motility: The Direction of Cell Movement and Cancer Invasion.* Nature, 208, 1183-187 (1965).

[122] Quigley J. P., Lacovara J., Cramer E. B., *The Directed Migration of B-16 Melanoma-cells in Response to a Haptotactic Chemotactic Gradient of Fibronectin.* J. Cell Biol., 97, A450-A451 (1983).

[123] Lacovara J., Cramer E. B., Quigley J. P., *Fibronectin Enhancement of Directed Migration of B16 Melanoma Cells.* Cancer Res., 44, 1657-1663 (1984).

[124] McCarthy J. B., Furcht L.T., *Laminin and Fibronectin Promote the Directed Migration of B16 Melanoma Cells in Vitro.* J. Cell Biol., 98, 1474-1480 (1984).

[125] Klominek J., Robert K. H., Sundqvist K.-G., *Chemotaxis and Haptotaxis of Human Malignant Mesothelioma Cells: Effects of Fibronectin, Laminin, Type IV Collagen, and an Autocrine Motility Factor-like Substance.* Cancer Res., 53, 4376-4382 (1993).

[126] Lawrence J. A., Steeg P. S., *Mechanisms of Tumour Invasion and Metastasis.* World J. Urol., 14, 124-130 (1996).

[127] Debruyne P.R., Bruyneel E.A., Karaguni I.-M. et al., *Bile Acids Stimulate Invasion and Haptotaxis in Human Corectal Cancer Cells through Activation of Multiple Oncogenic Signalling Pathways.* Oncogene, 21, 6740-6750 (2002).

[128] Stetler-Stevenson W. G., Hewitt R., Corcoran M., *Matrix Metallo-proteinases and Tumour Invasion: From Correlation to Causality to the Clinic.* Cancer Biol., 7, 147-154 (1996).

[129] Chambers A. F., Matrisian L. M., *Changing Views of the Role of Matrix Metalloproteinases in Metastasis.* J. Natl. Cancer Inst., 89, 1260-1270 (1997).

[130] Calabresi P., Schein P. S. (Eds.), *Medical Oncology.* 2nd Edn., McGraw-Hill, New York, 1993.

[131] Bray D., *Cell Movements.* Garland Publishing, New York, 1992.

[132] Melicow M. M., *The Three-steps to Cancer: A New Concept of Carcinogenesis.* J. Theor. Biol., 94, 471-511 (1982).

[133] Folkman J., Hochberg M., *Self-regulation of Growth in Three Dimensions.* J. Exp. Med., 138, 745-753 (1973).

[134] Casciari J. J., Sotirchos S. V., Sutherland R. M., *Variation in Tumour Cell Growth Rates and Metabolism with Oxygen-concentration, Glucose-concentration and Extracellular pH.* J. Cell. Physiol., 151, 386-394 (1992).

[135] Anderson A.R.A., Chaplain M.A.J., Newman E.L., Steele R.J.C., Thompson A.M., *Mathematical Modelling of Tumour Invasion and Metastasis.* J. Theor. Med., 2, 129-154 (2000).

[136] Terranova V.P., Diflorio R., Lyall R.M., Hic S., Friesel R., Maciag T., *Human Endothelial Cells are Chemotactic to Endothelial Cell Growth Factor and Heparin.* J. Cell Biol., 101, 2330-2334 (1985).

[137] Hotary K., Allen E.D., Punturieri A., Yana I., Weiss S.J., *Regulation of Cell Invasion and Morphogenesis in a 3-dimensional Type I Collagen Matrix by Membrane-type Metalloproteinases 1, 2 and 3.* J. Cell Biol., 149, 1309-1323 (2000).

[138] Zervoudaki A., Economou E., Pitsavos C., Vasiliadou K. et al., *The Effect of Ca2+ Channel Antagonists on Plasma Concentrations of Matrix Metallopro ases in Metastasis.* J. Natl. Cancer Inst., 89, 1260-1270 (1997).

[139] Johansson N., Ahonen M., Kahari V.-M., Matrix Metalloproteinases in Tumour Invasion. Cell. Mol. Life Sci., 57, 5-15 (2000).

[140] Sherwood L., *Human Physiology: From Cells to Systems.* 4th Ed., Pacific Grove, California, Brooks/Cole, 2001.

[141] Loudon R., *The Quantum Theory of Light.* 2-nd Ed., Oxford Univ. Press, Oxford, 1983.

[142] Sargent M. III, Scully M. O., Lamb W. E. Jr., *Laser Physics.* Addison Wesley Publ. Co., Reading, Massachusetts, 1977.

[143] Pantel R. H., Puthoff H. E., *Fundamentals of Quantum Electronics*. John Wiley & Sons, Inc., NewYork, 1969.

[144] Nussenzveig H. M., *Introduction to Quantum Optics*. Gordon and Breach Sci. Publ., London, 1973.

[145] Fowler A. C., Gibbon J. D., Mcguinness M. J., Physica, D4, 139 (1982).

[146] Haken H., Phys. Lett., A *53*, 77 (1975).

[147] Weiss B. O., Brock J., Phys. Rev. Lett., 57, 2804 (1986).

[148] Milonni P. W., Shih M.-L., Ackerhalt J. R., *Chaos in Laser-Matter Interactions*. World Sci. Publ. Co., Singapore, 1987.

[149] Newell A. C., Moloney J. W., *Nonlinear Optics*. Addison Wesley Publ. Co., Reading, Massachusetts, 1992.

[150] Ivancevic T. T., Bottema M. J., Jain L. C., *A Theoretical Model of Chaotic Attractor in Tumor Growth and Metastasis*. arXiv:0807.4272v1 [q-bio.CB] 27 Jul 2008.

[151] Olive L. C., Durand R. E., *Drug and Radiation Resistance in Spheroids: Cell Contact and Kinetics*. Cancer Metastasis Rev., 13, 121 (1994).

[152] Kozusko F., Bourdeau M., *A Unified Model of Sigmoid Tumour Growth Based on Cell Proliferation and Quiescence*. Cell Prolif., 40(6), 824-834 (2007).

[153] Chu I. M., Hengst L., Slingerland J. M., *The Cdk Inhibitor p27 in Human Cancer: Prognostic Potential and Relevance to Anticancer Therapy*. Nature Rev. Cancer, 8, 253-287 (2008).

[154] Castro M.A.A., Klamt F., Grieneisen V.A., Grivicich I., Moreira J.C.F., *Gompertzian Growth Pattern Correlated with Phenotypic Organization of Colon Carcinoma, Malignant Glioma and Non-small Cell Lung Carcinoma Cell Lines*. Cell Prolif. 36(2), 65-73 (2003).

[155] Ingber D. E., *Fibronectin Controls Capillary Endothelial Cell Growth by Modulating Cell Shape*. Proc. Natl. Acad. Sci., USA, 87 (1990).

[156] Boudreau N., Jones P. L., *Extracellular Matrix and Integrin Signaling: The Shape of Things to Come*. Biochem. J., 339 (1999).

[157] Aznavoorian M., Stracke L., Krutzsch H., Schiffman E., Liotta L. A., *Signal Transduction for Chemotaxis and Haptotaxis by Matrix Molecules in Tumour Cells*. J. Cell Biol., 110, 1427-1438 (1990).

[158] Vaidya V.G., Alexandro F.J. Jr., *Evaluation of Some Mathematical Models for Tumour Growth*. Int. J. Biomed. Comput., 13, 19-35 (1982).

[159] Schwartz M., *A Biomathematical Approach to Clinical Tumour Growth*. Cancer, 14, (1961).

[160] Nabeshima K., Lane W.S., Biswas C., *Partial Sequencing and Characterisation of the Tumor Cell-derived Collagenase Stimulatory Factor*. Arch. Biochem. Biophys., 285, 90-96 (1991).

[161] Xie B., Bucana C. D., Fidler I. J., *Density-dependent Induction of 92-kd Type Type-IV Collagenase Activity in Cultures of A431 Human Epidermoid Carcinoma Cells*. Am. J. Pathol., 144, 1958-1967 (1994).

[162] Werb Z., *ECM and Cell Surface Proteolysis: Regulating Cellular Ecology*. Cell, 91, 439-442 (1997).

[163] Rascle M., Ziti C., *Finite Time Blow-up in Some Models of Chemotaxis*. J. Math. Biol., 33, 388-414 (1995).

[164] Murray J. D., *Mathematical Biology*. Springer, Berlin, 1990.

[165] Hariharan P, *Basics of Holography*, Cambridge University Press, New York, 2002.